The Physics of Solids

The Physics of Solids

Richard Turton

OXFORD
UNIVERSITY PRESS

Great Clarendon Street, Oxford OX2 6DP
Oxford University Press is a department of the University of Oxford.
It furthers the University's objective of excellence in research, scholarship,
and education by publishing worldwide in

Oxford New York

Auckland Cape Town Dar es Salaam Hong Kong Karachi
Kuala Lumpur Madrid Melbourne Mexico City Nairobi
New Delhi Shanghai Taipei Toronto

With offices in

Argentina Austria Brazil Chile Czech Republic France Greece
Guatemala Hungary Italy Japan Poland Portugal Singapore
South Korea Switzerland Thailand Turkey Ukraine Vietnam

Oxford is a registered trade mark of Oxford University Press
in the UK and in certain other countries

Published in the United States
by Oxford University Press Inc., New York

First published 2000, reprinted 2002, 2004, 2005, 2006

A catalogue record for this book is available from the British Library

Library of Congress Cataloging in Publication Data

(Data applied for)

ISBN-13: 978-0-19-850352-1
ISBN 10: 0-19-850352-0

8

Typeset J&L Composition, Filey, N. Yorks
Printed in Great Britain
on acid-free paper by
CPI Antony Rowe,
Chippenham, Wiltshire

Preface

Is there really a need for another solid state physics textbook? Traditionally, solid state physics is introduced relatively late in an undergraduate physics course, and there are many textbooks which are designed for this level. A smaller number of texts are suitable for use at an earlier stage, but even these tend to assume a familiarity with quantum theory, and in particular, the Schrödinger equation.

However, since many aspects of solid state physics have important technological applications, there is an increasing tendency to introduce the subject in the first year of an undergraduate course. The main aim of this book is to meet the needs of these students by ensuring that a knowledge of quantum theory is not a prerequisite. Some basic ideas in quantum theory are introduced in Appendix A, but the majority of the book can be read without ever meeting Schrödinger's equation. (For those readers who do wish to apply quantum theory to solid state physics, relevant topics have been included in the optional sections of most chapters. These sections are marked with an asterisk.)

A second aim was to write a book which is both readable and yet encourages students to develop their ability to solve quantitative problems. I have tried to achieve this objective by integrating problem-solving questions into the fabric of the book. What I mean by this is that the questions are treated as an integral part of the process of reading the book. For example, instead of filling pages with derivations of equations, most of the key equations in this book are simply stated as 'fact'. If readers are interested in the origin of the equation then they are referred to a question in which they are encouraged to make the derivation for themselves. Similarly, students are encouraged to tackle problems in order to gain a feeling for the magnitudes of certain quantities. Some of these problems are dealt with as worked examples in the main body of the text, others are left for the student to perform. (In case any difficulties are encountered a comprehensive set of worked solutions are printed at the back of the book.) (A comprehensive set of solutions are provided at the back of the book to help with any difficulties, and worked examples are also included in the text.)

I would like to thank all of those people who read all or parts of this book during its development and for the many useful comments that they made. Any feedback from readers is always useful, so if you have any comments, please do let me know. Finally, I would like to thank the staff at Oxford University Press, in particular Michael Rodgers, who has once again given his unfailing support and encouragement throughout this project.

Contents

Table of physical constants

Quantity	Symbol	Value to three significant figures
Acceleration due to gravity (standard)	g	$9.81\ \mathrm{m\,s^{-2}}$
Atomic mass unit	u	$1.66\times10^{-27}\ \mathrm{kg}$
Avogadro's constant	N_A	$6.02\times10^{23}\ \mathrm{mol^{-1}}$
Bohr magneton	μ_B	$9.27\times10^{-24}\ \mathrm{J\,T^{-1}}$
Bohr radius	a_B	$5.29\times10^{-11}\ \mathrm{m}$
Boltzmann's constant	k_B	$1.38\times10^{-23}\ \mathrm{J\,K^{-1}}$
Electron charge	e	$1.60\times10^{-19}\ \mathrm{C}$
Electron rest mass	m_e	$9.11\times10^{-31}\ \mathrm{kg}$
Electron volt	eV	$1.60\times10^{-19}\ \mathrm{J}$
Molar gas constant	R	$8.31\ \mathrm{J\,mol^{-1}\,K^{-1}}$
Permeability of a vacuum	μ_0	$1.26\times10^{-7}\ \mathrm{T^2\,m^3\,J^{-1}}$*
Permittivity of a vacuum	ε_0	$8.85\times10^{-12}\ \mathrm{F\,m^{-1}}$
Planck's constant	h	$6.63\times10^{-34}\ \mathrm{J\,s}$
Planck's constant/2π	$\hbar$	$1.05\times10^{-34}\ \mathrm{J\,s}$
Speed of light in a vacuum	c	$3.00\times10^{8}\ \mathrm{m\,s^{-1}}$

* Note that μ_0 is usually given in units of $\mathrm{H\,m^{-1}}$, but in this book we have used units of $\mathrm{T^2\,m^3\,J^{-1}}$ for reasons described in Chapter 8. These units are equivalent, so the value of μ_0 in $\mathrm{H\,m^{-1}}$ is the same as that given above.

Chapter 1
Bonds between atoms

1.1 Introduction

Matter generally exists in one of three states—gas, liquid or solid. It is a well-known fact that we can change a substance from one of these states to another simply by altering the temperature. For example, water exists as a liquid at room temperature, but changes into a solid when the temperature drops below 0°C, or to steam when the temperature rises above 100°C. Because these temperatures are easily accessible, we are familiar with water in all three of these states. However, in many cases we associate a substance with one particular state. For example, in our normal daily experience air always exists as a gas and iron as a solid. However, both can be changed into a liquid, in one case by reducing the temperature to 77 K (−196°C) and in the other by raising the temperature to 1808 K (1535°C). Similar observations with a variety of different substances allow us to make the following statement: all (or nearly all) substances can exist in each of the three states of matter, but the temperature at which the material changes state varies enormously from one substance to another.

If you are content to accept this statement as simply a law of nature then read no further; this book is not for you. If, on the other hand, you wonder why a material changes from one state to another, or why some substances have melting points of several thousand kelvin, whereas others have boiling points as low as 4 K, then read on. The main aim of this book is not simply to catalogue the different properties of various materials, but to explain why certain substances have specific properties. In general, we shall find that in order to supply such an explanation we need to resort to a description of the material on the atomic scale. In particular, it is often the forces, or bonds, between the atoms which provide the initial clues. We can illustrate this approach by taking a more detailed look at the problem of melting and boiling.

The state of matter of any particular substance is governed by the interplay between two effects. Firstly, there is an attractive, or bonding, force between the atoms which has a tendency to pull the atoms as close together as possible. As we shall see in the next chapter, this is best achieved if the atoms form a regular, ordered arrangement. The second effect is a thermal one. Thermal energy causes random motion of the atoms, and as the temperature is increased, the atoms move about more. The two effects therefore act in opposition to one another. The interatomic forces try to produce regularity, attempting to impose law and order on the atoms, whereas thermal energy introduces an anarchic, chaotic element which does its best to disrupt this order.

Let us then compare the relative importance of these two effects in each of the three states of matter. Suppose we consider a substance which is initially in the solid state. In this state the interatomic forces are the dominant factor and the atoms are packed tightly together in a regular arrangement, or crystal, to form a rigid structure. Of course, thermal energy is also present, but since this effect is insufficient to overcome the relatively strong forces holding the atoms together, the only motion that it can produce is a vibration of the atoms about their fixed positions. As the temperature is increased, the vibration of the atoms increases in amplitude until we reach a stage where the interatomic forces are no longer strong enough to constrain the atoms to their positions in the crystal. At this point the crystal structure breaks up and the solid melts.

We now have an intermediate state—a liquid—in which neither one effect nor the other is clearly dominant. We can demonstrate this by considering a couple of well-known properties of liquids. The thermal motion of the molecules, or at least a consequence of it, can be observed in a simple experiment using an ordinary microscope. All that is necessary is to sprinkle a fine powder on the surface of a liquid. When examined through the microscope the particles are found to be moving, evidence of the constant buffeting they receive from the random motion of the molecules in the liquid. This effect is called Brownian motion. The importance of the interatomic forces is evidenced in many other phenomena. For example, surface tension, which causes water to form roughly spherical droplets when placed on a waxed surface, is directly related to the cohesive forces between the molecules in the liquid.

As the temperature increases further, the thermal energy of the atoms becomes more significant. When the boiling point is reached, the energies of the atoms become so large that they can escape completely from the liquid, and so form a gas. In this state, the interatomic forces become very small. Indeed, the concept of an ideal gas is based on the premise that the interatomic forces are negligible and that all the properties of the gas can be determined by considering only the thermal motion of the molecules.

The above provides at least a brief explanation of why materials change from one state to another as the temperature is altered. So why is it that the melting (and boiling) points of various substances differ so much? We can explain this in terms of the strength of the bonds between the atoms. If the forces between the atoms in the material are strong, then it is difficult to move the atoms far from their ideal position in the crystal, and so for a given temperature the vibrations of the atoms are relatively small. Conversely, if the forces are weak, then at the same temperature the vibrations are much larger, and so the melting point is likely to be significantly lower. Consequently, we can postulate a relationship between the melting point of a substance and the strength of the bonds between the atoms in the material. We will see how well this model holds up in Section 1.10.

It turns out that many other properties of solids can also be related to the strength or type of bond present in the material. This is a good enough reason to explain why we should begin our study of the solid state by examining the various types of bond that exist—because for relatively little effort we can determine quite a lot about the different properties of solids. However, first of all we need to introduce some basic concepts concerning the atoms themselves.

1.2 Bohr's model of the atom

The intention in this and the following section is not to give a detailed account of the structure of the atom. Instead I have tried to concentrate on just a few essential concepts which we will use in this and subsequent chapters. We will show that the electrons are arranged in a series of shells and subshells, each characterized by a particular energy, with specific limits imposed on the number of electrons allowed in each subshell. You may be familiar with such a concept as it is introduced at an early stage in many courses and is vital to an understanding of the bonding process. In Section 1.3 we use Pauli's exclusion principle to show why the electrons are arranged in this fashion.

Let us consider the ionization energy of an atom. This is the amount of energy required to remove one electron from an atom, thereby converting it into a positively charged ion. By studying the change in ionization energy from one element to another we will be able to build up a picture of the way in which the electrons are arranged in the atom.

A good place to start is with the simplest element, the hydrogen atom, which consists of just two particles, a proton and an electron. Because the proton has a positive charge and the electron has a negative charge, the two are mutually attracted and so form a single unit. The ionization energy in this case is simply the amount of energy required to separate these two particles. This quantity can be measured experimentally and turns out to be 13.6 eV. However, merely knowing the magnitude of the ionization energy is not very satisfying. What we would like to know is why it is 13.6 eV and not some other value. This means taking a more detailed look at the hydrogen atom.

A popular picture of a hydrogen atom consists of an electron describing a circular orbit around the proton, rather like a satellite in orbit around the Earth. However, in order to determine the energy of the electron we need to exploit the wavelike nature of the electron (see Appendix A). If we treat the electron as a de Broglie wave, rather than as a particle, then a more appropriate picture is to think of the electron orbit as a flexible plastic ring. The only waves which can be contained on this ring are standing waves, as shown in Fig 1.1(a), in which the circumference of the ring corresponds to an integral number of wavelengths. For all other wavelengths there is a condition of destructive interference and the wave cancels itself out (Fig 1.1(b)). Consequently, we can see that the electron is only allowed to occupy orbits of certain sizes. We will not go into further detail (a semi-classical derivation is given in Chapter 4 of Beiser—see Further reading—if you are interested), but merely quote the result that the radii of these allowed orbits are given by

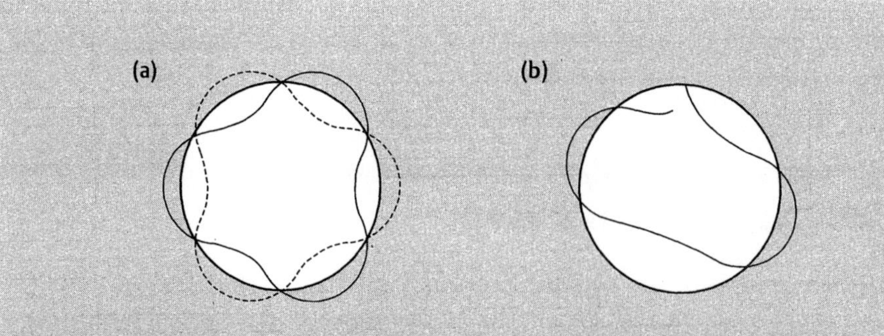

Figure 1.1 (a) A standing wave forms when the circumference of a loop is equal to an integral number of wavelengths. (b) In other cases the wave interferes destructively with itself.

$$r_n = \frac{n^2 h^2 \varepsilon_0}{\pi m_e e^2} \tag{1.1}$$

Here m_e denotes the mass of the electron, e is the electron charge, ε_0 is the permittivity of a vacuum, h is Planck's constant and n is an integer. Each value of n corresponds to one of the allowed orbits. It can also be shown that the corresponding energies of the electrons in these orbits are

$$E_n = -\frac{m_e e^4}{8\varepsilon_0^2 h^2 n^2} \tag{1.2}$$

If we substitute in the appropriate values for the physical constants and set n equal to 1, then we obtain $r_1 = 0.0529$ nm (see Example 1.1). This length is referred to as the Bohr radius. The corresponding energy is −13.6 eV, precisely the same value as is found experimentally for the ionization energy of hydrogen. (Notice that the energy of the electron in a hydrogen atom as given by eqn (1.2) is a negative value. This makes sense if we consider the reference level, or zero of energy, to be that of an isolated electron and an isolated proton. The minus sign then simply indicates that a hydrogen atom is a stable system, since an input of energy is required to separate the electron from the proton.)

Bohr's formula in eqn (1.2) gives good agreement with experiment for $n = 1$, but what meaning do we attach to other values of n? It follows from eqns (1.1) and (1.2) that higher values of n correspond to smaller binding energies and are associated with orbits of larger radius (see Question 1.1). Since electrons tend to find the lowest possible energy levels, we can conclude that the electron in a hydrogen atom will almost always occupy the $n = 1$ orbit. For hydrogen, at least, we can ignore the presence of the other allowed orbits.

Having considered the hydrogen atom in some detail, let us now examine the ionization energies of some of the other elements. How do we expect the ionization energy to vary from one element to another? It would seem a reasonable assumption to suppose that the ionization energy is larger for elements with a higher atomic number simply on the grounds that the energy required to drag an electron away from an atom should increase in proportion to the number of positive charges in the nucleus. We can obtain a mathematical expression for the binding energy based on this hypothesis simply by extending Bohr's theory. If the number of protons in the nucleus of a particular element is Z, then following Bohr's approach we can show that the electron energy should be Z^2 times larger than in hydrogen, i.e.

$$E_n = -\frac{Z^2 m_e e^4}{8\varepsilon_0^2 h^2 n^2} = -13.6\frac{Z^2}{n^2}\ \text{eV} \tag{1.3}$$

Example 1.1 Determine the energy and radius of the electron orbit in a hydrogen atom for $n = 1$.

Solution Setting $n = 1$ in eqn (1.1) we obtain a radius

$$r_1 = \frac{h^2 \varepsilon_0}{\pi m_e e^2} = \frac{(6.63\times10^{-34}\,\text{J s})^2(8.85\times10^{-12}\,\text{F m}^{-1})}{(3.14)(9.11\times10^{-31}\,\text{kg})(1.60\times10^{-19}\,\text{C})}$$

$$= 5.29\times10^{-11}\,\text{m}$$

and from eqn (1.2)

$$E_1 = -\frac{m_e e^4}{8\varepsilon_0^2 h^2} = -\frac{(9.11\times10^{-31}\,\text{kg})(1.60\times10^{-19}\,\text{C})^4}{(8)(8.85\times10^{-12}\,\text{F m}^{-1})^2(6.63\times10^{-34}\,\text{J s})^2}$$

$$= -2.17\times10^{-18}\,\text{J}$$

Dividing by the magnitude of the electron charge, e, converts this value to electron volts, i.e.

$$E_1 = -\frac{2.16\times10^{-18}}{1.60\times10^{-19}} = -13.6\,\text{eV}$$

Although this argument may seem very plausible, it has one major failing: it does not agree, even qualitatively, with experiment. Let us consider the actual ionization energies for the first 20 elements in the periodic table, which are shown in Fig. 1.2. We can see that the ionization energy of helium (with an atomic number of 2) is certainly larger than that of hydrogen, although it is not four times as large as eqn (1.3) suggests. Similarly, there is generally a steady increase in ionization energy as we go from lithium (atomic number 3) to neon (atomic number 10), and also from sodium (atomic number 11) to argon (atomic number 18). However, there is a dramatic decrease in ionization energy going from helium to lithium, or neon to sodium, or argon to potassium (atomic number 19). What is happening in these cases? Why is it so much easier to remove an electron from a lithium atom than from a helium atom?

To answer these questions we need to make a brief excursion into quantum theory.

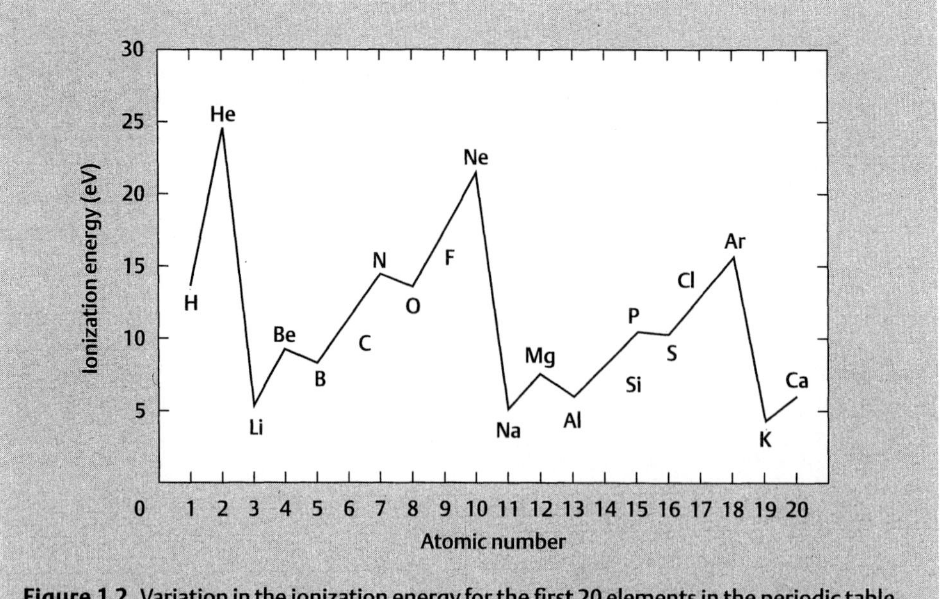

Figure 1.2 Variation in the ionization energy for the first 20 elements in the periodic table.

1.3 Pauli's exclusion principle and the shell model of the atom

We have seen that in Bohr's semi-classical theory of the atom the electrons are only allowed to occupy certain orbits, which are labelled using the integer n. This is often referred to as the principal quantum number. It turns out that in a more rigorous treatment three other 'labels', or quantum numbers, are needed to describe fully the electrons in an atom. These are

l, the orbital quantum number,
m_l, the orbital magnetic quantum number, and
m_s, the spin magnetic quantum number.

We do not need to worry about what these labels represent; our only interest is in what values these numbers can take. Fortunately, despite the mathematical complexities of quantum theory, the rules which prescribe the allowed values of these quantum numbers are remarkably straightforward and can be stated as follows. For any given value of the principal quantum number n, the orbital quantum number l can take only integer values between 0 and $(n - 1)$, and the orbital magnetic quantum number, m_l, is allowed integer values between $-l$ and $+l$. Finally, for each combination of n, l and m_l, the spin magnetic quantum number, m_s, takes values of $-\frac{1}{2}$ and $+\frac{1}{2}$.

Before we can put this into practice we need to introduce one further rule. This is the exclusion principle, which was introduced by Wolfgang Pauli in 1925 and turns out to be the key to deciphering the atomic code. The exclusion principle states that each electron in an atom must possess a unique set of quantum numbers. This statement is so simple that even a primary school child has the mathematical ability to put it into practice, and yet it is a statement that is so fundamental that without it we would be unable to understand the structure of the atom, or even the most basic properties of solids.

How do we use these rules of quantum theory to determine the arrangement of the electrons in an atom? We will begin by looking at those electrons where the principal quantum number, n, is equal to 1. We will refer to this as the first electron shell. Applying the above rules we find that the only permissible value of l is 0, and therefore m_l must also be 0. Since m_s takes values of $\pm\frac{1}{2}$ this means that there are two different combinations of quantum numbers with $n = 1$, and therefore, applying Pauli's exclusion principle, only two electrons in an atom can belong in this shell. This is a most profound statement. Since only two electrons are allowed to occupy the lowest energy level this means that all of the other electrons in an atom are forced into orbits which are further from the nucleus with correspondingly higher energies.

For the case with n equal to 2 there are more options available since l can now take values of 0 or 1. In the latter case there are three different possibilities for m_l; −1, 0 and +1. Since there are two values of m_s corresponding to each of these combinations, a total of eight electrons are allowed in this shell. This is shown in Table 1.1.

We now have the beginnings of a model of the atom in which the electrons are arranged in distinct groups. Can we use this shell model to explain the abrupt changes in the ionization energy shown in Fig. 1.2? Let us take a look at the difference between the elements helium and lithium where the first such change occurs. An

Table 1.1 Allowed combinations of quantum numbers for the first ($n = 1$) and second ($n = 2$) electron shells.

n	l	m_l	m_s	Total number of electrons
1	0	0	$-\frac{1}{2}$	
			$+\frac{1}{2}$	2
	0	0	$-\frac{1}{2}$	
			$+\frac{1}{2}$	
		−1	$-\frac{1}{2}$	
			$+\frac{1}{2}$	
2	1	0	$-\frac{1}{2}$	8
			$+\frac{1}{2}$	
		+1	$-\frac{1}{2}$	
			$+\frac{1}{2}$	

atom of helium has two electrons, which means that it has a full complement of electrons in the first shell. In comparison, lithium has three electrons, and so one of these electrons must be placed in the second shell. The amount of energy required to remove this electron from the atom is considerably smaller than that for the electrons in the inner shell for two reasons. Firstly, it is further away from the nucleus, and secondly the two electrons in the inner shell effectively cancel out some of the positive charge on the nucleus (see Fig. 1.3). This effect is known as screening. Consequently, it is not surprising that the ionization energy of lithium is considerably smaller than that of helium.

We encounter the same scenario when we look at the difference between neon, in which the electrons fill all the available states in the first two shells, and sodium, in which one electron is forced into the third shell.

Before we continue any further, it is helpful to introduce an abbreviated notation for the states corresponding to particular sets of quantum numbers. Since the energy of an electron is most affected by the quantum numbers n and l, a shorthand notation is often used in which each set of states is labelled by these two quantum numbers. The different values of l are represented by letters; thus the orbital with $l = 0$ is called an s level, and the letters p, d and f are used for $l = 1$, 2 and 3, respectively. (The reasons for this strange system of lettering are historical. With the benefit of hindsight it might have been easier to use labels a, b, c, etc., but the old system is with us whether we like it or not.) Following this scheme we can label the electrons in the first shell as 1s electrons, whilst those in the second shell are 2s and 2p electrons.

The third shell should follow a similar pattern, but now we appear to encounter a problem. The next sharp change in ionization energy occurs between the elements argon and potassium which have atomic numbers of 18 and 19, respectively. Since we have a total of 10 electrons in the first two shells, this implies that a further eight electrons can be placed in the third shell. However, Pauli's exclusion principle suggests that it should be possible for 18 electrons to occupy this shell, as is shown in Example

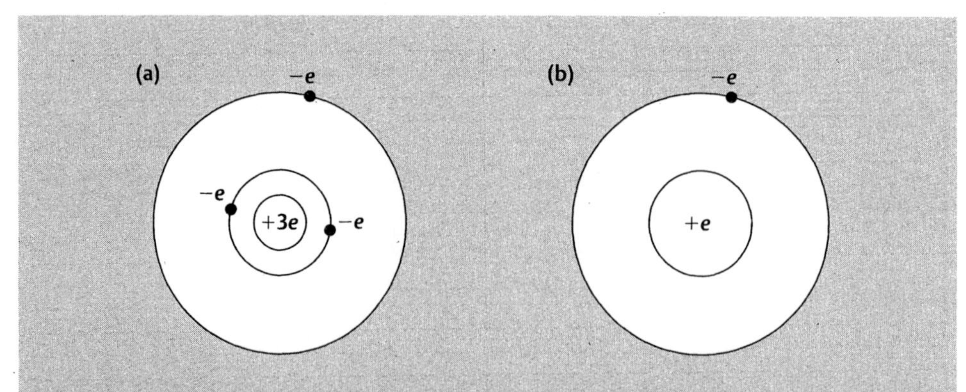

Figure 1.3 Schematic representation of a lithium atom showing (a) the arrangement of the electrons in shells, and (b) the effects of screening by the two electrons in the inner shell. From (b) it can be seen that the outer electron effectively experiences a nuclear charge of +e, not +3e.

Example 1.2 Determine how many electrons can be accommodated in the third electron shell by extending Table 1.1.

Solution When $n = 3$, l can take values of 0, 1 or 2. Applying the above rules we obtain the following table.

n	l	m_l	m_s	Label
3	0	0	$+\frac{1}{2}$	3s
			$+\frac{1}{2}$	
	1	−1	$-\frac{1}{2}$	3p
			$+\frac{1}{2}$	
		0	$-\frac{1}{2}$	
			$+\frac{1}{2}$	
		+1	$-\frac{1}{2}$	
			$+\frac{1}{2}$	
	2	−2	$-\frac{1}{2}$	3d
			$+\frac{1}{2}$	
		−1	$-\frac{1}{2}$	
			$+\frac{1}{2}$	
		0	$-\frac{1}{2}$	
			$+\frac{1}{2}$	
		+1	$-\frac{1}{2}$	
			$+\frac{1}{2}$	
		+2	$-\frac{1}{2}$	
			$+\frac{1}{2}$	

This gives a total of 18 electrons.

1.2. The cause of this discrepancy is that the energy of an electron depends not only on the value of n, but also to a lesser extent on l. If we look at the energies of the different states in potassium (Fig. 1.4) we find that the energy of the 4s electrons is actually slightly lower than that of the 3d electrons. Consequently, once the 3s and 3p subshells are filled the next electron is placed in the 4s shell, and so potassium has an electronic configuration which is similar to that of lithium and sodium.

The shell model of the atom not only allows us to explain the trends in the ionization energy, but also accounts for the structure of the periodic table. The elements were originally set out in this form by Dmitri Mendeleev in 1868, the arrangement being such that elements with similar chemical and physical properties were placed in the same column. This statement is still true in the modern form of the periodic table (see Fig. 1.5), although now it is the electronic configuration of the elements which determines their position in the table. For example, the elements in the far left hand column possess a single electron in the outer shell. The fact that this electron can easily be removed from the atom accounts for the extreme reactivity of these elements. In contrast, the elements in the far right hand column have filled shells or subshells. As a result, the ionization energies of these elements are comparatively large and so these elements are inert—they rarely react or form compounds with other elements. The elements in these columns are called group I and group 0 elements, respectively, the Roman numeral denoting the number of electrons present in the outer shell. Similarly, the elements in columns labelled group II to group VII have corresponding numbers of electrons in the outer shell. We will refer to these outermost electrons as the valence electrons.

We have introduced a lot of new ideas in this section, and so it is worth pausing for a moment to take stock of these findings. We have managed to explain the variation in the ionization energy of the elements and the structure of the periodic table by using a model of the atom in which the electrons are arranged in a series of shells and subshells. These we have labelled using numbers and letters, respectively. Electrons within a subshell are all at about the same distance from the nucleus and have the same energy, but electrons in different subshells generally have substantially different energies. What we now need to ask is how does this model help us to explain the process of bonding?

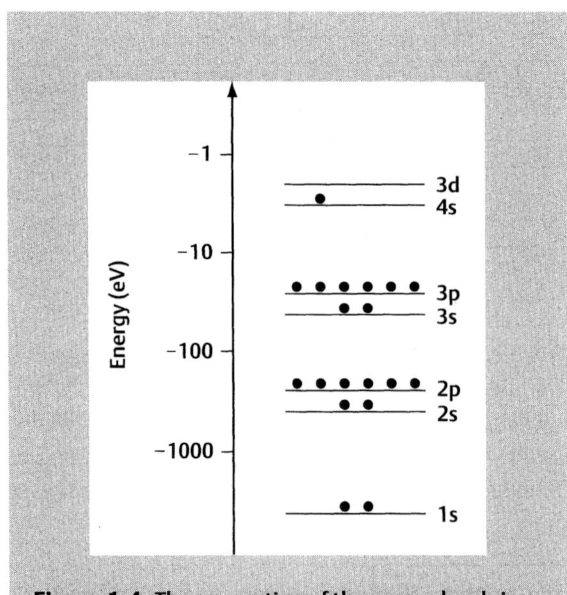

Figure 1.4 The occupation of the energy levels in potassium. Each electron is represented by a solid circle.

Transition metals

Post-transition metals

I	II											III	IV	V	VI	VII	0
1 H																	2 He
3 Li	4 Be											5 B	6 C	7 N	8 O	9 F	10 Ne
11 Na	12 Mg											13 Al	14 Si	15 P	16 S	17 Cl	18 Ar
19 K	20 Ca	21 Sc	22 Ti	23 V	24 Cr	25 Mn	26 Fe	27 Co	28 Ni	29 Cu	30 Zn	31 Ga	32 Ge	33 As	34 Se	35 Br	36 Kr
37 Rb	38 Sr	39 Y	40 Zr	41 Nb	42 Mo	43 Tc	44 Ru	45 Rh	46 Pd	47 Ag	48 Cd	49 In	50 Sn	51 Sb	52 Te	53 I	54 Xe
55 Cs	56 Ba	*	72 Hf	73 Ta	74 W	75 Re	76 Os	77 Ir	78 Pt	79 Au	80 Hg	81 Tl	82 Pb	83 Bi	84 Po	85 At	86 Rn
87 Fr	88 Ra	**															
		*	57 La	58 Ce	59 Pr	60 Nd	61 Pm	62 Sm	63 Eu	64 Gd	65 Tb	66 Dy	67 Ho	68 Er	69 Tm	70 Yb	71 Lu
		**	89 Ac	90 Th	91 Pa	92 U	93 Np	94 Pu	95 Am	96 Cm	97 Bk	98 Cf	99 Es	100 Fm	101 Md	102 No	103 Lr

Figure 1.5 The periodic table of the elements showing the atomic number and chemical symbol for each element.

1.4 Atoms in solids

So far we have considered the properties of isolated atoms, but in a solid the atoms are in close proximity to one another. What effect does this have on the electrons? To answer this it is helpful to classify the electrons into two categories: core electrons, which occupy the filled subshells, and valence electrons, which are those in the incomplete subshells. (It should be noted that there is no actual difference between these two types of electrons other than the fact that they occupy different states. If we moved a core electron into the incomplete subshell it would become a valence electron.)

Let us now consider the effect that a nearby atom exerts on these two sets of electrons. Since the core electrons are close to the nucleus, any external forces, such as those due to the presence of a nearby atom, are small in comparison with the attractive force from the parent nucleus. In contrast, we have seen that the valence electrons are not only further from the nucleus, but also see a smaller positive charge due to the screening effect produced by the core electrons. Consequently, the attractive force from the nucleus is greatly reduced and may only be comparable in magnitude with that from other nearby atoms. These situations are shown schematically in Fig. 1.6 for an atom of carbon. We conclude that the core electrons are essentially unaffected by their environment. In other words, the energies of the core electrons in a carbon atom which is part of a diamond crystal or a molecule of methane are virtually the same as they are in an isolated carbon atom. However, the energies of the valence electrons are significantly affected by the presence of other nearby atoms. This change in the energy of the valence electrons is the key to understanding the process of bonding.

Before we discuss different types of bonds in detail, we will take a general look at what happens when atoms bond together. For simplicity, let us consider a system of just two atoms. If we place the atoms in close proximity to one another then the valence electrons on the two atoms interact, and so the energies of these electrons are altered. If the change in energy is such that the total energy of the valence electrons in this composite system is lower than that in the isolated atoms, then we can see that it is energetically favourable for the atoms to form this composite system. This is precisely what we mean by a 'bond'. Furthermore, the difference in energy between the composite system and the isolated atoms is a measure of the strength of this bond.

So we have a simple prescription not only for determining whether a bond forms between two particular atoms, but also for finding the strength of this bond. All we have to do is calculate the total energy of the valence electrons in the isolated atoms and in the composite system. Unfortunately, calculating the energies of these valence electrons is a very complicated problem which in most cases can only be solved using a high-performance computer. Ideally, what we would like is a simple model which explains why elements combine in specific quantities to form molecules and crystals. It turns out that the majority of bonding processes can be understood simply by considering the number of valence electrons on each atom. Chemists often refer to this as the theory of valence, which originates from the work of Lewis in 1916. The basis for this theory is that the most stable elements are those in group 0 of the periodic

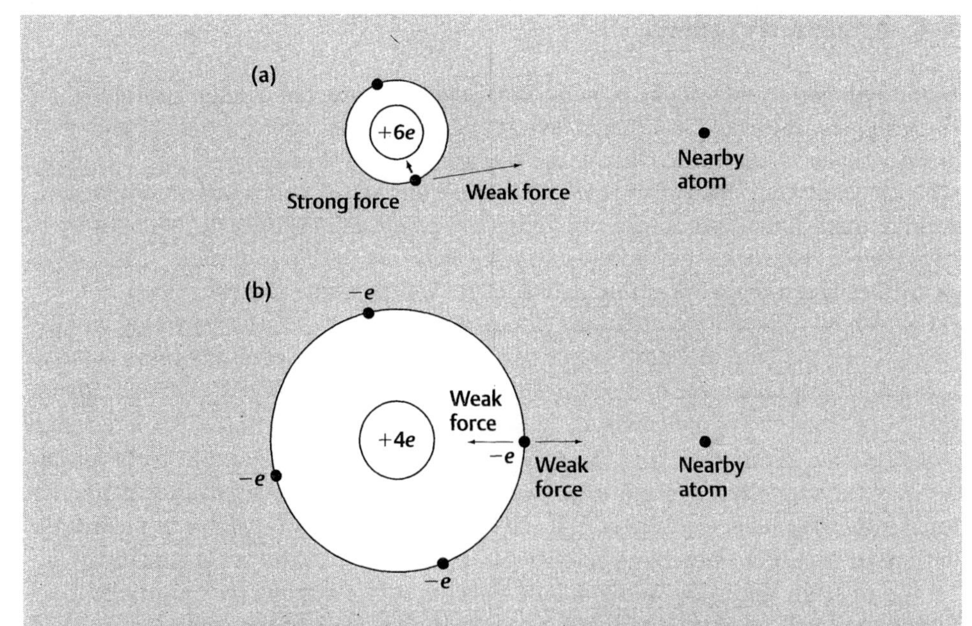

Figure 1.6 Schematic representation of the electrostatic forces experienced due to the parent nucleus and due to the presence of a nearby atom (which contains charged particles) for (a) the core electrons and (b) the valence electrons in an atom of carbon.

table which have filled shells or subshells. Bonding occurs in situations where atoms lose, gain or share electrons so that each atom achieves the electronic configuration of one of these elements. There are essentially three different mechanisms by which this can occur, corresponding to ionic, metallic and covalent bonding. We will examine each of these in turn in the following sections.

1.5 Ionic bonding

Conceptually the simplest type of bond to understand is that which occurs between atoms from group I and group VII of the periodic table (see Fig. 1.5). In order to obtain the electronic configuration of an inert atom the group I element needs to lose one electron and the group VII element needs to gain one. It is obvious that we can satisfy both criteria if an electron is transferred from the group I element atom to the group VII element atom. In doing so we end up with a positive and a negative ion—hence the term ionic bonding.

This is the picture we expect if we apply the theory of valence, but is this exchange of an electron energetically favourable? In other words, is the energy of this system lower than that of the two separate atoms? To answer this, let us take a look at a specific example in some detail.

We will consider sodium chloride, commonly known as rock salt. This substance is an obvious candidate for ionic bonding since sodium and chlorine belong to groups I and VII, respectively. The ionization energy of sodium is relatively small, as we can see from Fig. 1.2. In fact an energy of only 5.1 eV is required to detach the single valence electron from its parent nucleus. Chlorine, on the other hand, can actually reduce its energy if it gains an extra electron—the energy of a negative chlorine ion is lower than that of a neutral chlorine atom. The binding energy of this electron is referred to as the electron affinity. In the case of chlorine the electron affinity is 3.6 eV.

So it appears that we require a net input of $(5.1-3.6) = 1.5$ eV in order to persuade the electron to transfer from the sodium to the chlorine atom. However, there is one factor that we have not yet taken into account. If the sodium atom gives away an electron, it then has one more proton than it has electrons. This means that it has a net positive charge, so we now have a sodium ion, which we denote by Na^+. Similarly, the addition of an electron to a chlorine atom produces a negatively charged chlorine ion, Cl^-. These two ions therefore attract each other. The magnitude of the force F when the ions are a distance r apart is given by Coulomb's law

$$F_{Coulomb} = -\frac{e^2}{4\pi\varepsilon_0 r^2} \tag{1.4}$$

where the negative sign indicates that the force is attractive. However, what we really wanted to know is how much energy is saved when the ions attract one another. This is equal to the potential energy of the pair of ions, which from eqn (1.4) turns out to be

$$E = -\frac{e^2}{4\pi\varepsilon_0 r} \tag{1.5}$$

(If you are unsure how to obtain this result from eqn (1.4) this is explained in Appendix B.) The potential energy is negative, just as it is for an electron in an atom, because we need to supply energy in order to separate the two ions. If we calculate this energy for a pair of sodium and chlorine ions (see Example 1.3) we find that the reduction in energy which arises due to the Coulomb force between the ions is more than enough to allow the electron to transfer from the sodium to the chlorine atom.

Example 1.3 Calculate the Coulomb potential energy for a molecule of sodium chloride (consisting of one sodium and one chlorine ion). Assume that the distance between the nuclei is 0.236 nm.

Solution Substituting for the physical constants in eqn (1.5) we have

$$E = -\frac{e^2}{4\pi\varepsilon_0 r} = -\frac{(1.60\times 10^{-19}\,\mathrm{C})^2}{(4)(3.14)(8.85\times 10^{-12}\,\mathrm{F\,m^{-1}})\times(0.236\times 10^{-9}\,\mathrm{m})}$$

$$= -9.75\times 10^{-19}\,\mathrm{J}$$

$$= -6.10\,\mathrm{eV}$$

It is all very well being able to describe a molecule consisting of two ions, but a typical grain of salt contains somewhere in the region of a thousand billion billion (10^{21}) sodium and chlorine ions. Each positively charged sodium ion attracts all of the chlorine ions and repels the other sodium ions. Similarly, each chlorine ion attracts all of the sodium ions but repels the other chlorine ions. With a thousand billion billion ions to consider, the arrangement of the ions in a grain of salt would seem to be a major headache. How do we sort out this mess?

For simplicity let us first consider a one-dimensional chain of ions. The most satisfactory way to arrange the ions (i.e. to maximize the attractive force between the unlike ions whilst minimizing the repulsive force between like ions) is to cause them to alternate, as shown in Fig. 1.7. Extending this idea to three dimensions produces the arrangement shown in Fig. 1.8. Now we have to ask, what is the potential energy of an ion in the midst of such a structure? This may seem a daunting question, but in fact it can be solved with fairly basic algebra. We shall outline the procedure below. Those who are mathematically inclined should find it fairly easy to follow. Those who are not may skip the next couple of paragraphs to find out the result.

So, how do we determine the potential energy of an ion in a crystal of sodium chloride? Let us consider the chlorine ion at the centre of the cube in Fig. 1.8. We know

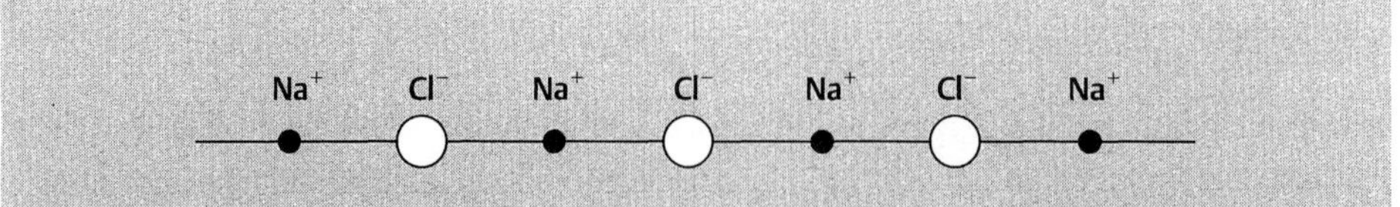

Figure 1.7 The arrangement of chlorine and sodium ions in an imaginary one-dimensional crystal.

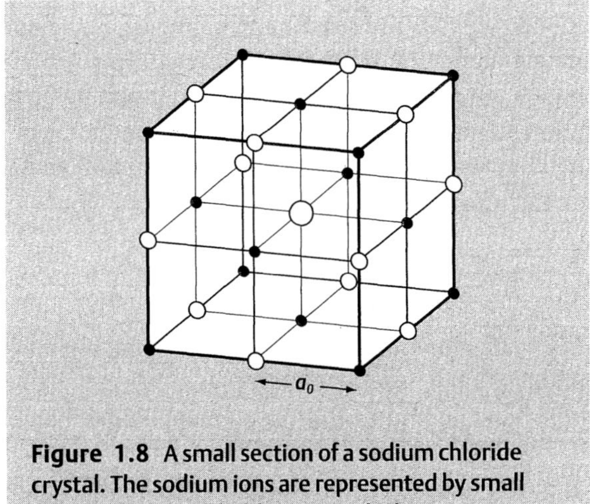

Figure 1.8 A small section of a sodium chloride crystal. The sodium ions are represented by small solid circles and the chlorine ions by large open circles.

that the potential energy due to a nearby ion decreases as the separation between the ions increases. Consequently, it is those ions which are closest to the central chlorine ion which most affect the energy of this ion. The nearest ions are the six sodium ions which lie at the centres of the cube faces in Fig. 1.8. Since these have opposite charge to the chlorine ion, they exert an attractive force and therefore produce a negative potential energy. If we denote the separation of the chlorine ion from each of these sodium ions as a_0, then using eqn (1.5) we can see that the potential energy of the chlorine ion due to these six sodium ions is

$$E = -6\frac{e^2}{4\pi\epsilon_0 a_0} \tag{1.6}$$

The next nearest ions to the centre are the 12 chlorine ions which lie at the midpoints of the outer edges of the cube in Fig. 1.8. Simple geometry shows that the separation of these ions from the central ion is $a_0\sqrt{2}$ and since the force is repulsive the potential energy is

$$E = +12\frac{e^2}{4\pi\epsilon_0 a_0\sqrt{2}} \tag{1.7}$$

Moving further outwards we come to the eight sodium ions at the vertices of the cube in Fig. 1.8 which are at a distance $a_0\sqrt{3}$ from the centre. We can therefore write the total potential energy due to all these ions as

$$E = -\frac{e^2}{4\pi\varepsilon_0 a_0} \times \left(6 - \frac{12}{\sqrt{2}} + \frac{8}{\sqrt{3}}\right) \tag{1.8}$$

Of course, this is not the end of the problem. We have considered only 24 of the 10^{21} or so ions in the grain of salt. However, we have dealt with the physics of the problem and are now concerned only with the mathematical details, which can be left for the mathematicians to solve. All we need to know is that the sum of this series does (eventually) converge to a value of 1.748. This value is known as the Madelung constant, M_d, and is the same for all crystals which have the same crystal structure as sodium chloride.

We therefore finally arrive at the following expression for the potential energy of an ion in a crystal of sodium chloride:

$$E = -M_d \frac{e^2}{4\pi\varepsilon_0 a_0} = -1.748 \frac{e^2}{4\pi\varepsilon_0 a_0} \tag{1.9}$$

All that remains now is to insert the appropriate values for the constants. Since the value of a_0 in sodium chloride is 0.281 nm we obtain a value for E of 1.43×10^{18} J or 8.95 eV (see Question 1.6).

How does this compare with experiment? Firstly, we should point out that when comparing with experimental measurements care should be taken to distinguish the lattice energy from the cohesive energy of the crystal. The former is the amount of energy required to separate the crystal into isolated ions, whilst the latter is the amount of energy required to separate the crystal into isolated atoms. The lattice energy is therefore given by eqn (1.9), whereas the cohesive energy is found by subtracting the amount of energy needed to convert the atoms into ions. For ionic crystals it is usually the lattice energy which is quoted.

Secondly, the experimental values are given in units of joules per mole. In order to convert our result into these units we should remember that 1 mole of sodium chloride contains Avagadro's number (6×10^{23}) of sodium ions and the same number of chlorine ions, giving a total of 12×10^{23} ions. However, if we multiply our result by this number we will be in error by a factor of 2. This is because in arriving at our expression for E in eqn (1.9) we have considered the interaction of one chlorine ion with every other ion in the crystal. If we now evaluate E for one of the other ions our summation will include this first ion again. Consequently, if we determine E separately for each ion in the crystal and add the values together, we are actually including each bond twice. The actual result is therefore obtained by multiplying eqn (1.9) by 6×10^{23} to give E = 861 kJ mol^{-1} (see Question 1.6). This can be compared with the experimental value of 776 kJ mol^{-1}. The agreement is surprisingly good considering that we have treated the ions as simple point charges and have completely neglected the fact that each is a complex system consisting of a nucleus surrounded by a swarm of electrons. However, as we shall see in the next section, far better agreement can be achieved if we include a simple term to take account of the finite size of the ions.

1.6 The repulsive force

We have seen that the ions of sodium and chlorine are brought together by the Coulomb force and that in a crystal of sodium chloride the separation between the ions is 0.281 nm. Considered together these two facts pose something of a puzzle.

Since the Coulomb force increases as the ions get closer together, we have to ask why it is that the ions do not approach any closer than this. The simplest solution is to suggest that the ions behave as hard spheres, like miniature billiard balls. However, this does not seem to agree with what we know about atoms. After all, an atom is just a disperse collection of particles. The only part of an atom which seems to have any substance is the nucleus, but the diameter of the nucleus is about a hundred thousand times smaller than the separation of the ions in sodium chloride.

To put this in perspective let us suppose that we are going to construct a large-scale model of an atom. We can use a golf ball to represent the nucleus and a handful of beads for the electrons. How far away should we place the beads from the golf ball? A few centimetres? Perhaps a few metres? No. In a scale model of the atom the beads will actually be a few kilometres from the golf ball. Can we seriously suggest that such a disperse collection of objects behaves as a solid sphere with a radius of a few kilometres? It seems improbable. And yet in an atom this is exactly what happens. If two atoms are moved towards one another we find that as the electron orbits begin to overlap it becomes virtually impossible to push them any closer together.

How can we understand this behaviour? Clearly it is difficult to comprehend from an intuitive point of view, and so it is perhaps not surprising that we need to invoke quantum theory. In particular, it turns out that this is another aspect of the exclusion principle. The argument goes as follows. We know that both the sodium and chlorine ions contain only filled shells of electrons, and according to the exclusion principle it is not possible to squeeze any more electrons into these shells. This means that if we push the two ions close together so that the outer electron orbits begin to overlap, then essentially we are trying to contravene the exclusion principle. So it appears that these ions (and indeed all atoms) behave rather like small hard spheres with a size which is comparable with the diameter of the electron orbits.

An alternative way to describe this behaviour is to introduce a repulsive force. The characteristics of this force are that it is virtually negligible at a separation of a few atomic diameters, but becomes exceedingly large if the separation becomes less than one atomic diameter. Such a force is referred to as a short-range force (see Question 1.7) and can be written as $F_{repel} \propto r^{-n}$ where n typically has a value of about 10. The total force can therefore be written as

$$F = F_{Coulomb} + F_{repel} = -\frac{Aa_0^2}{r^2} + \frac{Ba_0^{10}}{r^{10}} \qquad (1.10)$$

where A and B are constants. Here we have introduced factors of a_0^2 and a_0^{10} so that the constants A and B both have the dimensions of a force. (An alternative, and technically more accurate, expression for the repulsive force is $F_{repel} \propto e^{-r\alpha}$, where α is a constant;however, the simple power-law expression is adequate for our purposes.)

It is a useful exercise to sketch the repulsive and attractive forces on the same diagram, as shown in Fig. 1.9(a). At any given separation r the net force on the ions, as given by eqn (1.10), can be determined simply by finding the midpoint between these two curves. The equilibrium separation, a_0, occurs when the net force is zero, which, of course, is when the repulsive and attractive forces are equal in magnitude. An expression for the resultant potential energy can be obtained by integrating eqn (1.10)

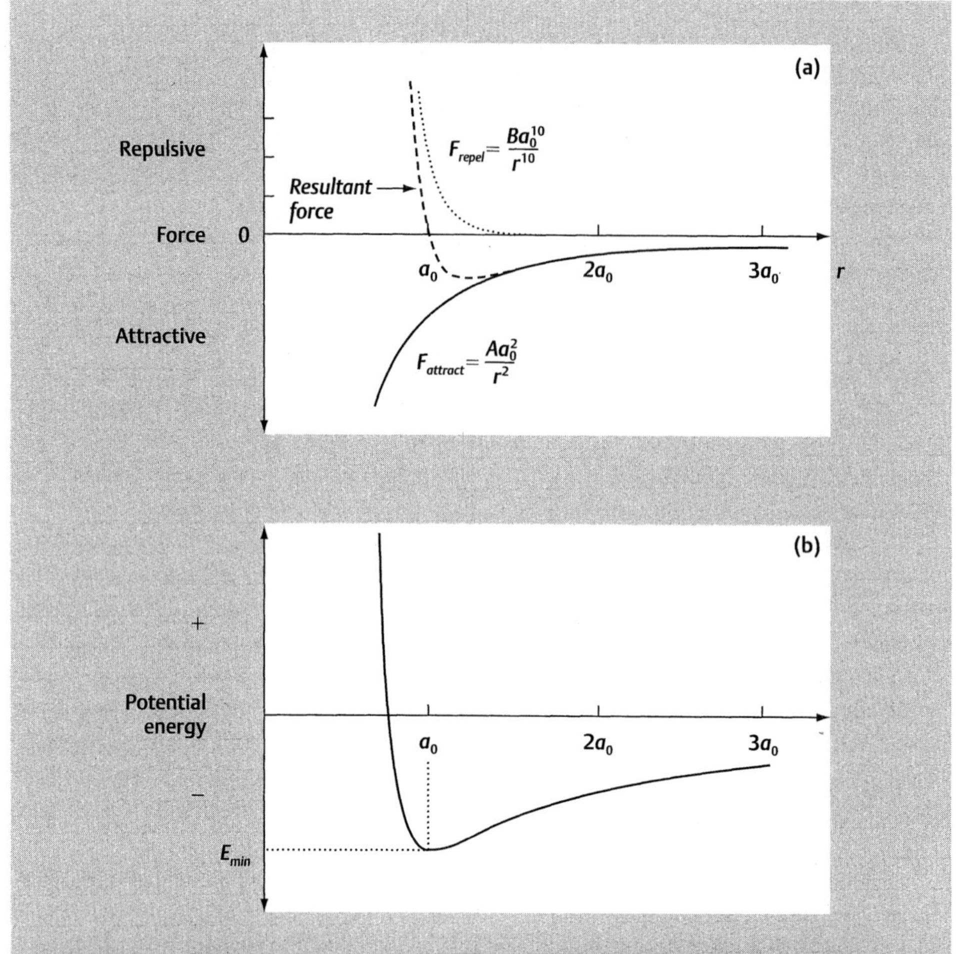

Figure 1.9 (a) The attractive, repulsive and resultant forces acting on a pair of dissimilar ions. (b) The resultant potential energy obtained by integrating the resultant force (see Example 1.4).

(see Example 1.4). The result is plotted in Fig. 1.9(b). From this curve we can see that the equilibrium separation a_0 corresponds to the position of minimum potential energy, E_{min}, as we might expect.

The question we should now ask is, what effect does the repulsive term have on the lattice energy of the crystal? Obviously it will reduce the energy, but by how much? It turns out that this correction is usually of the order of 10% (see Question 1.8). If we apply this correction to the value calculated previously we obtain a total lattice energy for sodium chloride which is very close to the experimental value.

Example 1.4 Obtain an expression for the potential energy of an ion in an ionic crystal assuming that the force is given by eqn (1.10). Also find the relationship between the constants A and B.

Solution The expression for the potential energy is found by integrating the expression for the force (see Appendix B), so

$$E(r) = -\int_{\infty}^{r} F(r)\,dr = \int_{\infty}^{r} \left(\frac{Aa_0^2}{r^2} + \frac{Ba_0^{10}}{r^{10}} \right) dr = -\frac{Aa_0^2}{r} + \frac{Ba_0^{10}}{9r^9}$$

Since we know that the resultant force at $r = a_0$ is zero, then from eqn (1.10) we have

$$0 = -\frac{Aa_0^2}{a_0^2} + \frac{Ba_0^{10}}{a_0^{10}}$$

which reduces to

$$A = B$$

1.7 Metallic bonding

Children learn at a very young age to distinguish objects which are made of metal from those which are not. The characteristic sheen of metal means that we can often identify that an object is metallic simply by looking at it. If there is any doubt, maybe because the object is painted or disguised in some other way, then touching the object or lightly tapping it with another piece of metal is usually sufficient to confirm our opinion. It seems that metals have a special significance, and their characteristic properties appear to be quite distinct from those of other solids.

It is perhaps not surprising then that if we look at the periodic table we find that the elements are divided into metals and non-metals. What is it that distinguishes a metallic element from a non-metallic one? We can find a clue by looking at the periodic table. Elements are generally classified as metals if the atoms possess one, two or three valence electrons. What makes these atoms so special is the way in which they bond together.

Let us consider an atom of the metal sodium, which we have already examined in some detail in Section 1.5. We know that the outer electron on the sodium atom can easily be removed, leaving a positive ion, Na^+. In the example of sodium chloride there were always plenty of chlorine atoms available to mop up these unwanted electrons. However, if there are only sodium atoms present then there is nowhere for these electrons to go—all of the atoms simply want to get rid of their electrons. This means that we are likely to end up with a collection of positively charged ions and a similar number of 'free', or delocalized, electrons which are not attached to any atom in particular. This appears to present a problem. We explained the presence of the bond in sodium chloride as being due to the attraction between the positive sodium ions and the negative chlorine ions. However, in this case all of the ions are positively charged, and so should repel one another.

This does not seem to be a very promising basis for the formation of a solid, but we have neglected the effect of the free electrons. Since there is one free electron corresponding to each ion, then, on average, at any instant in time there is likely to be an electron in the vicinity of each ion, as shown in Fig. 1.10. So although the sodium ions repel one another, they are each mutually attracted to the nearby electrons. These electrons therefore act as a sort of 'glue' which holds the collection of ions together.

Can we calculate the strength of this bond? Since the ions occupy fixed positions, we could estimate the strength of the repulsive potential between the ions. However, the attractive potential is far more difficult to deal with because the electrons are constantly on the move. So a simple calculation seems to be out of the question. The best that we can do is to say that the binding energy in a metal is likely to be considerably smaller than that in an ionic solid. This is based on the fact that in the latter case each ion is at least attracted by its nearest neighbours, whereas in the metal each ion is repelled by all the other ions. For this reason we also expect the spacing between the ions to be significantly larger in a metal than in a similar ionically bonded material. This is confirmed by experimental measurements. For example, the distance between the centres of two neighbouring sodium ions in a metal is found to be 0.382 nm, whereas sodium ions in sodium chloride and other ionically bonded materials behave as hard spheres with a diameter of only 0.194 nm. This is shown schematically in Fig. 1.11.

This describes the situation in simple or sp metals. These are the elements which are found in groups I and II of the periodic table, with a few other additions such as aluminium. The characteristic property of these elements is that they possess only s and p valence electrons. However, a large number of technologically important metals fall into the categories of transition and post-transition metals, which are found in the middle columns of the periodic table, as indicated in Fig. 1.5. These elements have quite different properties to the simple metals and this is largely a result of differences in the electronic configuration.

To illustrate this let us consider the fourth row of the periodic table which starts with the element potassium. We have already examined the electronic configuration

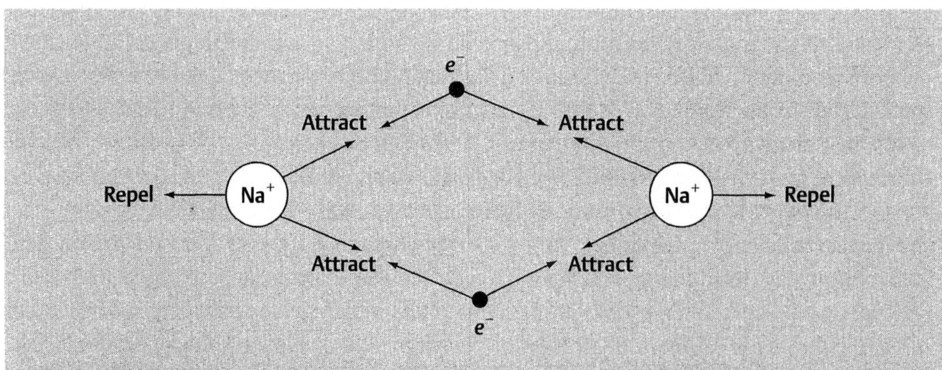

Figure 1.10 Metallic bonding—the repulsive and attractive forces between two sodium ions and the free valence electrons.

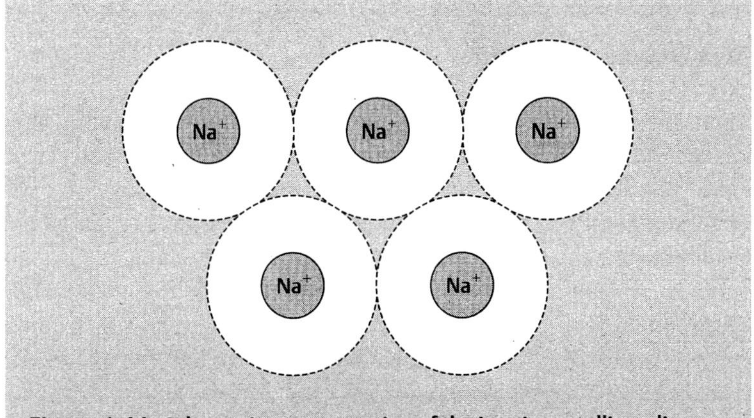

Figure 1.11 Schematic representation of the ions in metallic sodium. The dashed circles indicate the effective size of the ions and the shaded circles show the size of the ion core (i.e. the size of the ion in an ionic material such as NaCl).

of potassium in some detail. Figure 1.4 shows that it has a single electron in the 4s shell whilst the 3d states are at slightly higher energy and are unoccupied. The next element, calcium, has two electrons in the 4s level, so that this subshell is now filled. As we go further to the right along this row the electrons begin to fill the 3d subshell. These elements, from scandium to nickel, are the transition metals. The difference between the transition metals and the simple metals is therefore the presence of the partially filled shell of d electrons.

What is the significance of the d electrons and why do they have such an affect on the properties of the metal? To explain this let us consider the example of iron which possesses 26 electrons. Eighteen of these are in the subshells up to 3p making up the inert core of argon, and a further two are in the 4s subshell, leaving six to occupy the 3d states. The 4s and 3d electrons have very similar binding energies and are also at approximately the same distance from the nucleus. This means that if a couple of these outer electrons are removed, we are still left with an ion which is of approximately the same size as the original atom. This is quite different to the simple metal shown in Fig. 1.11. There is consequently a much stronger interaction between the neighbouring ions in the metal and this leads to higher cohesive energies. In fact, the cohesive energies for transition metals are generally comparable with those for ionic crystals, whereas the value for the simple metals is typically a factor of 5 times smaller.

The post-transition metals are those which have a filled d subshell with either one or two s valence electrons. These also have a large ion core like the transition metals, but the cohesive energies tend to be somewhat smaller.

1.8 The covalent bond

The covalent bond is synonymous with the idea of sharing electrons. It occurs between non-metallic elements, with the exception of the inert elements, and is the main form of bonding which exists between atoms in molecules. We can illustrate this concept by examining the hydrogen molecule, H_2. We have seen that each hydrogen molecule can be represented by supposing that the electron performs a circular orbit around the nucleus. How does this change if we have two nuclei and two electrons? We find that the electrons can now be pictured as performing a modified orbit around the two nuclei (Fig. 1.12). This means that it is no longer possible to associate the electrons with a particular atom—instead they are now shared by the two nuclei.

This picture seems straightforward enough, but there are several questions that it raises. For instance, why is it that hydrogen gas consists of H_2 molecules rather than individual atoms? And why are there just two hydrogen atoms in a molecule, not three or four? The answer must be that the H_2 molecule has a lower energy than that of any of the other possibilities. Unfortunately, although the H_2 molecule appears to be a very simple system, a calculation of the cohesive energy is far too complex to attempt here. However, we can provide a couple of simple arguments to explain why this should be the case.

Firstly, we will address the formation of the H_2 molecule. Why should two hydrogen atoms bond together? We can explain this in the same way as the metallic bond. If we look again at Fig. 1.12 we can see that there is a similarity with the picture of metallic bonding shown in Fig. 1.10. There is a repulsive force between the two nuclei, but the mutual attraction between the nuclei and the two electrons lowers the energy of the system. We can represent the energy levels schematically using the diagram in Fig. 1.13.

Why then are we limited to a molecule consisting of just two hydrogen atoms? To answer this we need to make use of Pauli's exclusion principle again. If we have two hydrogen atoms, each with a single 1s electron, then both of these electrons can be accommodated in the lowest energy level of the molecule (see Fig. 1.13). However, if we try to add a third hydrogen atom to the system, the exclusion principle tells us that this electron must go into a higher energy state. Consequently, it is not energetically favourable for this atom to join the molecule.

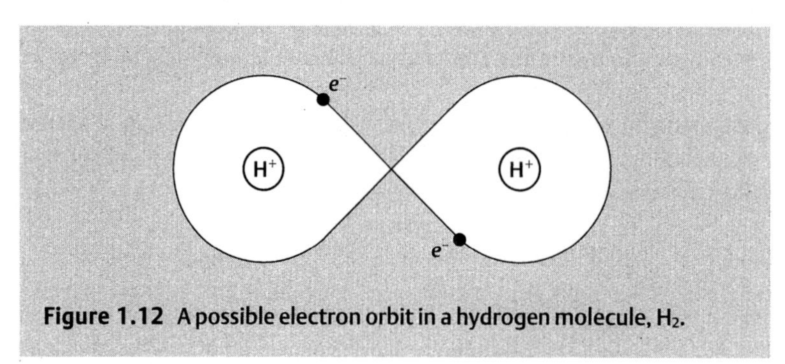

Figure 1.12 A possible electron orbit in a hydrogen molecule, H_2.

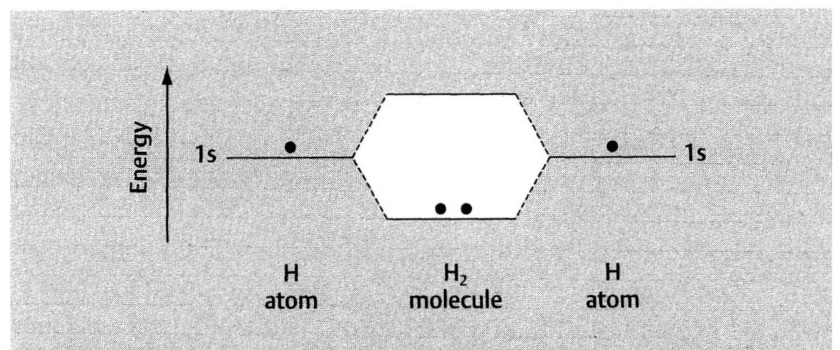

Figure 1.13 Schematic representation of the energy levels in a hydrogen molecule compared with the 1s levels in two isolated hydrogen atoms. The electrons are represented by solid circles.

We can follow similar arguments for other molecules, but the discussions become increasingly complex. A much simpler model that enables us to predict which combinations of atoms form molecules can be obtained by picturing the atoms as building blocks. These blocks can connect to one another at a certain number of points, with the number of these connection points corresponding to the number of vacant spaces in the outer subshell of the atom. For example, hydrogen has a single connection point (Fig. 1.14(a)) and so can connect with precisely one other hydrogen atom (Fig. 1.14(b)). Similarly, chlorine is one electron short of an inert gas configuration and so also has one connection point. We can therefore form HCl (Fig. 1.14(c)) and Cl_2 (Fig. 1.14d).

There is obviously not much scope for molecules between atoms with just one connection point, so let us turn our attention to the carbon atom which requires four electrons to complete its outer shell. The multitude of ways in which these extra four electrons can be provided gives rise to the entire field of organic chemistry. We will consider just a few of the vast number of carbon-based molecules. One of the simplest

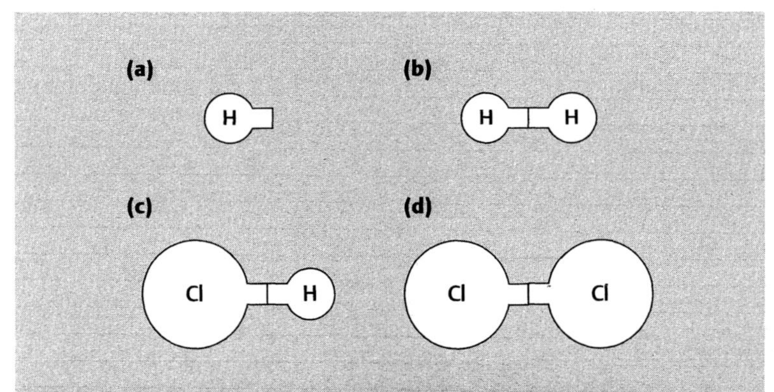

Figure 1.14 Building block models for (a) a single hydrogen atom, (b) a hydrogen, H_2, molecule, (c) a hydrogen chloride, HCl, molecule and (d) a chlorine, Cl_2, molecule.

carbon-based molecules is methane, CH_4, which consists of a single carbon atom and four hydrogen atoms. The building block diagrams become rather tiresome to draw as the molecules become larger, so we will introduce a shorthand notation in which each atom is represented by its chemical symbol and the connection points (or bonds) are shown by a short straight line linking the two symbols. Figure 1.15 shows the methane molecule in this format.

The methane molecule consists of precisely five atoms, but far larger structures can be produced by this same process. Imagine what would happen if we took a methane molecule and removed two of the hydrogen atoms. This leaves a group of atoms with the chemical formula CH_2 (see Fig. 1.16(a)). We will refer to this unit as a monomer. What happens to the two unpaired electrons? One possibility is that both of these unpaired electrons are shared with another CH_2 monomer to form the molecule ethylene shown in Fig. 1.16(b). Since there are two pairs of electrons involved in forming the bond between the carbon atoms we refer to this as a double bond. Alternatively, if we have just a single bond between the carbon atoms then we can form a long chain of CH_2 monomers, as shown in Fig. 1.16(c). The generic name for this type of molecule is a polymer. In this instance, since the polymer is based on the molecule ethylene, it is called polyethylene. In principle, there is no limit to the number of monomers which make up a polymer. In practice a typical polyethylene polymer consists of about twenty to thirty thousand CH_2 monomers.

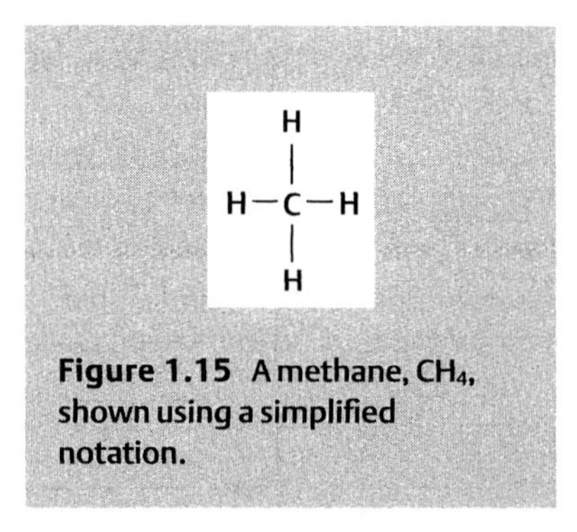

Figure 1.15 A methane, CH_4, shown using a simplified notation.

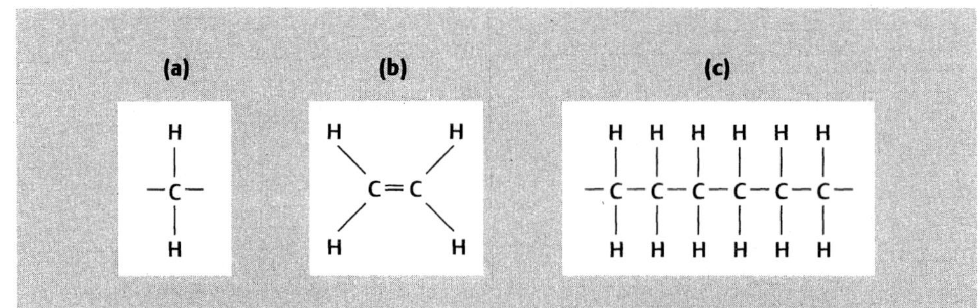

Figure 1.16 (a) A CH_2 monomer, (b) an ethylene molecule, C_2H_4 and (c) a short section of a polyethylene polymer.

Although polymers and other so-called macromolecules can contain many thousands of atoms, this number pales into insignificance when we consider a small crystal. For instance, a 1 carat diamond weighing 0.2 g contains in the region of 10^{21} carbon atoms (see Question 1.12). Diamond is a form of pure carbon in which each carbon atom is bonded covalently to four other carbon atoms. Although a diamond crystal is obviously a three-dimensional structure, we can show that all of the bonds are satisfied by constructing a two-dimensional model (Fig. 1.17).

Our model of the covalent bond is very successful at explaining why elements combine in specific quantities to form molecules, but it tells us nothing about the shape of the molecule. To find out what the molecules look like let us examine the methane molecule in more detail. We know that carbon has four valence electrons. Two of these are in the 2s subshell and the other two are in the 2p subshell. Our model suggests that two of the hydrogen atoms form bonds with the carbon 2s electrons whilst the other two bond with the carbon 2p electrons. This suggests that there should be some difference between the bonds. However, experiments have revealed that all four bonds are identical. It is as though each of the outer electrons on the carbon atom behaves as though it contains a mixture of all eight possible states. Since there are two 2s states and six 2p states, each valence electron essentially has one part s-like character to three parts p-like character. This is known as a hybrid sp^3 (pronounced 's p three') bond. Since all of the bonds are identical, the hydrogen atoms must be arranged symmetrically around the carbon atom. This is achieved if we picture the molecule as a cube with the carbon atom at the centre and the hydrogen atoms placed at alternate vertices, as shown in Fig. 1.18. It is then just a simple exercise in trigonometry (see Question 1.13) to show that the angles between the bonds are equal to 109.5°.

The characteristic shape of the methane molecule also appears in other carbon structures. In a polymer, such as polyethylene, the orientation of one monomer

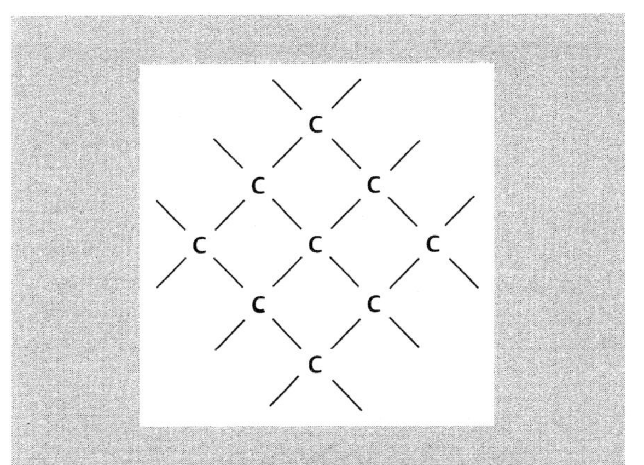

Figure 1.17 A two-dimensional representation of a diamond crystal. By forming covalent bonds with four other carbon atoms, each carbon atom obtains a filled outer shell.

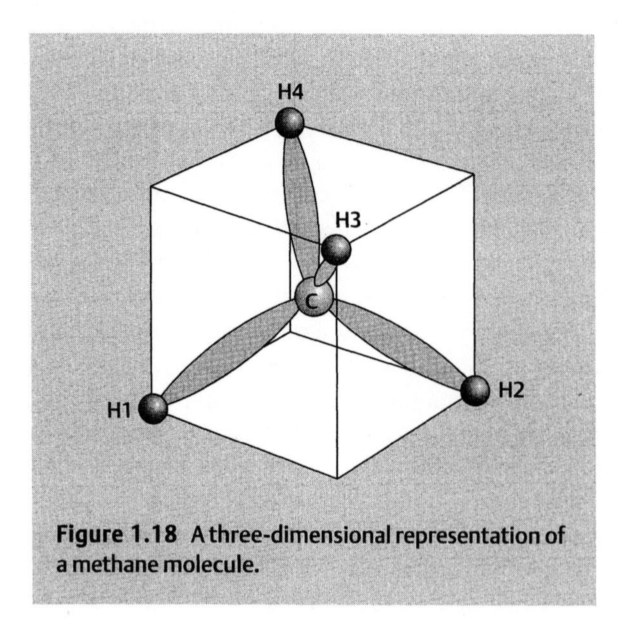

Figure 1.18 A three-dimensional representation of a methane molecule.

relative to its neighbour is more or less random. Consequently, a polymer typically has a convoluted shape as in Fig. 1.19. However, in a diamond there is only one way in which the atoms can be arranged so that each atom has four equidistant neighbours (Fig. 1.20). The resulting structure is highly regular and when extended over millions of atomic layers produces a very beautiful and much sought after crystal.

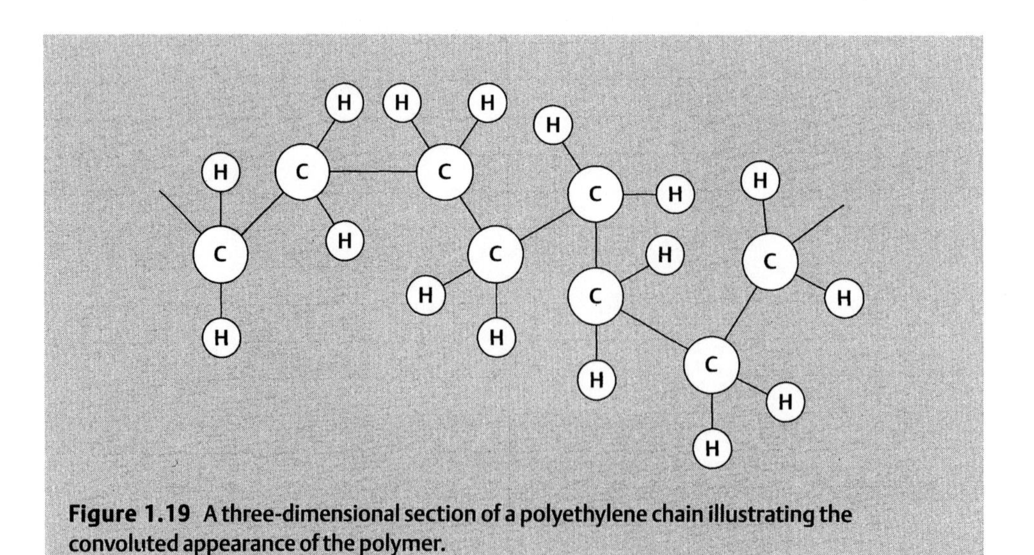

Figure 1.19 A three-dimensional section of a polyethylene chain illustrating the convoluted appearance of the polymer.

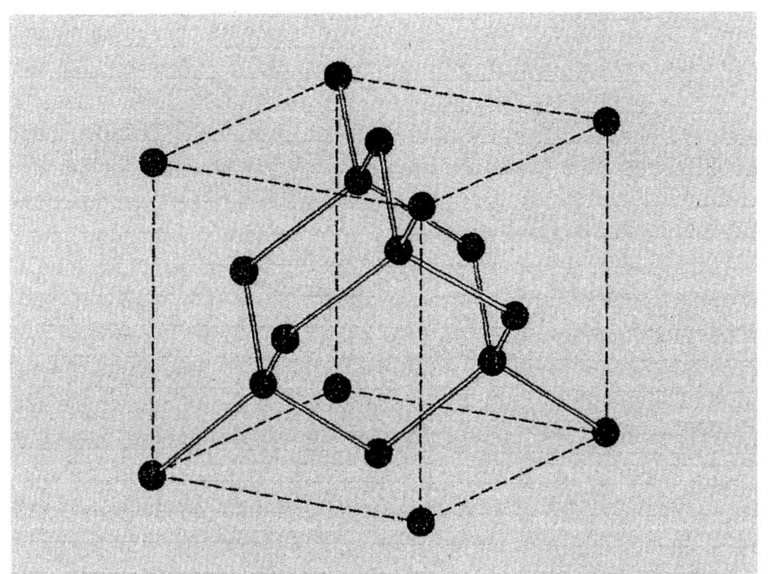

Figure 1.20 A diamond crystal. Each carbon atom forms the centre of a tetrahedron with the same shape as the methane molecule shown in Fig. 1.18. There is only one way in which these tetrahedra can be arranged so that each carbon atom forms bonds with four others.

1.9 Bonds between molecules

So far we have examined three ways in which atoms can attach themselves to one another and so form solids. In these materials the basic units are atoms or ions—that is to say, there are no molecules present. To illustrate this point let us return to the familiar example of sodium chloride. A sodium chloride molecule consists of just a single pair of sodium and chlorine ions, but a sodium ion in a crystal of sodium chloride exerts equal attractive forces on each of its six neighbouring chlorine ions. Consequently, we must say that the crystal is made up of individual sodium and chlorine ions, as opposed to sodium chloride molecules.

However, in the vast majority of solids the basic building block is the molecule. Familiar examples are available all around us—ice, organic materials and plastics are all made up of molecules. What we need to determine is how and why one molecule attracts another. As in the other types of bonding, the answer is to do with electrostatic forces, but the mechanism is rather more complex and relies on the electric dipole moment of the molecules. (If you are unfamiliar with this term, a dipole is simply a system of two equal and opposite electrical charges. The dipole moment measures the strength of the dipole and is equal to the magnitude of one of the charges multiplied by the distance between the charges.)

The molecular bond is best demonstrated by considering a couple of examples. We will begin by looking at a water molecule.

Let us consider a simple experiment. It is well known that if an ebonite rod is rubbed with a piece of fur it becomes negatively charged. If we now hold this charged rod next to a column of water, then the column of water is deflected, as shown in Fig. 1.21. We can explain this result by assuming that each water molecule behaves as a tiny electrical dipole and so is attracted to the charged ebonite rod. But why should a water molecule behave as a dipole? To answer this question we need to consider the bonds between the atoms in the molecule.

We know that a water molecule consists of two atoms of hydrogen and one of oxygen. Since oxygen is two electrons short of achieving an inert gas configuration and each of the hydrogen atoms has a single electron, we might expect the atoms to be joined by covalent bonds (see Fig. 1.22(a)). But a covalent bond means that each atom remains electrically neutral, so this process does not give rise to an electric dipole. However, the bonding process is not quite so straightforward because an oxygen atom is more electronegative than a hydrogen atom, i.e. it has a greater affinity for electrons. Consequently, we can suggest an alternative picture in which the oxygen atom steals the electrons from the hydrogen atoms, rather than sharing them. This means

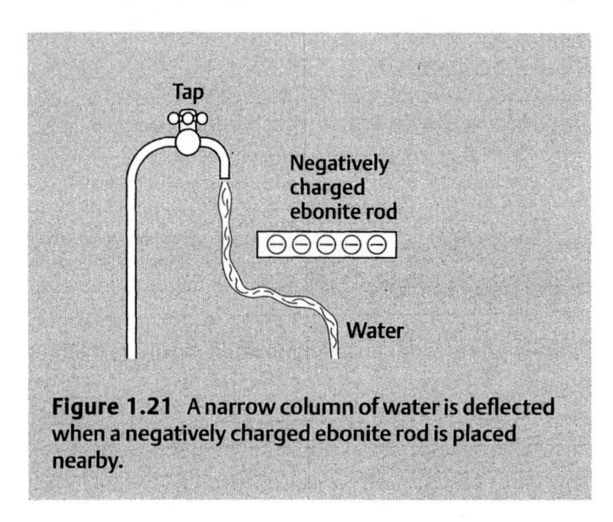

Figure 1.21 A narrow column of water is deflected when a negatively charged ebonite rod is placed nearby.

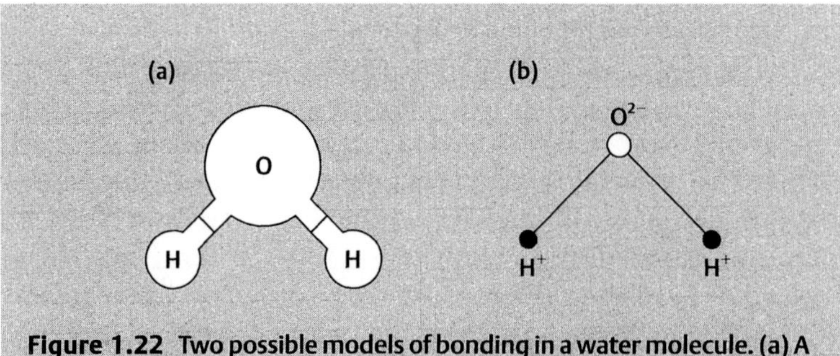

Figure 1.22 Two possible models of bonding in a water molecule. (a) A covalent representation using the building block model of Fig. 1.14, and (b) an ionic model.

that effectively we end up with a doubly charged oxygen ion, O^{2-}, and two bare hydrogen nuclei, H^+, as shown in Fig. 1.22(b). Owing to the shape of the molecule and the uneven distribution of the electrical charges, this molecule does exhibit a dipole moment. In fact, we can calculate that the dipole moment of this molecule is 1.91×10^{-29} C m (see Example 1.5). Experimental measurements show that the dipole moment is actually only 0.64×10^{-30} C m, so our ionic model is not quite correct. The hydrogen atoms are not completely ionized, but the electrons certainly spend more time in the vicinity of the oxygen atom than they do near the hydrogen atoms. So the bond between these atoms is really a mixture of ionic and covalent bonding. However, the important point is that the molecule has a finite dipole moment.

Example 1.5 Using Fig. 1.22(b), determine the dipole moment associated with the ionic model of the water molecule. (The length of the O—H bond is 0.097 nm and the angle between the bonds is 104.5°.)

Solution The water molecule has a net negative charge of 2e centred on the oxygen nucleus, and a net positive charge of 2e centred on the point A, midway between the two hydrogen ions (see diagram below).

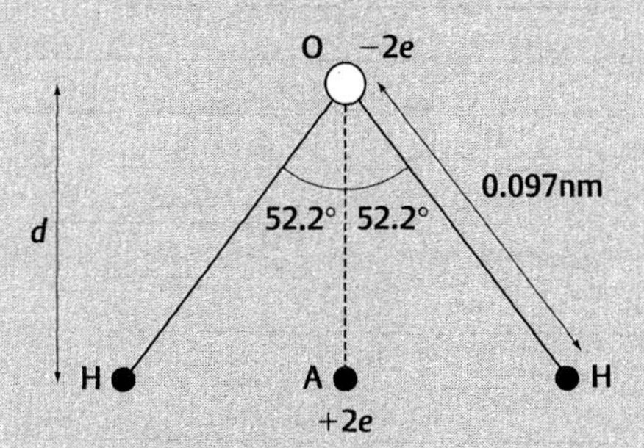

The dipole moment is therefore 2ed, where d is the distance from the oxygen nucleus to the point A.

Using trigonometry, we have

$$d = (0.097\,\text{nm})(\cos 52.25^\circ) = 0.0595\,\text{nm}$$

The dipole moment is therefore

$$2ed = (2)(1.6 \times 10^{-19}\,\text{C})(0.0595 \times 10^{-9}\,\text{m}) = 1.91 \times 10^{-29}\,\text{C m}$$

How does this information help to explain the formation of bonds between water molecules? We have seen that the oxygen and hydrogen atoms have an excess of negative and positive charge, respectively, so the oxygen atom from one molecule is attracted to the hydrogen atoms on other molecules, and vice versa. This is often referred to as a hydrogen bond. In an ice crystal there are twice as many hydrogen atoms as oxygen atoms, so this suggests that there should be two hydrogen atoms directed towards each oxygen atom. Figure 1.23 shows how this can be achieved in a small portion of an ice crystal.

The hydrogen bond occurs in many other instances besides water. For example, hydrogen bonds play an important role in determining the properties of polymers and are also responsible for binding the two chains of amino acids together in DNA to form the well-known double-helix structure.

The above model provides a bonding mechanism so long as there is some degree of ionicity in the bond, but what happens in molecules which do not exhibit an electric dipole? As an example let us consider an atom of an inert gas, such as helium. (You may wonder why we are now talking about atoms again rather than molecules. The reason is that since a helium atom has only a filled orbital of electrons it can be thought of as a molecule which consists of just a single atom.) In order to see how one helium atom exerts a force on another, let us imagine that we can take a sort of snapshot photograph of the atom at some instant in time and so determine the positions of the electrons. (Note that according to quantum theory it is not possible to determine the exact positions of small particles such as electrons at a specific moment in time. Strictly speaking we can only talk about the probability of finding an electron at a given point. However, this does not significantly affect the following argument, and so for simplicity we will assume that we know the exact positions of the electrons at any instant.) The result of our snapshot photograph might look something like Fig. 1.24(a). At this particular instant both of the electrons are to the right of the nucleus. Consequently, the centroid of the negative charge does not coincide with the nucleus, and so the atom behaves as an electric dipole. This imbalance of charge on one atom

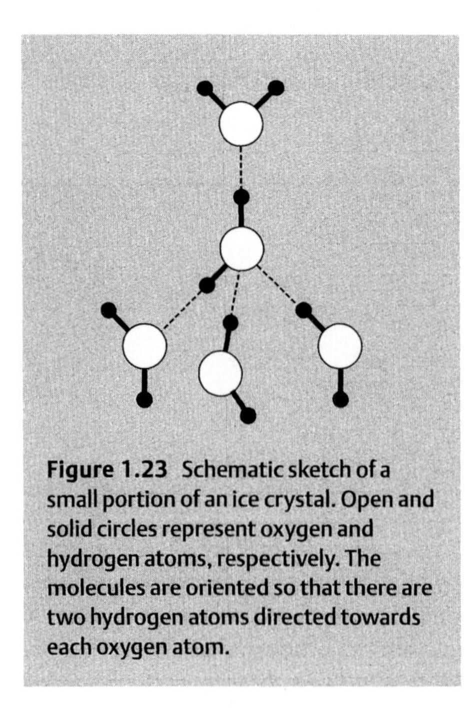

Figure 1.23 Schematic sketch of a small portion of an ice crystal. Open and solid circles represent oxygen and hydrogen atoms, respectively. The molecules are oriented so that there are two hydrogen atoms directed towards each oxygen atom.

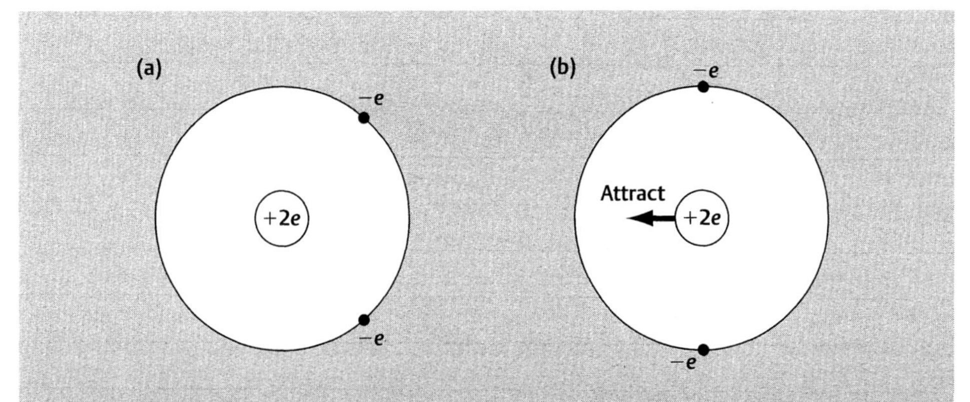

Figure 1.24 (a) A 'snapshot' view of a helium atom at a particular instant in time. (b) The instantaneous dipole on the first helium atom produces an attractive force on the the nucleus of a neighbouring helium atom.

tends to produce a similar disturbance on neighbouring atoms, and so gives rise to an attractive force between the atoms (Fig. 1.24(b)).

The result appears to be similar to that obtained for the water molecule, but there is one important difference. Whereas the dipole that exists on the water molecule can be considered to be a permanent feature of this structure, the dipole possessed by the helium atom is merely a result of the instantaneous arrangement of the electrons. If we take many similar snapshot pictures, the arrangement of the electrons, and therefore the dipole moment, will be different each time. Since there is no reason to suppose that the electrons should spend more of their time on one side of the nucleus than on the other, the average of all these dipole moments should be zero. However, the force exerted by a dipole is proportional to the square of the dipole moment. Since the square of a quantity must always be positive, the force does not average to zero.

Although the process that we have described may seem to be rather contrived, it does produce a very real effect. Admittedly, it is very weak in comparison with the other bonding forces that we described previously, but the existence of this force explains how non-polar molecules, such as carbon dioxide, solidify.

To summarize, we have seen that attractive forces arise owing to instantaneous or permanent distortions in the distribution of the electrons. These forces are referred to as van der Waals forces. The main point to note is that these forces between the molecules are considerably weaker than those between the atoms within the molecule, and therefore the molecules exist as distinct entities within the solid.

1.10 The relationship between the type of bond and the physical properties of a solid

Having considered the various ways in which bonds can form, let us now turn our attention to the question of how the type of bonding influences the properties of a material. In other words, assuming that we know what type of bond is present in a particular material, what can we deduce about the physical properties of that material?

Many properties of solids depend on the strength of the bonds which hold the constituent particles together. (In this section we will use the term 'atoms' to refer to these particles with the understanding that this refers to ions in the case of ionic and metallic bonding and to molecules in the case of molecular bonding.) For example, we started this chapter with a discussion of the variation in the melting point of different materials and postulated that the melting point is related to the strength of the bonds in the material. If we look at Fig. 1.25 we can see that indeed there is a good correlation between the melting points and the cohesive energies for the selection of materials shown. In Fig. 1.26 we present the information in a different manner, showing how the melting point varies according to the type of bonding present. We can see that the transition metals and the materials with covalent and ionic bonding form the strongest bonds, and the melting points of these materials are noticeably the highest. Accordingly, those materials which utilize the relatively weak molecular bond have very low melting points.

Since the bonds are what hold the atoms together, it is not surprising that the physical strength of a solid is also closely related to the strength of the bonds. There are several ways of quantifying the strength of a material and we will look at this subject in more detail in Chapter 3.

Some other physical properties of solids depend on certain characteristics of the bond type, rather than on the strength of the bond. A good example is the electrical conductivity of a material. What do we mean when we say that a material is a good

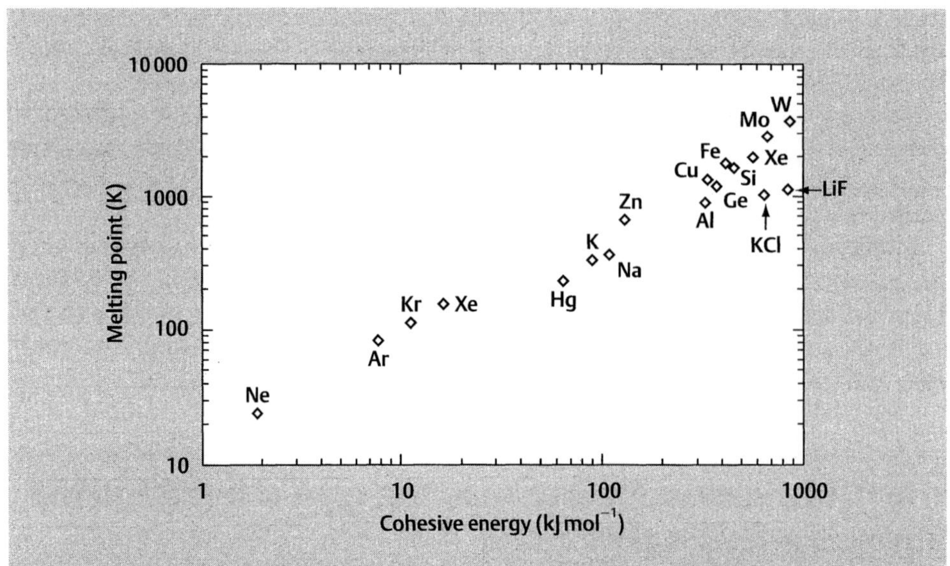

Figure 1.25 The melting point of a variety of materials plotted against the cohesive energy of the crystal. Note that log axes are used.

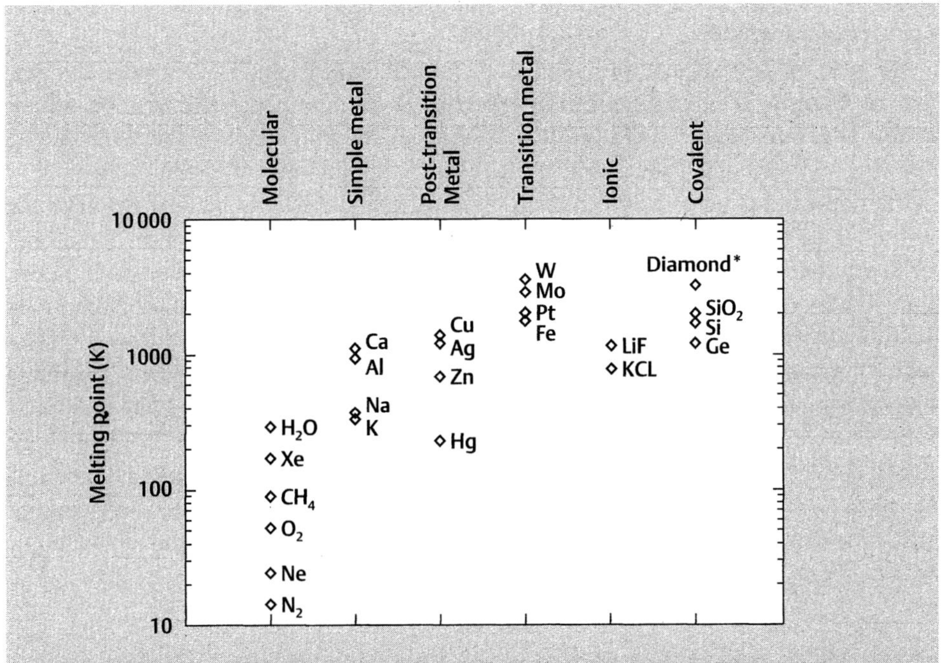

Figure 1.26 The melting point of a number of materials grouped according to the type of bonding. *Diamond is transformed to graphite at about 3300 K.

conductor of electricity? An obvious explanation is that if we place the material in an electrical circuit then a current is able to flow through it. However, this does not explain what is happening on the atomic scale. An electrical current only flows through a material if we can persuade the electrons which are already present in the material to move in the desired direction. Consequently, we can say that if an electric field is applied to the sample, then in a good conductor at least some of the electrons in the sample are free to move in response to this field. Now we have to ask, which types of bond will allow this to happen? We have seen that in metals the valence electrons are delocalized, but in other solids this is not the case. In ionic and covalent crystals all of the electrons are either strongly bound to a specific ion or shared between two atoms. Either way, they are not at liberty to stray from their assigned location. Similarly, in molecular crystals, the bonds which hold the molecule together account for all of the valence electrons. We conclude that it is a property of the type of bond which gives us the well-known result that metals conduct electricity whereas non-metals do not.

Thermal conductivity is also closely related to electrical conductivity, and so we again find that materials with a metallic bond tend to be good thermal conductors, whereas other materials are generally poor in comparison. (We shall see later, however, that it is not just the electrons which are responsible for redistributing heat through a crystal.)

It is less obvious, but equally true, that the optical properties of a material are also related to the presence of free electrons. We will not go into a detailed discussion at this stage, but just make an observation that metals reflect light, whereas many non-metals are transparent.

We have used these examples to give a brief flavour of the importance of the bonding mechanism in determining the properties of solids and we will mention others during the course of the book. We close this section by showing how the difference in bonding between diamond and graphite leads to dramatically different properties.

Diamond and graphite are both forms of pure carbon. In diamond the carbon atoms are covalently bonded to four other carbon atoms, but in graphite the carbon atoms form approximately hexagonal planes as shown in Fig. 1.27. Each carbon atom therefore bonds with just three other carbon atoms—the fourth valence electron from each atom behaves more like the delocalized electrons in a metal. It is these electrons which form the bonds between the planes of atoms. The presence of these unbound electrons means that graphite is a fairly good electrical conductor and like a metal it reflects light. However, graphite has little physical strength because of the weak bonds between the planes. In contrast, diamond is a very poor electrical conductor, it is transparent, and it is one of the hardest materials known. Since both materials consist of the same type of atom, these differences must be attributed to the type of bond within the crystal.

1.11 Summary

What have we learned from this first chapter? We have described the different types of bond which exist between atoms, or in one case between molecules, and shown that certain physical properties of solids can be understood simply by referring to the type of bond present in the material. It would seem, therefore, that the type of bonding forms a useful way of categorizing solids. Whilst this is true to some extent, and in this book we will often refer to one of these categories in order to explain some

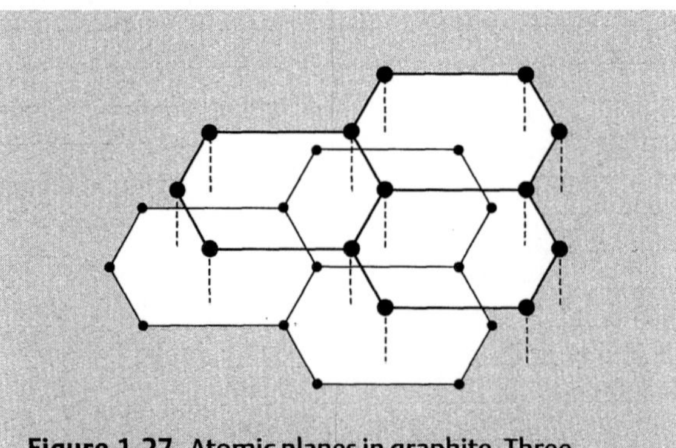

Figure 1.27 Atomic planes in graphite. Three covalent bonds exist between the carbon atoms in the same plane. The fourth electron forms a metallic bond between the planes.

general characteristic of a range of materials, we should realize that not all solids conform to such a simple classification scheme. For example, we have seen that graphite appears to possess both covalent and metallic bonds, and that the bonds in a water molecule are partially covalent and partially ionic.

In short, solids exhibit a vast variety of different properties and it is too much to expect that we can sort them into four neat categories. It is very useful if we can use a simple model to explain similar properties of many materials, but we should also remember that it is often those solids which do not behave according to the simple rules which often turn out to be the most useful materials.

Questions

1.1 The Bohr atom. Calculate the energy levels and radii of the corresponding orbits for the $n = 2$ and $n = 3$ states of the hydrogen atom.

1.2 The Bohr atom. Use eqn (1.3) to estimate the binding energy of an electron with principal quantum number $n = 1$ in a helium atom ($Z = 2$) and a lithium atom ($Z = 3$). Explain why the ionization energy of lithium is actually much lower than the value for helium and use this equation to estimate the binding energy of an $n = 2$ electron in lithium. (Remember to include the effects of screening by the $n = 1$ electron in your calculation.)

1.3 The shell model of the atom. Calculate the number of electrons required to fill all of the energy levels up to and including the 4p subshell. (Note: the 4p electrons have a higher energy than the 3d electrons but have a lower energy than the 5s electrons.)

1.4 Atoms in solids. Explain why the electrons in the filled electron subshells of an atom do not take part in the bonding process.

1.5 Ionic bonding. If the force between two ions is given by eqn (1.4), show that the potential energy is given by eqn (1.5).

1.6 Ionic bonding. Calculate the potential energy of an ion in a sodium chloride crystal in units of joules and electron volts. Hence obtain a value for the lattice energy of a sodium chloride crystal in units of joules per mole. (Ignore the effects of the repulsive force and assume that the separation of neighbouring ions is 0.281 nm.)

1.7 The repulsive force. Calculate the fractional change in the magnitude of the repulsive force between two ions as the separation is increased from a_0 to $2a_0$. Compare your result with the fractional change in the magnitude of the Coulomb force for the same increase in r. Hence explain why the repulsive force is a short-range force in comparison with the Coulomb force. (Assume that the repulsive force is as given in eqn (1.10).)

1.8 The repulsive force. (i) If the force between two ions is given by eqn (1.10), show that the potential energy is a minimum when $r = a_0$. (ii) Calculate the percentage reduction in the lattice energy of an ionic crystal due to the inclusion of the repulsive term in eqn (1.10).

1.9 The repulsive force. If the binding energy of an ion in a sodium chloride crystal is 1.29×10^{-18} J and the equilibrium separation a_0 is 0.281 nm, determine the value of the constant A in eqn (1.10).

1.10 **Metallic bonding.** The separation between neighbouring Na^+ and Cl^- ions in a crystal of NaCl is 0.281 nm, whilst the separation between neighbouring Na^+ ions in metallic sodium is 0.372 nm. Given that the ionic radius of Cl^- is 0.181 nm, determine the radius of the Na^+ ions in each case. Explain why the values are different.

1.11 **Covalent bonding.** If the length of a covalent bond between two carbon atoms is 0.154 nm, estimate the length of a polyethylene molecule which contains 25 000 CH_2 monomers.

1.12 **Covalent bonding.** Estimate the number of carbon atoms in a diamond of mass 0.2 g. (Carbon has a molar mass of 12 g mol^{-1}.)

1.13 **Covalent bonding.** Calculate the angles between the bonds in a methane molecule (as shown in Fig. 1.18).

1.14 **Molecular bonding.** (i) The potential energy between two simple, neutral molecules at a separation r is given by

$$E(r) = -\frac{Aa_0^6}{r^6} + \frac{Ba_0^{12}}{r^{12}}$$

By obtaining an expression for the force, determine the relationship between the constants A and B.

(ii) The potential energy can also be written in an alternative form:

$$E(r) = 4\epsilon\left[\left(\frac{\sigma}{r}\right)^{12} - \left(\frac{\sigma}{r}\right)^{6}\right]$$

Show that $\sigma = a_0 2^{-1/6}$ and that ϵ is the binding energy.

Chapter 2
Crystals and crystalline solids

2.1 Introduction

How are the atoms arranged in a solid? The first clues were obtained by mineralogists over 200 years ago. They noticed that crystals of a given material exhibit certain characteristic shapes. This discovery lead the Frenchman Rene-Just Häuy to predict that crystals are ordered structures made up from many small, identical building blocks arranged in a regular fashion, as shown in Fig. 2.1. The faces of the crystal are then determined by the shape of the building blocks and the way in which these blocks are arranged to form the crystal. (This model also explains how it is possible to cleave diamonds and other gemstones since the crystal will break most easily along the planes between the building blocks.)

We now know that these building blocks, which we refer to as unit cells, consist

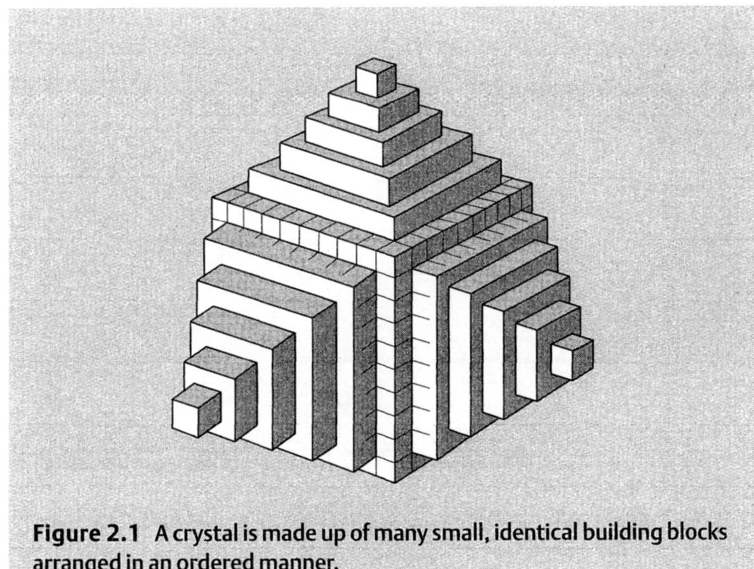

Figure 2.1 A crystal is made up of many small, identical building blocks arranged in an ordered manner.

typically of just a few atoms or molecules. This means that the shape of a crystal, which may be several centimetres across, directly reflects the arrangement of the atoms on the scale of a few tenths of a nanometre! So next time you see a diamond in a jeweller's shop window, just remember that what you are looking at is the result of a few thousand billion billion carbon atoms arranged with uncanny precision in an almost perfect geometrical array. Maybe then the price tag will not seem so exorbitant!

Of course, crystals tend to be the exception rather than the rule in nature. In many solids the regular arrangement of the atoms is not so obvious. For instance, metals can be fashioned into an infinite variety of different shapes. This would seem to suggest that the atoms in a metal are distributed in some sort of random pattern. However, if we examine the material under a high-powered microscope we find that a metal consists of many small crystals, or 'grains', as shown in Fig. 2.2. These materials are referred to as polycrystalline solids. Within each grain the atoms are arranged in a regular manner, just like a small crystal, but the orientation of the atomic planes in each grain is different from that in the neighbouring grains (see Fig. 2.3). At the grain boundaries (i.e. the regions where neighbouring grains meet) the crystal structure is disrupted and the ordered arrangement of the atoms is destroyed. We therefore have a material in which the arrangement of atoms is ordered in some regions and disordered in others. For a typical grain size of 50 μm we find from Example 2.1

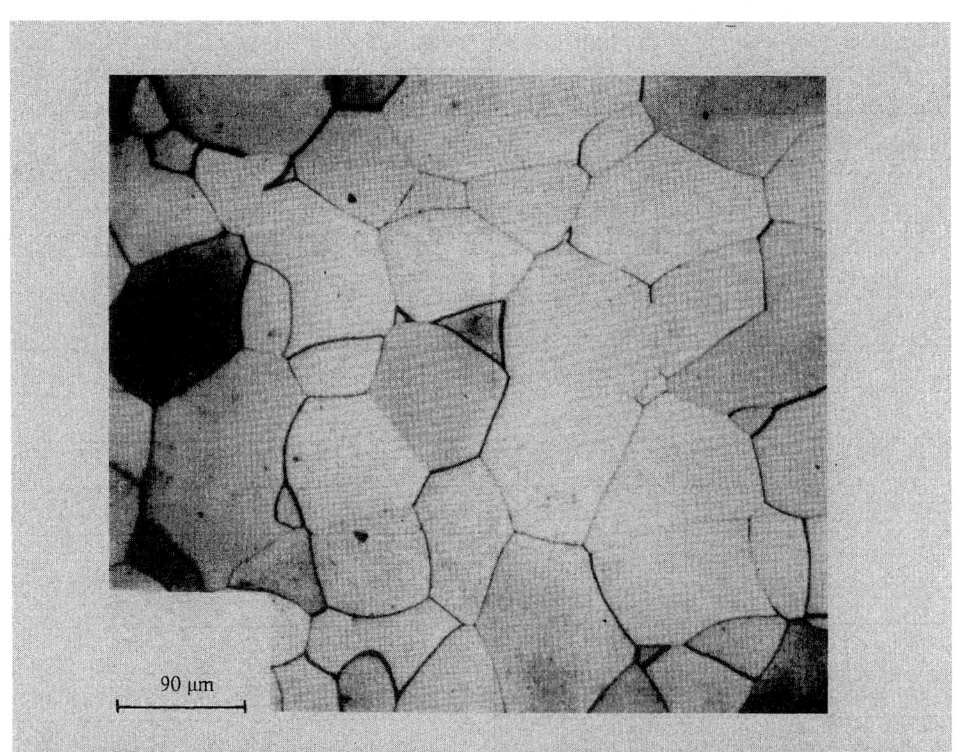

Figure 2.2 An image of the polished surface of a sample of aluminium obtained using an optical microscope. The dimensions of the grains range from about 10 μm to 100 μm.

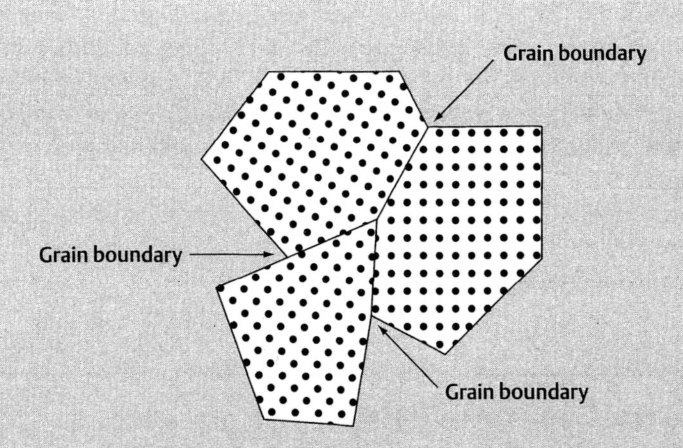

Figure 2.3 A schematic diagram showing the ordered arrangement of atoms within each grain and the random way in which the planes of atoms in neighbouring grains are aligned. (Note that the number of atoms in a typical grain is far larger than shown here.)

that only about one in every 30 000 atoms is close to a grain boundary—the remaining 99.998% of the atoms are just as they would be in a perfect crystal. We are therefore justified in describing such a material as an ordered structure.

Example 2.1 Given that the grains in a polycrystalline metal are typically 50 μm across and that metal ions have a radius of 0.15 nm, estimate the average number of ions in a grain and the proportion of these ions which are adjacent to a grain boundary.

Solution Assume the grain is roughly cubic in shape.

Volume of grain is $\approx (50.0 \times 10^{-6}\ \mathrm{m})^3 = 1.25 \times 10^{-13}\ \mathrm{m}^3$
Volume of ion is $\approx (0.30 \times 10^{-9}\ \mathrm{m})^3 = 2.70 \times 10^{-29}\ \mathrm{m}^3$
Therefore, number of ions per grain is

$$\approx \frac{(1.25 \times 10^{-13}\ \mathrm{m}^3)}{(2.70 \times 10^{-29}\ \mathrm{m}^3)} = 4.63 \times 10^{15}$$

Surface area of grain is $\approx (6)(50.0 \times 10^{-6}\ \mathrm{m})^2 = 1.5 \times 10^{-8}\ \mathrm{m}^2$
Area corresponding to one ion is $\approx (0.30 \times 10^{-9}\ \mathrm{m})^2 = 9.00 \times 10^{-20}\ \mathrm{m}^2$
Therefore, number of ions adjacent to surface of grain is

$$\approx \frac{(1.5 \times 10^{-8}\ \mathrm{m}^2)}{(9.00 \times 10^{-20}\ \mathrm{m}^2)} = 1.67 \times 10^{11}$$

Proportion of ions adjacent to grain boundary is

$$\approx \frac{1.67 \times 10^{11}}{4.63 \times 10^{15}} = 3.60 \times 10^{-5}$$

I should mention at this point that there are some solids which have no recognizable crystal structure on any scale. These are called amorphous solids and are most definitely disordered structures. In this chapter, and indeed in most of this book, we will be concerned only with the properties of ordered solids, i.e. single crystals or polycrystalline materials. We will, however, take a more detailed look at amorphous solids in Chapter 11.

In this chapter we will consider first the theoretical aspects which determine how the atoms are arranged in a crystal and then look at experimental methods of probing the structure of crystals.

2.2 Close-packed structures

What factors determine how the atoms are arranged in a crystal? The main criterion is that the total potential energy of the crystal is minimized, i.e. the binding energy is a maximum. Can we solve this problem theoretically?

For a diatomic molecule the problem would seem to be relatively straightforward, at least in principle, since we need to consider only how the potential energy varies as a function of the separation between the two atoms. For molecules with three atoms the calculation becomes more complicated because there are now three distances (between each pair of atoms) to consider. Since it is unlikely that all of the atoms will lie in the same plane, we have the added complication of dealing with a three-dimensional problem in this case. If this sounds difficult, then the prospect of trying to optimize the positions of about 10^{23} atoms in a crystal seems like a hopeless task. However, we can make use of the fact that a crystal is a regular structure made up of unit cells consisting of typically just a few atoms. This means that we need to solve the minimum energy problem for just a small number of atoms. In general, such calculations are still far from trivial, but there are many cases for which we can make useful qualitative predictions based on very simple models.

Let us begin by looking at the solids of the inert elements, such as neon and argon. Admittedly, these are not materials which we are likely to encounter very often, if at all—neon and argon melt at temperatures of 25 K (−248°C) and 84 K (−189°C), respectively. Our reason for examining these materials is that the atoms have only filled shells of electrons. This makes the problem considerably easier because we can treat the atoms as identical, electrically neutral, hard spheres (see Section 1.6). In order to minimize the energy of the crystal we simply have to find the most efficient way of arranging the spheres so as to minimize the total volume occupied. In other words, what started out as a physics problem has ended up as a purely mathematical one.

So what is the most efficient way to pack spheres together? Let us approach this in a trial-and-error manner. As a first attempt we could try a simple cubic structure which is formed by arranging the spheres so that their centres lie on a cubic lattice, as shown in Fig 2.4. The calculation in Example 2.2 shows that only just over half of the total volume of the cube is actually occupied by the spheres in this case—the rest is empty space. Surely, we can do better than this. An improvement is to place

Example 2.2 Determine the actual volume occupied by the spheres in the simple cubic structure as a percentage of the total volume.

Solution In order to solve this problem we need to determine the volume of a single cube, V_c, and the total volume of the parts of the spheres within this cube, V_s.
To calculate V_s: From Fig. 2.4 we can see that there are parts of eight spheres (one centred at each vertex) within the cube. Since each sphere is shared equally between eight cubes (because each vertex is common to eight cubes), the total volume occupied by these parts of the spheres is equivalent to the volume of one sphere, i.e.

$$V_s = (8)\left(\frac{1}{8}\right)\left(\frac{4\pi r^3}{3}\right) = \frac{4\pi r^3}{3}$$

where r is the radius of a sphere.

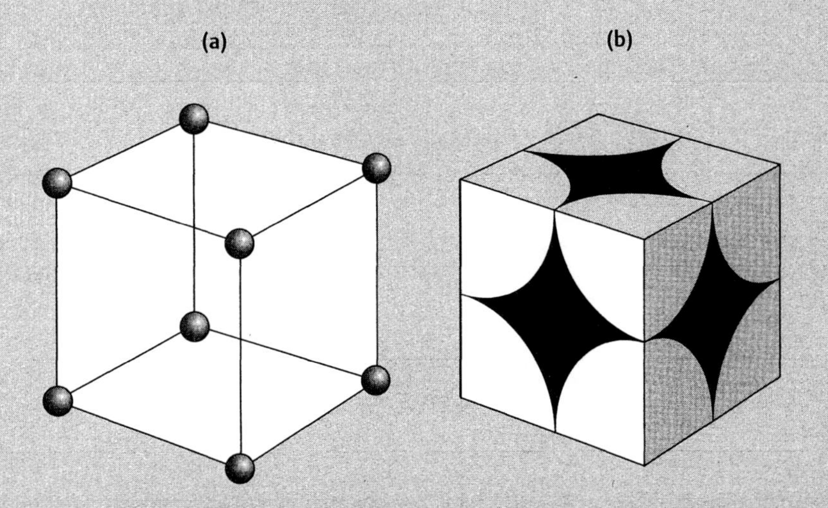

Figure 2.4 (a) A simple cubic structure is formed by placing an atom at each vertex of a cubic lattice. (b) The minimum volume is achieved when the spheres just touch along each edge of the cube.

To calculate V_c: From Fig. 2.4(b), each edge of the cube is of length $2r$. Therefore the volume of the cube is

$$V_c = (2r)^3 = 8r^3$$

The percentage volume occupied by the spheres is therefore

$$\frac{V_s}{V_c} \times 100\% = \frac{4\pi r^3/3}{8r^3} \times 100\% = \frac{\pi}{6} \times 100\% = 52.4\%$$

another atom at the centre of each cube to form a body-centred cube (see Fig. 2.5). In order to do this we have to increase the length of each side of the cube by about 15%, but since the cube now contains an extra sphere, this leads to a significant improvement in the packing efficiency. A calculation (see Question 2.1) shows that the spheres now occupy 68% of the total volume. A further refinement can be made by constructing a face-centred cubic structure. As the name suggests, this arrangement is achieved by taking a simple cubic structure and placing a sphere at the centre of each face of the cube, as shown in Fig. 2.6. (Note that there is no sphere at the centre of the cube.) In this case the spheres account for 74% of the total volume (see Question 2.2). It turns out that the face-centred cubic structure is the most efficient way to pack spheres, and so we refer to this arrangement as a close-packed structure.

Although it is not possible to pack spheres into a smaller volume than in the face-centred cubic structure, there are other equally efficient ways of arranging spheres. To see how this can be achieved we will take a more systematic approach to the problem. Let us first of all restrict ourselves to a two-dimensional problem. What is the most efficient way of arranging spheres on a plane? It is quite easy to convince yourself that this is achieved by arranging the spheres in a hexagonal pattern, as shown in Fig. 2.7. In this case there are six outer spheres which each touch the central sphere (marked 'X') and the two neighbouring spheres on either side. Since we can tessellate a plane using hexagons, this pattern can be repeated to cover an area of any size. The result is shown in Fig. 2.8(a).

How do we extend this argument to three dimensions? Let us try to picture the upper surface of the structure in Fig. 2.8(a). It must look something like an inverted

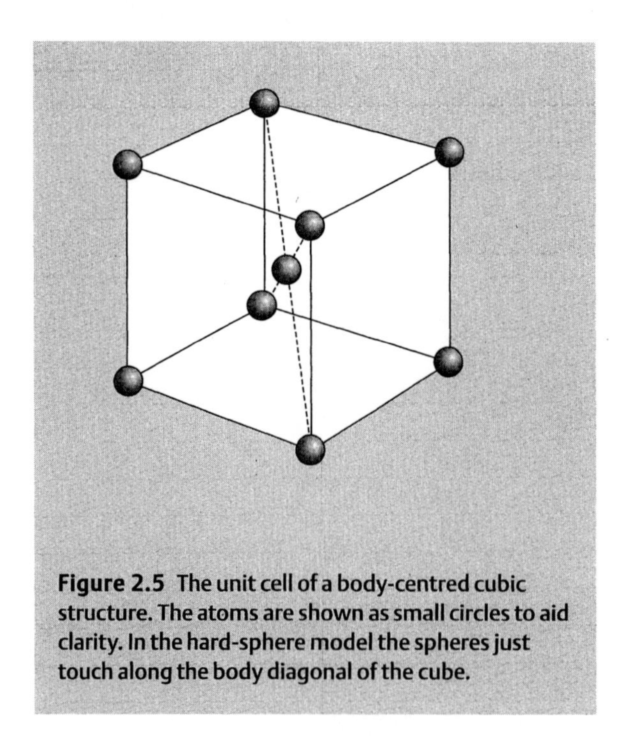

Figure 2.5 The unit cell of a body-centred cubic structure. The atoms are shown as small circles to aid clarity. In the hard-sphere model the spheres just touch along the body diagonal of the cube.

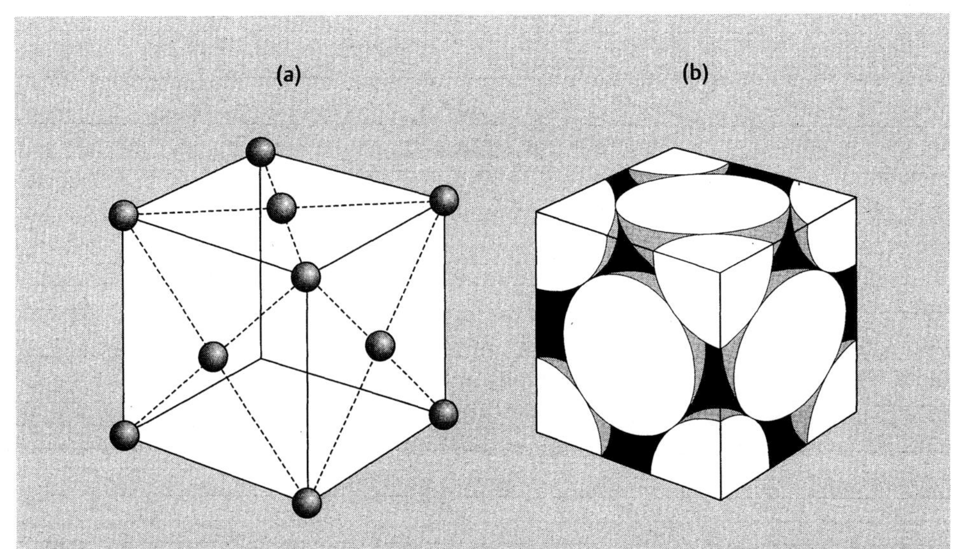

Figure 2.6 (a) A face-centred cubic structure is achieved when an atom is placed at each vertex and at the centre of each face of a cubic lattice. The atoms on the hidden faces are not shown for clarity. (b) When the atoms are represented by solid spheres, the spheres are in contact along the face diagonals of the cube.

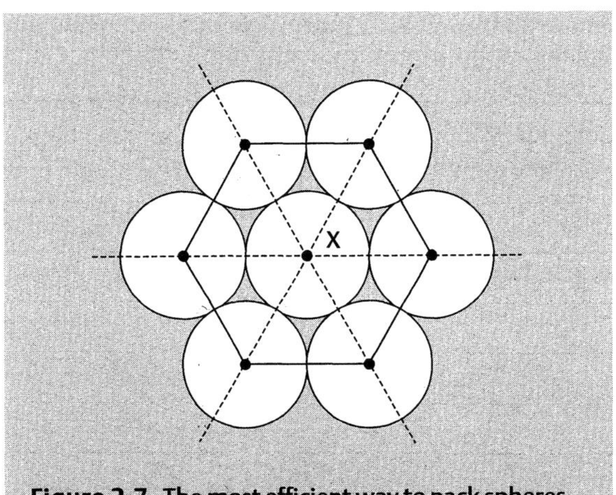

Figure 2.7 The most efficient way to pack spheres on a plane is to arrange them in a hexagonal pattern. When the pattern is repeated to cover the plane, each sphere is in contact with six others.

eggbox with the deep depressions corresponding to the gaps between the spheres. If we are going to build upwards from the plane then these depressions are the obvious places to put the next layer of spheres. In the figure we have labelled these sites alternately 'B' and 'C'. If we place spheres on all the points marked 'B', as in Fig.

2.8(b), we obtain a new plane of spheres which is identical to the first. A similar result is obtained if the spheres are placed at the the points labelled 'C'. It makes no difference which set of points is used, so long as we do not try to put spheres at both 'B' and 'C' in the same plane. What do we do with the third plane of spheres? There are two choices. If we assume that the spheres in the second plane occupy sites 'B', then the third plane of spheres can be placed either at the points 'C', or directly above the spheres in the first plane (points 'A'). Does it matter which we choose? Mathematically, the spheres occupy precisely the same volume in either case, but in practice with real atoms the total energy of the two arrangements is likely to differ slightly, and it is therefore necessary to distinguish between the different alternatives. One possibility is that the spheres alternate in the sequence ABABAB . . . and so on. This is referred to as the hexagonal close-packed structure. Another possibility is that all three possible sites are used in rotation, so that we obtain the sequence ABCABC . . . . It turns out that this is a structure that we have met already—it is the face-centred cubic structure. This is not immediately obvious, but if we slice the corner off a face-centred cubic structure, as in Fig. 2.9, we can see that the atoms in the exposed plane are arranged in the same hexagonal pattern as in Fig. 2.8. Consequently, we can conclude that both the face-centred cubic and the hexagonal close-packed structures have exactly the same packing efficiency.

Having analysed the minimum energy of a crystal from the purely mathematical viewpoint of packing spheres with maximum efficiency, let us now take a look at how well this idealized model compares with real crystals. Do the atoms in a crystal really form a close-packed structure? The answer is'yes', at least for the inert solids. It is found that all of the inert solids form face-centred cubic crystals with the exception of helium which has a hexagonal close-packed structure. What about other materials? Are there any other solids which crystallize in a close-packed structure?

Let us take a look at pure metals. (By 'pure' we mean in this context that all of the atoms are identical.) We know that in a metal each atom gives up its outer (valence) electrons, leaving an ion which has only filled shells or subshells of electrons, rather

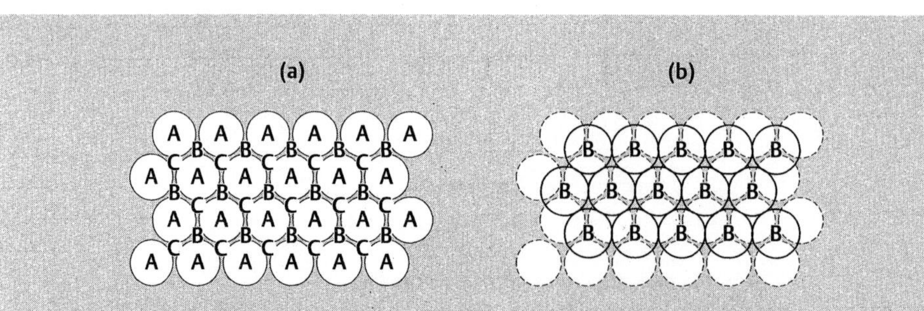

Figure 2.8 (a) An extension of the hexagonal packing of spheres to three dimensions can be achieved by first considering a plane of atoms. The centres of the spheres are labelled 'A' and the spaces between the spheres are labelled alternately 'B' and 'C'. (b) A second plane of close-packed spheres is formed by placing the centres of the spheres directly above each space marked 'B'. (An equivalent structure is formed by using the sites marked 'C' instead.)

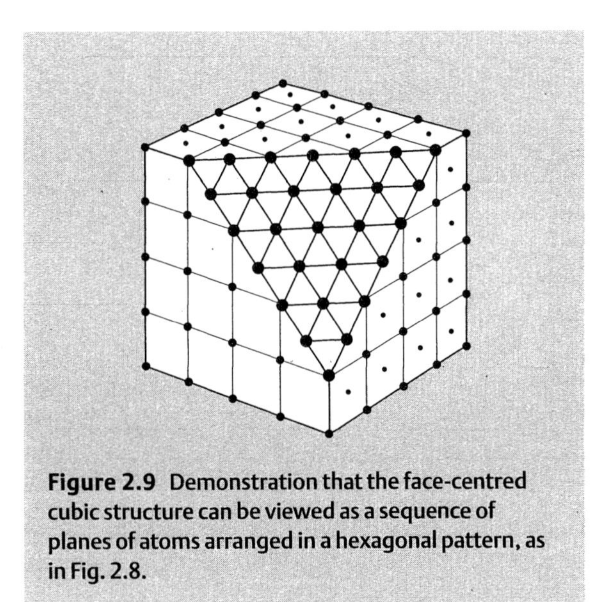

Figure 2.9 Demonstration that the face-centred cubic structure can be viewed as a sequence of planes of atoms arranged in a hexagonal pattern, as in Fig. 2.8.

like an inert element. So we might reasonably expect the ions in a pure metal to form a close-packed arrangement. How well does this agree with the experimental evidence? It turns out that roughly three-quarters of the metallic elements crystallize in either the face-centred cubic or the hexagonal close-packed structures. (Of the other metallic elements, the majority crystallize in the body-centred cubic structure, which is a slightly less efficient method of packing spheres. Obviously there are other factors which we need to take into account in order to determine the minimum energy configuration in these cases.)

Can we apply similar arguments to alloys—metals consisting of more than one type of element? The simplest type of alloy occurs when we have two types of atom, A and B, with similar atomic radii. Suppose that the atoms of type A form a close-packed structure. Since the atoms of type B are of similar size, it may be possible to substitute some of these atoms for those of type A without altering the structure of the crystal. This means that we still have a close-packed structure, but not all of the atoms are of the same type. An example is brass, which is an alloy of copper and zinc (see Fig. 2.10). Pure copper forms a face-centred cubic crystal, and since the zinc ions are only 4% larger than the copper ions, it is possible to achieve concentrations of up to 38% zinc at room temperature without altering the crystal structure. Alloys of this type are referred to as substitutional solid solutions. (The term 'solution' merely implies that there is no particular order to the way in which the ions A and B are distributed in the crystal.)

Another type of alloy is obtained when the sizes of the atoms are quite different. In this case it may be possible to squeeze the smaller atoms (B) into the gaps between the larger atoms (A) without significantly distorting the crystal structure, as shown in Fig. 2.11. This is referred to as an interstitial solid solution. What is the maximum size for an interstitial atom in a close-packed structure? Intuitively we would expect it to be very small because we are already making optimum use of the space available.

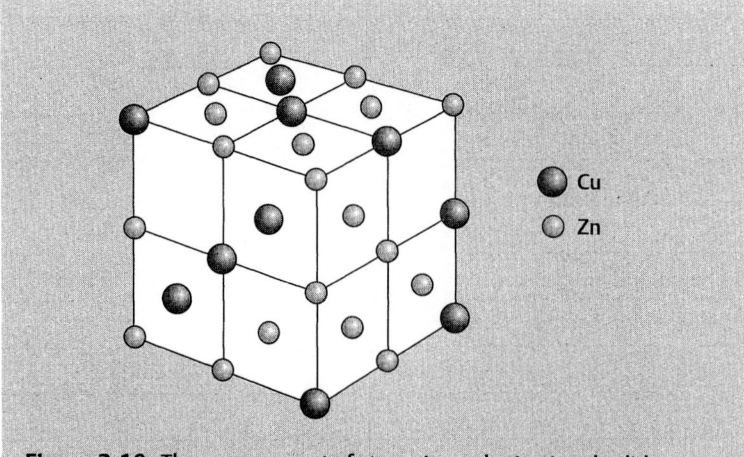

Figure 2.10 The arrangement of atoms in a substitutional solid solution, in this case a section of a brass crystal. The copper and zinc ions are placed at random on a face-centred cubic lattice. (Only atoms on the visible surfaces are shown for clarity.)

However, the results are surprising. Example 2.3 shows that the radius of an interstitial atom in a face-centred cubic structure can be as much as 41% of that of the larger atoms without affecting the positions of these atoms. In practice, substantially larger interstitial atoms can be accommodated by allowing the lattice to distort slightly. In

Example 2.3 Determine the maximum radius of the sphere *B* in Fig. 2.11 which can be placed into a face-centred cubic structure without affecting the positions of the other spheres.

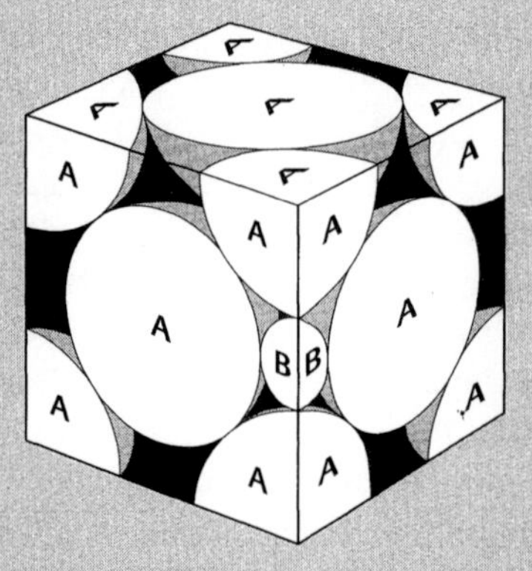

Figure 2.11 In an interstitial alloy a small atom, B, is accommodated in the spaces between the larger atoms, A. In this instance the spheres of type A form a face-centred cubic structure.

Solution If the radii of the spheres *A* and *B* are denoted *a* and *b*, respectively, we can redraw a section of Fig. 2.11 as follows:

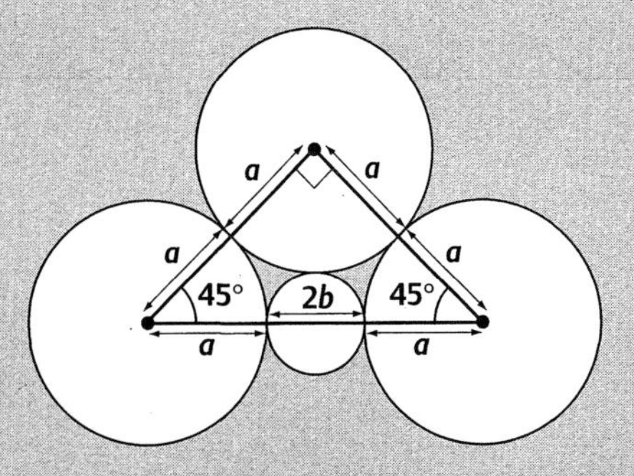

It follows that

$$\cos 45^\circ = \frac{2a}{2a + 2b}$$

Rearranging, we obtain

$$b = a\sqrt{2} - a = 0.41a$$

contrast, although the packing of spheres is less efficient in the body-centred cubic structure, the largest interstitial atom in this case which does not disrupt the lattice has a radius of only 16% of that of the larger atoms (see Question 2.3). This result is relevant in the manufacture of steel, as we shall see in Section 2.8.

In ionic materials the ions also achieve filled electron shells, but the situation is rather more complicated than in metals because we now have negative as well as positive ions. However, we have already seen (in Section 1.5) that the potential energy can be minimized by arranging the ions so that each negative ion is surrounded by positive ions (and vice versa). The other important point is that the ions differ in size—the negative ions are generally much larger than the positive ions. (This is because a positive ion has more protons than electrons, and so the electrons experience a greater attractive force than in the uncharged atom. The converse applies for a negative ion.) On the basis of this information we would expect the ionic crystal to form a crystal similar to that shown in Fig. 2.11 with the negative ions forming a close-packed structure and the positive ions occupying the interstitial sites. This is indeed the case for sodium chloride and for many other ionic compounds, although in most cases the positive ions are rather too large to occupy the interstitial sites without disrupting the positions of the negative ions. In the case of sodium chloride, for example, the sodium ions have a radius which is 54% of that of the chlorine ions. Consequently, the chlorine ions are pushed slightly further apart than they would be in an ideal face-centred cubic structure, as shown in Fig. 2.12.

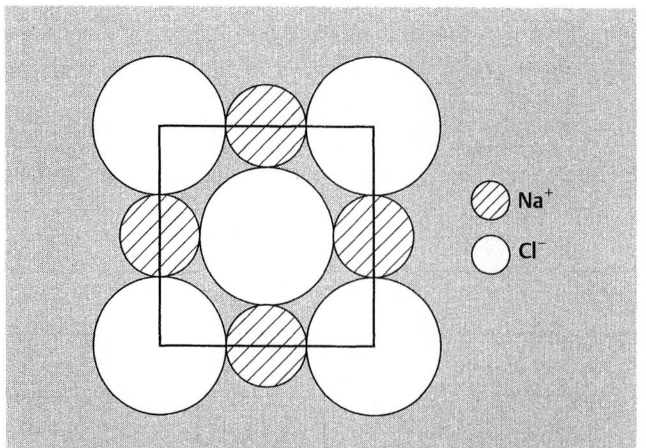

Figure 2.12 A sodium chloride crystal can be described approximately as a face-centred cubic chlorine crystal with the sodium ions occupying the interstitial positions. In this diagram, one face of a unit cell is shown schematically with the ions drawn to scale. Note that because the sodium ions are slightly larger than the maximum size for an interstitial, the chlorine ions are not in contact across the face diagonal of the cube.

2.3 Non-close-packed structures

Although the idea of packing spheres into the minimum possible volume gives us a good idea of the crystal structure in many cases, there are some instances in which this approach is not applicable.

In a covalent material the bonds form specifically between pairs of atoms, and we have seen that the angles between the bonds are fixed. They are sometimes referred to as directed bonds. Consequently, the lowest energy structure of a covalent crystal is determined by the angles between the bonds. In a diamond each carbon atom bonds to four other carbon atoms and we obtain the structure shown in Fig. 1.20. We say that this is an open structure, meaning that a relatively small percentage of the total volume is occupied. If we assume that the atoms are hard spheres with a radius equal to half the bond length (i.e. the distance between the centres of adjacent carbon atoms) it can be shown that the spheres occupy just 34% of the total volume.

It is interesting to note that graphite has a rather more compact structure than diamond, and accordingly has a larger cohesive energy than diamond. This means that graphite is the most stable form of carbon at normal temperatures and pressures. This might seem to be a rather worrying statement if you have recently invested a large amount of money in a diamond, but there is no need to panic. Your diamond will not turn into graphite overnight (or even in the next thousand years!). The structures of diamond and graphite are so different that it requires a large amount of energy to change from one state to the other. This amount of thermal energy is only available at very high temperatures. Consequently, there is no danger of a diamond turning into graphite—unless it is kept at a temperature in excess of 3300 K for any length of time!

Molecular crystals are another example of materials in which packing of spheres does not provide a reliable model of the crystal structure. In this case the reason is obvious: although individual atoms and ions are approximately spherical in shape, the majority of molecules are far from spherical.

A second reason why some molecules do not pack together quite as efficiently as we might expect is due to the distribution of electrical charges in the molecule. We have already seen an example of this in the previous chapter when we considered the water molecule. Since the oxygen and hydrogen atoms acquire slight negative and positive charges, respectively, each molecule in a crystal of ice orients itself so that the hydrogen atoms are directed towards the oxygen atoms in nearby molecules (and vice versa). The result is shown in Fig. 1.23. Each water molecule has just four nearest neighbours, which suggests that the molecules occupy only a relatively small percentage of the total volume. In fact, the random arrangement of the molecules in liquid water turns out to be more efficient (in terms of the percentage volume occupied) than the ordered arrangement in an ice crystal, which is why ice is less dense than water.

2.4 The crystal lattice

So far we have been concerned with the crystal structure—in other words, with the way in which the atoms are arranged in a crystal. In some circumstances it is more useful to consider the crystal from a slightly different viewpoint, by describing it in terms of a lattice. In this and the following section we will look at the crystal lattice and see how this picture helps up to label the faces of a crystal and the atomic planes within a crystal.

A lattice is a mathematical concept which was studied in particular by the nineteenth century French crystallographer Auguste Bravais. It is defined as an infinite array of points which appears to be exactly the same no matter which of the points we choose as the origin. This is the same as saying that all points in a lattice are equivalent to one another. An example of a section of a two-dimensional Bravais lattice is shown in Fig. 2.13. The vectors **a** and **b** shown in the diagram are the primitive vectors of the lattice, so called because we can generate all the other points in the lattice by using combinations of these two vectors. For example, in the case shown it is easy to convince yourself that all of the points in the lattice can be described by the position vectors

$$\mathbf{R}_{mn} = m\mathbf{a} + n\mathbf{b} \tag{2.1}$$

where m and n can take all integer values.

In three dimensions the simplest Bravais lattice is based on the simple cubic structure (Fig. 2.4). The primitive vectors in this case are mutually perpendicular and are of

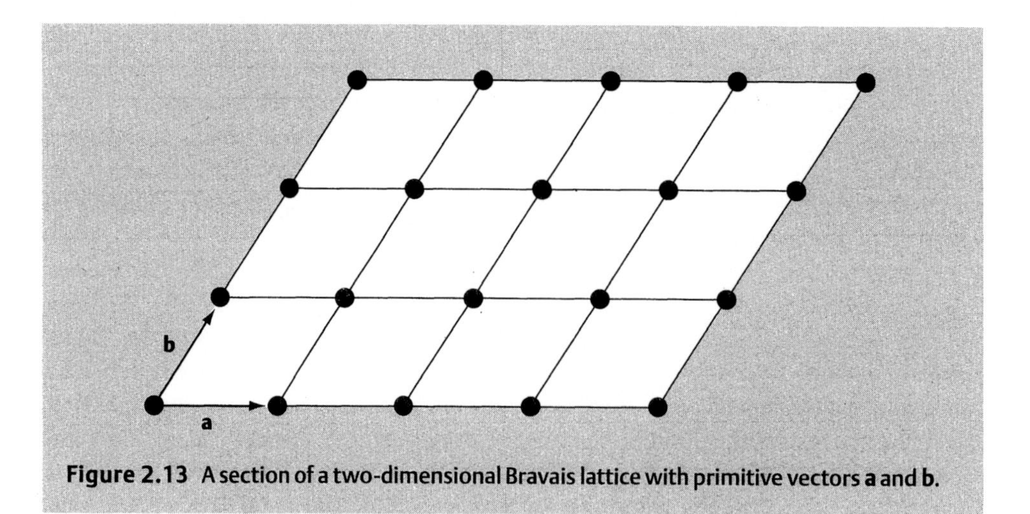

Figure 2.13 A section of a two-dimensional Bravais lattice with primitive vectors **a** and **b**.

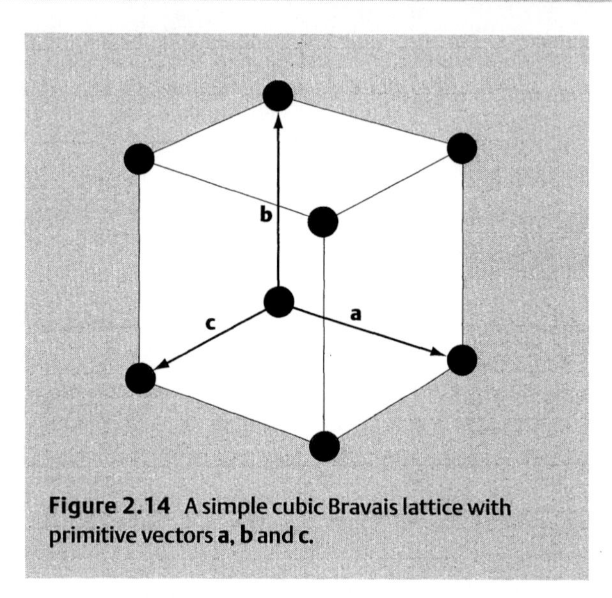

Figure 2.14 A simple cubic Bravais lattice with primitive vectors **a**, **b** and **c**.

equal length, as shown in Fig. 2.14. It is perhaps less obvious, but it is also possible to construct lattices based on the body-centred cubic and face-centred cubic structures. Figure 2.15 demonstrates that all the sites in a body-centred cubic structure are equivalent. You may wish to construct a similar diagram to convince yourself that the same is true for the face-centred cubic structure.

It follows that for the inert solids (with the exception of helium) and for some pure metals (those in which the atoms are arranged in a face-centred cubic or body-centred cubic structure), the positions of the atoms correspond directly to the lattice points. This is not true for all crystals. For example, the atomic sites in a sodium chloride crystal cannot constitute a lattice because not all of the sites are equivalent to one another—half of them are occupied by sodium ions and half by chlorine ions. However, if we consider the sodium ions separately, we can see from the sketch of the sodium

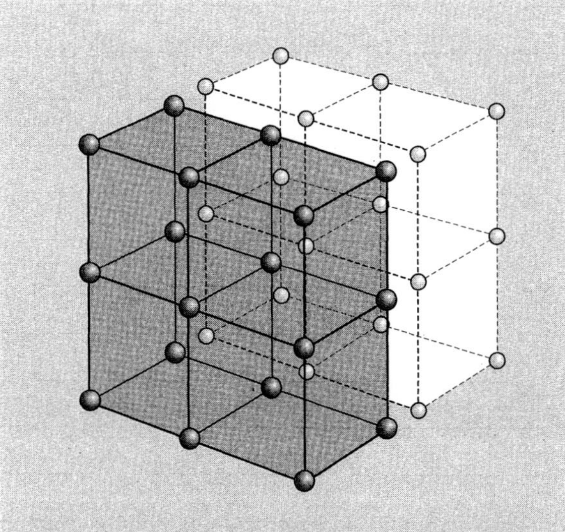

Figure 2.15 A demonstration that a body-centred cubic structure forms a Bravais lattice. We can view the structure in two ways. The solid lines form a cubic lattice with the solid circles at the vertices and the open circles at the centres of the cubes. Alternatively, the dashed lines form a cubic lattice with the open circles at the vertices and the solid circles at the centres of the cubes. Consequently, the structure looks the same regardless of whether the origin is on a solid circle or an open circle.

chloride crystal in Fig. 1.8 that they are arranged in a face-centred cubic structure. Similarly, it is found that the chlorine ions also form a face-centred cubic structure. Consequently, the structure of sodium chloride can be represented by a face-centred cubic lattice in which there are two ions, one sodium and one chlorine, associated with each lattice site. We therefore can see that the crystal lattice is related to, but is not necessarily the same as, the crystal structure.

2.5 Labelling crystal planes

Having described what is meant by the crystal lattice, how can we use this concept to label a crystal plane, such as the one shown in Fig. 2.16? A general prescription, first suggested by William Miller in 1839, is as follows:

(1) determine the intercepts on the axes in units of the primitive vectors **a**, **b** and **c**;
(2) take the reciprocal of each number;
(3) reduce these numbers to the three smallest integers h, k and l, having the same ratio;

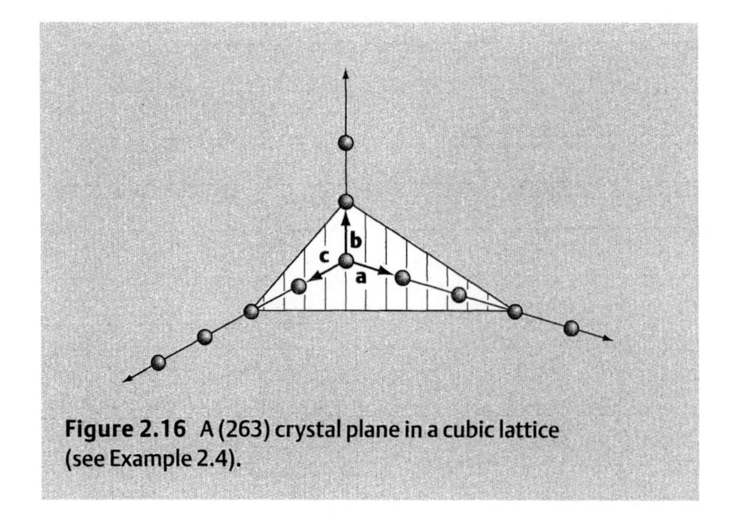

Figure 2.16 A (263) crystal plane in a cubic lattice (see Example 2.4).

(4) the values h, k and l are called the Miller indices and are enclosed in parentheses (hkl) to denote a crystal plane.

This procedure may seem unnecessarily complicated, but the beauty of it is that it can be applied to any lattice, not just cubic ones.

Example 2.4 Determine the Miller indices for the plane shown in Fig. 2.16.

Solution Following the procedure outlined above:

(1) The intercepts on the axes in units of **a**, **b** and **c** are 3, 1 and 2, respectively.
(2) The reciprocals are $\frac{1}{3}$, 1 and $\frac{1}{2}$, respectively.
(3) The smallest set of integers with the same common ratio are 2, 6, 3.
(4) Consequently, we can label the plane a (263) plane.

By following the above prescription we find that the plane in Fig. 2.16 is a (263) plane (see Example 2.4). Some other examples are shown in Fig. 2.17 for a cubic lattice.

A few points concerning Miller indices are worth noting. I will simply list them here for reference:

- If a plane is parallel to a particular axis, the corresponding Miller index is zero. (In effect, we are saying that the intercept with the axis occurs at infinity and the reciprocal of infinity is zero.)
- If a plane intercepts an axis at a negative value, this is denoted by placing a bar over the corresponding Miller index, i.e. $(\bar{1}11)$.
- The Miller indices (*hkl*) do not actually denote a single plane but a set of parallel planes. For example, if we apply the above procedure to the planes parallel to the one shown in Fig. 2.17(a) with intercepts at 2**c**, 3**c**, and so on, we find that they are all (001) planes.

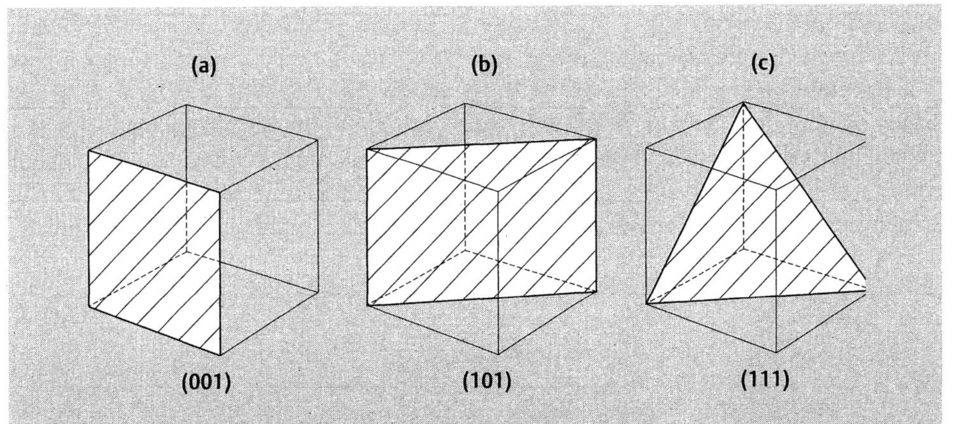

Figure 2.17 Some common crystal planes shown in a cubic lattice using the primitive vectors defined in Fig. 2.14.

- The set of equivalent planes (i.e. those which have the same symmetry) is denoted by using braces $\{hkl\}$, e.g. in a cubic lattice $\{100\}$ refers to the (100), (010) and (001) planes.
- A direction in a lattice is specified using square brackets $[hkl]$. This is defined as the direction perpendicular to the plane (hkl).

2.6 X-ray diffraction

So far in this chapter we have examined the structure of crystals from a purely theoretical viewpoint. How do we know that these ideas are correct? Can we determine experimentally how the atoms are arranged in a crystal? We certainly cannot see the atoms in a crystal using a conventional optical microscope. This is not a problem of inadequate technology, it is far more fundamental. The reason is due to the fact that a light wave does not have a definite position in space, but is smeared out over a region corresponding to the wavelength of the light. Consequently, it is not possible to observe any features which are much smaller than the wavelength of the light that we are using to examine the structure. Since the wavelength of visible light is at least 380 nm, and the separation of atoms in a crystal is only about 0.2 nm, we cannot hope to see any atoms even with the most sophisticated optical microscope.

This suggests that if we want to 'see' the atomic structure of a crystal we should use electromagnetic radiation with a wavelength which is comparable with the spacing of the atoms, i.e. the X-ray region of the spectrum. However, we cannot build an X-ray microscope because there is no equivalent of the optical lens which works at X-ray wavelengths. Instead we must use a different approach.

X-rays were first used to probe the atomic structure of crystals in 1912 following a suggestion by the German physicist Max von Laue that a crystal might be capable of strongly diffracting X-rays. This indeed proved to be the case. When X-rays are directed onto a crystal, as in Fig. 2.18(a), a characteristic diffraction pattern is observed on the photographic plate (see Fig. 2.18(b)).

A simple explanation of this result was provided the following year by W. L. Bragg based on the idea that the X-rays are weakly reflected by the lattice planes in the crystal so that the resulting image is due to a combination of reflected beams from many parallel lattice planes. The main assumption in Bragg's interpretation of the experiment is that the surface from which the X-rays are reflected is flat. This means

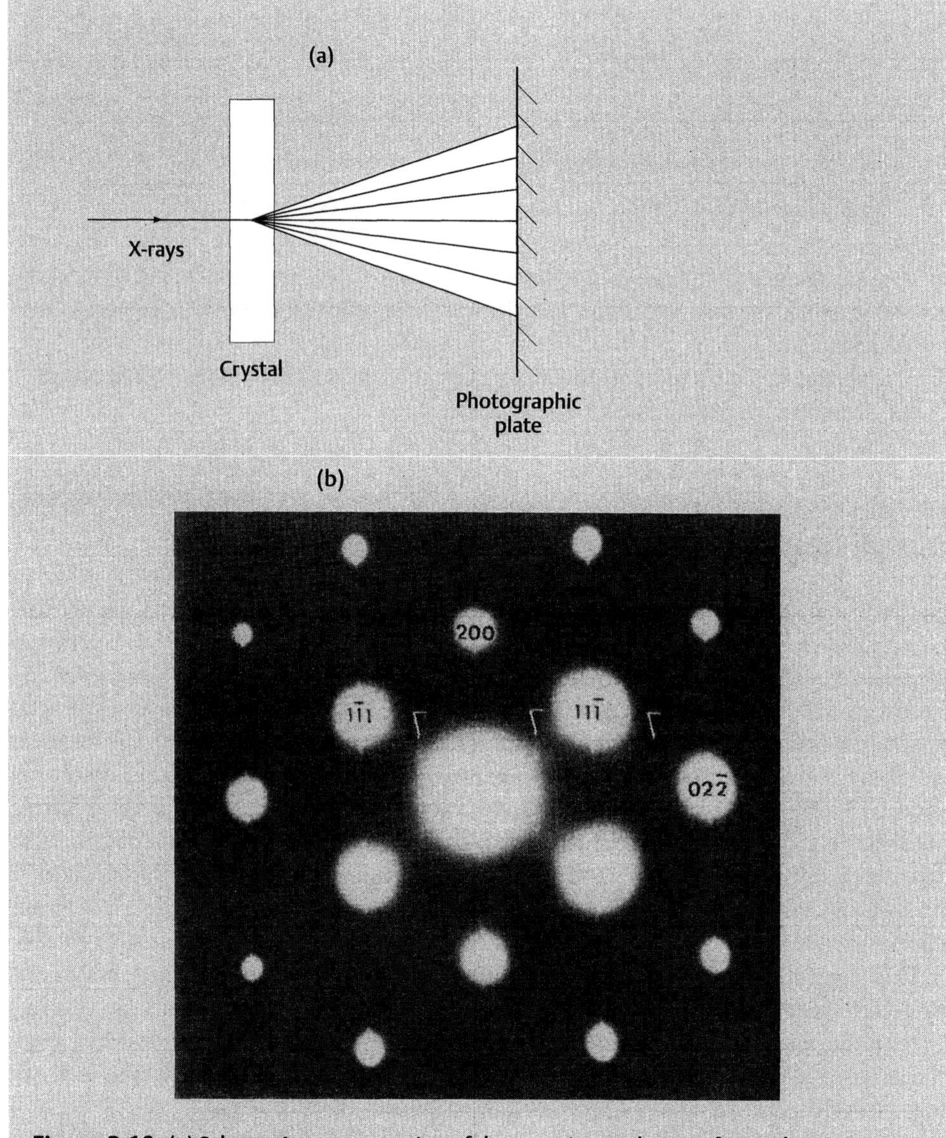

Figure 2.18 (a) Schematic representation of the experimental set-up for studying diffraction of X-rays by a crystal. (b) An example of a diffraction pattern from an SiGe crystal.

that, as in conventional optics, the angle between the incident beam and the plane is equal to that between the reflected beam and the plane. Let us then consider the reflections from two parallel planes within the crystal, as shown in Fig. 2.19. We can see from the diagram that the ray which is reflected from plane B travels slightly further than that which is reflected from plane A. The extra distance travelled by the ray reflected from plane B is equal to the sum of the distances from U to V and from V to W. Using basic trigonometry we find that this distance is equal to $2d \sin \theta$ (see Question 2.7), where θ is the angle between the incident beam and the reflecting plane and d is the separation of the atomic planes. This path difference means that in general the reflected waves from adjacent planes will have different phases and so will partially cancel one another. However, if the path difference is equal to one wavelength (or any integral number of wavelengths), then the reflected waves are in phase. This is known as the Bragg condition and is expressed by the relation

$$n\lambda = 2d \sin \theta \tag{2.2}$$

where n is an integer and λ is the wavelength of the X-rays.

The above discussion suggests that if we keep the wavelength constant and rotate the sample (so that the angle θ changes) we should observe a series of bright and dark regions corresponding to conditions of constructive and destructive interference, respectively. You may have encountered similar interference patterns in other situations, such as Young's double-slit experiment. However, there is a major difference in this case. So far we have only considered the reflections from two lattice planes, but experimental measurements suggest that only about 0.01% of the wave is reflected by each lattice plane. This means that we need to consider contributions from about 10 000 lattice planes. How does this affect our argument? If the Bragg condition is exactly satisfied for reflections from two parallel planes, then the result is valid for reflections from any number of planes because all of the parallel planes are equally

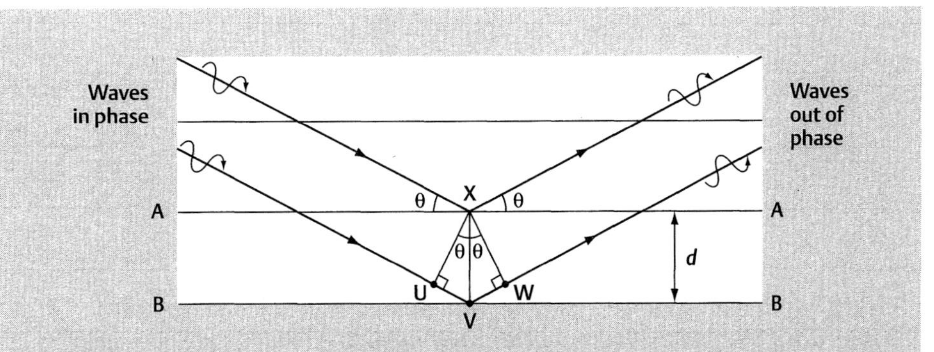

Figure 2.19 The Bragg construction used to explain the reflection of X-rays from a crystal. Reflections from two adjacent atomic planes are shown. It is assumed, as in standard optics, that the angle of incidence is equal to the angle of reflection. (Notice, however, that the angle θ is measured from the reflecting plane, not from the normal to the plane.)

spaced. In other words, we will still get a condition of constructive interference. However, a small deviation from the Bragg condition means that each reflected wave has a slightly different phase. The result of combining 10 000 such waves is that destructive interference occurs. We therefore observe a very sharp peak, as shown in Fig. 2.20, from which the separation of the lattice planes can be determined with considerable accuracy.

Whilst the above analysis leads to the correct result, I should point out that the basic assumption of this argument is incorrect. Since the sizes of the atoms are comparable with the wavelength of the X-rays, the surface from which the X-rays are diffracted is certainly not flat. How then do we explain the ray diagram in Fig. 2.19? The way to proceed is to treat the crystal as a collection of atoms and consider how the X-rays are scattered by the individual atoms. It can be shown, after some mathematical gymnastics, that this approach gives the same result as eqn (2.2). (Further details of this analysis are given in Chapter 2 of Rosenberg.)

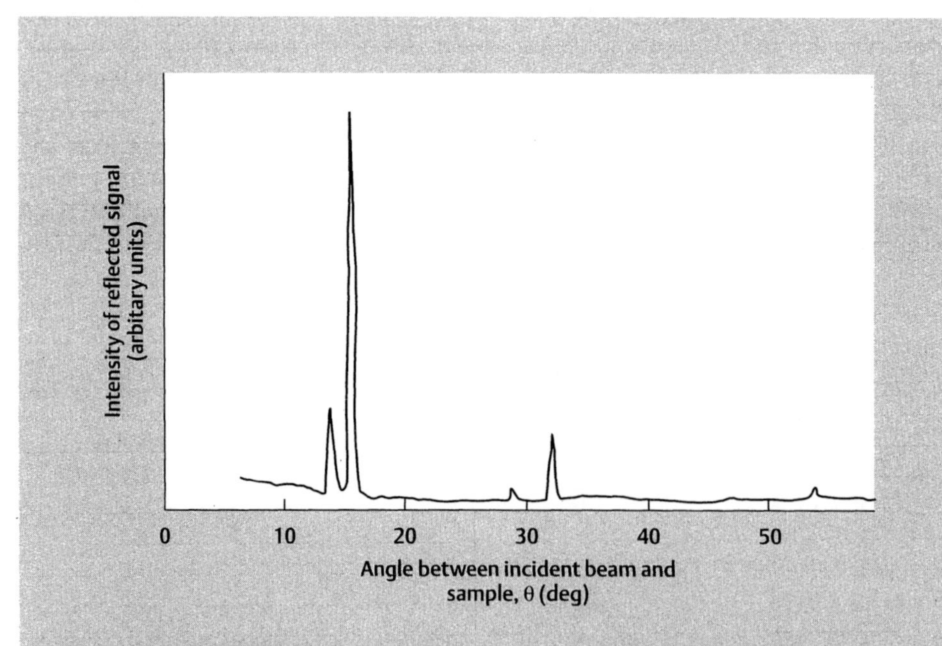

Figure 2.20 Results obtained from a simple undergraduate laboratory experiment involving X-ray diffraction from the (100) planes of a sodium chloride crystal. The graph shows the intensity of the reflected signal as a function of the incident angle θ. Note that the X-ray source in this instance generates two wavelengths of X-rays which explains the appearance of the double peaks. Three sets of peaks can be seen, corresponding to $n = 1, 2$ and 3. (Data supplied by Margaret Evans.)

2.7 Electron microscopes

X-rays are not the only means of probing the atomic structure of solids. Electrons and neutrons are also capable of exhibiting wavelike behaviour with a wavelength which is comparable with, or smaller than, the separation of the atoms in a crystal. Neutron scattering produces a diffraction pattern which is similar to that obtained using X-rays, but since neutrons have a magnetic moment this can provide additional information when studying magnetic materials. In contrast, electrons can be used in a quite different way because it is possible to focus the electron beams using electric or magnetic fields. This means that we should be able to build an instrument analogous to an optical microscope which allows the atomic structure of a crystal to be viewed directly. Does this happen in practice?

According to quantum theory, the wavelength of a particle of mass m is given by (see Appendix A)

$$\lambda = \frac{h}{\sqrt{2mE}} \tag{2.3}$$

where E is the kinetic energy of the particle and h is Planck's constant. Using this formula, the calculation in Example 2.5 suggests that electrons with an energy of a few electron volts should be suitable for observing the atomic structure in a crystal. However, if we direct a beam of electrons with this sort of energy at a crystal, it is unlikely that we will discover anything about the arrangement of the atoms, because most of the electrons will be absorbed by the crystal! To avoid this problem it is necessary to use electrons with a much higher energy—typically several hundred keV—and the crystal samples must be at most a few hundred nanometres thick so that a large proportion of the electrons pass through the specimen.

Example 2.5 Find the kinetic energy of an electron with a wavelength of 0.5 nm.

Solution Using Eqn (2.3)

$$E = \frac{h^2}{2m\lambda^2} = \frac{(6.63\times 10^{-34}\,\mathrm{J\,s})^2}{(2)(9.11\times 10^{-31}\,\mathrm{kg})(0.5\times 10^{-9}\,\mathrm{m})^2}$$

$$= 9.65\times 10^{-19}\,\mathrm{J} = 6.03\,\mathrm{eV}$$

What is the smallest object that we can see using an electron microscope? Since an electron with an energy of 100 keV has a wavelength of about 0.004 nm, we might suppose that such an instrument would easily be capable of resolving individual atoms. Unfortunately, there are other factors which limit the resolution of an electron microscope to about 0.5 nm. (The reasons are discussed in Young—see Further reading.) This is just about sufficient to observe individual atoms in a crystal, as can be seen from Fig. 2.21. The image is certainly not good enough to be used for making any quantitative measurements, but clearly demonstrates the ordering of the atoms

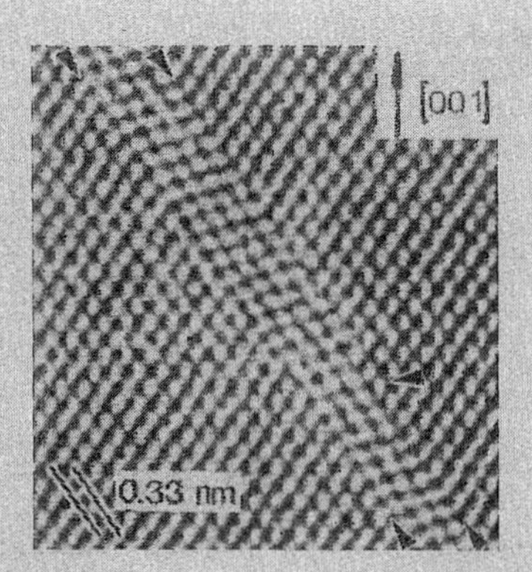

Figure 2.21 An electron microscope image of an SiGe crystal. This level of resolution allows faults in the atomic structure of the crystal to be observed directly. The fault lines run from the top left to the bottom right corner of the image. They can be seen most clearly by studying the image at a shallow angle from the lower left corner.

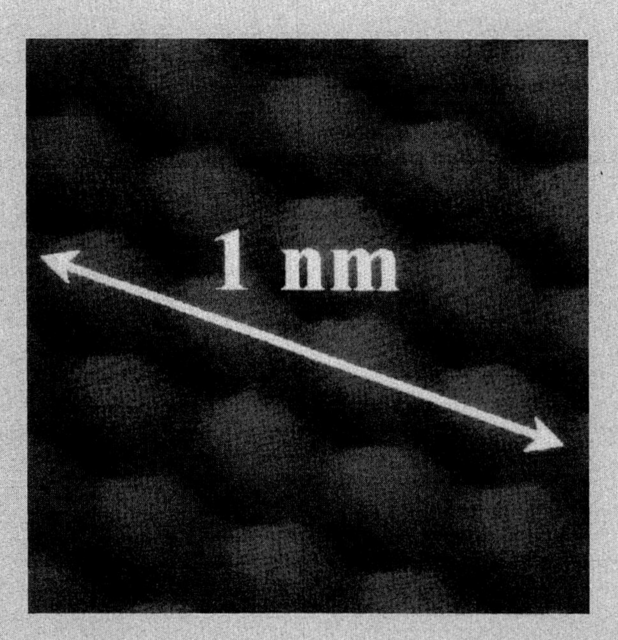

Figure 2.22 An image of a graphite surface obtained using a scanning tunnelling microscope. The individual carbon atoms are clearly visible. (By kind permission of Ken Snowdon.)

in the crystal. It is also possible to notice disturbances in the ordering such as the two fault lines in this crystal which run from near the bottom right to the top left corner of the image.

Whilst we are on the subject it is worth mentioning that other microscopes have been developed which have far greater sensitivity than the electron microscope. Scanned probe microscopes use a fine probe—with a tip which may be just a single atom across—to scan the surface of an object. The first such instrument, the scanning tunnelling microscope, was invented by Gerd Binnig and Heinrich Rohrer at IBM Zurich in 1981. Since then, numerous variations on this theme have been produced. The principle of the scanned probe microscope relies on keeping the tip of the probe at a constant distance (typically less than 1 nm) from the surface. In this way a contour map of the surface is produced, which allows the individual atoms to be clearly distinguished. Figure 2.22 shows an image of a graphite surface obtained using a scanning tunnelling microscope.

2.8 Allotropic phase transitions—changing the crystal structure*

We are all familiar with the idea that a change in temperature or pressure can bring about a phase change. For example, we know that at atmospheric pressure pure ice melts and turns into a liquid at a temperature of 273 K (0°C). However, it is less well known that a phase change can occur in which the material remains solid both before and after the phase change. This is referred to as an allotropic phase change and in this case it is the crystal structure which changes.

When a material goes from a solid to a liquid an obvious physical change takes place, but it is not always so easy to tell when an allotropic phase change occurs. In some instances there is an abrupt change in one or more properties of the solid, but the sure way to know that an allotropic phase change has occurred is if latent heat is involved. Latent heat is a quantity of heat which is either absorbed or released without any corresponding change in the temperature of the substance. In an allotropic phase change the latent heat can be thought of as the energy required to rearrange the atoms into the new crystal structure.

Why does an allotropic phase change occur? The answer lies in thermodynamics. In previous discussions we have assumed that the most stable structure is the one with the lowest total potential energy, i.e. the greatest binding energy. However, thermodynamics tells us that the most stable structure at a temperature T is actually the one in which the free energy is minimized. The free energy, F, is defined as

$$F = U - TS \tag{2.4}$$

where U is the total potential energy (which in thermodynamics is usually referred to as the internal energy) and S is the entropy of the system. (The entropy can simply be thought of as a measure of the disorder in the system.)

We can see from this definition that at a temperature of 0 K the free energy is exactly equal to the total potential energy, and so our previous assumption should certainly predict the most stable structure at absolute zero. However, for many materials the total potential energy is quite similar for two or more crystal structures. For example, in several metals the differences between the total potential energies of hypothetical face-centred cubic, body-centred cubic and hexagonal close-packed arrangements are relatively small. Therefore, variations in the entropies of these structures can mean that at a high temperature the structure with the lowest free energy is not necessarily the same as the structure with the minimum potential energy.

To be more specific, let us consider two possible structures, A and B, of a particular material. Suppose that the potential energy and the entropy of structure A are lower than those of structure B. This means that the free energy of the two structures varies as shown in Fig. 2.23. At absolute zero the crystal has structure A, but at the critical temperature T_c there is an allotropic phase change. (This of course assumes that T_c is lower than the melting point of the material.)

Several metals such as aluminium, copper, iron and titanium exhibit allotropy. For example, iron forms a body-centred cubic crystal at room temperature, but when heated above 1185 K (912°C) it changes to a face-centred cubic structure. This allotropic phase change is of considerable importance in the manufacture of steel. Let us consider the behaviour of a small quantity of carbon in an iron crystal. Since the radius of a carbon atom is 62% of that of an iron atom, the carbon atoms are just about small enough to occupy an interstitial position in the face-centred cubic crystal. Consequently, at temperatures greater than 1185 K it is possible to incorporate several per cent of carbon into an iron crystal without significantly distorting the crystal structure. However, as the temperature is lowered, the crystal transforms to the body-centred cubic structure and the interstitial sites become too small to accommodate the carbon atoms (see Section 2.2). The way in which the excess carbon atoms are incorporated into the room temperature crystal depends on various factors, such as how rapidly the material is cooled, and has a significant effect on the mechanical properties of the material. This means that two pieces of steel which have the same carbon content may vary enormously in strength depending on how they have been formed.

2.9 Summary

In this chapter we have seen that the structure of a solid is determined by energy considerations. In particular, the arrangement of the atoms is generally such that the total potential energy of the system is minimized. In many cases this corresponds to a close-packed structure, meaning that the atoms are arranged so as to occupy the minimum volume.

We also know that the atoms in a solid form an ordered structure which we can picture as a regular lattice with an atom, or group of atoms, associated with each lattice

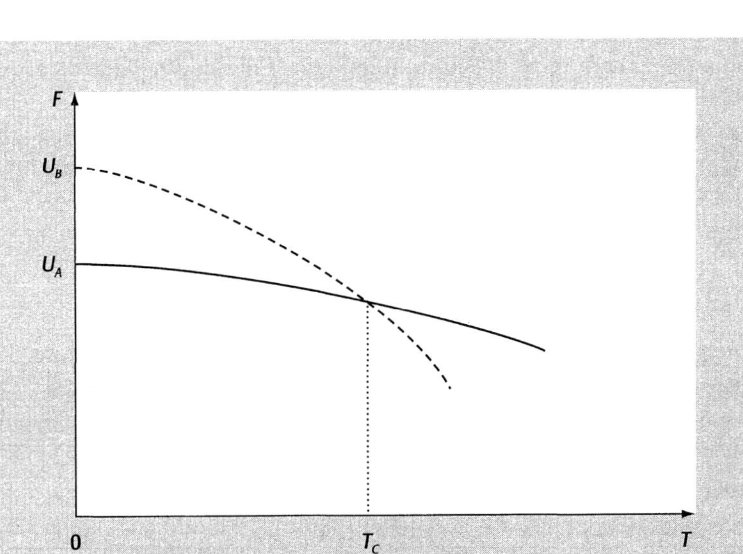

Figure 2.23 The free energies of two hypothetical structures A and B for a particular material are shown by the solid and dashed lines, respectively. Since the most stable structure is the one with the lowest free energy, there is an allotropic phase change at the critical temperature T_C.

site. This ordering is simply another example of the atoms achieving a minimum energy configuration since the total potential energy of such a structure is lower than that of a disordered arrangement.

The regular structure of a crystal is central to our understanding of the behaviour of solids. Indeed, more advanced texts on this subject rely on the periodicity of a solid in order to make quantitative calculations of many of these properties. Having said that, I should point out that in some circumstances it is the irregularities in the crystal structure which ultimately determine the properties of a real solid. For example, in the next chapter we shall see that even when only a tiny fraction of a per cent of the atoms are associated with imperfections in the crystal structure, the strength of the solid may be altered by several orders of magnitude.

Questions

*Questions marked * indicate that they are based on the optional sections.*

2.1 Close-packed structures. Determine the actual volume occupied by the spheres in the body-centred cubic structure shown in Fig. 2.5 as a percentage of the total volume.

2.2 Close-packed structures. Determine the actual volume occupied by the spheres in the face-centred cubic structure shown in Fig. 2.6 as a percentage of the total volume.

2.3 Close-packed structures. Determine the maximum radius of sphere which can be placed into a body-centred cubic structure without affecting the positions of the other spheres.

2.4 The crystal lattice. The primitive vectors for a face-centred cubic structure, **a**, **b** and **c**, are shown in Fig. 2.24. Write down expressions for **a**, **b** and **c** in terms of the length of one edge, d, and the unit vectors $\hat{\mathbf{x}}$, $\hat{\mathbf{y}}$ and $\hat{\mathbf{z}}$. Show that the points A, B, C and D can be specified by position vectors of the form

$$\mathbf{R}_{lmn} = l\,\mathbf{a} + m\,\mathbf{b} + n\,\mathbf{c}$$

where l, m and n are integers. Determine the values of l, m and n in each case.

2.5 Miller indices. A plane in a lattice with primitive vectors **a**, **b** and **c** has intercepts at 2**a**, −**b** and **c**. Use Miller indices to label the plane and the direction perpendicular to this plane.

2.6 Miller indices. Draw a sketch of a (012) plane in a simple cubic lattice.

2.7 X-ray diffraction. Show that the path difference between the two rays shown in Fig. 2.19 is equal to $2d\sin\theta$.

2.8 X-ray diffraction. Determine the maximum wavelength for which Bragg reflection can be observed from a crystal with an atomic separation of 0.2 nm.

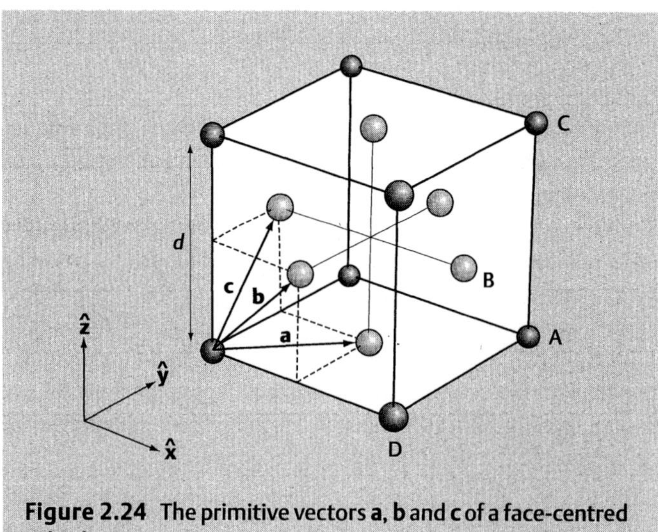

Figure 2.24 The primitive vectors **a**, **b** and **c** of a face-centred cubic lattice.

2.9 X-ray diffraction. The X-ray wavelengths used to obtain the data in Fig. 2.20 are 0.153 93 nm (denoted K_α) and 0.139 02 nm (K_β). Decide which of the peaks correspond to the K_α radiation. Given that $n = 1$ K_α peak at $\theta = 15.80°$, determine a value of d for the sodium chloride crystal used in this experiment. Assuming that there is no error in the measurement of θ and that the width of the peak at half maximum intensity is 0.1°, estimate the uncertainty in the value of d.

2.10 X-ray diffraction. A simple cubic crystal is illuminated with X-rays of wavelength 0.09 nm at a glancing angle. The crystal is rotated and the angles at which Bragg reflection occurs are measured. Which set of crystal planes will give the smallest angle for first-order reflection? If this angle is 8.9° determine the spacing between these crystal planes. At what angle will first-order reflection be obtained from the (110) crystal planes?

2.11 Electron microscopes. Determine the wavelength of an electron with a kinetic energy of 100 keV.

2.12 Electron microscopes. Discuss the advantages and disadvantages of using electron microscopes in comparison with X-ray diffraction for determining the structure of a crystal.

2.13 *Allotropic phase change. Iron is observed to undergo a transformation from body-centred cubic to face-centred cubic structure at 1185 K. Assuming that in each case the atoms behave as hard spheres and that the size of these spheres is not affected by the transformation, determine the percentage change in volume that occurs.

Chapter 3
Mechanical properties of solids

3.1 Introduction

The basic mechanical properties of solids—how a material reacts when it is subject to an applied stress—are something that we become familiar with at a very young age. As infants we discover that most solids do not change shape no matter how hard we hit them or pull them, a few change shape quite dramatically, and some have the alarming habit of breaking into several pieces when we bend or drop them.

A more scientific approach reveals that no solids are completely rigid. All materials deform to some extent, although in many cases the deformation is so small that it can only be observed with sophisticated measuring apparatus. If the stress is below a certain limit, known as the yield stress, the deformation is elastic, which means that the object returns to its original shape once the stress is removed. If the stress exceeds the yield stress, the deformation is no longer reversible and there are essentially two types of behaviour. The material either breaks as the yield stress is reached (this is obviously not a reversible process!) or it undergoes plastic deformation, in which case the change in shape is permanent. A typical stress–strain curve for the latter case is shown in Fig. 3.1. It can be seen that the magnitude of the plastic deformation is much larger than is achieved in the elastic regime.

The mechanical behaviour of solids therefore falls into two categories. A material is described as ductile if it undergoes plastic deformation beyond the yield stress, or as brittle if there is no region of plastic deformation. It should be noted that, contrary to the normal usage of the word, the scientific term 'brittle' does not necessarily imply that a material breaks easily. For example, a diamond is brittle because it does not undergo any plastic deformation, but it is certainly not easy to break!

Having briefly described the basic mechanical properties of solids, there are several questions that we might be tempted to ask. For example, why are some solids brittle and others ductile? What is the difference between elastic and plastic deformation? And, can we predict the value of the yield stress for a given material? To answer these questions we need to examine how a stress affects a solid on the atomic scale.

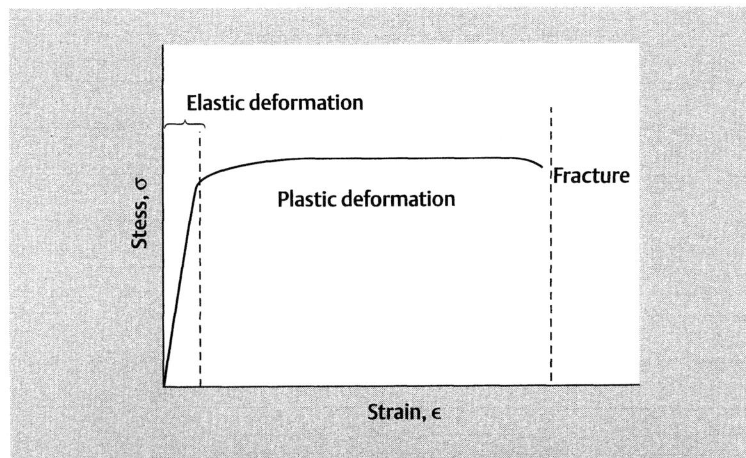

Figure 3.1 A typical stress–strain characteristic for a ductile material. Note that the strains produced during plastic deformation are typically several hundred times larger than those produced by elastic deformation.

3.2 Elastic deformation—macroscopic approach

Let us begin by looking at the behaviour of a material in the elastic regime. In this section we will examine the observable changes that take place when a stress is applied to a solid object, and in the next section we will explain these characteristics from an atomic viewpoint.

The behaviour of a wire subject to a tensile stress can be investigated by using a metal wire which is fixed at its upper end so that it hangs freely, as shown in Fig. 3.2. A stress can be applied to the wire by attaching a mass to its lower end. If a force F produces an extension Δl in a wire which is originally of length l and cross-sectional area A, then we can define the stress, σ, and the strain, ε, by the equations

$$\sigma = \frac{F}{A} \tag{3.1}$$

$$\epsilon = \frac{\Delta l}{l} \tag{3.2}$$

By varying the force and measuring the corresponding extension we find that the strain is directly proportional to the stress, as shown in Fig. 3.3. This means that we can define a constant of elasticity, Young's modulus Y, which is equal to the gradient of the stress-strain line. Thus

$$Y = \frac{\sigma}{\epsilon} = \frac{F}{A}\frac{l}{\Delta l} \tag{3.3}$$

For a given material it is found that the gradient of the curve is independent of the shape or size of the wire; therefore the value of Young's modulus is a characteristic of the material.

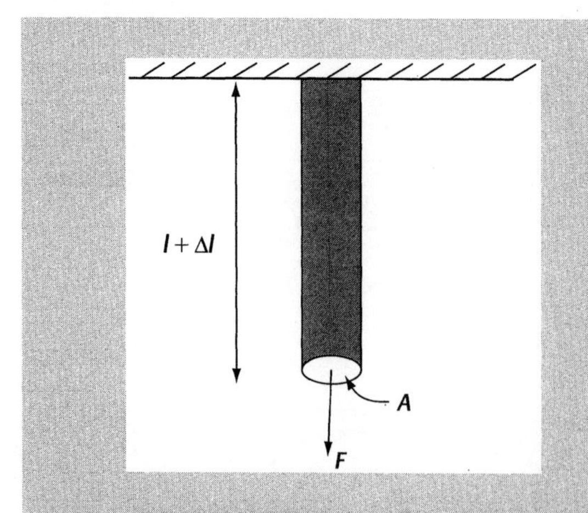

Figure 3.2 A wire of length l and cross-sectional area A is extended in length by Δl when a force is applied along the axis of the wire.

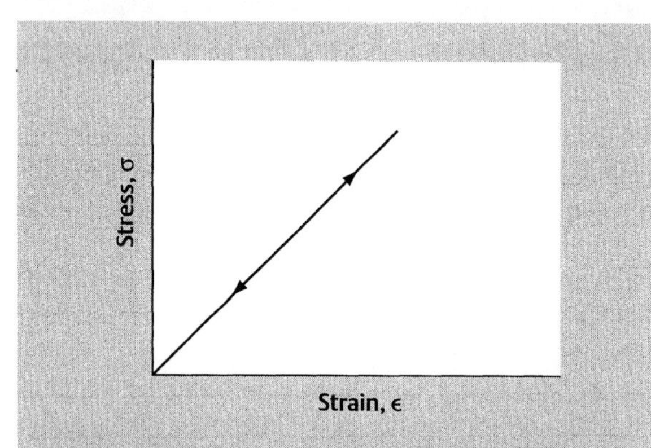

Figure 3.3 A typical stress–strain graph obtained using the apparatus in Fig. 3.2. The arrows on the line indicate that the curve is the same whether the wire is being loaded or unloaded, i.e. the deformation is reversible.

The extension of the wire is not the only change that takes place when we apply a tensile stress. There are also changes in the perpendicular directions. If we consider a cuboid with sides of length l_1, l_2 and l_3, as shown in Fig. 3.4, we find that the effect of a stress parallel to the sides of length l_1 produces changes in the perpendicular directions given by

$$\frac{\Delta l_2}{l_2} = \frac{\Delta l_3}{l_3} = -\nu \frac{\Delta l_1}{l_2} \tag{3.4}$$

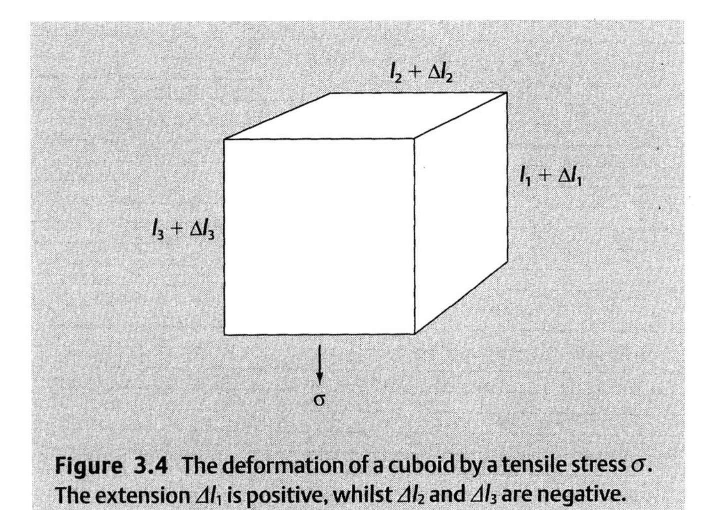

Figure 3.4 The deformation of a cuboid by a tensile stress σ. The extension Δl_1 is positive, whilst Δl_2 and Δl_3 are negative.

where ν is a dimensionless quantity called Poisson's ratio. The minus sign indicates that the change in length in the directions perpendicular to the applied stress is of the opposite sense to that in the direction parallel to the stress, i.e. a tensile stress which produces an increase in length along l_1 leads to a decrease in l_2 and l_3. Since the value of ν is always found to be less than 0.5, we can show that the volume of a solid always increases under a tensile stress (see Question 3.2).

A solid can also be deformed in other ways. In particular, we can apply a shear stress, which results in a twisting action, or a hydrostatic pressure, which means that a uniform pressure is applied to all faces of the object. The corresponding elastic moduli are the modulus of rigidity, G, and the bulk modulus, K, respectively. These are defined in a similar way to Young's modulus. The modulus of rigidity is given by the shear stress, τ, divided by the angle of shear, α

$$G = \frac{\tau}{\alpha} \tag{3.5}$$

(see Fig. 3.5) and the bulk modulus is equal to the change in the applied pressure, ΔP, divided by the resulting fractional change in volume—hence

$$K = -\Delta P \frac{V}{\Delta V} \tag{3.6}$$

The minus sign is included to make the value of K positive, because increasing the pressure leads to a decrease in volume and therefore ΔV is negative.

The mechanical properties of a solid in the elastic regime are therefore characterized by four quantities: Young's modulus, Poisson's ratio, the modulus of rigidity and the bulk modulus. If we assume that the solids are isotropic (i.e. that they have the same properties in every direction), it can be shown that these moduli are connected by the relationships

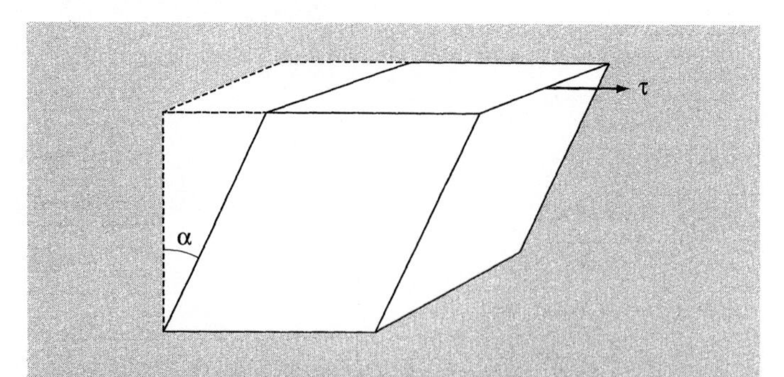

Figure 3.5 The application of a shear stress τ to a cuboid produces a deformation which can be characterized by the angle of shear α.

$$G = \frac{Y}{2(1+\nu)} \tag{3.7}$$

$$K = \frac{Y}{3(1-2\nu)} \tag{3.8}$$

(A proof is given on pages 178–182 of the book by Guinier and Jullien—see Further reading.) Typical values of these quantities for a selection of materials are given in Table 3.1. The fact that the elastic moduli are very large explains why most solids do not deform significantly under stress. For example, a mass of 1 tonne suspended from the end of a steel cable of 1 cm diameter produces an extension of only 0.06% (see Question 3.1).

Table 3.1 Typical values of Young's modulus Y, the modulus of rigidity G, the bulk modulus K and Poisson's ratio ν for a selection of solids

	Y ($10^9\,\mathrm{N\,m^{-2}}$)	G ($10^9\,\mathrm{N\,m^{-2}}$)	K ($10^9\,\mathrm{N\,m^{-2}}$)	ν
Diamond	950	390	540	0.21
Al	70	24	72	0.33
Cu	130	48	140	0.35
Fe	120	70	170	0.17
Pb	15	6	43	0.40
W	350	150	320	0.28
Brass	100	37	110	0.35
Glass	75	23	41	0.22
Steel	210	84	170	0.29

3.3 Elastic deformation—microscopic approach

How do we explain the elastic deformation of a solid from an atomic viewpoint? As a starting point, let us begin by reminding ourselves of the forces between the atoms in a crystal. In Chapter 1 we found that the combination of repulsive and attractive forces leads to a resultant force which varies with atomic separation as shown in Fig. 1.9(a). The equilibrium separation, a_0, corresponds to the position at which the net force is zero. This is a position of stable equilibrium because if the atoms are moved apart there is a net attractive force which pulls them back together again, and if they become too close together the repulsive force pushes them apart. Consequently, if a tensile force, F_T, is applied to the crystal, the atoms move apart until the attractive force balances the external force, as shown in Fig. 3.6. This results in a new equilibrium separation, a_1. The process is reversible because if the force is removed, the net attractive force pulls the atoms back to their original separation, a_0. A similar argument shows that a compressive force, F_C, leads to a reduction in the atomic spacing.

This argument not only explains why the deformation is reversible, but also demonstrates why the stress is proportional to the strain. If we look closely at Fig. 3.6, we can see that in the region around the equilibrium separation, a_0, the curve approximates to a straight line. This means that the increase in the separation of the atoms is directly proportional to the external force, and therefore on a macroscopic scale the strain is linearly related to the stress. This is only true in the vicinity of a_0, i.e. for very small values of strain. As the atoms move further from the equilibrium separation we can see that the curve is definitely non-linear. This is rather academic in most cases because the elastic limit generally corresponds to a strain of much less than 0.01 (i.e. a change in dimensions of less than 1%), for which the curve deviates only very slightly from a straight line. However, larger values of elastic strain can be

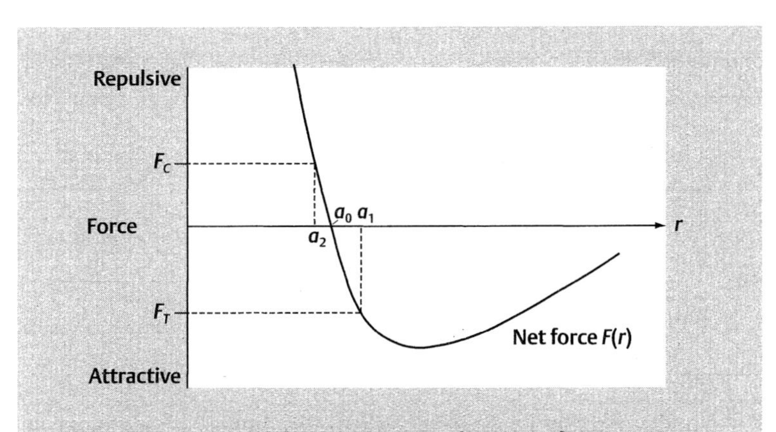

Figure 3.6 The net interatomic force as a function of separation, r. With no external force the equilibrium separation is at a_0. If a tensile force F_T or a compressive force F_C is applied, the separation changes to a_1 or a_2, respectively.

achieved with very fine filaments of material known as whiskers. In this case there is clear evidence that the stress–strain curve becomes non-linear for strains of a few per cent, in accordance with the atomic model (see Fig. 3.7). This means that the value of Young's modulus is not quite constant, but decreases slightly as the strain increases (see Question 3.4).

An understanding of shear stress on an atomic scale can be achieved by considering a plane of atoms in a face-centred cubic or hexagonal close-packed structure (see Section 2.2). When the material is unstressed, we can line up the centres of the atoms in adjacent rows, as in Fig. 3.8(a), but when a shear stress is applied the atoms are moved slightly from their idealized positions and we end up with a distorted arrangement, as in Fig. 3.8(b). Again, we can understand the reversible nature of this process, because if the stress is removed, the atoms naturally fall back into the arrangement in Fig. 3.8(a) which is a position of lower energy. We can also use the following argument to relate the macroscopic value of the modulus of rigidity to the displacement of the atoms. If the separation between the centres of the atoms in alternate rows is a and the difference in the displacement of the atoms in these rows is x, then the angle α is given by

$$\alpha = \tan^{-1}\left(\frac{x}{2a}\right) \approx \frac{x}{2a} \tag{3.9}$$

(The approximate form on the far right is due to the fact that for elastic deformation, the displacement x is always much smaller than $2a$.) Since the angle α shown in Fig. 3.8(b) is the same as the macroscopic angle of shear in a bulk crystal shown in Fig. 3.5, we can use eqns (3.5) and (3.9) to write

$$\tau = G\alpha \approx \frac{Gx}{2a} \tag{3.10}$$

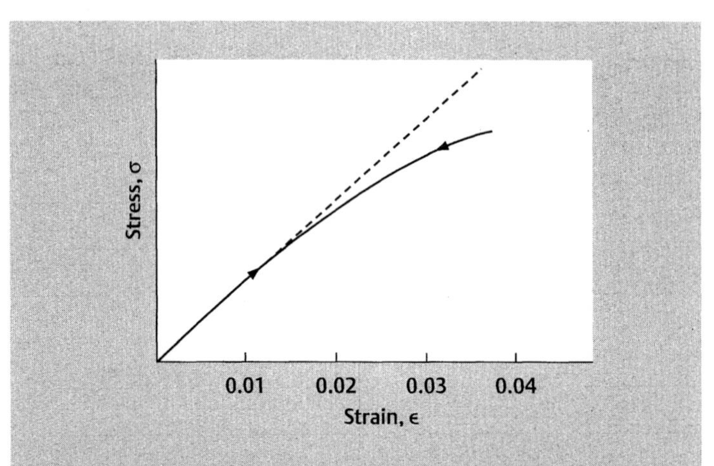

Figure 3.7 The solid line shows a typical stress–strain curve for a whisker; the dashed line corresponds to the criterion that stress is proportional to strain. Note that the deformation remains elastic (i.e. reversible) beyond the point at which it ceases to be linear.

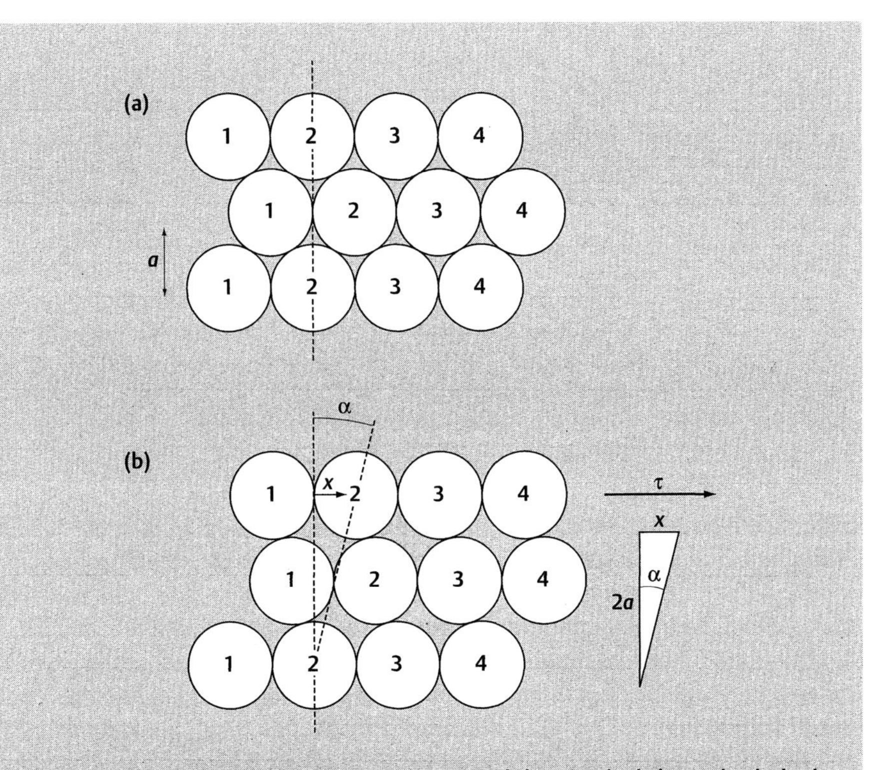

Figure 3.8 (a) The positions of the atoms in a hexagonal close-packed plane. The dashed line demonstrates that the centres of atoms in alternate rows are aligned. (b) When a shear stress, τ, is applied the rows of atoms are displaced with respect to one another, producing an angle of shear α. Note that the angle α is greatly exaggerated in this diagram. For elastic deformation α is generally less than 0.05°.

The above discussion shows that the mechanical deformation which occurs on a macroscopic scale in a crystal is a direct result of the changes which take place in the atomic structure, i.e. the effect of applying a stress is to stretch, compress or rotate the bonds between the atoms. Since the strength of these bonds is large compared with the magnitudes of forces that we experience in everyday life, this explains why most solids appear to be more or less rigid. However, there are a few notable exceptions, such as rubber, in which large deformations can be achieved with relatively small stresses. The Young's modulus of rubber is about 10^5 times smaller than that of metals and most other common materials. Consequently, a stress which produces an extension of only 0.001% in a metal wire is sufficient to double the length of a rubber band.

Why does rubber exhibit such extraordinary elastic behaviour? To discover the answer we again need to examine the material on an atomic scale. Rubber is made up of extremely long molecules called polymers (see Section 1.8) which consist of a chain of carbon atoms joined by covalent bonds. Since the covalent bond is very strong, the elastic properties of rubber cannot be attributed to the stretching of these bonds.

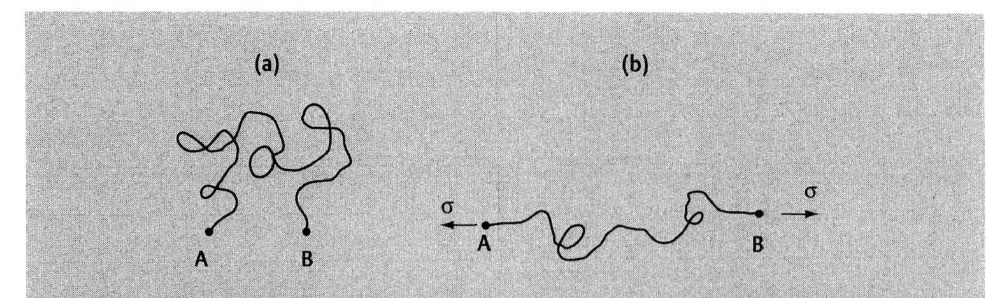

Figure 3.9 (a) The typical arrangement of a molecule in a sample of unstressed rubber. A and B denote the ends of the polymer. (b) When a small tensile stress σ is applied, the molecule extends significantly in length.

Instead, it is found to be due to the shape of the molecules. In fact, the behaviour is analogous to that of a steel spring. When we apply a tensile stress to the ends of the spring, we can easily achieve a change in length of 50% or more, but this is because the shape of the spring changes, not because the atoms in the steel wire are moving further apart. Similarly, the molecules in a sample of unstressed rubber tend to form convoluted shapes, as shown in Fig. 3.9(a), but when a stress is applied the molecules uncoil (Fig. 3.9(b)). This requires some weak molecular bonds between neighbouring polymers to be broken, but the strong covalent bonds between the carbon atoms are unaffected by this process. Consequently, a sample of rubber can undergo a dramatic change in shape for a relatively small applied stress.

3.4 The elastic limit

As the stress applied to a ductile material is increased, there is a transition point at which the behaviour changes from elastic to plastic deformation, i.e. from reversible and (usually) linear deformation to permanent and distinctly non-linear deformation. How can we explain this behaviour?

In the previous section we showed that the effect of a shear stress causes adjacent planes of atoms to move slightly with respect to one another. If the shear stress is sufficiently large, it should be possible to move the atoms to the position shown in Fig. 3.10(a), i.e. so that they occupy a position directly on top of those in the plane below. This corresponds to a position of unstable equilibrium—in principle the atoms could remain in this position with no applied stress, but in practice the slightest stress will cause the atoms to move one way or the other. If the atoms move to the left they fall back into their original position (Fig. 3.8(a)), but if they move to the right, they end up in a new position of stable equilibrium, as shown in Fig. 3.10(b). In the latter case the solid is permanently distorted because the atoms will remain in this new configuration when the external stress is removed.

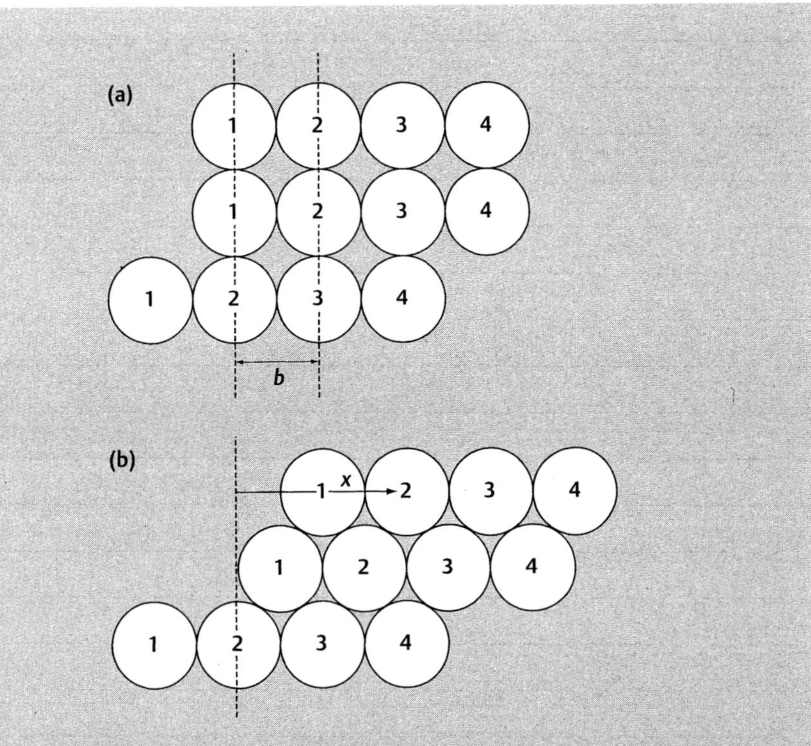

Figure 3.10 (a) When the displacement between the atoms in alternate planes is equal to b, where b is the distance between the centres of adjacent atoms in the plane, the atoms occupy a position of unstable equilibrium. (b) When the displacement x is equal to $2b$, the atoms occupy a new position of stable equilibrium. They will not return to their original position if the stress is removed.

This picture gives a qualitative explanation of the process of plastic deformation, but we can also use it to make an estimate of the yield stress. We can see from Figs 3.8(a) and 3.10(b) that the atoms are in a position of stable equilibrium when the displacement is equal to zero or $2b$, where b is the distance between the centres of adjacent atoms measured along the row. It follows that the same situation applies when the displacement is $4b$, $6b$, and so on. We can also see from Fig. 3.10(a) that the stress may be zero for a displacement of b (or $3b$, $5b$, etc.). This suggests that there is a periodic relationship between the shear stress τ and the displacement x. For example, let us postulate that these two quantities are related by an equation of the form

$$\tau = C\sin\left(\frac{\pi x}{b}\right) \tag{3.11}$$

where C is a constant. This obviously satisfies the above conditions because τ is equal to zero when x is equal to any multiple of b.

Example 3.1 Obtain an expression for the yield stress using eqns (3.10) and (3.11).

Solution The yield stress, τ_Y, is given by the maximum value of eqn (3.11), which occurs when the sine term is equal to 1.0, i.e.

$$\tau_Y = C$$

If we assume that eqn (3.11) is valid for all values of strain, then for small displacements (i.e. when $x \ll d$) we obtain

$$\tau \approx C\frac{\pi x}{b}$$

Equating this with eqn (3.10) (which is also valid for small displacements) gives

$$\frac{Gx}{2a} \approx C\frac{\pi x}{b}$$

Rearranging, we obtain

$$C \approx \frac{bG}{2\pi a}$$

Therefore,

$$\tau_Y \approx \frac{bG}{2\pi a}$$

Since the atomic spacings in each direction are roughly the same, we can put $a \approx b$, and so the yield stress is given approximately by

$$\tau_Y \approx \frac{G}{2\pi} \approx \frac{G}{6}$$

By using eqns. (3.10) and (3.11) we arrive at a theoretical value for the yield stress of about $G/6$ (see Example 3.1). Using the values of G in Table 3.1, this analysis predicts that the yield stress is about 10^{10} N m^{-2} for most materials. However, the measured values of yield stress (as shown in Table 3.2) are much smaller than this. Typically the values are between about 10^6 N m^{-2} and 10^9 Nm^{-2}. This suggests that our result is in error by a factor of between 10 and 10 000, depending upon the material in question. Although our calculations are fairly crude, more sophisticated methods only alter our result by a factor of about 5, i.e. to give a theoretical yield stress of $G/30$. This is still nowhere near the experimental values for most materials.

How do we explain this discrepancy? It seems that there must be some fundamental error in our reasoning. The problem is that we have assumed the crystal to be an idealized structure in which the atoms are arranged with perfect regularity. Of course, in a real solid this is not quite true. There are always defects present, and in particular there is a type of extended defect called a dislocation which has a dramatic effect on the mechanical properties of a solid.

Table 3.2 Values of stress and strain corresponding to the yield point (elastic limit) and breaking point in three common materials. (Note that this data is given only as an indication of 'typical' values. The values may differ considerably depending on many criteria such as the number and type of impurities present in the sample.)

	Yield stress (10^6 N m^{-2})	Yield strain	Breaking stress (10^6 N m^{-2})	Breaking strain
Al	26	0.0004	30	0.25
Glass	600	0.008	600	0.008
Steel	200	0.001	450	0.30

3.5 Dislocations

The concept of a dislocation in a crystal was first proposed in 1934 by Geoffrey Taylor to explain the discrepancy between the measured and calculated values of the elastic limit. (However, it was not until 1955 that dislocations were actually observed experimentally.)

The simplest form of dislocation, called an edge dislocation, is caused by an extra partial plane of atoms, as shown in Fig. 3.11. This produces a fault line which runs through the crystal corresponding to the edge along which the partial plane terminates. The core of the dislocation is the region in which the crystal lattice is distorted. In the plane shown in Fig. 3.11 this distortion is significant over a distance of only a few atomic spacings, but the distortion extends over a distance of several thousands (or even millions) of atomic spacings in the direction perpendicular to this plane (i.e. into and out of the page).

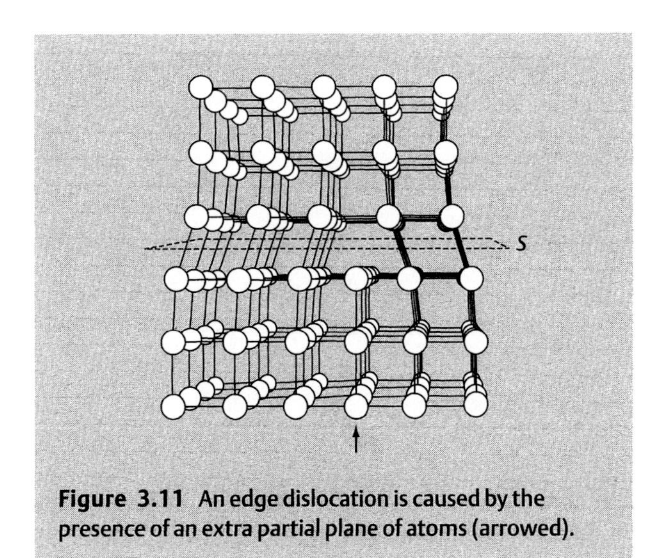

Figure 3.11 An edge dislocation is caused by the presence of an extra partial plane of atoms (arrowed).

How does the presence of a dislocation explain the low measured value of the yield stress? To answer this, let us examine the close-up view of the core of the dislocation in Fig. 3.12(a). The plane labelled S is called the slip plane (for reasons which will soon become apparent). If we look at the rows of atoms shown in the diagram we can see that the rows below the slip plane contain one more atom than the rows above this plane. Consequently, if we pair up the atoms on either side of the slip plane we are left with one atom (B) which does not have a partner. Suppose that we now apply a shear stress along a direction parallel to the slip plane and perpendicular to the dislocation. Because the atoms immediately above the slip plane are more widely spaced than those below this plane, only a small force is required to achieve the situation shown in Fig. 3.12(b). We can see that atom X is now paired up with atom B, leaving C unpaired. However, atom Y can now move across and pair up with atom C, and so on. In this way we can picture the dislocation moving one atomic step at a time through the crystal.

We can therefore see why our previous estimate of the yield stress was far too large—because we assumed that it was necessary to move all of the atoms in a plane simultaneously. The presence of the dislocation means that the atoms can shuffle along one at a time, and so the stress required is far smaller. This model also shows that there is another difference between elastic and plastic deformation. In the former case the deformation is distributed uniformly throughout the crystal, i.e. each plane of atoms moves the same distance with respect to the plane immediately beneath it. In contrast, in the latter case the majority of the deformation is concentrated in just a few planes (the slip planes).

We should be able to use this model to produce a revised estimate of the yield stress, since this is just the stress required to cause a dislocation to move. However, calculations give answers of about 10^5 N m^{-2}, which is considerably smaller than the observed values. So it seems that the presence of dislocations allows us to understand

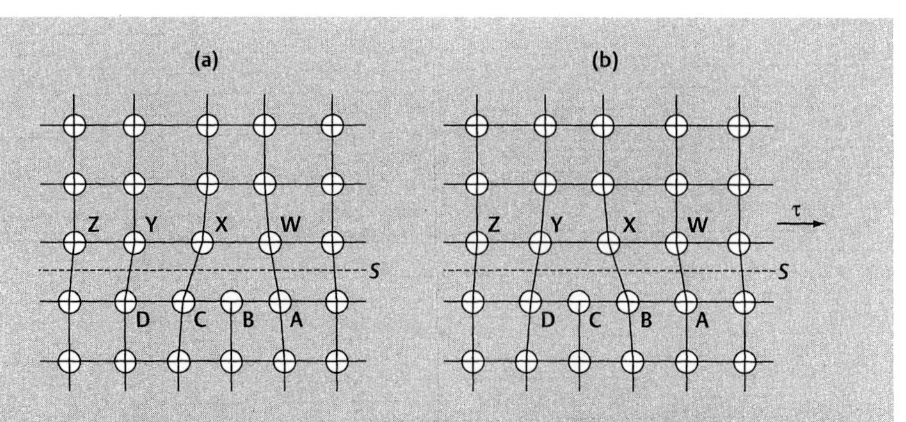

Figure 3.12 (a) A more detailed view of the region around the core of the dislocation. (b) When a small shear stress τ is applied parallel to the slip plane, the atoms above the plane are able to move slightly in the direction of the applied stress, and the dislocation moves one atomic spacing in the opposite direction.

the mechanism of plastic deformation, but we are still no closer to being able to make a reliable calculation of the yield stress.

The main reason why it is more difficult to move a dislocation than expected is because of the presence of impurities. Since the atoms near the dislocation on one side of the slip plane are slightly further apart than normal, this is the ideal place for large impurity atoms to collect. Similarly, interstitial impurities can also take advantage of this extra space (see Fig. 3.13). The presence of these impurities in the core of the dislocation obstructs the movement of the slip plane, and so leads to a higher value of yield stress. As a consequence it is found that alloys tend to have higher elastic limits than pure metals. For example, the yield stress of mild steel, which in its simplest form is a crystal of iron with carbon atoms occupying a small proportion of the interstitial positions, is about 10 times larger than that of pure iron.

Another factor which affects the yield stress is the temperature. In particular, the yield stress is generally observed to decrease as the temperature increases. We can explain this behaviour because the increased thermal vibration of the atoms at higher temperatures makes it easier for a dislocation to move past an obstructing impurity atom. This process is also important in the phenomenon of creep, which is the slow deformation of a material when exposed to high temperatures (i.e. greater than 50% of the melting point in kelvin) for any length of time. Creep occurs because the movement of a dislocation past an impurity is a statistical event which has a small probability of occurring even when the stress is well below the expected yield stress. Consequently, the material deforms slowly over a period of hours or days. This is a particularly serious problem in jet engines which operate at temperatures often in excess of 750 K. A rate of strain as low as 0.01% per hour can be important because the deformation is permanent. This means that the effects are cumulative: successive exposure to high temperatures leads to increasing levels of deformation and eventually to failure.

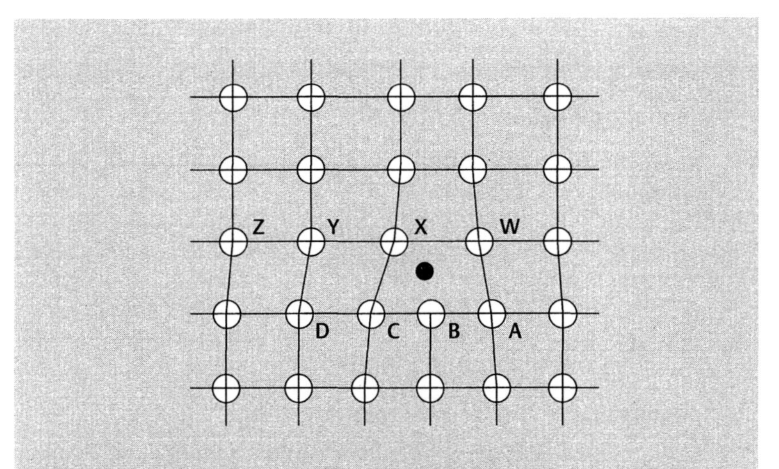

Figure 3.13 The presence of an interstitial atom (shown by the solid circle) at the core of the dislocation obstructs the movement of the dislocation.

3.6 Plastic deformation

Once the yield stress is exceeded, the dislocations are usually able to move quite freely, resulting in a large plastic deformation for a relatively small change in stress. This is the region of easy glide. In some cases this behaviour continues until the material eventually fractures. A typical stress–strain curve for such a material is shown in Fig. 3.1. At the point of fracture the strain may be 100% or more. This is over a thousand times larger than the strains usually produced by elastic deformation.

However, not all materials behave in this fashion. In some instances there is a noticeable increase in stress as the strain gets larger, as shown in Fig. 3.14. This effect is known as work hardening and has an important consequence, as we can see if we consider the process shown in Fig. 3.15. We suppose that a stress σ_1, which is greater than the yield stress σ_Y, is applied to the material (as shown by the single arrows). When the stress is removed, the material remains permanently deformed. If the material is now stressed again, it is found to exhibit elastic deformation up to the higher stress σ_1 (double arrows). It therefore appears that the yield stress in these materials depends on the past history of stress that has been applied to the material. This has many useful practical applications—in fact, most metal wires that are used in load-bearing applications are usually pre-stressed in this way in order to raise the elastic limit of the sample.

Having described the phenomenon of work hardening, let us look at why it occurs. The reason is that as the applied stress is increased, the dislocations in the sample multiply. We will not discuss how this occurs (a description is given in Rosenberg if you are interested) but we can see evidence of this process in the electron microscope

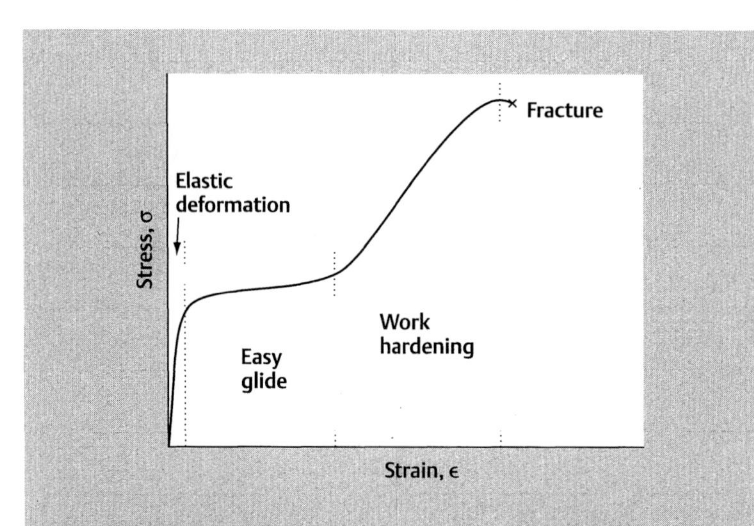

Figure 3.14 A typical stress–strain curve for a material which exhibits work hardening. The magnitude of the strain produced in the region of easy glide varies enormously from one material to another, and in some cases is almost non-existent.

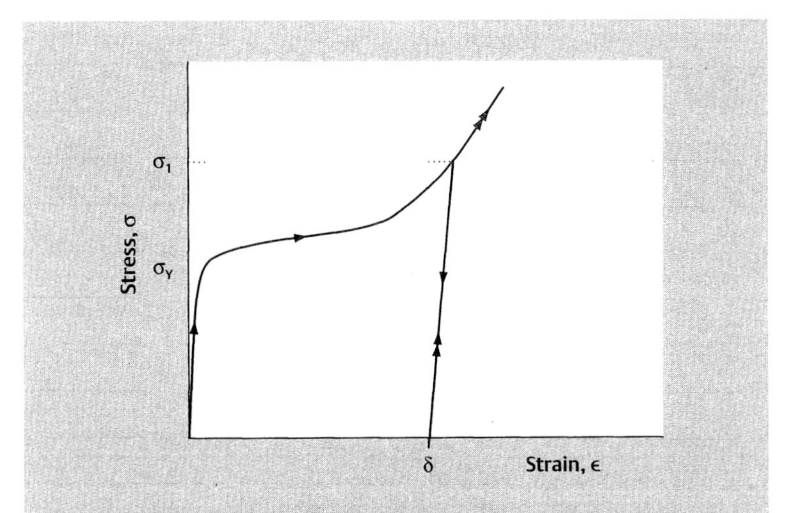

Figure 3.15 The curve described by the single arrows shows the effect of applying and removing a stress of magnitude σ_1 (where $\sigma_1 > \sigma_Y$). If a stress is reapplied (double arrows) the deformation remains elastic until a stress σ_1 is reached.

images of Fig. 3.16. At relatively low levels of strain (Fig. 3.16(a)) the dislocation density is low, but at higher levels of strain (Fig. 3.16(b)), the dislocations form a tangled mass and it becomes increasingly difficult for the dislocations to move past one another.

From Figs 3.1 and 3.14 we can see that the final stage of plastic deformation is accomplished with a decrease in stress. We can explain this behaviour by considering a wire subject to a tensile stress. As the wire is stretched, the diameter of the wire

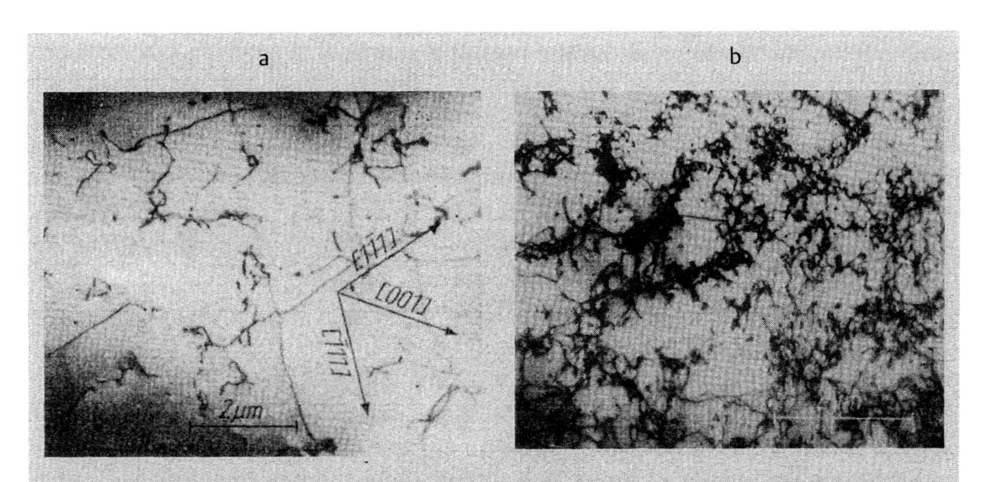

Figure 3.16 Dislocation lines in a single crystal of molybdenum subjected to strains of (a) 1% and (b) 4%.

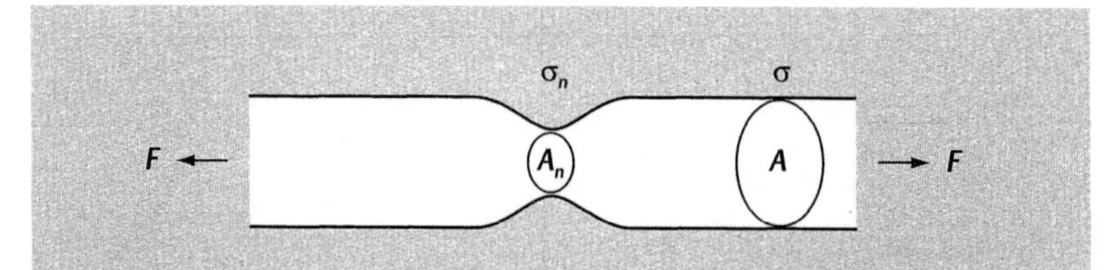

Figure 3.17 Necking occurs in a ductile material close to the point of fracture. The reduction in cross-sectional area in the necked region leads to an increase in the stress σ_n in comparison with the stress σ in the rest of the material.

decreases slightly, but we can assume that the diameter remains uniform along the length of the wire. However, just before the wire fractures it is found that the cross-section becomes significantly smaller at one point on the wire (see Fig. 3.17). This is termed 'necking'. Since the force in this region is concentrated over a smaller area, A_n, the stress σ_n is considerably larger than that in the rest of the wire. Consequently, even if the overall stress, σ, is reduced, plastic deformation will continue in the necked region so long as σ_n is larger than the critical value.

3.7 Brittle fracture

Not all materials exhibit plastic deformation. In brittle materials fracture occurs at the elastic limit and the stress–strain diagram looks something like Fig. 3.18.

Theoretically, the breaking stress of a brittle material is determined by the force required to pull two planes of atoms apart. There are various ways of calculating this stress (for examples, see Tabor, pp. 222–224), but in general the result is about 0.1*Y*, where *Y* is Young's modulus. This result is similar in magnitude to the predicted value of yield stress in ductile materials, and again it is found to be a gross overestimate of the measured value. However, the cause of the discrepancy must be different—it cannot be due to the presence of dislocations because the material does not undergo plastic deformation. So why is the breaking stress so small?

To answer this, let us consider a familiar example. We know that a sheet of glass can be cut to size by scoring a shallow groove across the surface. When a small amount of pressure is applied, the glass sheet breaks cleanly along the groove. (At least, that is what should happen. Whenever I have tried this the results have been rather less successful!) Although the glass sheet may be several millimetres thick and the groove is only a fraction of a millimetre deep, it appears that the presence of the groove on the surface substantially weakens the structure along this line. Why does this happen? If we look at Fig. 3.19(a) we can see that the stress lines are concentrated immediately below the groove, and therefore the stress in this region is larger than in other places. To be precise, if the radius of curvature of the tip of the groove is r and the groove is of depth l, then the stress at the tip of the groove is increased by a factor $2\sqrt{l/r}$. If this

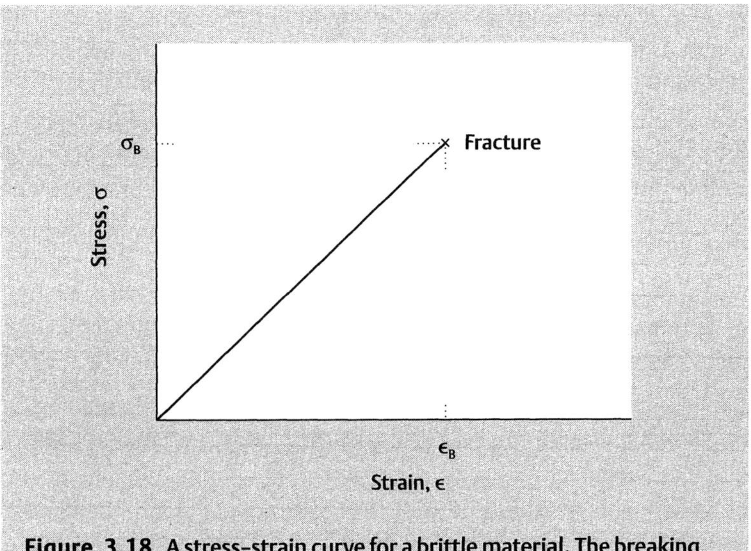

Figure 3.18 A stress-strain curve for a brittle material. The breaking strain ϵ_B is generally less than 1%.

stress exceeds the breaking stress of the solid, then the crack opens up further, as shown in Fig. 3.19(b). This in turn leads to an even greater concentration of stress at the tip of the crack, and so the crack propagates through the material. All of this happens very quickly. In fact, the crack travels at close to the speed of sound—roughly 300 m s^{-1}—so it appears that the fracture occurs instantaneously.

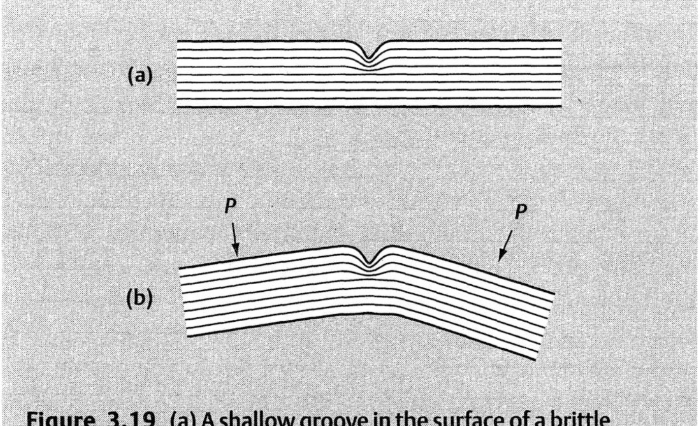

Figure 3.19 (a) A shallow groove in the surface of a brittle material leads to a concentration of the stress lines at the tip of the groove. (b) When a pressure *P* is applied on either side of the groove, the groove becomes deeper and the stress concentration increases.

In 1920 Alan Griffith suggested that a similar mechanism accounts for the low breaking stress of brittle solids. Even when there are no cracks visible to the naked eye, most surfaces possess many small surface cracks which are typically a few micro metres deep. We can see from Example 3.2 that the stress concentration at the tip of such a crack is sufficient to lower the breaking stress by a factor of about 100, which is roughly the amount required to bring the results in line with the measured breaking stress.

Example 3.2 Estimate the stress required to open up a surface crack in a glass rod given that the crack has a depth of 2.0 μm and a radius of curvature at the tip of 1.0 nm.

Solution From Table 3.1, Young's modulus for glass is $75 \times 10^9\,\mathrm{N\,m^{-2}}$. Therefore the theoretical breaking stress is

$$\sigma_B \approx 0.1Y \approx 7.5 \times 10^9\,\mathrm{N\,m^{-2}}$$

The stress concentration factor at the tip of the crack is

$$2\left(\frac{l}{r}\right)^{1/2} = 2\left(\frac{2.0\times 10^{-6}\,\mathrm{m}}{1.0\times 10^{-9}\,\mathrm{m}}\right)^{1/2} = 89$$

Therefore the actual breaking stress is

$$\frac{\sigma_B}{89} \approx 8\times 10^{7}\,\mathrm{N\,m^{-2}}$$

Having identified the cause of failure in brittle materials, can we do anything to increase the breaking stress? One solution is to polish the surface in order to remove as many of the surface cracks as possible. This increases the breaking stress, but only until the material becomes scratched again. A more long-term solution is to find a way of stopping the cracks spreading through the material. For example, a crack can propagate only if it is on a surface which is under a tensile stress. If a compressive stress is applied, as in Fig. 3.20, the crack does not open up any further. The breaking stress is therefore increased by ensuring that all of the surfaces are subject to a compressive stress. This can be achieved by applying a hydrostatic pressure to the sample, but this is not a practical proposition in most circumstances. However, a similar effect can be achieved in glass by rapidly cooling the sample during manufacture. This has the effect of causing the outer surfaces to harden whilst the central layer of the material is still relatively hot. As the inner region cools down, it contracts and so exerts a compressive stress on the outer surfaces. This principle is used in the manufacture of toughened glass. An alternative approach, which is used to make laminated windscreens for cars and other types of safety glass, involves sandwiching a layer of ductile material (such as a clear polymer) between two sheets of glass, as shown in Fig. 3.21. The presence of the polymer layer means that even if a crack propagates through one

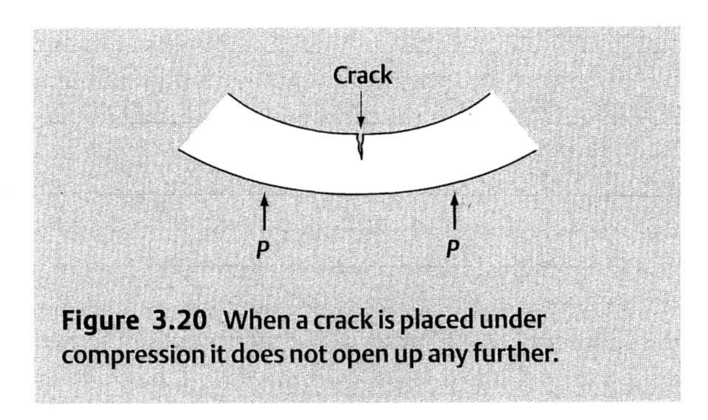

Figure 3.20 When a crack is placed under compression it does not open up any further.

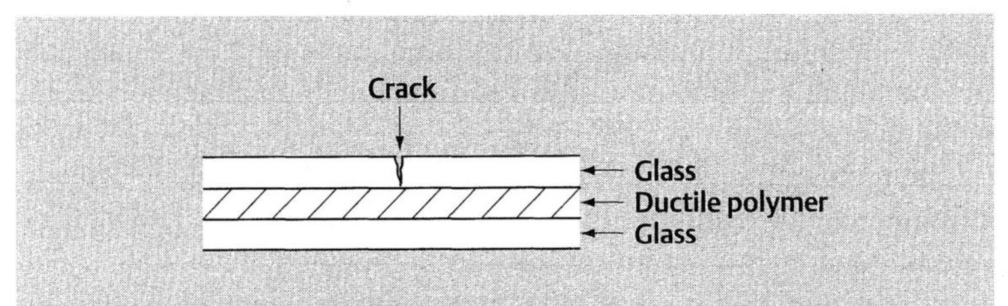

Figure 3.21 Laminated glass is formed by sandwiching a ductile polymer layer between two sheets of glass. Even if a crack appears in one of the glass sheets, it cannot propagate through the ductile layer and so does not cause a catastrophic failure.

layer of the glass, it is blocked by the ductile layer and so the windscreen does not break completely.

3.8 Brittle or ductile?

Why are some materials brittle and others ductile? Alternatively, why do some materials crack and others undergo plastic deformation?

A simple answer is that the behaviour depends on the type of bond present in the material. Metals are ductile because the atoms are not particular about which atom they bond with, and therefore the dislocations can move easily. Similarly, some molecular solids, such as polymers, are ductile because the molecules can slide over one another. In comparison, most other materials are brittle because the dislocations are not mobile. For example, in an ionic crystal the movement of a dislocation would upset the distribution of electrical charges. In a covalent crystal the presence of strong, directed bonds means that the dislocations cannot move easily. And in amorphous solids, such as glass, there are no dislocations because there is no regular crystal structure.

This explanation appears to agree with our experience of solids at room temperature, but the brittle or ductile nature of a solid is also found to depend on the temperature and pressure. For example, at high temperatures (but well below the melting point), glass becomes so amazingly ductile that it can be shaped simply by blowing it. At the other end of the scale, iron and many types of steel become brittle at temperatures around 0°C. How do we explain this behaviour?

The answer is that all materials are capable of exhibiting either ductile or brittle behaviour. The mechanism which occurs in practice is the one which has the lowest critical stress. Therefore, if a material is ductile, the stress needed to initiate the movement of a dislocation is smaller than that required to open up a crack. In this instance the dislocations are able to redistribute the stress at the tip of a crack before it becomes sufficient to allow the crack to propagate. Alternatively, the material is brittle if the critical stress for moving a dislocation exceeds the stress necessary to propagate a crack. The type of fracture therefore depends on the competition between these two processes. In general, the mobility of a dislocation increases with temperature and pressure, whereas the brittle breaking stress remains roughly constant. This means that materials tend to be ductile at high temperatures (and pressures), but undergo a transition to brittle behaviour as the temperature (or pressure) is reduced. Consequently, when we classify materials as being either brittle or ductile, what we really mean is that the transition temperature is above or below room temperature, respectively.

3.9 Summary

Despite the fact that the mechanical properties of solids are exploited in a wide range of applications, our understanding of these properties is still incomplete. For example, it is not yet possible to make quantitative predictions of the elastic limit or the breaking point of a material with any degree of certainty. However, the atomic models that we have developed in this chapter at least give us a qualitative understanding of the criteria which determine the yield stress and the way in which the material behaves beyond this point. This means that a scientific approach can be taken in designing new materials to meet specific criteria. For example, a knowledge of the effects of impurities and other factors on the yield stress has enabled alloys and plastics to be developed which offer a lightweight alternative to steel whilst providing a similar mechanical strength and rigidity.

Our discussion in this chapter has been restricted to simple models. In particular, we have assumed that the materials are isotropic—in other words, their mechanical properties are the same in every direction. Although this is quite a good approximation for many polycrystalline solids, we should be aware that there are other instances in which this assumption is not valid. A familiar example is wood. It is well known that the strength of a piece of wood depends on whether the stress acts perpendicular or parallel to the grain. This is because wood is made up of many long fibres which run more or less parallel to one another (see Fig. 3.22). The individual fibres are hard to break, but it is relatively easy to split the wood by separating the fibres. Many artificial composite materials are based on a similar principle using glass or carbon or other types of fibres.

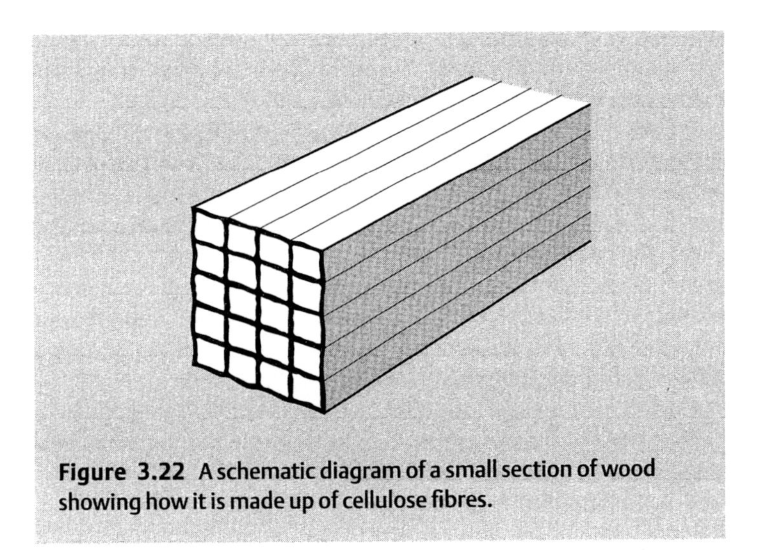

Figure 3.22 A schematic diagram of a small section of wood showing how it is made up of cellulose fibres.

Having dealt exclusively in this chapter with the way in which the shape of a solid is altered by an applied stress, it should be pointed out that this is not necessarily the only property that is affected. For instance, the optical properties may be affected as a result of the movement of the atoms when a material is subject to a stress. This explains the interference patterns that can be seen in a toughened glass windscreen which are due to the internal stresses (see Section 3.7). In some materials the stress may also give rise to a voltage across the sample. This is the piezoelectric effect which we will discuss in more detail in Chapter 10.

Questions

3.1 Elastic deformation. Using the data in Table 3.1, determine the strain produced when a mass of 1000 kg is suspended from a steel cable of diameter 1 cm.

3.2 Elastic deformation. Show that a tensile stress applied to a cuboid (as in Fig. 3.4) results in an increase in volume if Poisson's ratio, ν, is less than 0.5.

3.3 Elastic deformation. Use the values for Y and ν in Table 3.1 and eqns (3.7) and (3.8) to calculate values of G and K for aluminium and iron. Compare your results with the values in Table 3.1 and give a possible explanation for the discrepancies between the figures.

3.4 Elastic deformation. Measurements on an iron whisker have shown that in the elastic regime the stress and strain are related by the equation

$$\sigma = C(\epsilon - 2\epsilon^2)$$

where C is a constant. Determine to what extent C differs from the value of Young's modulus for strains of 0.001 and 0.04.

3.5 Elastic limit. A lift cable of diameter 2 cm has a yield stress of 2×10^8 N m^{-2}. Assuming that the empty lift has a mass of 150 kg and that the average mass of a person is 70 kg, determine the maximum number of people that can be carried in the lift given that the safe operating limit is 20% of the elastic limit.

3.6 Elastic limit. A steel bar of length 2 m and diameter 3 cm projects horizontally from a building. Given that the elastic limit is reached when a mass of 2000 kg is suspended from the free end of the bar, determine the yield shear stress. Give an estimate of the maximum shear stress based on a theoretical approach and explain why the values differ. Calculate the angle of shear corresponding to the elastic limit.

3.7 General. Calculate the stress required to break a bar of aluminium by brittle fracture assuming that the most serious surface cracks are of depth 1 μm and have a tip radius of curvature of 1.0 nm. Hence, using the data in Table 5.2, explain why aluminium undergoes plastic deformation at room temperature.

3.8 General. Given that the yield stress of glass is considerably higher than that of steel (see Table 3.2), explain why glass cables are not used in load-bearing applications.

3.9 General. Describe the types of deformation that occur for each of the materials listed in Table 3.2 and sketch stress–strain diagrams in each case.

Chapter 4
Electrical properties of metals

4.1 Introduction

How does an electric current flow through a material? A current is a flow of charged particles, which implies that either ions or electrons are involved. In liquids, both of these types of particles take part in electrical conduction, but in solids the ions occupy fixed positions in the crystal lattice and so are not free to move. (A few solids do exhibit some ionic conduction, but the magnitude of this type of conductivity is usually very small.) Consequently, we can conclude that the electrical conductivity of solids is due almost entirely to the movement of the electrons. This implies that metals should be good electrical conductors because they have a plentiful supply of essentially free electrons, whereas all other materials should be insulators because there are no free electrons. This picture is actually rather oversimplified, and in the next chapter we will look at how some non-metals can be reasonably good electrical conductors. However, in this chapter we will concentrate on the process of electrical conduction in metals.

What happens when we apply a voltage to a metal object? The negatively charged electrons in the sample are attracted towards the positive terminal and are repelled by the negative terminal. We can represent this by tilting the energy bands to indicate that the electrons at the positive end have a lower energy than those at the negative end, as shown in Fig. 4.1. Since the valence electrons in a metal are essentially free, we would expect them all to rush down the slope towards the lower energy states near the positive terminal. However, we will see that this is far from the truth.

4.2 Drude's classical theory of electrical conduction

The first attempt to explain the behaviour of electrons in a metal was made by P. Drude in 1900, only three years after the discovery of the electron. The theory is incorrect in many respects, despite the fact that it gives several rather convincing results, but it is worth examining in detail because it introduces some useful

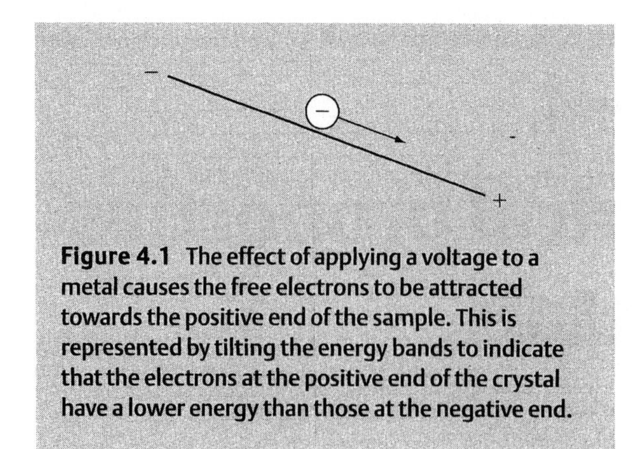

Figure 4.1 The effect of applying a voltage to a metal causes the free electrons to be attracted towards the positive end of the sample. This is represented by tilting the energy bands to indicate that the electrons at the positive end of the crystal have a lower energy than those at the negative end.

concepts which we will use later in a more sophisticated treatment of electrical conduction.

Drude assumed that a metal is composed of ions, which are stationary, and valence electrons, which are free to move. He treated the electrons as small solid objects which behave rather like the ball in a pinball machine, colliding with one ion after another so that a typical trajectory looks something like Fig. 4.2. If there is no voltage applied to the metal then at each collision the electron is deflected in a different direction so that the overall motion is quite random. Of course, this is just what we expect when there is no applied voltage, because any net movement of the electrons in a particular direction would constitute a flow of current.

Since the random motion of the valence electrons appears to be similar to that of the molecules in an ideal gas, Drude suggested borrowing some of the concepts used in the kinetic theory of ideal gases to describe the properties of the electron 'gas'. Let us see what we can determine by this approach. (If you are unfamiliar with kinetic theory, it may be worth consulting an introductory general physics text—see Further reading.)

We can begin by estimating the velocity of the electrons. In kinetic theory the

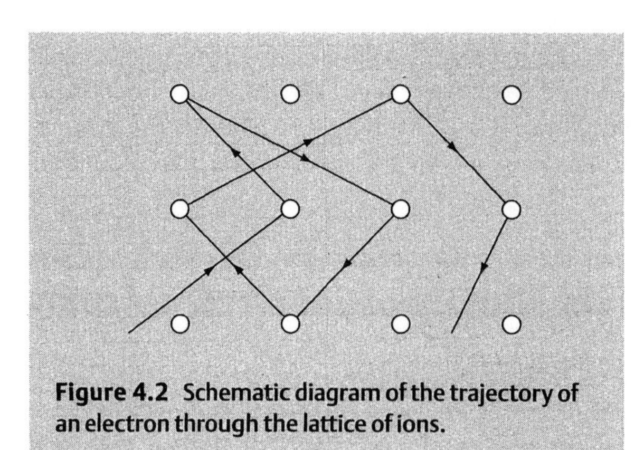

Figure 4.2 Schematic diagram of the trajectory of an electron through the lattice of ions.

average thermal velocity, v_t, of a molecule with mass m at temperature T (measured in kelvin) is given by the equation

$$\frac{1}{2}mv_t^2 = \frac{3}{2}k_B T \tag{4.1}$$

where k_B is Boltzmann's constant. If we apply the same formula to the electrons in a metal and replace m by the electron mass m_e then we obtain a value for the speed of the electrons at room temperature of about 10^5 m s^{-1} (see Question 4.1).

Using the analogy with kinetic theory we can also define the mean free path λ—which is the average distance that an electron travels between collisions—and the relaxation time τ—which is the average time duration between collisions. These quantities are related by the equation

$$v_t = \frac{\lambda}{\tau} \tag{4.2}$$

If we assume that the mean free path is of the order of a few atomic spacings, i.e. about 1 nm, then the relaxation time is typically about 10^{-14}s.

Now let us consider what happens when we apply an electric field (i.e. a voltage) to the sample. The electrons are attracted towards the positive end of the sample, and so we expect a net flow of electrons in this direction. Using some simple arguments we can make an estimate of the magnitude of this effect. If the electric field is $\mathcal{E}$, then the force on each electron is $e\mathcal{E}$. Dividing by the electron mass, m_e, we find that the acceleration on each electron in the direction of the field is

$$a = \frac{e\mathcal{E}}{m_e} \tag{4.3}$$

This acceleration produces a change in velocity Δv along the direction of the electric field. Since the acceleration is constant, the value of Δv at time t after a collision is given by

$$\Delta v = at = \frac{e\mathcal{E}\,t}{m_e} \tag{4.4}$$

If the average time between collisions is denoted by τ, then the average velocity in the direction of the field is

$$\overline{\Delta v} = \frac{e\mathcal{E}\,\tau}{m_e} \tag{4.5}$$

This quantity is referred to as the drift velocity and represents the average velocity of the valence electrons in the direction of the electric field. If we substitute typical values into this equation we find that in a moderate electric field of 10 V m^{-1} the drift velocity is only about 0.02 m s^{-1} (see Question 4.2). This is of the order of 10 million times smaller than the thermal velocity of the electrons.

Why is the drift velocity so small? It is because even when an electric field is applied to the metal, the electrons still collide with the ions and so they continue to follow a more or less random trajectory (as in Fig. 4.2) with just a very slight tendency to move towards the positive end of the sample. Consequently, we arrive at the rather surprising result that when a current flows in a metal, the valence electrons are not all moving towards the positive terminal. In fact, at any instant in time there are nearly as many electrons travelling in the opposite direction!

What evidence do we have that Drude's theory is correct? Well, first of all we can use it to provide a qualitative explanation of the phenomenon of electrical resistance. Let us assume that when an electron collides with an ion, the excess energy that the electron has gained from the electric field is transferred to the ion. This increases the vibration of the ion, and so leads to an increase in the temperature of the sample. In other words, the kinetic energy of the electrons is converted into thermal energy. This is a familiar effect because it is used in electric kettles, tóasters and light bulbs: if we pass an electric current through a metal wire it gets hot. If the resistance of the wire is increased, the wire gets hotter. Consequently, resistance can be thought of as a measure of the amount of electrical energy that is converted into heat energy, which in turn can be attributed to the collisions between the electrons and the ions.

We can also show that Drude's model of electrical conduction is consistent with Ohm's law. First of all we will define a quantity called the electron mobility μ which is a measure of the ease with which the electron moves through the lattice of ions. If we write

$$\overline{\Delta v} = \mu \mathcal{E} \tag{4.6}$$

then by comparing eqns (4.5) and (4.6) we can see that

$$\mu = e\tau / m_e \tag{4.7}$$

If we assume that the relaxation time has a fixed value (i.e. independent of temperature and electric field strength) for a particular material, then the mobility is also a constant. (In practice this is not quite true. It is found that τ, and therefore μ, vary slightly with temperature and electric field strength, but these effects can usually be neglected.)

Let us then consider the behaviour of the valence electrons in a metal wire in which the electric field is directed along the axis of the wire. The analysis is simplest if we neglect the random thermal motion of the electrons—we can do this because we know that the average thermal velocity in any particular direction is zero—and assume that each electron travels at a velocity $\overline{\Delta v}$ along the wire, as shown in Fig. 4.3. This means that in a time interval t each electron moves a distance $\overline{\Delta v}\,t$ along the wire. As a result, the number of electrons passing through the cross-section A in this time is equal to $n\,\overline{\Delta v}\,t\,A$, where n is the number of valence electrons per unit volume. Since the current is defined as the amount of charge passing through the wire per unit time, and as each electron carries a charge e, the current I flowing in the wire is $n\,\overline{\Delta v}\,Ae$. So, finally, we arrive at an expression for the current density J:

$$J = \frac{I}{A} = n\overline{\Delta v}e = n\mu\mathcal{E}e \tag{4.8}$$

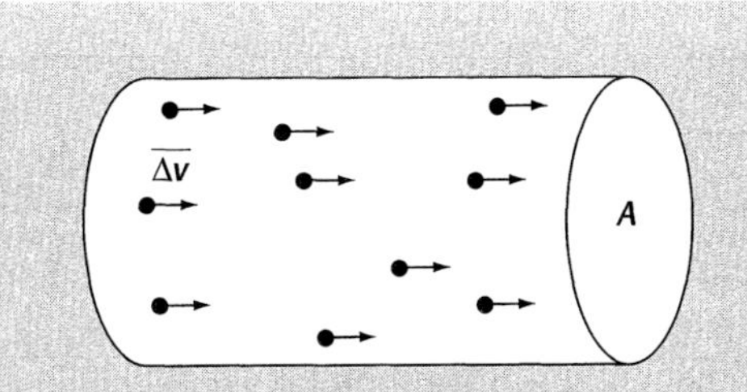

Figure 4.3 A simple picture of conduction in a wire in which it is assumed that all of the valence electrons have the same drift velocity $\overline{\Delta v}$. The random thermal component of the electron velocity is omitted.

where we have substituted for $\overline{\Delta v}$ from eqn (4.6). Since n, μ and e are constants for a given material, we can express the product of these three terms as a single constant

$$\sigma = n\mu e \tag{4.9}$$

where σ is referred to as the conductivity. Alternatively, we can define the resistivity, ρ, which is the inverse of the conductivity, i.e.

$$\rho = \frac{1}{\sigma} = \frac{1}{n\mu e} \tag{4.10}$$

Consequently, we can write eqn (4.8) as

$$J = \sigma\mathcal{E} = \frac{\mathcal{E}}{\rho} \tag{4.11}$$

This is Ohm's law, or at least the microscopic form of Ohm's law. We can put it in a more familiar form (in terms of current, voltage and resistance) by substituting $J = I/A$, $\mathcal{E} = V/L$ and $R = \rho L/A$ (see Question 4.3).

Table 4.1 Conductivity and resistivity data at 273 K for a selection of metals. *n* is the number of valence electrons per unit volume, which is equal to the number of atoms per unit volume multiplied by the valence of the atom.

	Valence	σ ($10^7\ \Omega^{-1}\ m^{-1}$)	ρ ($10^{-8}\ \Omega\ m$)	n ($10^{28}\ m^{-3}$)
Sodium	1	2.38	4.20	2.65
Copper	1	6.45	1.55	8.50
Silver	1	6.80	1.47	5.86
Gold	1	4.88	2.05	5.90
Magnesium	2	2.54	3.94	8.60
Zinc	2	1.82	5.50	13.2
Aluminium	3	4.00	2.50	18.1
Tin	4	0.87	11.5	14.5
Lead	4	0.52	19.2	13.2

Example 4.1 Estimate the typical conductivity of a metal at 295 K assuming that the mean free path is about 1 nm and that the number of valence electrons is about 10^{29} m^{-3}.

Solution From eqns (4.1) and (4.2), the thermal velocity and the relaxation time of the electrons are 1.16×10^5 m s^{-1} and 8.62×10^{-15} s, respectively (see Question 4.1).

Combining eqns (4.7) and (4.9) we have

$$\sigma = \frac{ne^2\tau}{m_e} = \frac{(10^{29}\,\mathrm{m}^{-3})(1.60\times 10^{-19}\,\mathrm{C})^2(8.62\times 10^{-15}\,\mathrm{s})}{9.11\times 10^{-31}\,\mathrm{kg}}$$

$$= 2.42\times 10^{7}\,\Omega^{-1}\mathrm{m}^{-1}$$

Finally, if we use the expression for the conductivity from eqn (4.9) and assume that the mean free path is of the order of a few atomic spacings then we find that the predicted value of conductivity is in good agreement with the experimental data in Table 4.1 (see Example 4.1).

So we have some quite convincing evidence that Drude's model of electrical conduction is correct. We have managed to show that Drude's model is:

- consistent with Ohm's law;
- qualitatively explains the phenomenon of electrical resistance; and
- gives good values for the conductivity.

However, as we shall see in the next section, Drude's model has some severe limitations.

4.3 Failures of the classical model

Despite the successes of Drude's theory, there are many features of the electrical conductivity of solids that cannot be explained by this model.

For instance, let us first of all compare the conductivities of different metals. Equation (4.9) shows that the conductivity is proportional to the number density of valence electrons n and to the mobility μ and Table 4.1 gives values of n for various different metals, but how does the mobility vary from one metal to another? From eqns (4.7) and (4.2) we can see that the mobility varies inversely with the mean free path, and we would expect the mean free path to be dependent on the size of the ions and on the percentage of the total volume that is occupied by the ions. Since most metallic ions tend to be of a more or less uniform size (with an ionic radius of about 0.15 nm) and form close-packed crystal structures, we can assume that the mean free path—and therefore the mobility—does not vary substantially between different metals. This suggests that the conductivity should be directly proportional to the valence electron concentration, n. However, this is not in good agreement with the experimental data plotted in Fig. 4.4. In fact, it appears that those metals which have the highest electron concentration tend to have the lowest conductivities (with the exception of aluminium) and vice versa!

Another problem occurs if we consider the electrical properties of alloys (which are mixtures of two or more metallic elements). The classical model seems to suggest that the resistivity of an alloy should be intermediate between the values for the corresponding pure materials. However, many alloys have resistivities which are considerably larger than those of either of the pure constituents. For example, nichrome, which is an alloy of 80% nickel and 20% chromium, has a resistivity which is 10 times higher than that of either nickel or chromium (see Fig. 4.5).

The dependence of resistivity on temperature is also at odds with experimental evidence. If we assume that the mean free path is approximately constant with temperature, then we can use Drude's model to predict that the resistivity should be proportional to $T^{1/2}$ (see Question 4.8). But experimental measurements show that the resistivity is actually proportional to T over a wide range of temperatures.

Finally, let us consider the molar specific heat capacity, which is the amount of energy required to raise the temperature of 1 mole of solid by 1 K. If we again use the analogy with an ideal gas, then we can see from Example 4.2 that Drude's theory predicts a value for a monovalent metal of $9R/2$. A similar argument suggests values of $6R$ for a divalent metal, $15R/2$ for a trivalent metal, and so on (see Question 4.9). However, experimental results show that the molar specific heat capacity at room temperature is approximately $3R$, regardless of the valency of the metal.

There are many other characteristics that cannot be explained using Drude's model, but I am sure that by now you are more than convinced that we need to improve this model if we want to obtain a better understanding of electrical conduction.

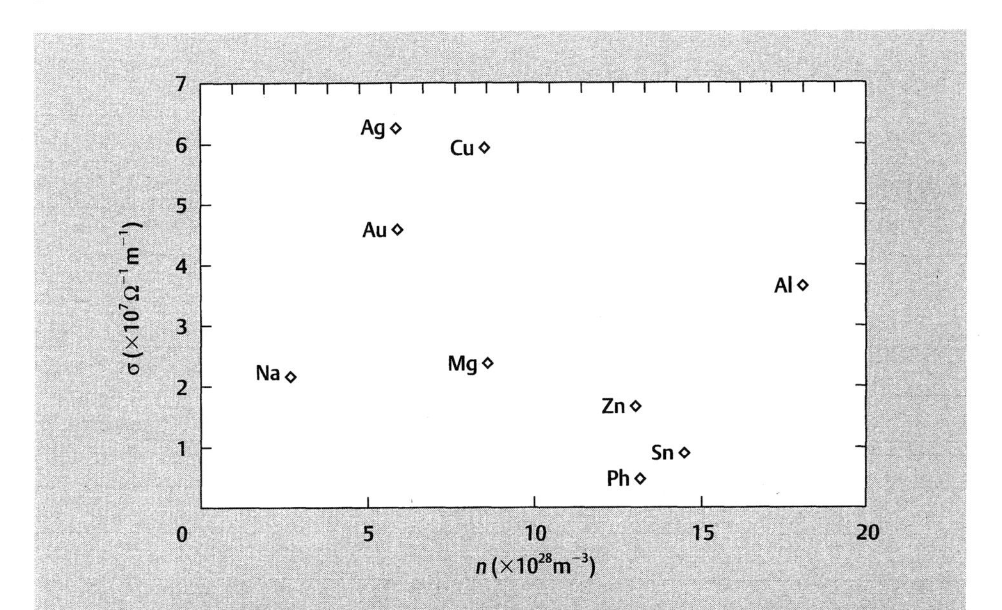

Figure 4.4 A plot of conductivity σ against the number density of valence electrons n for a number of common metals at 295 K. (Data is taken from Table 4.1.)

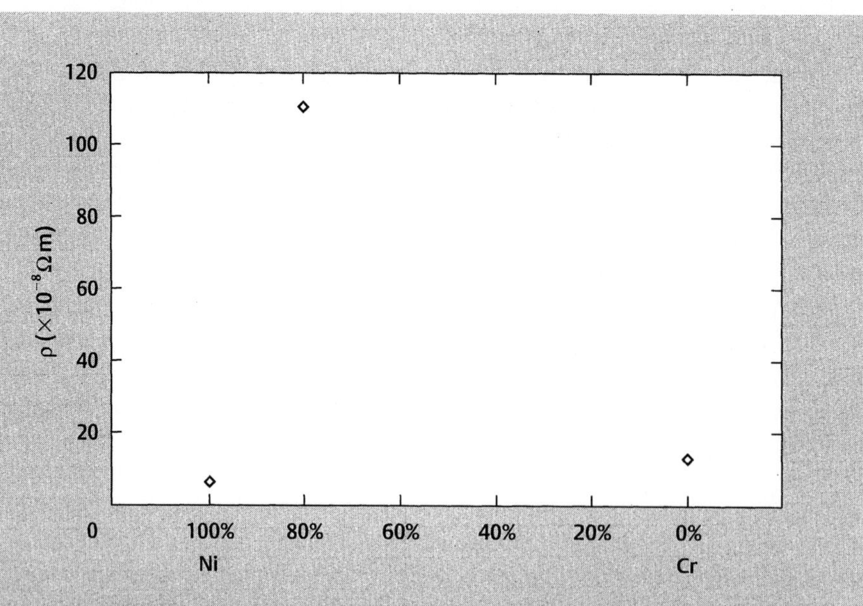

Figure 4.5 The resistivity ρ for nickel–chromium alloys at 295 K. The x axis indicates the percentage of nickel in the alloy. Data points are shown for 100% Ni (i.e. pure nickel), 80% Ni (nichrome) and 0% Ni (i.e .pure Cr).

Example 4.2 Determine the molar specific heat capacity of a monovalent metal.

Solution We will make use of the theory of equipartition of energy in which it is assumed that each atomic degree of freedom has an energy $\frac{1}{2}k_BT$. (You may be familiar with this result from the kinetic theory of gases. If not, consult one of the introductory texts listed in Further reading.)

Each valence electron has three translational degrees of freedom with a total energy of $\frac{3}{2}k_BT$.

Each ion vibrates in three mutually perpendicular directions, each of which is associated with both kinetic and potential energy giving six degrees of freedom (see Chapter 7) with a total energy of $3k_BT$.

In 1 mole of a monovalent metal there are N_A ions and N_A valence electrons, giving a total energy of

$$3N_A k_B T + \frac{3}{2} N_A k_B T = \frac{9}{2} RT$$

where R is the molar gas constant (which is the product of Avagadro's constant with Boltzmann's constant.)

In order to increase T by 1 K, we therefore need to supply an energy of $\frac{9}{2}R$.

4.4 Bloch's quantum theory of electrical conduction

We should not be too surprised that Drude's theory does not give us a complete picture of electrical conduction in solids. After all, we have no justification for assuming that eqn (4.1) can be applied to the electrons in a metal. Also we have treated the ions as inert objects that simply get in the way of the electrons, ignoring the electrostatic interactions that occur between the valence electrons and the ions (electron–ion interactions) and between the valence electrons themselves (electron–electron interactions). The latter turns out to be fairly insignificant in most cases, so we can safely ignore this effect, but the interactions between the electrons and the ions are of great importance and substantially alter our view of electrical conduction.

Let us first consider the behaviour of an electron approaching an isolated positive ion. The electron is attracted to the ion when the separation between them is relatively large, but as the electron gets close to the ion it is repelled by the outer filled shell of electrons (Fig. 4.6). We are therefore not dealing with 'collisions' between the electrons and ions, as we assumed in the previous sections, rather the electrons are deflected or 'scattered' by the ions. To describe the motion of an electron in a solid we need to extend this idea to a large number of ions in close proximity.

Another failing of Drude's model is that it treats the electron as a classical particle. However, we have already seen in Chapter 1 that we cannot explain the behaviour of an electron in an atom using classical arguments. It follows that in order to describe the interactions between the valence electrons and the ions we must use a quantum theory treatment.

How does this affect our understanding of electrical conduction? The first person to tackle this problem was F. Bloch in 1925. By solving the quantum mechanical equations for an electron in a perfect crystal lattice, Bloch showed that an electron moves through the crystal without being deflected at all. We can picture this behaviour by imagining the electron weaving through the lattice of ions, as shown in Fig. 4.7.

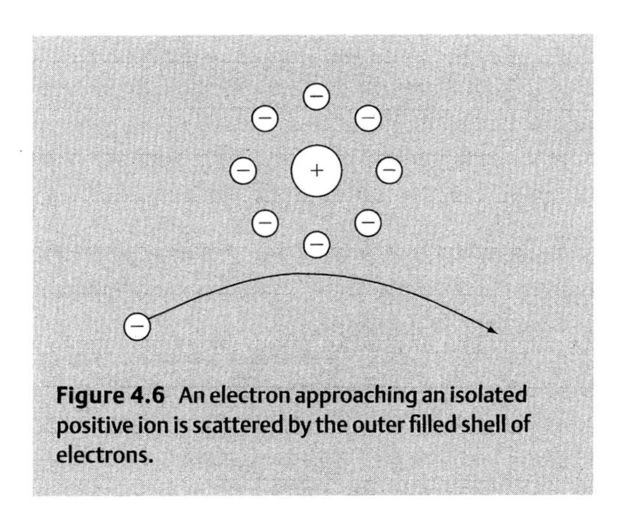

Figure 4.6 An electron approaching an isolated positive ion is scattered by the outer filled shell of electrons.

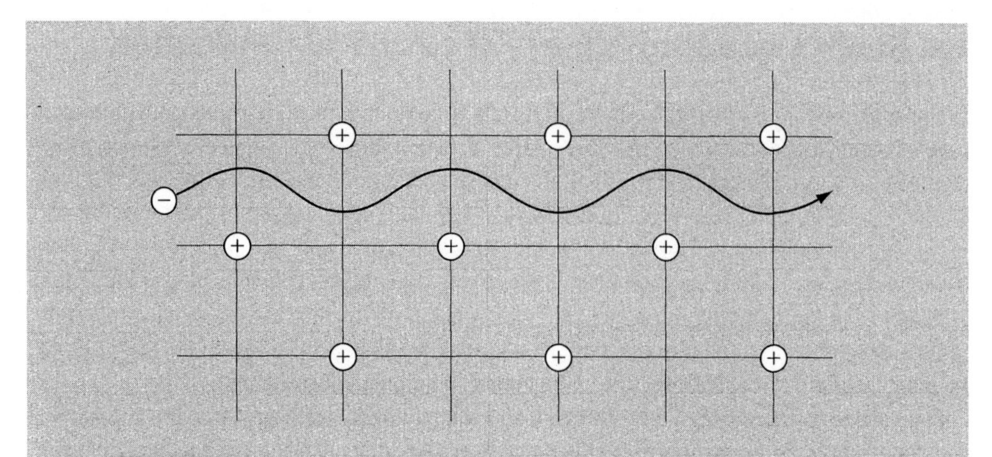

Figure 4.7 Quantum theory predicts that an electron is able to pass through a perfect crystal without being scattered.

However, this result poses a riddle. Since it is the scattering of the electrons by the ions that gives rise to an electrical resistance, if the electrons are not scattered then a metal should offer no resistance to an electric current. Since this is clearly at odds with our everyday observations, we might be inclined to scrap this model and look for an alternative approach. It turns out that the model is quite correct. The flaw in our reasoning is that we have assumed that the crystal is perfect. However, any imperfections in the crystal will cause the electrons to be deflected. Consequently, the mean free path of an imperfect crystal is not infinite, and so the material has a non-zero resistance.

There are essentially three different types of imperfection which produce electron scattering, and therefore can be thought of as the origin of electrical resistance in a metal. Firstly, the ions are not stationary, but are in a state of continual thermal vibration about their equilibrium positions. Consequently, at any instant in time the ions do not occupy the positions of a perfect lattice. This leads to scattering of the electrons, as shown schematically in Fig. 4.8, which in turn gives rise to a resistance.

A second cause of scattering is the presence of impurities. This is particularly important if the impurity ion is significantly larger or smaller than the host ions, or if it has a different valence. For example, in brass, which is a copper–zinc alloy, the zinc ions are about 9% larger in diameter than the copper ions and have a charge of $+2e$ compared with $+1e$ on the copper ions. The effect on scattering is shown schematically in Fig. 4.9.

Thirdly, imperfections in the crystal structure also disrupt the lattice and so cause scattering. We illustrate this effect in Fig. 4.10 by considering one of the simplest imperfections—a missing atom, or vacancy.

To summarize, we have shown that the processes which result in the scattering of the electrons are very different to the simple picture of electrons colliding with fixed ions that we assumed in Drude's model. However, the trajectories of the electrons are quite similar in the two cases: the electrons are frequently deflected and so follow an almost random path. Consequently, the concepts that we introduced in the classical

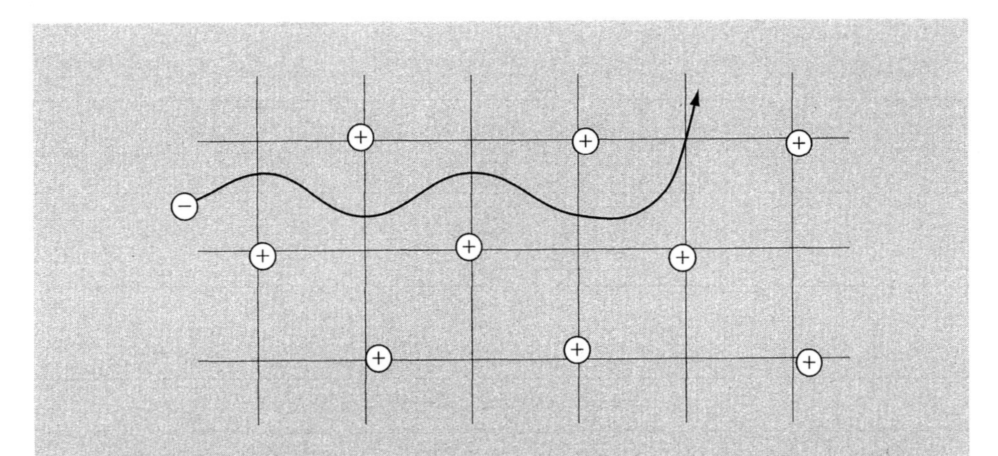

Figure 4.8 Thermal vibrations of the ions destroy the periodicity of the lattice, leading to scattering of the electrons.

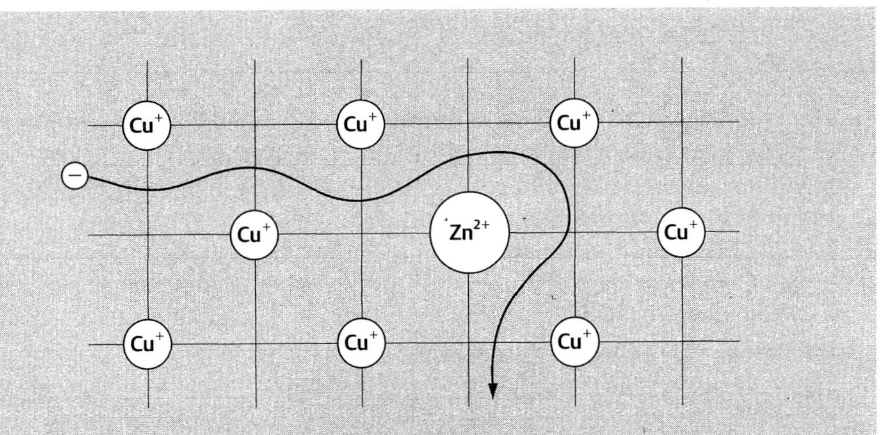

Figure 4.9 Impurities may have a substantial effect on the scattering of electrons, particularly if they are of a different size or valence to the host ions.

theory (such as the mean free path, the drift velocity and the mobility) are all relevant in a quantum theory of conduction. However, the values of these parameters may be quite different, as we shall see in Section 4.10.

4.5 Band theory of solids

We now have a qualitative description of electrical conduction from a quantum viewpoint. If we wish to obtain quantitative results then we also need to consider the energies of the electrons in a solid.

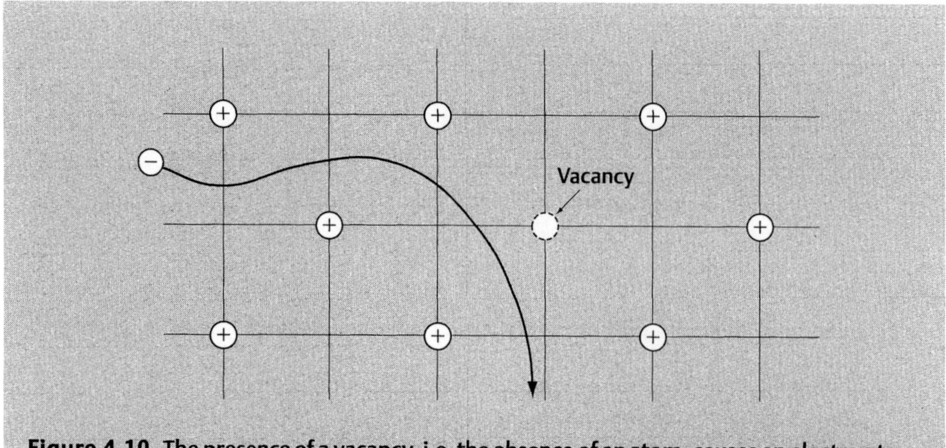

Figure 4.10 The presence of a vacancy, i.e. the absence of an atom, causes an electron to be scattered.

To begin with let us remind ourselves of the restrictions which govern the energy levels in an isolated atom (see Chapter 1). These can be summarized by the following statements:

1. The properties of the electrons in an atom are determined by four parameters, or quantum numbers, which are denoted n, l, m_l and m_s. n is the principal quantum number, l takes integer values from 0 to $(n-1)$, m_l is allowed integer values from $-l$ to $+l$ and m_s can be $-\frac{1}{2}$or $+\frac{1}{2}$.
2. The electrons are only allowed to occupy certain discrete energy levels, which we label using the quantum numbers n and l (since the energy of an electron does not usually depend on m_l or m_s). The letters s, p, d and f are used to denote the states $l = 0, 1, 2$ and 3, respectively.
3. The occupancy of these levels is determined by the exclusion principle which states that each electron must possess a unique set of quantum numbers.

If we apply these rules to a sodium atom, which has 11 electrons, we find that two electrons occupy each of the 1s and 2s levels, six occupy the 2p level, leaving one electron in the 3s level, as shown in Fig. 4.11(a). However, in a solid the atoms are not isolated. In fact they are in close proximity to a large number of other atoms. What happens in this case?

Let us first of all consider the situation with just two sodium atoms. When the atoms are far apart they behave like isolated atoms, but as we move them closer together, the outer (i.e. 3s) electrons begin to interact, which affects the energies of these electrons. If we plot the positions of the energy levels as a function of the separation between the atoms we obtain a picture as in Fig. 4.12. At the equilibrium separation, a_0, we therefore have two states, one of which is at a lower energy than the 3s state in an isolated atom, and one which is at a higher energy. We have to be very careful how we interpret this result. The energy levels can no longer be considered to belong to a specific atom, rather they are a property of the pair of atoms considered

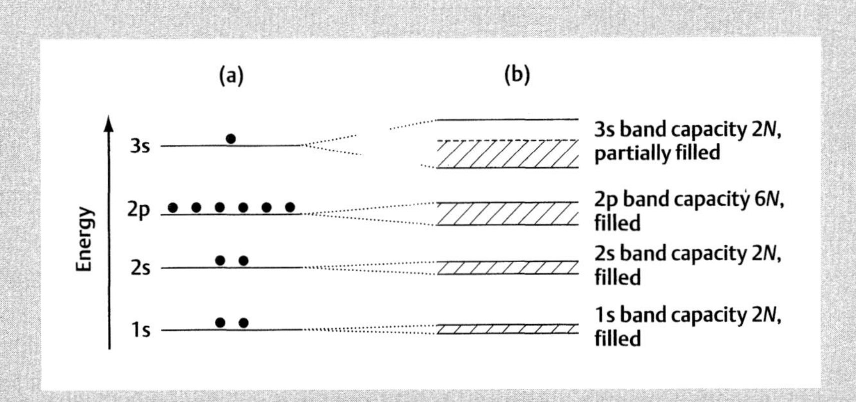

Figure 4.11 (a) The occupation of the electron energy levels in an isolated sodium atom. (b) The occupation of the energy bands in a sodium crystal.

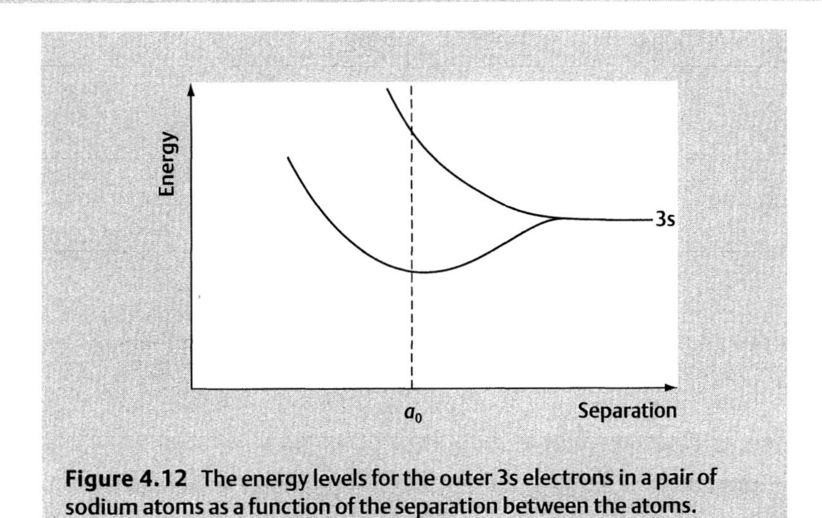

Figure 4.12 The energy levels for the outer 3s electrons in a pair of sodium atoms as a function of the separation between the atoms.

as a whole. Consequently, this picture does *not* imply that the 3s electron attached to one atom has a higher energy than that on the other atom. In fact, since there are two allowed values of m_s for each state, both of the electrons can occupy the lowest energy level (Fig. 4.13).

If we extend this same argument to a system of N sodium atoms, we should get N discrete 3s energy levels, as shown in Fig. 4.14(a). However, in a macroscopic crystal the number of atoms, N, is typically of the order of 10^{24} and the energy levels span a range of only a few electron volts. Consequently, the spacing between adjacent levels is so small (see Question 4.11) that we effectively have a continuous band of energies, as shown in Fig. 4.14(b). Similar interactions occur for the 1s, 2s and 2p electrons, but the interactions are much weaker because these electrons are closer to their respective nuclei, and so the corresponding bands cover a smaller range of energies (Fig. 4.15).

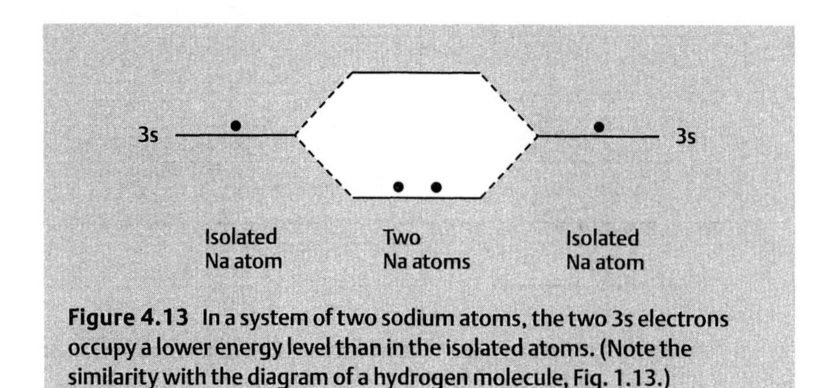

Figure 4.13 In a system of two sodium atoms, the two 3s electrons occupy a lower energy level than in the isolated atoms. (Note the similarity with the diagram of a hydrogen molecule, Fig. 1.13.)

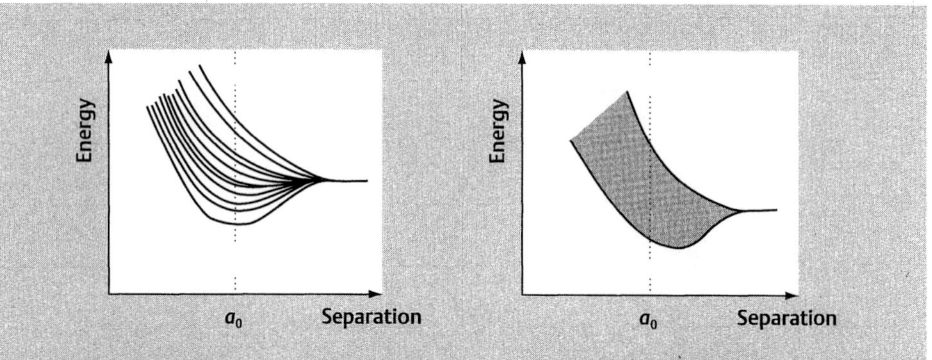

Figure 4.14 (a) The energy levels for the 3s electrons in a group of N sodium atoms as a function of the separation between the atoms. (b) For large values of N the states are so close together that we effectively have a continuous band of allowed energies.

Although the quantum states in a solid are so closely spaced that we can treat them as a continuous band of energies, we still need to be aware of the existence of the discrete energy levels when we consider how many electrons can occupy each band. If the crystal contains N atoms then in the 1s band we have N discrete states, which are capable of accommodating a total of $2N$ electrons. Since there are two 1s electrons in each sodium atom, we have precisely the right number of 1s electrons needed to occupy all of the states in the 1s band. Similarly, we find that all $2N$ and $6N$ available positions in the 2s and 2p bands are occupied. Now we come to the 3s band. This also has a capacity to accommodate $2N$ electrons, but since there is only a single 3s electron in each sodium atom, there are only N electrons available to occupy these spaces. Consequently, the 3s band is just partially occupied. We therefore obtain an overall picture for the energy bands of sodium as shown in Fig. 4.11(b).

Before we consider how the presence of the energy bands affects our model of electrical conduction in a metal we need to examine this partially filled band in more detail. In particular, we want to know which quantum states are filled and which are vacant. We would also like to have some information about the energies of these quantum states. We will address these questions in the following sections.

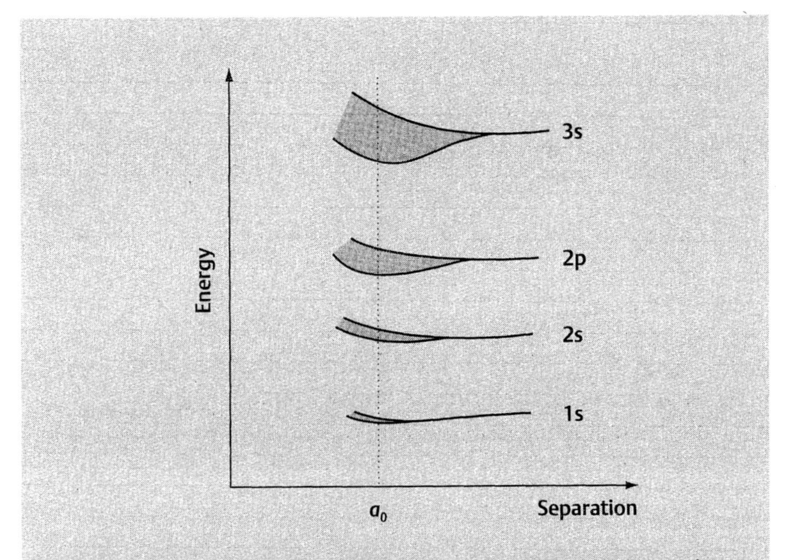

Figure 4.15 The formation of 1s, 2s, 2p and 3s energy bands in sodium. The bands become progressively narrower for electrons which are more tightly bound to the parent nucleus.

4.6 Distribution of the electrons between the energy states—the Fermi–Dirac distribution

If an energy band is only partially filled, such as the 3s band in sodium, which states do the electrons occupy? We know that electrons tend to occupy the lowest available energy levels, so if there are N electrons in the band then as a first approximation we can assume that the lowest $N/2$ states are occupied and the higher energy states are vacant. This suggests that there is an abrupt cut-off between those states which are occupied and those which are vacant. The corresponding energy is called the Fermi energy, E_F. We can state this another way by saying that the probability that a state is occupied is equal to 1.0 if the energy of the state is less than the Fermi energy, and equal to 0.0 if the energy is greater than the Fermi energy, as shown in Fig. 4.16.

Whilst the above analysis is correct at a temperature of absolute zero, at any finite temperature some of the electrons gain thermal energy and are excited into higher energy states. Consequently, at any instant in time there are a number of electrons which have energies greater than the Fermi energy, and a corresponding number of vacant spaces below the Fermi energy. This is shown schematically in Fig. 4.17.

The statistical distribution of the electrons in this case was first determined by Enrico Fermi and Paul Dirac in 1926. According to their theory, the probability $f(E)$ that a state with energy E is occupied at temperature T is given by

$$f(E) = \frac{1}{e^{(E-E_F)/k_B T}+1} \tag{4.12}$$

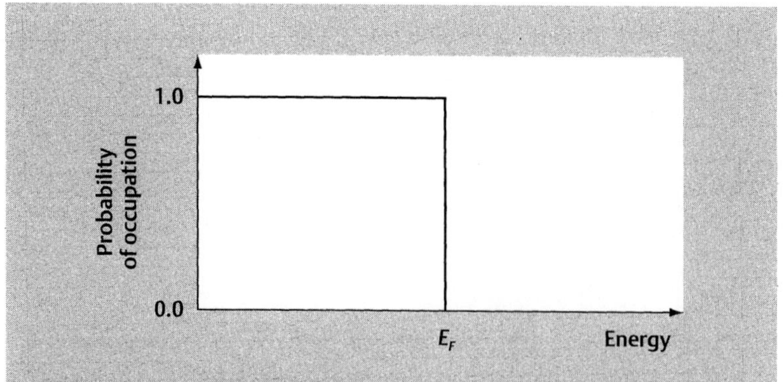

Figure 4.16 The probability distribution for a system of electrons at $T = 0$ K. The energy of the highest occupied state is the Fermi energy E_F.

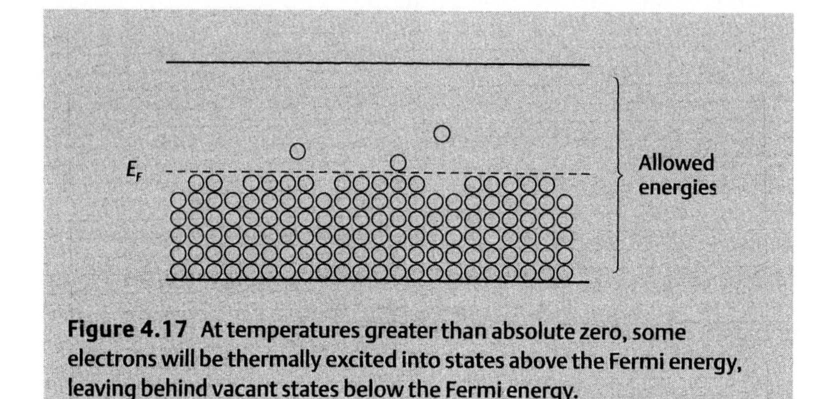

Figure 4.17 At temperatures greater than absolute zero, some electrons will be thermally excited into states above the Fermi energy, leaving behind vacant states below the Fermi energy.

This is the Fermi–Dirac distribution. We will not attempt to justify this relationship (if you want more details then refer to Beiser—see Further reading) but we are interested in what the distribution looks like. At $T = 0$ K the equation predicts the same distribution as shown in Fig. 4.16 (see Question 4.12), but at higher temperatures the abrupt step in the distribution becomes more gradual, as shown in Fig. 4.18. In this case it is no longer meaningful to refer to a maximum energy of the electrons, but the

Example 4.3 Show that the probability of occupation for an electron state at the Fermi energy is equal to 0.5 for all finite temperatures.

Solution When $E = E_F$, the exponent in eqn (4.12), $(E - E_F)/k_B T$, is equal to zero for all non-zero values of T.

Consequently,

$$f(E) = \frac{1}{e^0 + 1} = \frac{1}{1+1} = \frac{1}{2}$$

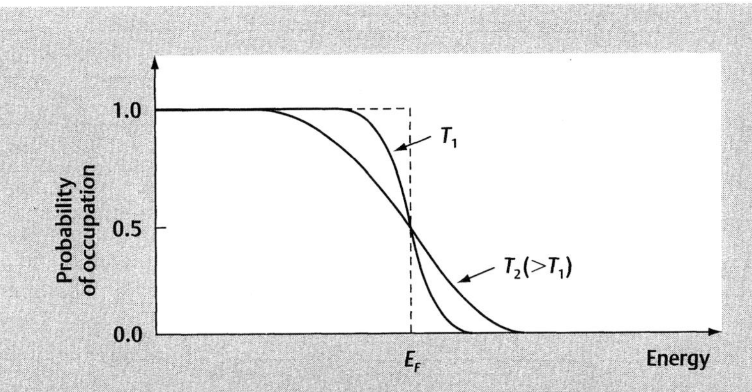

Figure 4.18 The Fermi–Dirac probability distribution for two temperatures T_1 and T_2 (where $T_2 > T_1$). The distribution at $T = 0$ K is shown by the dashed lines. Note that the Fermi energy can be defined as the energy at which the probability of occupation is equal to 0.5.

Fermi energy now represents a sort of average maximum energy. In fact, we can see from Example 4.3 that the Fermi energy can be defined as the energy for which the probability of occupation is equal to one-half.

4.7 The density of states

In the previous section we introduced the Fermi–Dirac equation. This describes how the electrons are distributed as a function of energy, but it does not tell us how many, if indeed any, electrons exist at a particular energy. To find out this information we need to determine how many quantum states exist at the relevant energy.

Since we are dealing with a continuous band of energies, we should really determine how many quantum states occur in a small energy range, e.g. between E_1 and $E_1 + \delta E$. By repeating this procedure over many such intervals we can construct a histogram showing how the number of quantum states varies as a function of energy (see Fig. 4.19). If we make the energy range δE sufficiently small, the histogram becomes a smooth curve. This curve is referred to as the density of states function and can be determined either by experimental methods or by calculation. The density of states for a typical energy band is shown in Fig. 4.20. A characteristic feature is that the majority of states occur in the middle of the band, with comparatively few states available near the extremes of the band.

A calculation of the density of states is in general rather complicated, but for a simple metal the density of states can be obtained, at least for the lower part of the energy band, without too much difficulty. We will consider this problem in the next section. Readers with no previous experience of quantum theory may wish to omit this section.

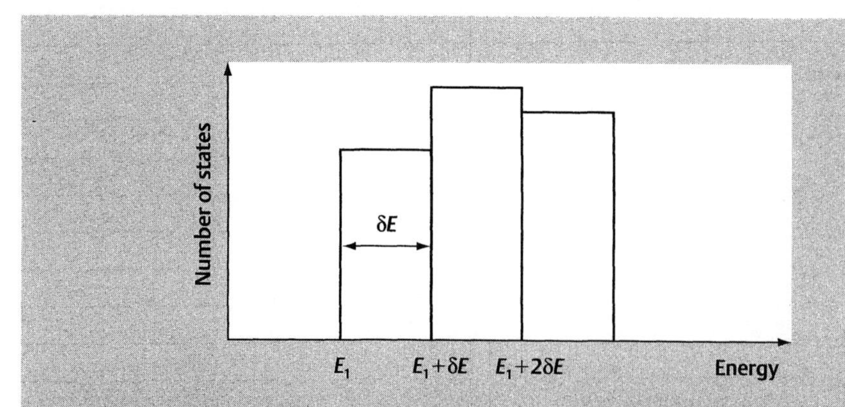

Figure 4.19 A histogram of the number of quantum states with energies between E_1 and $E_1 + \delta E$, $E_1 + \delta E$ and $E_1 + 2\delta E$, and so on.

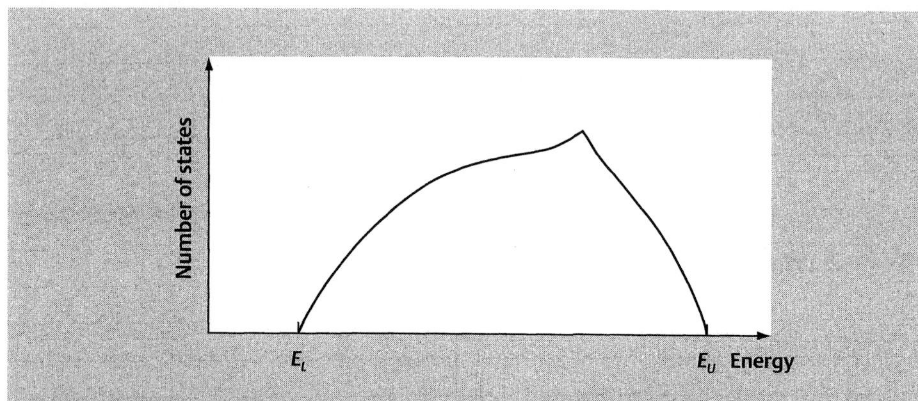

Figure 4.20 The density of states for a typical energy band. E_L and E_U indicate the lower and upper edges of the energy band.

4.8 The free electron model*

In order to calculate the density of states, we first of all need to determine the positions of the energy levels in the crystal. To do this we will use the so-called free electron model. In this model it is assumed that the average potential inside the metal due to the valence electrons and ions is constant throughout the sample, but that at the edge of the sample there is a large potential which stops the electrons escaping from the metal. The depth of the box is equal to the amount of energy required for an electron to escape from the metal. This quantity is known as the work function and can be determined from measurements of the photoelectric effect (see Serway in Further reading). The work function is usually of the order of a few electron volts. In comparison, the average thermal energy of the electrons is equal to k_BT, which at room temperature is approximately 0.025 eV. Consequently, the chances of an electron escaping from the metal under normal conditions are very small, and so for mathematical convenience we will assume that the box containing the electrons is of infinite depth.

To simplify matters further, let us begin by restricting our attention to a one-dimensional case. We will assume that the box is of length L, where L is the size of the sample. According to quantum theory, the energy, E, of an electron in such a system is given by the differential equation

$$-\frac{\hbar^2}{2m_e}\frac{d^2\psi}{dx^2} = E\psi \tag{4.13}$$

where ψ is known as the eigenfunction. (If you are familiar with quantum theory, you may recognize the above equation as the Schrödinger equation for a particle in a one-dimensional potential well of infinite depth.) From Example 4.4, we can see that Equation (4.13) is satisfied by eigenfunctions of the form

$$\psi = A\sin(k_x x) \tag{4.14}$$

where A is a constant. These eigenfunctions can be represented graphically, as shown in Fig. 4.21, and the corresponding energies are given by

$$E = \frac{\hbar^2 k_x^2}{2m_e} = \frac{\hbar^2 \pi^2 n_x^2}{2m_e L^2} \tag{4.15}$$

where n_x is an integer.

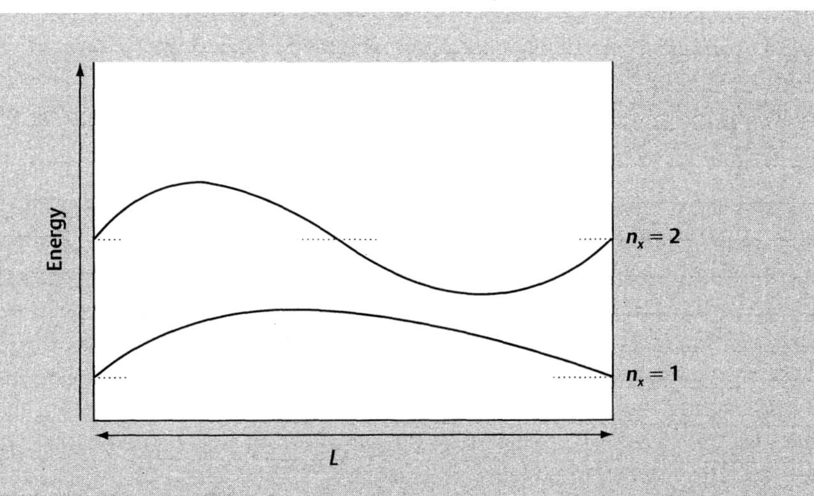

Figure 4.21 In the free electron model it is assumed that the electrons are trapped in a deep potential well. The figure shows the form of the first few eigenfunctions for a one-dimensional case. The corresponding energies are given by eqn (4.15).

Example 4.4 Solve eqn (4.13) to obtain the allowed energies for an electron in an infinitely deep one-dimensional box of width L.

Solution A knowledge of second-order differential equations suggests that a suitable form for ψ is

$$\psi = A\sin(k_x x)$$

where A is a constant. Differentiating this expression twice gives

$$\frac{d^2\psi}{dx^2} = -k_x^2 A\sin(k_x x) = -k_x^2\psi$$

Substituting in eqn (4.13) we obtain

$$E = \frac{\hbar^2 k_x^2}{2m_e}$$

We also need to ensure that the solution satisfies the boundary conditions. Because the box is infinitely deep, the probability of finding the electron outside the box is zero. Consequently, the value of ψ should be zero at $x = 0$ and at $x = L$.

The first condition is automatically satisfied by our choice of ψ, but the condition that ψ is zero at $x = L$ is only satisfied if $k_x = n_x\pi/L$ where n_x is an integer.

Consequently, the allowed energies can be written as

$$E = \frac{\hbar^2\pi^2 n_x^2}{2m_e L^2}$$

If we now extend this treatment to the three-dimensional case of an infinitely deep box with dimensions L in each direction we find (see Question 4.14) that the allowed energies are given by

$$E = \frac{\hbar^2 k^2}{2m_e} = \frac{\hbar^2}{2m_e}\left(\frac{\pi}{L}\right)^2 (n_x^2 + n_y^2 + n_z^2) \tag{4.16}$$

Now that we have obtained an expression for the allowed energy levels, the next step is to determine which of these states are occupied by electrons. To keep matters simple, let us assume a temperature of absolute zero. In this case we know that all of the states up to the Fermi energy are occupied and all the states of higher energy are vacant, so what is the energy of the highest occupied state? We know from the exclusion principle that each quantum state can hold two electrons, so we can answer this question by starting with the lowest energy level and putting two electrons in each state until we have used up all of the electrons. Let us try this. The lowest energy state given by eqn (4.16) occurs when

$$n_x = n_y = n_z = 1$$

so we can put two electrons in this state. Next we have three states

$n_x = 2, n_y = 1, n_z = 1$

$n_x = 1, n_y = 2, n_z = 1$

$n_x = 1, n_y = 1, n_z = 2$

all of which have the same energy. These states accommodate a further six electrons.

We could continue in this way, but as we are dealing with about 10^{23} electrons it would be extremely tedious! However, we can achieve the same result by recognizing that each combination of n_x, n_y and n_z corresponds to a point on a 3-D grid. For a particular grid point the value of $(n_x^2 + n_y^2 + n_z^2)^{1/2}$, is equal to the distance of that point from the origin. If we construct a sphere with this radius, centred on the origin, then all points lying on the surface of the sphere have the same value of $(n_x^2 + n_y^2 + n_z^2)^{1/2}$, and so, according to eqn (4.16), have the same energy. Consequently, we can determine the energy of the highest occupied state if we determine the radius n_{max} of the surface which just encloses sufficient grid points to accommodate all of the N electrons in a crystal (see Fig. 4.22). As only integer values of n_x, n_y and n_z are allowed, each grid point

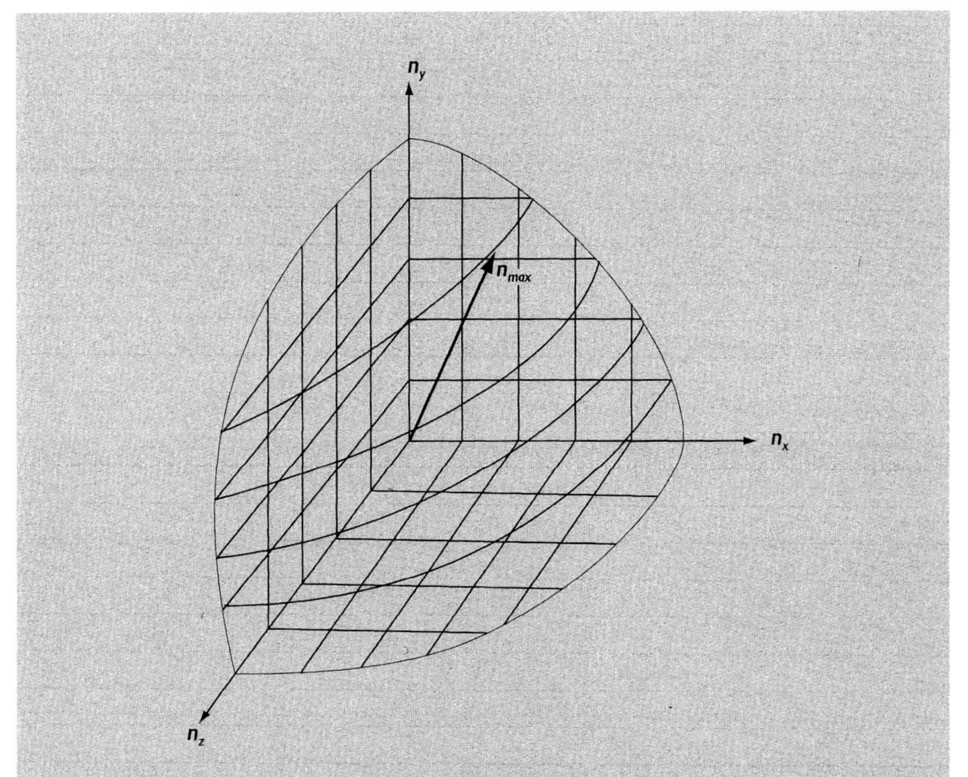

Figure 4.22 Each allowed quantum state given by eqn (4.16) can be represented by a grid point (n_x, n_y, n_z) as shown. Grid points with the same energy are connected by a surface of constant radius. In order to determine the Fermi energy, we need to find the value of the radius n_{max} which just contains sufficient states to accommodate all of the valence electrons in the crystal.

in Fig. 4.22 corresponds to a cube of unit volume, and so the number of grid points is equal to the volume under the surface. By pursuing this argument further, we find (see Example 4.5) that the energy of the highest occupied state (i.e. the Fermi energy) is given by

$$E_F = \frac{\hbar^2}{2m_e}\left(\frac{3\pi^2 N}{V}\right)^{2/3} \tag{4.17}$$

where V is the volume of the crystal.

All that remains now is to determine the density of states, $g(E)$. How do we do this? Let us suppose first of all that we know the form of the density of states function. If we integrate the density of states function up to the Fermi energy then we are actually performing a sum over all of the occupied states. In other words, the number of electrons N is given by

Example 4.5 By determining the number of grid points contained beneath the surface of radius n_{max} (shown in Fig. 4.22), show that the Fermi energy (i.e. the energy of the highest occupied state at $T = 0$ K) is given by

$$E_F = \frac{\hbar^2}{2m_e}\left(\frac{3\pi^2 N}{V}\right)^{2/3}$$

Solution The volume contained beneath the surface in Fig. 4.22 corresponds to one-eighth of a sphere of radius n_{max}, i.e. the volume is

$$\frac{1}{8}\cdot\frac{4}{3}\pi n_{max}^3 = \frac{\pi n_{max}^3}{6}$$

Since each grid point corresponds to a cube of unit volume, this expression also gives the number of grid points beneath the surface.

As each grid point corresponds to a state which can accommodate two electrons, for a crystal containing N electrons we require $N/2$ states. Therefore

$$\frac{N}{2} = \frac{\pi n_{max}^3}{6}$$

Rearranging, we obtain

$$n_{max} = \left(\frac{3N}{\pi}\right)^{1/3}$$

By writing $n_{max} = (n_x^2 + n_y^2 + n_z^2)^{1/2}$ and substituting in eqn (4.16) we obtain an expression for the energy of the highest occupied state (i.e. the Fermi energy) as follows:

$$E_F = \frac{\hbar}{2m_e}\left(\frac{\pi}{L}\right)^2\left(\frac{3N}{\pi}\right)^{2/3}$$

$$= \frac{\hbar^2}{2m_e}\left(\frac{3\pi^2 N}{V}\right)^{2/3}$$

where $V = L^3$.

$$N = \int_0^{E_F} g(E)\, dE \tag{4.18}$$

(see Fig. 4.23). By rearranging the above equation we can see that an expression for $g(E)$ can be found by differentiating N with respect to E. Consequently, using eqn (4.17) we can show that the density of states function is given by (see Question 4.15)

$$g(E) = \frac{dN}{dE} = \frac{V}{2\pi^2}\left(\frac{2m_e}{\hbar^2}\right)^{3/2} E^{1/2} \tag{4.19}$$

So our calculations have lead us to predict that the density of states varies as $E^{1/2}$. Comparing this with the true density of states (as shown in Fig. 4.20), we find that the agreement is good for the lower part of the band (see Fig. 4.24), but we obviously do not have a complete picture because our model suggests that there is no maximum energy for the band. To obtain the true density of states curve we need to include further refinements. For instance, the potential within the metal is not constant as we

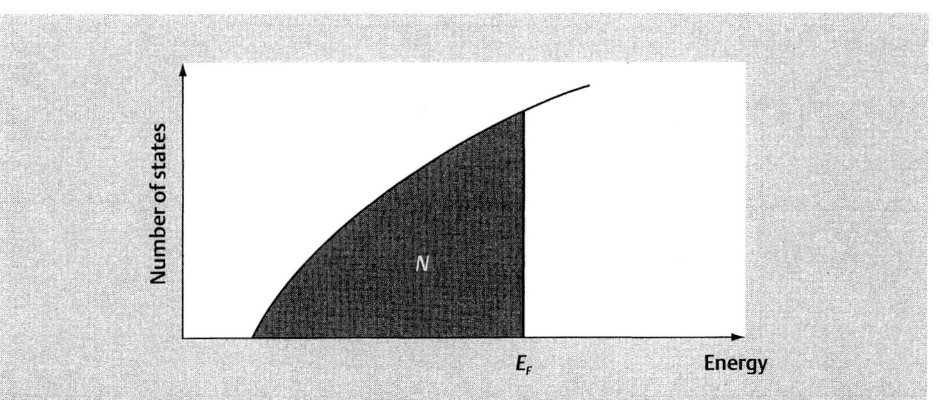

Figure 4.23 The number of single electron states N is equal to the area under the density of states curve $g(E)$ up to the Fermi energy.

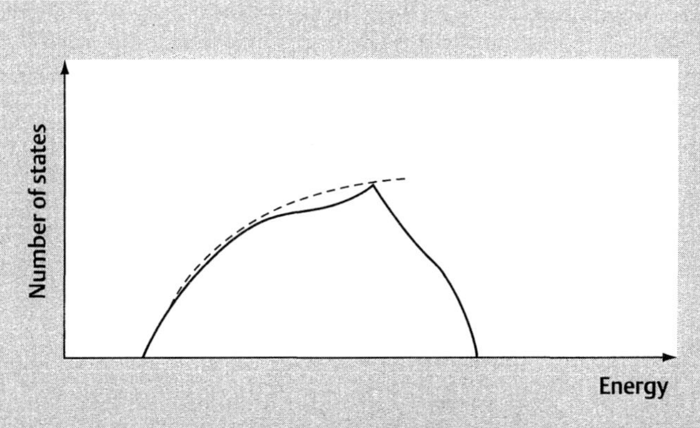

Figure 4.24 A comparison of the true density of states for a typical metal (solid line) with the density of states predicted by eqn (4.19) (dashed line).

have assumed, but varies periodically with the spacing of the atoms. Such a model is considered in more advanced texts, such as Kittel (see Further reading). However, for our purposes the free electron model is more than adequate.

4.9 The density of occupied states

In the preceding sections we have investigated the number of quantum states that exist at a particular energy and also looked at the probability that a state with a given energy is occupied. Our final step is to combine these two quantities in order to find out how many electrons have a particular energy. This is called the density of occupied states. We can determine this quantity graphically by superimposing the curve

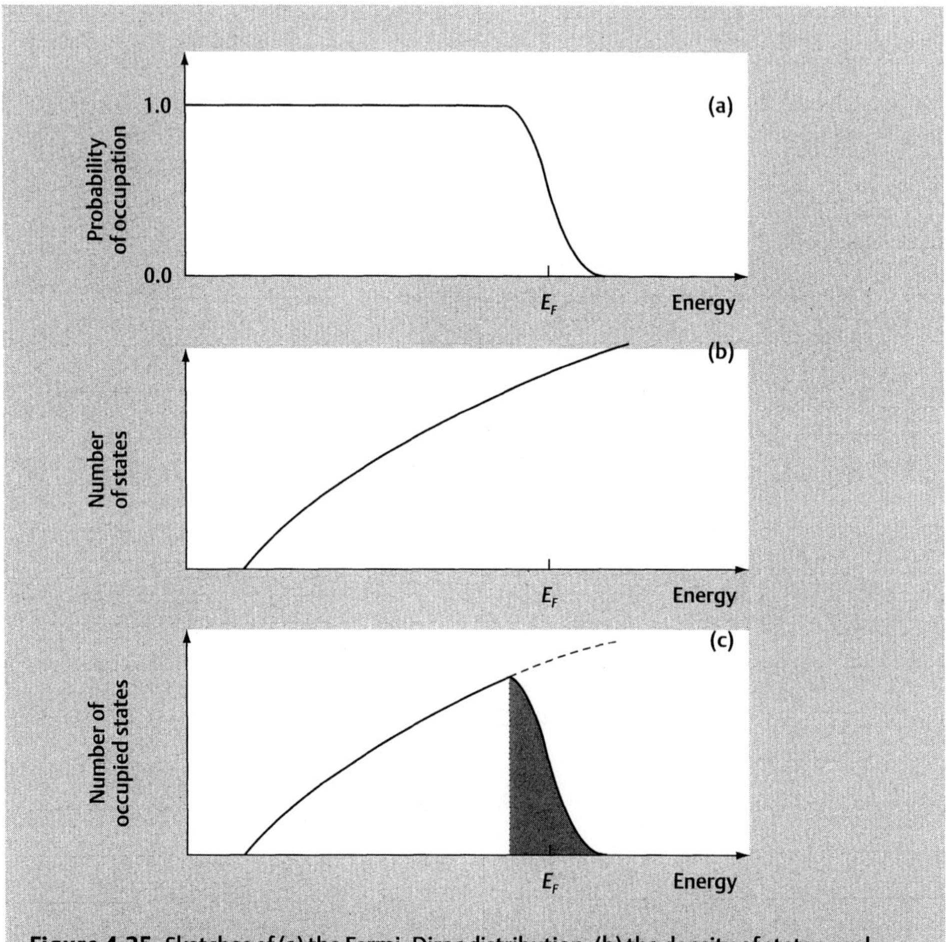

Figure 4.25 Sketches of (a) the Fermi–Dirac distribution, (b) the density of states, and (c) the density of occupied states for a simple metal at a finite temperature.

for Fermi–Dirac distribution in Fig. 4.18 on top of the density of states curve shown in Fig. 4.20. The result is shown in Fig. 4.25. Alternatively, if the density of states function is denoted by $g(E)$ and the Fermi–Dirac distribution by $f(E)$, then the density of occupied states is equal to the product $g(E).f(E)$.

4.10 Band theory of electrical conduction

Now that we have discussed the implications of band theory in some detail, we can round off this chapter by estimating the values of some of the parameters relevant to the theory of electrical conduction and compare our results with the values that we assumed in Drude's theory.

Let us begin by looking at the number of electrons that take part in electrical conduction. In the classical theory we assumed that all of the valence electrons are free to move in the direction of an applied electric field. This would seem to be a common-sense assumption, but do we still believe that this is true? If an electron is capable of being accelerated by an electric field then it must be able to move into a vacant state of slightly higher energy. By inspecting Fig. 4.25(c) we can see that this criterion rules out electrons which are well below the Fermi energy because there are no

Example 4.6 Using the Fermi-Dirac distribution, determine the values of energy corresponding to $f = 0.9$ and $f = 0.1$ at a temperature of 300 K. This corresponds approximately to the shaded region in Fig. 4.25(c). Hence estimate the proportion of valence electrons that can participate in electrical conduction. (Assume the Fermi energy is 3.22 eV, the value for sodium.)

Solution Rearranging eqn (4.12) we have

$$E - E_F = k_B T \ln\left(\frac{1}{f(E)} - 1\right)$$

When $f(E) = 0.1$,

$$E - E_F = k_B T \ln\left(\frac{1}{0.1} - 1\right) = 2.20\, k_B T$$

Similarly, when $f(E) = 0.9$, $E_F - E = 2.20\ k_B T$.

Therefore the width of the shaded region in Fig. 4.25(c) is approximately $4.40\ k_B T$. At 300 K this is 0.11 eV.

If we assume that the density of states is constant across the energy band, then the number of electrons in this region as a proportion of the total number of valence electrons is roughly equal to

$$\frac{\text{energy range of shaded region}}{\text{energy range of occupied states}} = \frac{0.11\ \text{eV}}{3.22\ \text{eV}} = 0.034$$

appropriate vacant states available. Consequently, the only electrons that can participate in electrical conduction are those that are close to the Fermi energy, indicated by the shaded region in Fig. 4.25(c). We can make a crude estimate of the proportion of electrons that fall into this category by comparing the energy range spanned by the shaded region with the range of energies covered by all the occupied states in the valence band—see Example 4.6. For a more sophisticated approach to this problem see Question 4.17. Either way, it turns out that at room temperature only a few per cent of the valence electrons occupy the states near to the Fermi energy. In other words, the number of electrons that participate in electrical conduction is of the order of a hundred times smaller than we assumed in the classical model.

What about the velocity of the electrons? In Drude's theory (Section 4.2) we assumed that the velocities are given by the kinetic theory of ideal gases, and so we predicted a value of about 10^5 m s^{-1} at room temperature. In the band theory approach we have seen that it is only those electrons which have an energy close to the Fermi level that participate in conduction. If we equate the Fermi energy with the kinetic energy of the electrons then we can write

$$\frac{1}{2} m_e v_F^2 = E_F \tag{4.20}$$

where v_F is the velocity of the electrons at the Fermi energy.

From Table 4.2, we can see that the Fermi energy in a metal is typically of the order of a few electron volts. Consequently the velocity of electrons at the Fermi energy is about 10^6 m s^{-1}(see Question 4.18), about 10 times larger than predicted by Drude's theory at room temperature. We should also note that in the band theory model the velocity is independent of temperature, whereas Drude's theory predicts that the velocity goes to zero as the temperature drops to zero kelvin.

The mean free path of the electrons is also likely to be substantially different from the classical assumption of a few nanometres. As we have seen in Section 4.4, the mean free path depends on the temperature (because of the thermal vibrations of the ions) and on the concentration and type of impurities and imperfections. In very

Table 4.2 Values of the Fermi energy E_F for a selection of metals. (These values can be obtained by substituting $n = N/V$ from Table 4.1 in eqn (4.17).)

	E_F (eV)
Sodium	3.22
Copper	7.00
Silver	5.46
Gold	5.49
Magnesium	7.05
Zinc	9.38
Aluminium	11.58
Tin	9.99
Lead	9.38

nearly perfect crystals at temperatures of less than 1 K, mean free paths of several centimetres have been measured. However, even under normal circumstances the mean free path in a metal is typically 1 μm (see Question 4.19), about 1000 times larger than the classical value.

We can also use the band theory model to explain the specific heat capacity anomaly that we discussed in Example 4.2. According to Drude's theory, each electron has an energy $\frac{3}{2}k_BT$, producing a total contribution of $\frac{3}{2}N_AK_BT$ for 1 mole of a monovalent metal. However, we have seen from Example 4.6 that only a very small percentage of the valence electrons are in a position to increase their energy by a small amount. Consequently, the contribution of the electrons to the specific heat capacity is much smaller than the classical value. In fact, it turns out that the electronic contribution is negligible in comparison with the contribution from the ions (see Question 4.20), so our band theory model predicts a molar specific heat capacity of almost exactly $3R$, which is in good agreement with the experimental values for most metals at room temperature. (We will discuss the specific heat capacity again, and in more detail, in Chapter 7.)

4.11 Summary

In this chapter we have investigated the process of electrical conduction in metals using both a classical and a quantum theory approach.

In the classical model we assumed that electrical resistance is due to collisions between the valence electrons and the ions. Despite some rather convincing results, there are many properties of metals that cannot be explained by this approach.

In the quantum model the resistance of a metal is found to be due to interactions of the valence electrons with

(1) the thermal vibrations of the ions;
(2) impurities; and
(3) imperfections in the crystal.

These three factors are important in many other circumstances as we shall see in subsequent chapters.

The quantum and classical models therefore give very different explanations for the origin of electrical resistance. However, there are some noticeable similarities between these models. For example, the overall trajectory of the valence electrons in the quantum picture is not unlike the haphazard motion that we assumed in the classical model. Consequently, many of the concepts that we developed in the classical model—such as mean free path, mobility and drift velocity—are also relevant in the quantum model, although the values associated with these parameters can be quite different from the classical values.

In the following chapter we will consider the electrical behaviour of semiconductors and compare these characteristics with those of metals.

Questions

*Questions marked * indicate that they are based on the optional sections.*

4.1 Drude's theory. Use eqn (4.1) to determine the average thermal velocity of a conduction electron in a metal at a temperature of 295 K. Assuming that the mean free path of the electrons is about 1 nm, calculate the average time between collisions.

4.2 Drude's theory. Use the values obtained in Question 4.1 to determine the mobility and drift velocity of an electron in a metal in an electric field of 10 V m^{-1} at 295 K. Explain why the drift velocity is much smaller than the thermal velocity calculated in Question 4.1.

4.3 Drude's theory. Show that the microscopic form of Ohm's law, as given by eqn (4.11), is equivalent to the more familiar form $V = IR$.

4.4 Drude's theory. Calculate the number density of valence electrons in lithium given that the density is 530 kg m^{-3}, the molar mass is 7 g mol^{-1} and that lithium is monovalent (i.e. each atom has one valence electron). Assuming that the mean free path of the electrons is 1 nm, calculate the conductivity and the mobility of lithium at 295 K.

4.5 Drude's theory. Using the data from Table 4.1, calculate the relaxation time and mean free path for aluminium at 273 K.

4.6 Drude's theory. A potential difference of 0.3 V exists between the ends of a copper wire of length 5 m. If the cross-sectional area of the wire is 2.5 mm^{-2}, use the data in Table 4.1 to calculate the mobility and the drift velocity of the electrons in the wire. Also determine the net number of electrons that pass through a given cross-section of the wire in 1 second.

4.7 Failures of the classical model. Assuming that the mean free path is about 1 nm, use the values of n given in Table 4.1 to calculate values of conductivity for silver and lead at 273 K. Compare your results with the actual values in Table 4.1 and comment on your findings.

4.8 Failures of the classical model. Show that Drude's theory predicts that the resistivity ρ is proportional to $T^{1/2}$ where T is the temperature. (Assume that the mean free path is constant with temperature.)

4.9 Failures of the classical model. Show that Drude's theory predicts that a divalent metal (i.e. a metal which has two valence electrons per atom) has a specific heat capacity of $6R$.

4.10 Quantum theory of conduction. Using the quantum model of conductivity (see Section 4.4) describe what happens to the resistivity of a metal as the temperature is lowered towards absolute zero. In particular, does this theory predict that the resistivity goes to zero or remains finite at $T = 0$ K? Give reasons for your answer.

4.11 Band theory. For a crystal of sodium with a volume of 1 cm^3, estimate the average spacing between the energy levels in the 3s band given that the 3s electron energies span a range of 3.22 eV. (You can use the value for the electron concentration, n, from Table 4.1.)

4.12 The Fermi–Dirac distribution. Show that the Fermi–Dirac equation predicts the same distribution as shown in Fig. 4.16 at $T = 0$ K.

4.13 The Fermi–Dirac distribution. Using the Fermi–Dirac distribution, determine the values of energy (as a function of the Fermi energy) corresponding to $f = 0.95$ and $f = 0.05$ at a temperature of 300 K.

4.14 *Free electron model. For a 3-D infinitely deep quantum well with dimensions L in each direction, use the trial function.

$$\psi = A \sin(k_x x) \sin(k_y y) \sin(k_z z)$$

and show that the allowed values of k_x, k_y and k_z are $n_x\pi/L$, $n_y\pi/L$ and $n_z\pi/L$, respectively. Hence show that the allowed energy levels are given by

$$E = \frac{\hbar^2 k^2}{2m_e}$$

where

$$k^2 = \left(\frac{\pi}{L}\right)^2 (n_x^2 + n_y^2 + n_z^2)$$

4.15 ***Free electron model.** Using the expression for the Fermi energy in eqn (4.17), show that the density of states function $g(E)$ is given by

$$g(E) = \frac{dN}{dE} = \frac{V}{2\pi^2}\left(\frac{2m_e}{\hbar^2}\right)^{3/2} E^{1/2}$$

4.16 ***Free electron model.** Calculate the Fermi energy and the Fermi velocity (i.e. the velocity of an electron at the Fermi energy) for copper using the data in Table 4.1. Use the density of states function to determine the average spacing between energy levels near the Fermi energy in a sample of copper of volume 1 cm^3.

4.17 ***Band theory of electrical conduction.** (In this question we estimate the proportion of valence electrons that participate in electrical conduction in a metal. The method is more accurate than that used in Example 4.6 because the density of states function is included in the calculation.)

From Example 4.6 we can see that a probability of occupation of 0.9 corresponds to an energy of $E_F - 2.2\,k_BT$. The number of electrons with energies greater than this is given by

$$N_c = \int_{E_F - 2.2\,k_BT}^{\infty} f(E) . g(E)\, dE$$

As this integral is not easy to perform, we will assume that the number of electrons with energies greater than the Fermi energy is approximately equal to the number in the shaded region in Fig. 4.26. Consequently, we can write

$$N_c \approx \int_{E_F - 2.2\,k_BT}^{E_F} g(E)\, dE$$

Solve this equation and use the data in Tables 4.1 and 4.2 to determine the proportion of

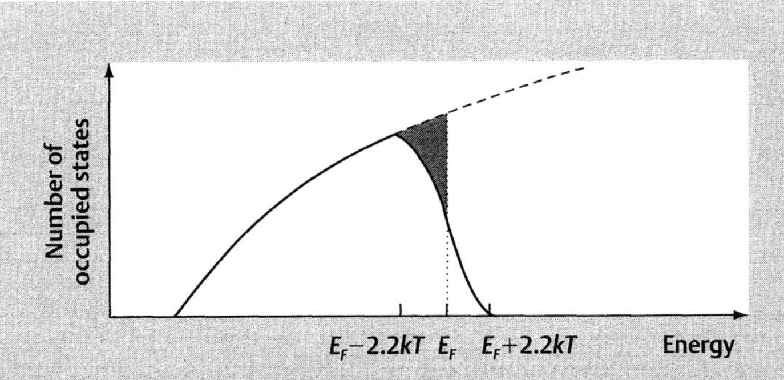

Figure 4.26 Sketch to show that the number of electrons with energies greater than the Fermi energy is approximately equal to the number in the shaded area.

valence electrons that participate in electrical conduction in a sample of sodium at 300 K.

4.18 **Band theory of electrical conduction.** Use the data in Table 4.2 to determine the velocity of an electron at the Fermi energy in sodium, copper and aluminium.

4.19 **Band theory of electrical conduction.** Assuming that the velocity of a conduction electron is given by eqn (4.20) and that at room temperature only 3.0% of the valence electrons take part in electrical conduction, use the data in Tables 4.1 and 4.2 to estimate values for the relaxation time τ and the mean free path λ in copper at 273 K.

4.20 **Band theory of electrical conduction.** We have seen in Example 4.6 that at room temperature only about 3.0% of the valence electrons in a metal take part in electrical conduction, and so we can assume that it is only this small percentage of electrons that are capable of being thermally excited. Further calculations show that these electrons gain an average thermal energy k_BT. Using this information, estimate the contribution of the valence electrons to the molar specific heat capacity of a monovalent metal (i.e. one valence electron per atom) at 300 K. Compare your result with the contribution to the heat capacity from the ions (assuming that the heat capacity of the ions can be described by the classical approach described in Example 4.2.)

Chapter 5

Semiconductors

5.1 Introduction

One of the most remarkable features about the electrical conductivity of solids is that the range of values is so vast. From Fig. 5.1 we can see that the conductivity of the best conductors, such as copper and silver, is over 10^{24} times higher than that of the best insulators, such as sulphur. Such a ratio is hard, if not impossible, to comprehend. It is similar, for example, to comparing the diameter of a planetary orbit to that of an electron in orbit around an atom. However, if it was not for this large range of values, we would have great difficulty in making safe use of electricity. For example, if the electrical conductivity varied by a factor of only 10^5 (which is typical of the range of

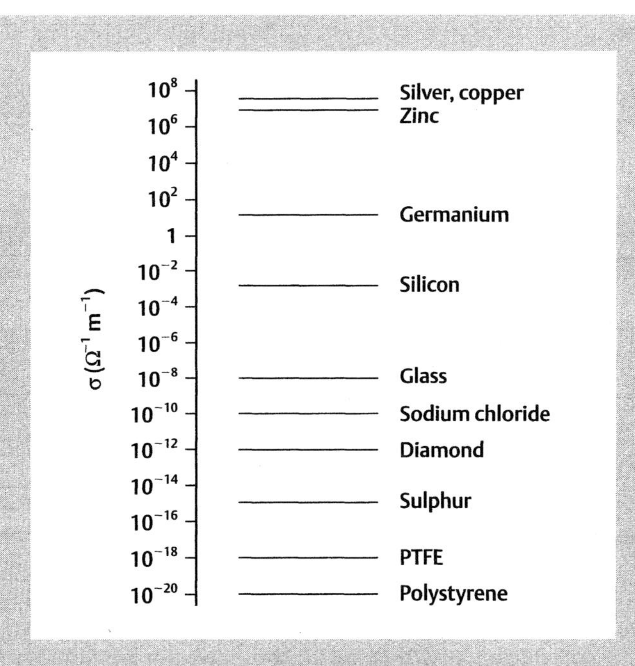

Figure 5.1 The conductivities of a selection of solids at room temperature. Note that a logarithmic scale is used because of the extremely large range of values.

values of mechanical strength or thermal conductivity of solids), you would get electrocuted every time you switched on an electric light!

We can account for this enormous range of electrical conductivities quite simply by considering the type of bond present in the material. Metallic bonding produces delocalized electrons which are able to move in response to an electric field. Therefore these materials form electrical conductors. In contrast, other types of bonding involve the valence electrons being strongly bound to a particular atom or ion, or to a pair of atoms in the case of a covalent bond. These materials should therefore be insulators.

This simple model explains why there are are two extremes of behaviour—conductors and insulators—but it does not explain why there is a continuous range of conductivities in between these extremes. Another puzzling feature that cannot be explained by our simple bonding model is that materials which employ similar types of bonds may have very different electrical properties. For example, silicon and diamond are both covalent solids with identical crystal structures, and yet diamond is a good insulator, whereas silicon is a reasonably good conductor, particularly at high temperatures. In order to resolve these problems we need to develop further the energy band theory that we introduced in Chapter 4.

In this chapter we will concentrate on semiconductors—materials which have electrical conductivities intermediate between those of metals and insulators—as these materials have many important technological applications. In the first part of the chapter we will look at the electrical properties of pure, or intrinsic, semiconductors. These properties can be dramatically altered by very small concentrations of impurity atoms. In particular, certain impurities are often added on purpose to control the conductivity of a given material. This process is known as doping, and we will consider these doped, or extrinsic, semiconductors in the second part of the chapter.

5.2 Band theory of solids—again

In Chapter 4 we found that the interactions between the outermost electrons in a solid lead to the formation of a continuous band of energies. In the case of sodium, this band is only partially occupied by electrons. However, sodium is a particularly simple case because it has only a single valence electron on each atom. In this section we will look at what happens when there is more than one valence electron per atom. In particular, we will study diamond, which of course is a crystal consisting purely of carbon atoms.

An isolated carbon atom contains six electrons, two each in the 1s, 2s and 2p levels, as shown in Fig. 5.2(a). If we plot the energies of the 2s and 2p energy levels for a group of N carbon atoms as a function of separation we find that the two sets of energy levels interact, as shown in Fig. 5.3. In particular, at the equilibrium separation the states are split into two bands which are each a mixture of 2s- and 2p-like states. The lower energy band, which is known as the valence band, corresponds to electrons which are involved in forming covalent bonds and can accommodate $4N$ electrons. The upper band is called

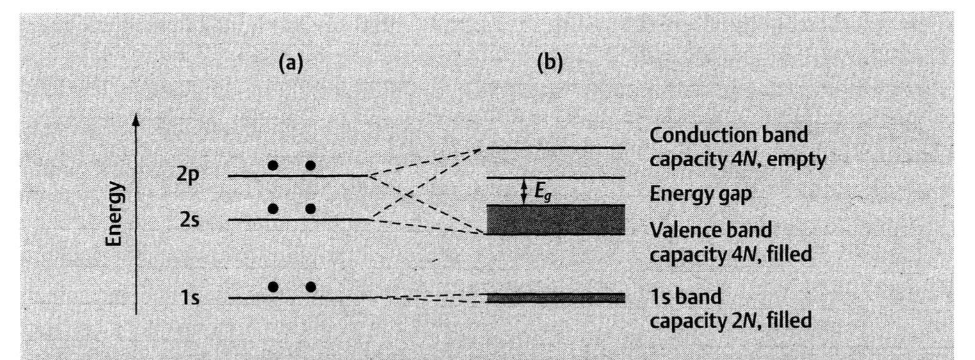

Figure 5.2 The occupation of (a) the energy levels in an isolated carbon atom, and (b) the energy bands in a diamond crystal. Notice that there is an energy range E_g separating the highest occupied states (in the valence band) from the lowest vacant states (in the conduction band). This is a characteristic feature of all insulators.

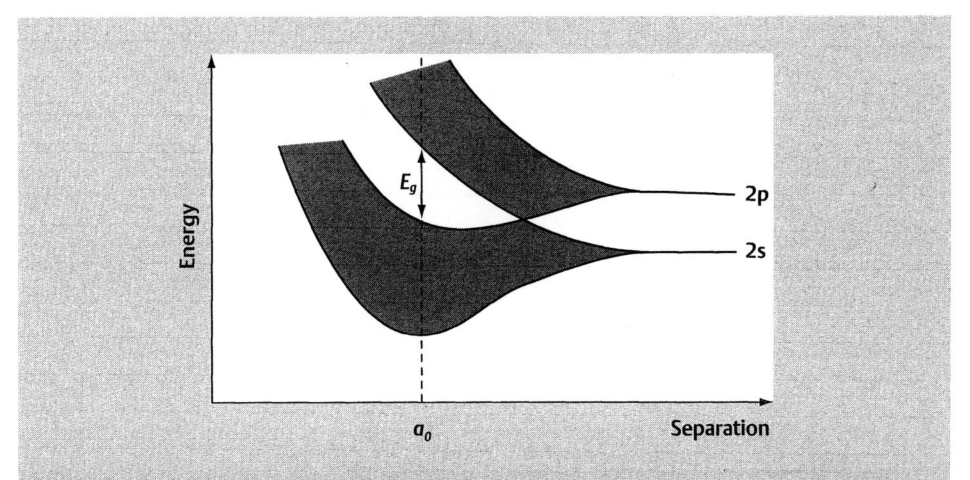

Figure 5.3 The energy levels of the 2s and 2p states for a group of N carbon atoms as a function of the separation of the atoms.

the conduction band because the electrons in these states are relatively free and so can easily take part in electrical conduction. This band also has a capacity of 4 N electrons. Since each carbon atom has four valence electrons, this is just sufficient to fill all of the states in the valence band, leaving the conduction band empty. (At least, this is the case at zero kelvin. We will consider the effects of temperature in a moment.) The valence and conduction bands are separated by an energy range in which there are no allowed electron energy levels. This is referred to as the forbidden band gap (or often simply as the 'band gap'), and has a magnitude which we will denote by E_g (see Fig. 5.2(b)).

As we shall see, the band gap is crucial to our understanding of the electrical properties of semiconductors and insulators. Unfortunately, there is no simple method of calculating the magnitude of the band gap, or indeed whether such a gap exists at all.

This information must be obtained from sophisticated computer calculations or from experimental measurements. These results reveal an important difference between metals and non-metals. It is found that the energy band diagram in Fig. 4.11(b), in which the highest occupied band is only partially filled, is characteristic of metals, whereas non-metals have an arrangement similar to that shown in Fig. 5.2(b) in which the highest occupied states (at absolute zero) are separated from the lowest vacant states by a region of forbidden energies.

We can use this finding to explain the difference in electrical conductivity between metals and non-metals. The electrons in a partially filled band can gain an arbitrarily small amount of energy by moving into a nearby unoccupied state. This means that when an electric field is applied to the material, the electrons can increase their kinetic energy by being accelerated in the direction of the electric field. Conversely, if there is an energy gap between the filled and vacant states then this process cannot occur, and so these materials are electrical insulators.

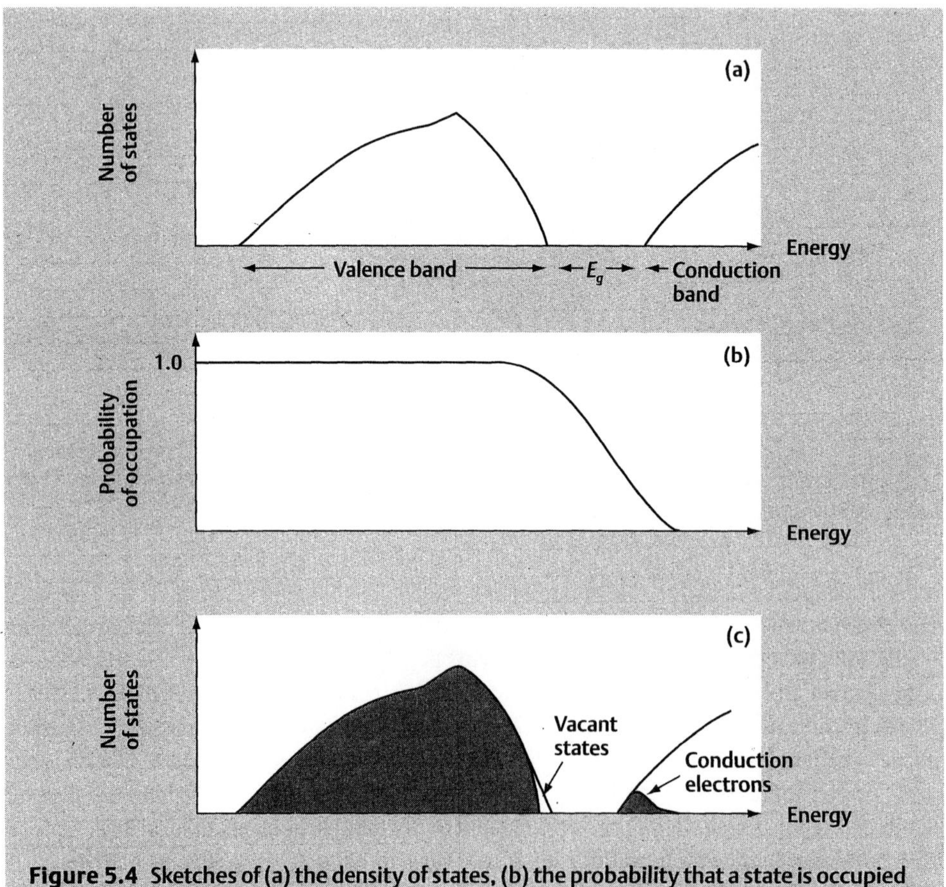

Figure 5.4 Sketches of (a) the density of states, (b) the probability that a state is occupied (as given by the Fermi–Dirac distribution), and (c) the density of occupied states for a semiconductor at finite temperature.

Once again we appear to have a clear distinction between the electrical properties of metals and insulators, this time based on the energy band structure. So how do we account for the behaviour of semiconductors? The key to understanding this problem is to remember that at a temperature above absolute zero the electrons do not necessarily occupy the lowest possible energy levels (see Section 4.6). This means that if the band gap is sufficiently small, there is a possibility of finding some electrons in the conduction band (see Fig. 5.4), and consequently the material is able to conduct electricity to some degree. We will explore this idea in more detail in the next section.

5.3 The difference between insulators and semiconductors

In the previous chapter we found that the probability that a state with energy E is occupied at temperature T is given by the Fermi–Dirac distribution

$$f(E) = \left[e^{(E-E_F)/k_B T} + 1\right]^{-1} \tag{5.1}$$

However, before we can apply this formula to a semiconductor or an insulator, we first of all need to determine the position of the Fermi energy. In the previous chapter we defined the Fermi energy in a metal as the energy for which the probability of occupation is equal to one-half. If we try to apply the same definition to a non-metal we obtain a rather ambiguous result. For example, at a temperature of absolute zero the probability that a valence state is occupied is equal to 1.0 and the probability that a conduction state is occupied is equal to 0.0 (see Fig. 5.5). So at what energy is the probability equal to 0.5? A common-sense argument suggests that we place the Fermi en-

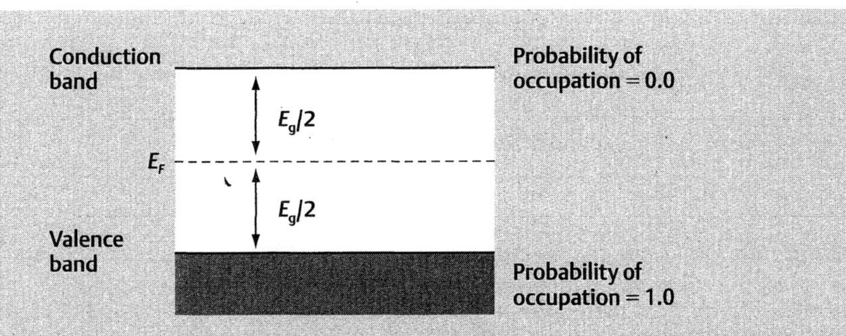

Figure 5.5 At a temperature of absolute zero the probability of a state being occupied is equal to 1.0 for all valence states and 0.0 for all conduction states. The Fermi energy, which corresponds to a probability of occupation of 0.5, is therefore placed in the centre of the band gap.

ergy at the centre of the band gap. (We will see later in this chapter that this is a remarkably good approximation in most circumstances.)

By making this assumption we can now use eqn. (5.1) to determine the probability of finding an electron in the conduction band. If we take the value of E in eqn. (5.1) to be the energy of the lowest energy state in the conduction band, then we can see from Fig. 5.5 that

$$E - E_F = \frac{E_g}{2}$$

Consequently, the probability of finding an electron at this energy becomes

$$f(E) = \left[e^{E_g/2k_B T} + 1\right]^{-1} \tag{5.2}$$

We can further simplify this expression if we note that at room temperature the magnitude of the band gap is usually of the order of 1 eV or more (see Table 5.1), whereas $k_B T$ is only about 0.025 eV. Consequently, the exponential term in eqn. (5.2) is much larger than 1, and so to a good approximation (see Question 5.1) we can write

$$f(E) \approx \left[e^{E_g/2k_B T}\right]^{-1} = e^{-E_g/2k_B T} \tag{5.3}$$

Since we expect the concentration of electrons in the conduction band, n, to be proportional to the probability in eqn. (5.3), we can write

$$n = Ce^{-E_g/2k_B T} \tag{5.4}$$

where C is a constant. We will see in Section 5.11 how to obtain an expression for C, but for now we will simply assume that $C \approx 10^{25}$ m^{-3}. This value gives good results at room temperature for a wide range of semiconductors and insulators (see Question 5.2).

Table 5.1 Values of the band gap E_g at 300 K for a selection of semiconductors and insulators

	Eg (eV)
Indium antimonide	0.18
Indium arsenide	0.36
Germanium	0.67
Silicon	1.11
Gallium arsenide	1.43
Silicon carbide	2.3
Zinc sulphide	3.6
Diamond	5.5

Example 5.1 Estimate the number of conduction electrons per cubic metre in a sample of silicon at 300 K. (Assume that the constant C is equal to $2 \times 10^{25}\ \mathrm{m}^{-3}$.)

Solution From Table 5.1 the band gap of silicon is $E_g = 1.11\ eV = 1.78 \times 10^{-19}\ J$, so using eqn (5.4) with $C = 2 \times 10^{25}\ \mathrm{m}^{-3}$ gives a carrier concentration at 300 K of

$$n = Ce^{-E_g/2k_B T} = (2\times 10^{25}\,\mathrm{m}^{-3})\exp\left(-\frac{1.78\times 10^{-19}\,\mathrm{J}}{(2)(1.38\times 10^{-23}\,\mathrm{J\,K}^{-1})(300\mathrm{K})}\right)$$

$$= 9.22\times 10^{15}\,\mathrm{m}^{-3}$$

Now that we have dealt with the theory, let us put some numbers into these equations. If we substitute for the band gap of silicon in eqn (5.4) we find that at room temperature there are about 10^{16} conduction electrons per cubic metre (see Example 5.1). Therefore we should not be surprised that silicon behaves as an electrical conductor. However, the concentration of conduction electrons is about 13 orders of magnitude smaller than in a metal (see Table 4.1), which explains why the conductivity of silicon at room temperature is much lower than that of metallic conductors.

In comparison, if we perform the same calculation for diamond, which has a band gap only about five times larger than that of silicon, we find that there are virtually no electrons in the conduction band (see Question 5.3). Consequently, we expect diamond to be a perfect insulator.

The behaviour of a material as a semiconductor or as an insulator also depends on the temperature. As we can see from eqn (5.4), the concentration of electrons in the conduction band increases exponentially as the temperature is raised, leading to a corresponding increase in the conductivity (see Question 5.4). For example, at low temperatures silicon makes an excellent insulator, but at very high temperatures it has a conductivity approaching that of a metal (see Fig. 5.6). An interesting demonstration of the effect of temperature on the electrical conductivity of an insulator can be performed by connecting a glass bar in series with a light bulb in an electrical

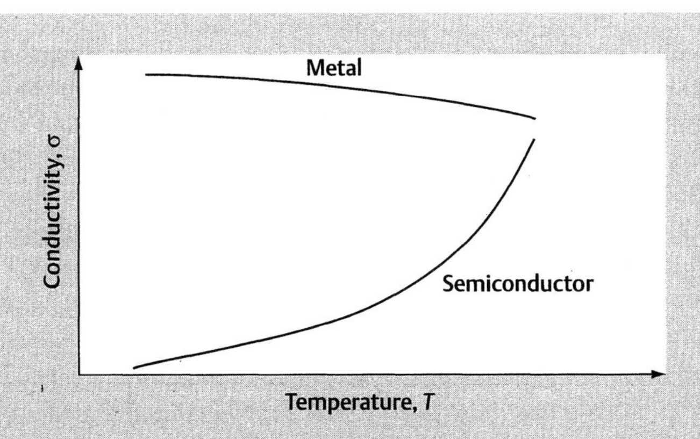

Figure 5.6 Variation of conductivity with temperature for a typical metal and semiconductor.

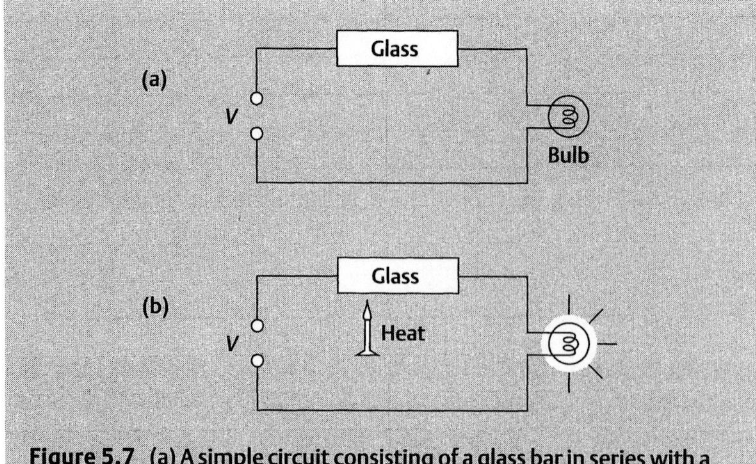

Figure 5.7 (a) A simple circuit consisting of a glass bar in series with a light bulb. At room temperature the bulb does not light because glass is an insulator. (b) But if the glass bar is heated (e.g. using a Bunsen burner), the conductivity of the glass increases and the bulb lights.

circuit (Fig. 5.7). At room temperature the glass is an insulator and so the bulb does not light, but if the glass bar is heated it begins to conduct and the bulb lights.

So we can see that the magnitude of the band gap and the temperature are both very important factors in determining the electrical properties of a semiconductor or an insulator. In practice materials are generally referred to as semiconductors if the band gap is less than 3 eV, and as insulators if the band gap is greater than 3 eV, but it should be obvious from the above discussion that in fact there is no clear-cut distinction between an insulator and a semiconductor.

5.4 Holes

It is not only the conduction electrons in a semiconductor which affect the conductivity. As we can see from Fig. 5.4(c), at a finite temperature we also have the possibility of finding vacant states in the valence band. In this section we will look at how these vacant states contribute to the conductivity of the material.

We have already seen that in a covalent crystal, such as silicon or diamond, there are just enough states in the valence band to accommodate all of the electrons that are involved in forming covalent bonds with the neighbouring atoms. Consequently, if there is a vacant state in the valence band, this means that one of the bonds must be incomplete, as shown in Fig. 5.8(a). If we apply an electric field to the crystal, this vacant state offers an opportunity for one of the electrons to move closer to the positive end of the crystal (Fig. 5.8(b)). In doing so, it leaves behind a new vacant state. This process can continue again and again, as in Fig. 5.8(c), so that we have a sequence of

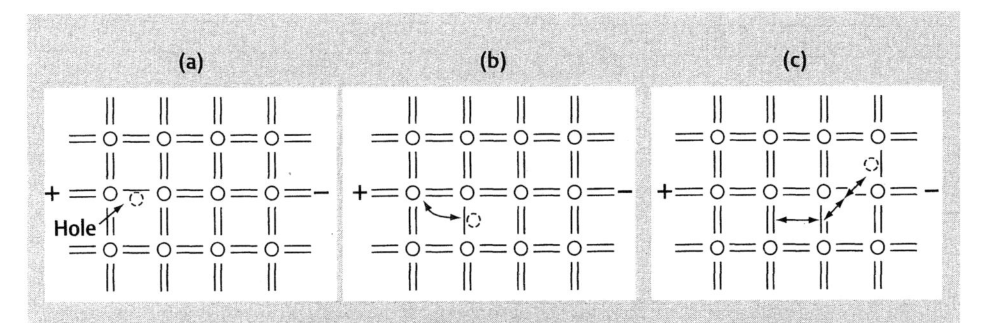

Figure 5.8 (a) A vacant state (or hole) in the valence band corresponds to an incomplete bond. (b) When an electric field is applied, an electron is able to move into this vacant state and the hole moves to the position previously occupied by the electron. (c) As this process is repeated, the hole is observed to move towards the negative end of the crystal.

events in which an electron moves towards the positive end of the crystal. An alternative description is to say that the vacant state moves towards the negative end of the crystal. Since it is far easier to describe the motion of this single vacant state than that of the large number of valence electrons, we will give the vacant state a name—we will call it a 'hole'. Since this hole appears to be attracted towards the negative end of the crystal we can also assume that it has a positive charge. In fact, it has precisely the same amount of charge as an electron, but of the opposite sign.

So how does the presence of these holes affect the conductivity of the material? When we apply an electric field to a semiconductor we find that the conduction electrons move in one direction and the holes move in the opposite direction. However, since these two sets of particles also have charges of opposite polarity, the total current is equal to the sum of the electron current and the hole current. Using eqn (4.9) we can therefore write the total conductivity as

$$\sigma = |e|(n\mu_e + p\mu_h) \tag{5.5}$$

where p is the concentration of holes and μ_e and μ_h are the mobilities of the electrons and holes, respectively.

One final point in this section concerns the concentration of holes, p. As we can see from Fig. 5.9, each time we create a conduction electron, it leaves behind a hole in the valence band. This implies that the number of holes should be exactly equal to the number of conduction electrons. Although this is certainly true if we have a very pure sample of material, we will see shortly that it does not necessarily apply if there are impurities present. Consequently, we will express this relationship as

$$n_i = p_i \tag{5.6}$$

where the subscript i shows that this applies only to intrinsic (i.e. pure) semiconductors.

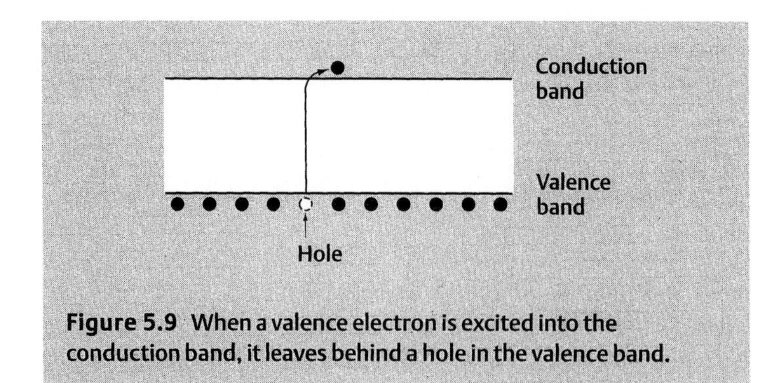

Figure 5.9 When a valence electron is excited into the conduction band, it leaves behind a hole in the valence band.

5.5 Optical properties of semiconductors

We have considered the probability of electrons being thermally excited into the conduction band. Another possibility is that we can optically excite an electron across the band gap. The process is shown schematically in Fig. 5.10(a). The energy of the photon is transferred completely to an electron, giving the electron enough energy to enter the conduction band and leaving behind a hole in the valence band. When this occurs we say that the photon has been absorbed by the material. The main consideration that governs whether or not this process will occur is that the energy of the photon must be at least as large as the band gap of the semiconductor. Since the energy of a photon with frequency ν and wavelength λ is given by

$$E = h\nu = \frac{hc}{\lambda}$$

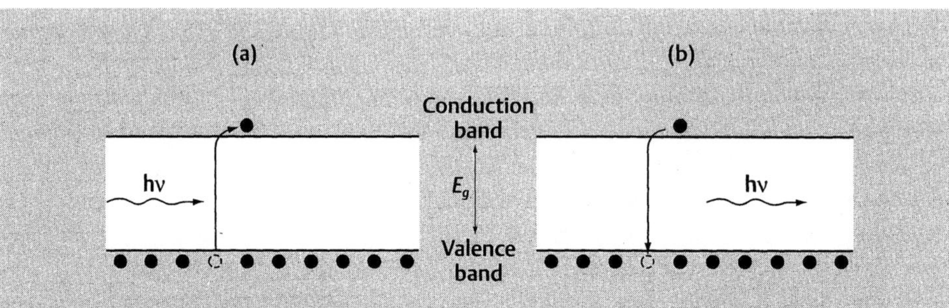

Figure 5.10 (a) A single photon is capable of exciting an electron from the valence band into the conduction band provided that the energy of the photon $h\nu$ is greater than the band gap energy of the semiconductor. (b) Conversely, a conduction electron can recombine with a hole and produce a photon of energy $h\nu$.

(see Appendix A), the minimum frequency, ν_{min}, and corresponding maximum wavelength, λ_{max}, at which absorption can occur is given by

$$\nu_{min} = \frac{c}{\lambda_{max}} = \frac{E_g}{h} \tag{5.7}$$

Note that c refers to the speed of light.

This process is known as photoconductivity, because a beam of light of the appropriate frequency produces a large number of conduction electrons and holes, and so significantly increases the conductivity.

The reverse process can also occur; in other words, an electron in the conduction band can recombine with a hole in the valence band with the excess energy emitted as a photon (see Fig. 5.10(b)). This is the basis for the light-emitting diode (LED) and semiconductor laser which we will consider in more detail in the next chapter.

It is worth pointing out at this stage that not all semiconductor materials are capable of producing light. The reason is due to the fact that the properties of electrons and holes in a crystal are specified by the energy E and the wavevector k. An electron and hole can only recombine if both of these quantities are conserved. (Since the wavevector is often referred to as the 'crystal momentum', we can say that energy and momentum are the conserved quantities.) A photon of the correct frequency can take care of the conservation of energy, but since the wavevector of a photon is negligible at the appropriate frequencies (see Question 5.10), it cannot affect the wavevector of the electron. Consequently, we conclude that the process of light emission occurs only if the electrons in the lowest energy conduction states have the same wavevector as the holes in the highest energy valence states. Semiconductors which satisfy this condition include gallium arsenide, indium antimonide and indium arsenide. These materials are called direct gap semiconductors. However, materials such as silicon and germanium do not satisfy this criterion. Consequently, these materials, which are known as indirect gap semiconductors, do not exhibit light emission and so cannot be used to fabricate light-emitting diodes or semiconductor lasers.

5.6 The effective mass

What is the mass of an electron in a semiconductor crystal? Surely the mass is the same as that of a free electron. However, certain experiments suggest that the electrons in a semiconductor behave as though they have a quite different mass to that of a free electron. For example, let us consider the behaviour of an electron in a uniform magnetic field B. Theory suggests (see Example 5.2) that the electrons move in circular orbits with a frequency given by

$$\nu_c = \frac{Be}{2\pi m_e} \tag{5.8}$$

where ν_c is known as the cyclotron frequency. Although this result is well established for free electrons, in general the measured cyclotron frequency of an electron in a semiconductor does not agree with eqn (5.8).

What is wrong? The problem is that an electron in a solid experiences other forces (e.g. due to the ions and other valence electrons) and so it does not respond in the same way as a free electron. However, we can obtain agreement with eqn (5.8) if we assume that the mass of an electron in a crystal is somehow different from the mass of a free electron.

This may seem a rather drastic solution. After all, the electron mass is normally considered to be a fundamental constant of nature. However, we are not suggesting that the mass of an electron in a crystal *is* different from that of a free electron, merely that it *behaves* as though it is different. In other words, if we take an electron out of a crystal it will have the same mass as any other electron, but when we apply a magnetic field to the crystal, the net effect of this field combined with all of the other forces in the crystal can be described by assuming that the electron has an 'effective mass', which we will denote by m_e^*. The cyclotron frequency for an electron in a crystal is therefore given by

$$\nu_c = \frac{Be}{2\pi m_e^*} \tag{5.9}$$

Example 5.2 By considering the forces acting on an electron in a uniform magnetic field B, show that the electrons move in circular orbits with a frequency given by eqn (5.8).

Solution The force on an electron (with charge e) due to a magnetic field B is

$$F_{\text{magnetic}} = Bev$$

where v is the velocity of the electron.

If the electron moves in a circular orbit of radius r, then the centripetal force

$$F_{\text{centripetal}} = \frac{m_e v^2}{r}$$

must exactly balance the magnetic force.

By equating these two expressions we find that the velocity of the electron is given by

$$v = \frac{Ber}{m_e}$$

In order to obtain an expression for the cyclotron frequency, ν_c, we note first that the angular frequency ω is related to the velocity by $v = \omega r$, and that the frequency ν is related to the angular velocity by $\nu = \omega/2\pi$. Therefore

$$\nu_c = \frac{\omega_c}{2\pi} = \frac{v}{2\pi r} = \frac{Be}{2\pi m_e}$$

Values of effective masses for a selection of metals and semiconductors are given in Table 5.2. We can see that the holes also have an effective mass, m_h^*, which may be quite different to the electron effective mass in a particular material. We should also

Table 5.2 Values of the effective masses of electrons and holes, m_e^* and m_h^*, for a selection of semiconductors and metals. The values are given as a ratio of the free electron mass, m_e. Note that in some semiconductors, such as silicon and germanium, there is more than one effective mass associated with the electrons. In these cases an average value has been given. It is also usual to assign up to three different masses to the states at the top of the valence band. The values quoted here are for the so-called heavy hole states.

	m_e^*/m_e	m_h^*/m_e
Indium antimonide	0.014	0.4
Indium arsenide	0.022	0.4
Germanium	0.60	0.28
Silicon	0.43	0.54
Gallium arsenide	0.065	0.5
Sodium	1.2	
Copper	0.99	
Zinc	0.85	

note that the effective masses for the metals in this table are not very different to the free electron mass, and similar results are obtained for a large number of other metals. This explains why we did not need to introduce the concept of effective mass in the previous chapter when we were discussing metals. However, for semiconductors the difference between the effective mass and the free electron mass can be as much as a factor of 70, so it is vital that we use the effective mass in these materials if we wish to obtain reasonably accurate results.

The importance of the effective mass is not just that it gives the correct cyclotron resonance frequency, but that the same correction factor is applicable in a wide range of circumstances. For instance, the acceleration on an electron in an electric field $\mathcal{E}$ should really be written as

$$a = \frac{e\mathcal{E}}{m_e^*} \tag{5.10}$$

(cf. eqn (4.3)). In the next section and in Section 5.11 we will see examples of other instances in which the effective mass is important.

5.7 n-type semiconductors

So far in this chapter we have considered the electrical properties of pure, or intrinsic, semiconductors. However, the most important property of semiconductors is the fact that we can control the conductivity of the material by deliberately introducing impurities. In fact, a surprisingly small concentration of impurities—less than 0.0001% or one in every million atoms—leads to a dramatic change in the conductivity.

For the rest of this chapter we will look at the properties of these extrinsic semiconductors. In this section we will consider the effects of n-type doping in which donor impurities are used to increase the number of negatively charged carriers (i.e. conduction electrons) and in the next section we will look at p-type doping which uses acceptor impurities to control the concentration of positively charged carriers (i.e. holes).

A donor impurity is an atom that has one more valence electron than the atom that it replaces. An example is a phosphorus atom in silicon since phosphorus has five valence electrons and silicon has four. To determine the effect of this impurity, let us consider a silicon crystal. Each silicon atom uses its four valence electrons to form covalent bonds with the four neighbouring atoms. If we substitute a phosphorus atom for one of the silicon atoms we would expect the phosphorus atom to form bonds with the four neighbouring silicon atoms, leaving one electron left over (see Fig. 5.11). What happens to this fifth electron?

To answer this question, let us consider the diagram in Fig. 5.12(a) in which the electrons are shown at their average distance from the nucleus. There is a total of 10 electrons in the two filled shells closest to the nucleus. Four more take part in the bonds with neighbouring silicon atoms and so we place these further from the nucleus. Finally, we have the fifth valence electron, which we will refer to as the donor electron. This electron is not involved in any of the bonds, so we assume that it is the furthest electron from the nucleus and has the smallest binding energy. The system depicted in Fig. 5.12(a) is obviously very complex, but we can simplify matters greatly if we assume that the donor electron is screened from most of the nuclear charge by the presence of the other fourteen electrons. This suggests that on average the donor electron 'sees' the nucleus and the collection of other 14 electrons as just a single

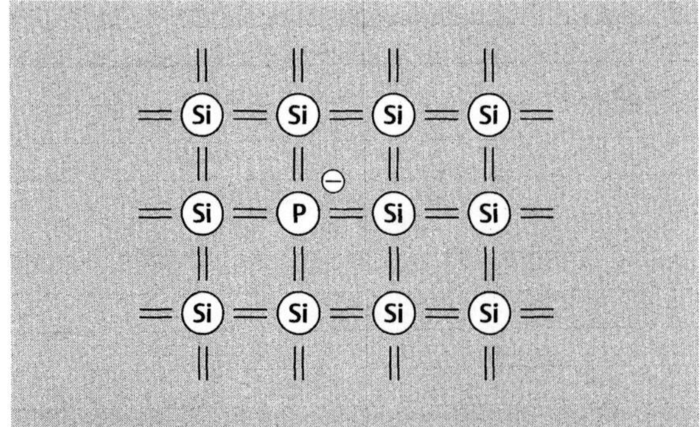

Figure 5.11 If we introduce a donor impurity, e.g. a phosphorus atom in a silicon crystal, there is one extra electron which does not participate in the covalent bonds.

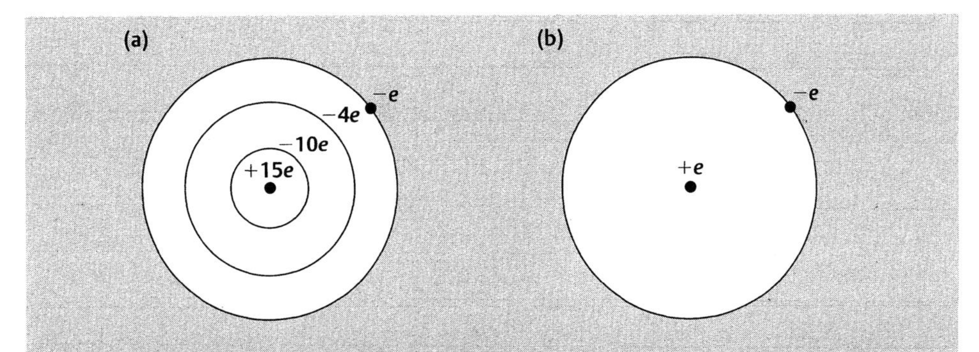

Figure 5.12 (a) A schematic diagram showing the arrangement of the electrons associated with a phosphorus impurity in silicon. (b) A simplified picture is obtained by assuming that the inner electrons screen most of the nuclear charge from the outer electron.

positive charge, as shown in Fig. 5.12(b). This now resembles a system with which we are very familiar—a hydrogen atom.

In Chapter 1 we calculated the binding energy of the electron in a hydrogen atom using Bohr's theory, so let us assume that we can apply the same model to determine the binding energy of the donor electron. There are two modifications that we need to make. Firstly, the mass of the electron should be replaced by the effective mass, m_e^*, as discussed in the previous section. Secondly, the electron in a hydrogen atom is assumed to move in a vacuum and so we used the dielectric constant for a vacuum, ε_0. In this instance the electron is in a silicon crystal and so we should use the dielectric constant for silicon, which can be written as $\varepsilon\varepsilon_0$, where ε has a value of 11.7. So the binding energy of the donor electron is given by

$$E_d = -\frac{m_e^* e^4}{8\varepsilon^2 \varepsilon_0^2 h^2} \tag{5.11}$$

Substituting in appropriate values for silicon we predict that the binding energy is 43 meV (see Example 5.3). Experimental measurements show that the binding energy of a donor electron in silicon is typically between about 40 meV and 50 meV depending on the type of donor impurity, so our simple model gives a reasonably accurate result. We can also use Bohr's model to determine the average distance of the donor electron from the phosphorus nucleus (see Question 5.12). The value is of the order of several atomic spacings, which confirms our assumption that this electron is not in close proximity to the phosphorus nucleus but rather is moving through the silicon crystal.

How does this donor electron fit into the energy band picture? The electron does not take part in bonding, so there is no room for it in the valence band. On the other hand, since it is weakly attached to the phosphorus atom, it cannot belong in the conduction band either. However, if we supply the small amount of energy required to liberate it from the phosphorus atom, then it will be able to take part in conduction. So we conclude that the donor electron must occupy a state at an energy E_d below the conduction band edge, as shown in Fig. 5.13. But this means that it is in the band gap

Example 5.3 Use Bohr's model of the hydrogen atom to estimate the binding energy of a donor electron in silicon.

Solution The binding energy of the electron in a hydrogen atom is given by eqn (1.2) with $n = 1$, *i.e.*

$$E_1 = -\frac{m_e e^4}{8\varepsilon_0^2 h^2} = -13.6\,\text{eV}$$

For the donor electron in silicon we substitute m_e^* instead of m_e and $\varepsilon\varepsilon_0$ instead of ε_0.

Using $m_e^* = 0.43 m_e$ and $\varepsilon = 11.7$ the binding energy of the donor electron is

$$E_d = -\frac{m_e^* e^4}{8\varepsilon^2\varepsilon_0^2 h^2} = -\frac{0.43}{(11.7)^2}\frac{m_e^4}{8\varepsilon_0^2 h^2} = -(3.14\times 10^{-3}).(13.6\,\text{eV})$$
$$= 0.043\,\text{eV}$$

where there are no allowed states! So we appear to have two contradictory statements. However, the second statement is not quite true. There are no allowed states in the band gap of a perfect crystal, but if an impurity is present then the crystal is no longer perfect and so it is quite reasonable for a state associated with the impurity to be located in the band gap.

Where is the Fermi energy in this case? We can answer this question most easily by considering two extremes of temperature. At a temperature of absolute zero all of the electron states are occupied up to and including the donor state, and the lowest vacant state is in the conduction band. Using the same argument as in Section 5.3 we can therefore place the Fermi energy midway between the donor state and the conduction band edge, i.e.

$$E_F = E_g - \frac{E_d}{2} \tag{5.12}$$

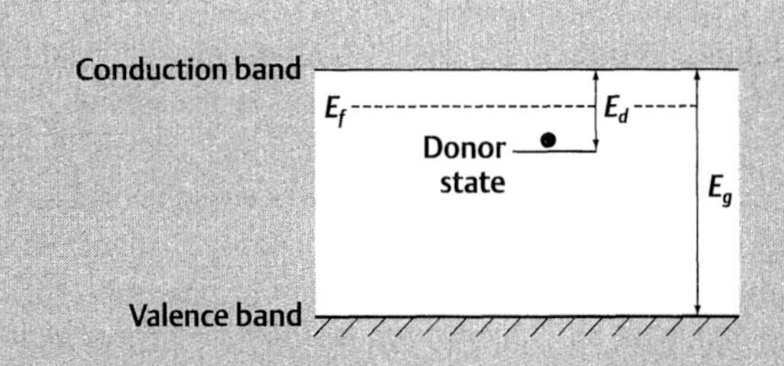

Figure 5.13 The energy band diagram for an n-type semiconductor. The donor state is at an energy E_d below the conduction band edge, where $E_d \ll E_g$. At $T = 0$ K the Fermi energy is midway between the donor state and the conduction band edge, i.e. typically about 20–30 meV below the conduction band edge.

On the other hand, at very high temperatures the number of intrinsic carriers becomes comparable with the number of donor electrons and so we have a material which behaves like an intrinsic semiconductor, i.e. the Fermi energy is close to the centre of the gap. We therefore predict that the Fermi level in an n-type semiconductor moves down in energy as the temperature is increased—from just below the conduction band edge at $T = 0$ K to near the centre of the band gap at very high temperatures.

The Fermi energy also depends on the concentration of donor impurities. Again, we can demonstrate this by considering two extremes of doping. When the donor concentration is very low then the Fermi energy is near the centre of the band gap at moderate temperatures because under these conditions the number of intrinsic carriers is comparable with the number of donor electrons. In contrast, when the doping concentration is high, then the intrinsic carrier concentration is negligible even at temperatures close to the melting point, and so the Fermi energy never moves far below the conduction band edge. The variation of the Fermi energy with temperature and doping concentration is shown in Fig. 5.14.

How does the presence of the donor impurity affect the conductivity of a semiconductor? From Example 5.4 we can see that under most circumstances the concentration of conduction electrons is given by

$$n = Ce^{-(E_g - E_F)/k_B T} \tag{5.13}$$

where C is a constant. Note that this equation is similar to eqn (5.4) which describes the concentration of conduction electrons in an intrinsic semiconductor, but the energy term in the exponent is now $(E_g—E_F)$ rather than $E_g/2$. From Fig. 5.14 we can see that for an n-type semiconductor the Fermi energy is generally much closer to the

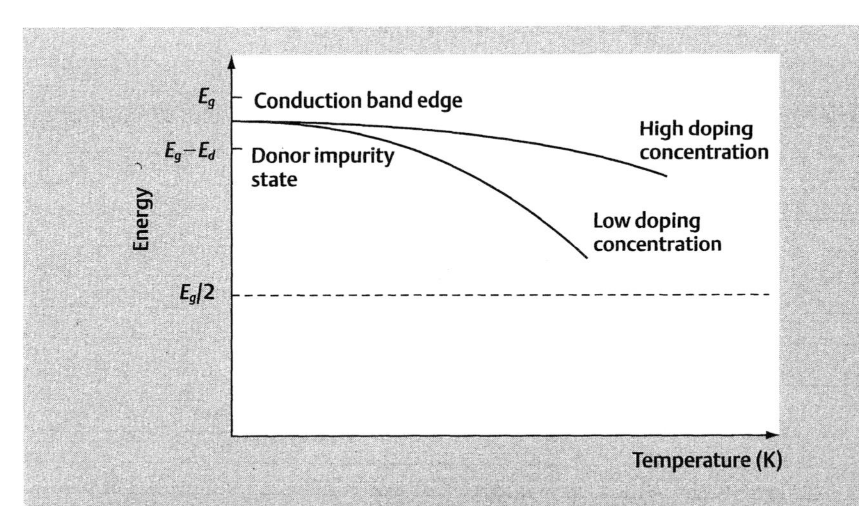

Figure 5.14 Variation of the position of the Fermi energy with temperature for two different levels of n-type doping. In each case the Fermi energy at $T = 0$ K is at an energy $E_d/2$ below the conduction band edge, where E_d is the binding energy of the donor impurity.

Example 5.4 By considering the probability of an electron occupying a state at the conduction band edge, show that the concentration of conduction electrons in a doped semiconductor is given by eqn (5.13)

Solution The Fermi-Dirac distribution (eqn (5.1)) gives the probability of finding an electron at the conduction band edge (i.e. with energy E_g) as

$$f(E_g) = \left[e^{(E_g - E_F)/k_B T} + 1\right]^{-1}$$

From Fig. 5.15 we can see that in most cases the difference in energy between the conduction band edge and the Fermi energy, E_g—E_F, is much larger than the average thermal energy, $k_B T$. (The exception is for very high doping concentrations at relatively high temperatures.) Consequently, in most circumstances we can make the same approximation as in eqn (5.3), i.e. the exponential term is much larger than 1, so

$$f(E_g) \approx \left[e^{(E_g - E_F)/k_B T}\right]^{-1} = e^{-(E_g - E_F)/k_B T}$$

If we assume that the concentration of conduction electrons is proportional to $f(E_g)$, then

$$n = Ce^{-(E_g - E_F)/k_B T}$$

where C is a constant.

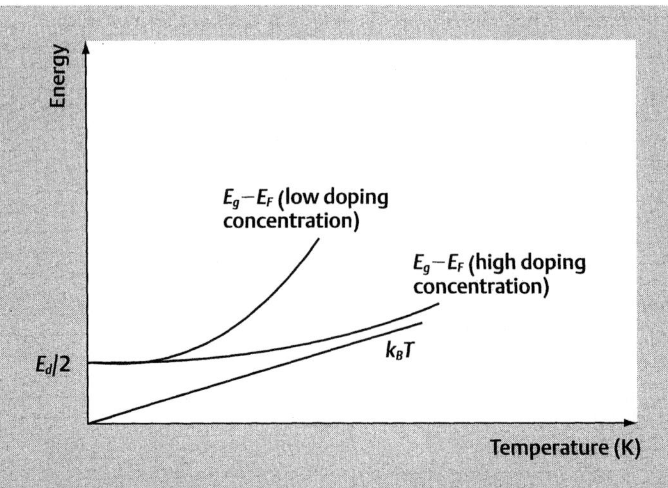

Figure 5.15 Sketch of the variation in energy between the conduction band edge and the Fermi energy, E_g—E_F, as a function of temperature and doping for an n-type semiconductor. The data is the same as that shown in Fig. 5.14. For comparison the thermal energy $k_B T$ is also shown.

conduction band edge than it is to the valence band edge, i.e. $(E_g—E_F) \ll E_g/2$, which results in an exponential increase in the number of carriers in the doped case.

An alternative way to address this problem is to consider the probability of finding an electron in the donor state. At room temperature we find that the proportion of donor electrons occupying these states is extremely small (see Example 5.5), and therefore we can assume that virtually all of the donor impurities are ionized at room temperature. Consequently, if the concentration of donor impurities is given by N_d, then the concentration of conduction electrons n is given by

$$n = N_d + n_i \tag{5.14}$$

where n_i is the intrinsic carrier concentration.

We can now see why only a small concentration of impurities results in a large change in conductivity. In a sample of intrinsic silicon the concentration of conduction electrons at room temperature is about $1 \times 10^{16}\text{m}^{-3}$, whereas if we replace one in every million silicon atoms with a donor impurity and all of these impurities are ionized, then the concentration of conduction electrons is about $5 \times 10^{22}\text{ m}^{-3}$, or 5 million times as many as in the intrinsic case (see Question 5.14).

Example 5.5 A silicon crystal is doped with phosphorus atoms which have a binding energy E_d of 45 meV. Assuming that the Fermi energy is 200 meV below the conduction band edge at a temperature of 300 K, determine the proportion of donor states that are occupied.

Solution The proportion of donor states that are occupied, N_{D0}, is equal to the probability of an electron having an energy equal to the donor state energy. Since the donor state is at an energy $E_g—E_d$, the probability is given by the Fermi-Dirac distribution with $E = E_g—E_d$, i.e.

$$N_{D0} = f(E_g - E_d) = \left[e^{(E_g - E_d - E_F)/k_B T} + 1\right]^{-1}$$

Since the Fermi energy E_F is at E_g—200 meV, the energy term in the exponent is

$$E_g - E_d - E_F = E_g - 45\,\text{meV} - (E_g - 200\,\text{meV}) = 155\,\text{meV}$$
$$= 2.48 \times 10^{-20}\,\text{J}$$

Therefore

$$N_{D0} = \left[\exp\left(\frac{2.48 \times 10^{-20}\,\text{J}}{(1.38 \times 10^{-23}\,\text{J K}^{-1})(300\,\text{K})}\right) + 1\right]^{-1} = 2.5 \times 10^{-3}$$

So only 0.25% of the donor states are occupied, i.e. virtually all of the donors are ionized.

The above result also demonstrates the need for obtaining extremely pure samples of semiconductors. If the concentration of accidental impurities is comparable with the dopant concentration, then the effects of doping a material could well be obscured. For example, if all of the accidental impurities happen to be acceptors, then a material which is intentionally doped with donors may turn out to have p-type characteristics! Since the doping densities used in many semiconductor devices are typically of the order of one per million, we need to have materials in which the level of accidental impurities is at least an order of magnitude lower. Techniques for obtaining such low levels of impurities are now routinely available. In fact, impurity levels as low as only one atom in every hundred billion (10^{11}) have been reported in semiconductors. In comparison, a metal is generally considered to be pure if the impurity concentration is of the order of a few atoms per million.

Finally, we should point out that although the addition of donor impurities leads to an increase in the number of conduction electrons, the semiconductor remains electrically neutral because the number of positively charged donor ions is exactly equal to the number of conduction electrons in the sample.

5.8 p-type semiconductors

If we introduce an acceptor impurity, which has one fewer valence electrons than the atom that it replaces, then we have an incomplete bond, as shown in Fig. 5.16. An example is an aluminium atom in a silicon crystal because aluminium has just three valence electrons.

The presence of the incomplete bond means that we have a hole in the valence band. Well, that is not quite true because the hole is bound to the acceptor impurity, just like the extra electron is bound to the donor impurity. In order to create a hole which can move through the valence band, we need to excite a valence electron into this incomplete bond and so ionize the acceptor impurity. So the acceptor state is actually just above the valence band edge, as shown in Fig. 5.17. For an aluminium impurity in silicon the binding energy, E_a, is 57 meV, so as in the case of the donor impurities, we can assume that all of the acceptor impurities are ionized at room temperature.

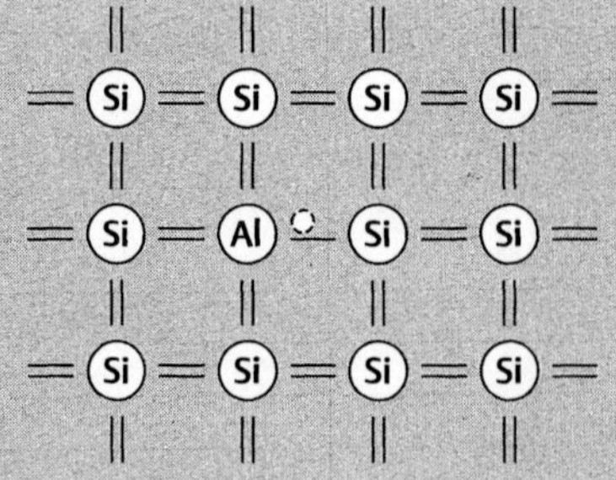

Figure 5.16 The effect of an acceptor impurity, e.g. aluminium in silicon, is to produce an incomplete bond, i.e. a hole in the valence band.

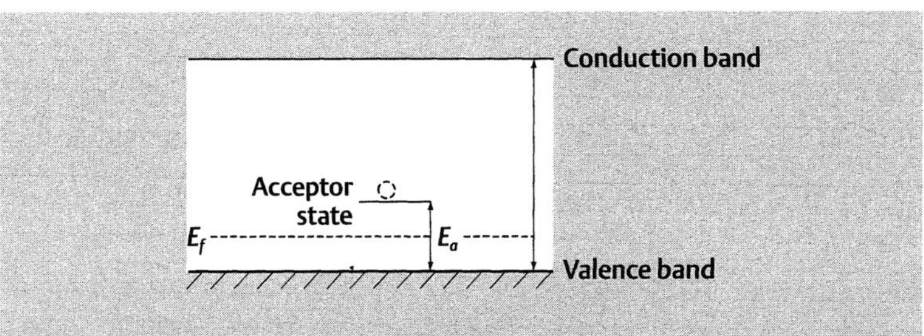

Figure 5.17 The energy band diagram for a p-type semiconductor. The acceptor state is at an energy E_a above the valence band edge, where $E_a \ll E_g$. At $T = 0$ K the Fermi energy is midway between the valence band edge and the acceptor state.

The position of the Fermi energy can also be obtained using a similar argument to that for n-type semiconductors. At absolute zero the highest energy occupied state is at the top of the valence band and the lowest vacant state is the acceptor level, so the Fermi energy can be assumed to be at an energy of $E_a/2$ above the valence band edge. At higher temperatures the effect of the intrinsic carriers is to push the Fermi energy upwards towards the middle of the band gap, but at room temperature we can assume that the Fermi energy is much closer to the valence band than to the conduction band.

Using an analysis similar to that in Example 5.5 we can also show that the concentration of holes in a doped semiconductor is given by (see Question 5.16)

$$p = Ce^{-E_F/k_B T} \tag{5.15}$$

5.9 Majority and minority carriers

In the previous two sections we have seen that by using a small concentration of impurities we can not only dramatically increase the conductivity of the material, but also determine whether the current is carried predominantly by electrons or by holes.

However, we should note that in an n-type semiconductor there is a small concentration of holes which are produced by thermal excitation. Despite the fact that these holes are greatly outnumbered by the conduction electrons, we cannot simply ignore them. In fact, they turn out to be of great importance in the operation of a transistor, as we shall see in the next chapter. We therefore need to distinguish between the majority and minority carriers. In an n-type semiconductor the conduction electrons are the majority carriers and the holes are the minority carriers, whereas in a p-type material it is the other way round.

One important result is that for a given type of semiconductor at a particular

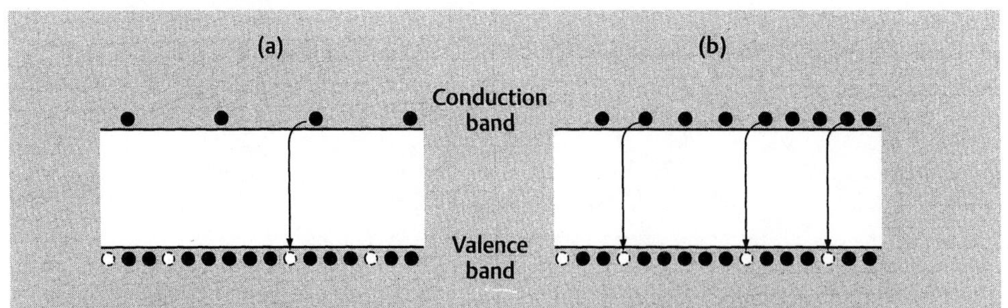

Figure 5.18 (a) In an intrinsic semiconductor, the concentrations of conduction electrons and holes are very low, and so the rate of recombination is relatively low. (b) In an n-type semiconductor, the high concentration of conduction electrons means that there are far more opportunities for a hole to recombine with a conduction electron, and so the concentration of holes is lower than in the intrinsic case.

temperature, the product of the concentration of minority carriers with the concentration of majority carriers is a constant, irrespective of the level or type of doping. Therefore we can write

$$np = n_i\, p_i = \text{constant} \tag{5.16}$$

where n_i and p_i are the concentrations of conduction electrons and holes, respectively, in the intrinsic material, and n and p are the corresponding quantities in a doped material. This is known as the law of mass action and is proved in Example 5.6. Physically, what this relationship means is that since the majority carrier concentration is greater than in the intrinsic case, the minority carrier concentration is lower than in the intrinsic material. We can also understand this relationship from a qualitative point of view. For example, in an n-type semiconductor the rate at which electron–hole pairs are created is the same as in the intrinsic case, but there are now far more possibilities for the holes to recombine. This is shown schematically in Fig. 5.18.

5.10 The Hall effect

We have concentrated on the effects of applying an electric field to a material, but an interesting effect occurs if there is also a magnetic field perpendicular to the electric field.

Let us consider the situation shown in Fig. 5.19(a) in which an electric field $\mathcal{E}_x$ is applied in the x direction and a magnetic field B_z is applied in the z direction. From electromagnetic theory we know that the electrons are deflected in the direction which is perpendicular to both the applied electric and magnetic fields (see e.g. Alonso and Finn in Further reading). This leads to an accumulation of negative charge on one side of the sample, and so produces an electric field in the y direction. The field

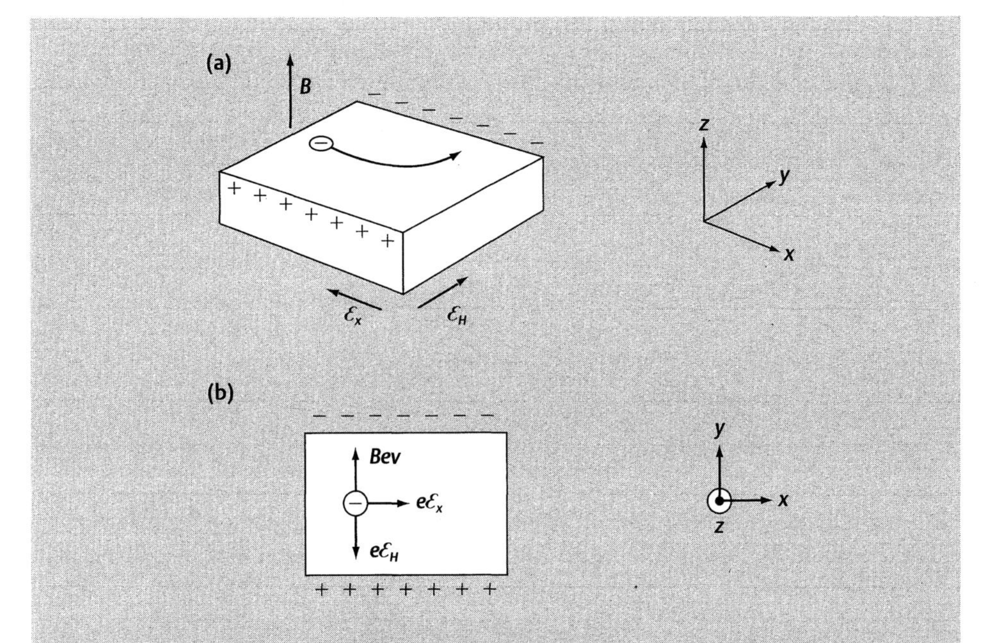

Figure 5.19 (a) The combined effect of an electric field in the x direction, $\mathcal{E}_x$, and a magnetic field in the z direction, B_z, is to cause the electrons to be deflected in the y direction. (b) Plan view of the sample. At equilibrium, the force on the electrons due to the magnetic field, $B_z ev$ (where v is the velocity of the electron), is balanced by the force due to the Hall field, $e\mathcal{E}_H$.

is referred to as the Hall field, $\mathcal{E}_H$, named after Edwin Hall who discovered this effect in 1879.

The Hall field is found to be proportional to the strength of the magnetic field, B_z, and the current density flowing in the x direction along the sample, J_x, so we can write

$$\mathcal{E}_H = R_H J_x B_z \tag{5.17}$$

where R_H is known as the Hall coefficient.

In the case where both types of dopant impurity are present, it is also possible to obtain another relationship between the Hall field and the perpendicular magnetic field. If we look at the plan view in Fig. 5.19(b), we can see that the electrons are deflected to the left by the magnetic field, but the Hall field causes the electrons to be deflected to the right. In equilibrium these two forces must balance. Since the force due to the magnetic field is $B_z ev$ (where v is the velocity of the electrons in the x direction) and the force due to the Hall field is $e\mathcal{E}_H$, this means that

$$e\mathcal{E}_H = B_z ev$$

Example 5.6 Show that the product of the concentrations of minority and majority carriers is a constant for a given semiconductor at a particular temperature.

Solution First of all we note that the expression for the concentration of conduction electrons in eqn (5.13) is equally valid for a p-type material as for an n-type material. (Of course, in this case the energy in the exponent, $E_g—E_F$, is greater than $E_g/2$, and therefore the concentration of conduction electrons is lower than in the intrinsic material.) Similarly, eqn (5.15) gives the hole concentration in n-type as well as p-type materials. Therefore, the product of the concentration of conduction electrons and holes in either type of material is

$$np = Ce^{-(E_g - E_F)/k_B T} Ce^{-E_F/k_B T} = C^2 e^{-E_g/k_B T}$$

Since this expression is independent of the position of the Fermi energy, we can deduce that it is also independent of the level of doping. Also note that in the intrinsic case the carrier concentration is given by eqn (5.4), i.e. $n_i = Ce^{-Eg/2kBT}$ and $n_i = p_i$. Therefore

$$n_i p_i = (Ce^{-E_g/2k_B T})^2 = C^2 e^{-E_g/k_B T}$$

and thus

$$np = n_i p_i = \text{constant}$$

Therefore

$$\mathcal{E}_H = B_z v \tag{5.18}$$

From the above equations we can show (see Example 5.7) that the Hall coefficient is also given by

$$R_H = \frac{1}{Ne} \tag{5.19}$$

where N is the concentration of charge carriers. Since we can measure the quantities J_x, $\mathcal{E}_H$ and B_z, we can use eqns (5.17) and (5.19) to calculate the value of N. So the Hall effect provides an experimental method of measuring the concentration of carriers in the sample.

For metals the current is usually carried by electrons, and since the electron charge is negative, the value of R_H is also negative. However, if we consider a current of holes, we can see from Fig. 5.20 that the holes are deflected to the same side as the electrons. Consequently, the Hall field is of opposite polarity, and so the Hall coefficient is positive. This means that the Hall effect not only determines the concentration of carriers,

Example 5.7 Show that the Hall coefficient R_H is given by eqn (5.19).

Solution From eqn (5.18) we have

$$\mathcal{E}_H = B_z v$$

and from eqn (4.8) the current density is given by

$$J_x = Nve$$

Substituting these expressions in eqn (5.17) gives

$$R_H = \frac{\mathcal{E}_H}{J_x B_z} = \frac{B_z v}{NevB_z} = \frac{1}{Ne}$$

but also gives us information about the type of charge carrier. So for n-type semiconductors we find that R_H is negative, and for p-type semiconductors it is positive. It is interesting to note that a number of metals such as zinc and aluminium also have positive values of R_H, indicating that the current is carried by holes in these materials.

We should also mention that a novel variation on the Hall effect occurs if the electrons are confined to a very thin layer (i.e. so that the thickness of the sample in the z direction is small). If we define the Hall resistance, ρ_H, as the product of the magnetic field strength and the Hall coefficient, i.e.

$$\rho_H = B_z R_H \tag{5.20}$$

then by substituting for R_H from eqn (5.19) we can see that ρ_H should increase linearly as the magnetic field strength increases. However, at very low temperatures and high magnetic fields the Hall resistance is found to increase by a series of steps, as shown in Fig. 5.21. This strange behaviour, which was first observed by Klaus von

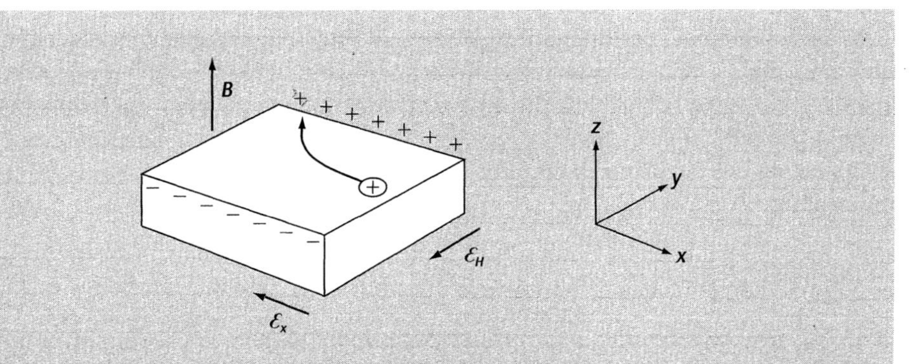

Figure 5.20 For a p-type semiconductor the holes are deflected in the same direction as the electrons in Fig. 5.19(a), and consequently give rise to a Hall field $\mathcal{E}_H$ of opposite polarity.

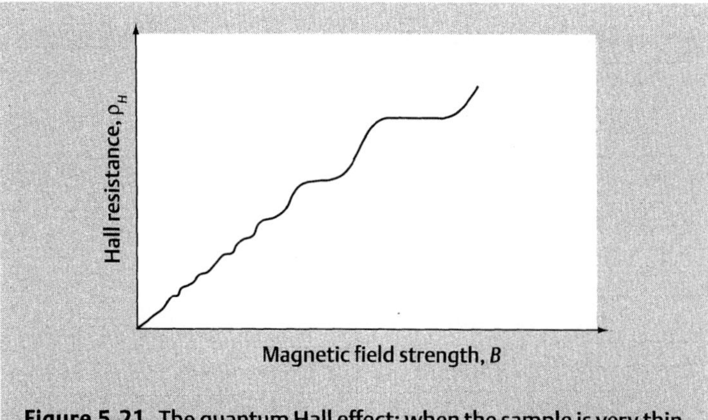

Figure 5.21 The quantum Hall effect: when the sample is very thin and subject to large magnetic fields, the Hall resistance ρ_H rises in a series of steps as the magnetic field strength is increased. Note that at low magnetic fields, the Hall resistance is proportional to the magnetic field strength, as we expect from the ordinary Hall effect.

Klitzing and coworkers in 1980, is known as the quantum Hall effect. (For an explanation of this phenomenon see Kittel or one of the articles mentioned in Further reading.)

5.11 The free electron model applied to semiconductors*

In this section we will use the free electron model (which was introduced in Section 4.8) to prove some of the formulae that we have introduced in the earlier sections of this chapter. We begin by taking a more detailed look at the effective mass.

In Section 5.6 we introduced the concept of effective mass in order to explain the cyclotron resonance measurements. This suggests that the effective mass is a quantity which must be determined empirically. However, it is also possible to develop a theoretical basis for this concept. By considering the group velocity of a electron, we can see from Example 5.8 that the effective mass can be defined by the equation

$$m_e^* = \hbar^2 \left(\frac{d^2 E}{dk^2} \right)^{-1} \tag{5.21}$$

Consequently, we can think of the effective mass as a measure of the curvature of the energy bands with respect to the wavevector.

If we apply this definition to the free electron model, for which the energies are given by

Example 5.8 By considering the group velocity v_g of an electron in a crystal, show that the effective mass of the particle is given by eqn (5.21).

Solution From wave theory we know that the group velocity is defined as

$$v_g = \frac{d\omega}{dk}$$

From quantum theory we also know that the energy of a particle is related to the angular frequency ω by

$$E = \hbar\omega$$

(see Appendix A), so we can write

$$v_g = \frac{1}{\hbar}\frac{dE}{dk}$$

Differentiating this expression, we find that the acceleration a is given by

$$a = \frac{dv_g}{dt} = \frac{1}{\hbar}\frac{d}{dt}\frac{dE}{dk}$$

$$= \frac{1}{\hbar}\frac{d^2E}{dk^2}\frac{dk}{dt}$$

Now consider an electron moving at speed v_g in an electric field. In a time dt the electron moves a distance $v_g\,dt$. Since the force on the electron is $e\mathcal{E}$, we can write the work done on the electron, dE, as

$$dE = e\mathcal{E}v_g\,dt$$

Therefore

$$e\mathcal{E}v_g = \frac{dE}{dt} = \frac{dE}{dk}\frac{dk}{dt}$$

Rearranging we get

$$\frac{dk}{dt} = e\mathcal{E}v_g\frac{dE}{dt}$$

and substituting in the expression for the acceleration gives

$$a = e\mathcal{E}v_g.\frac{1}{\hbar}.\frac{d^2E}{dk^2}.\frac{dk}{dE} = \frac{e\mathcal{E}}{\hbar^2}\frac{d^2E}{dk^2}$$

where we have also used the expression for v_g.

Finally, comparing this equation with eqn (5.10) we can see that

$$m_e^* = \hbar^2\left(\frac{d^2E}{dk^2}\right)^{-1}$$

$$E = \hbar^2 k^2 / 2m_e \tag{5.22}$$

(see Section 4.8) we find that the effective mass is exactly equal to the free electron mass, i.e.

$$m_e^* = m_e$$

(see Question 5.19). Of course, this is just what we would expect because we derived this expression for the energy of the electrons (in Section 4.8) by assuming that the potential inside the metal is constant.

The free electron model works well for most metals, but it does not appear that we can use this same model to describe the energy levels in a semiconductor. For example, we can see from Fig. 5.22 that the energy–wavevector relationship for a semiconductor is quite different to the free electron form given by eqn (5.22). However, we are only interested in the lowest energy conduction band states and the highest energy valence band states, because these are the states which are populated by conduction electrons and holes, respectively. (All of the states outside this region are either fully occupied or completely vacant and so do not contribute to the conductivity of the semiconductor.) This makes matters much simpler because we do not have to worry about obtaining a relationship which reproduces the entire energy–wavevector curve in Fig. 5.22(b). All we need is an expression which gives us the correct form near the edges of each band. Is it possible to achieve this by simply modifying the free electron model?

Let us consider the conduction band first of all. Can we alter eqn (5.22) to describe the lowest states in the conduction band? Firstly, we need to replace the mass m_e with the effective mass m_e^* since this will give us the correct curvature for the band. Also, if we arbitrarily take the zero of energy to coincide with the top of the valence band, then we note that there are no conduction states with an energy less than E_g.

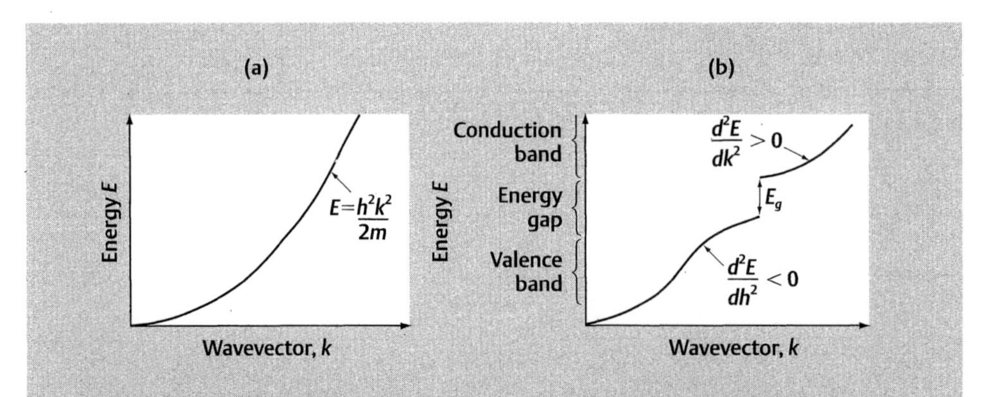

Figure 5.22 (a) In a metal the free electron model predicts that the energy E and wavevector k are related by the expression $E = \hbar^2 k^2 / 2m$. (b) In a semiconductor, the presence of the band gap requires a more complicated relationship between E and k.

Consequently, we propose that the energy–wavevector relationship can be written as

$$E = E_g + \frac{\hbar^2 k^2}{2m_e^*} \tag{5.23}$$

The density of states in the conduction band, $g(E)$, can now be found by following an argument similar to that used in Section 4.8 but with the electron mass, m_e, replaced by the effective mass, m_e^*, and with the energy scale shifted so that the lowest energy state in the conduction band is at an energy E_g. Comparing with eqn (4.19) the density of states is therefore given by

$$g(E) = \frac{V}{2\pi^2}\left(\frac{2m_e^*}{\hbar^2}\right)^{3/2}(E - E_g)^{1/2} \tag{5.24}$$

The concentration of electrons in the conduction band is then found by integrating over the occupied states that lie above the edge of the conduction band, i.e.

$$n = \frac{1}{V}\int_{E_g}^{\infty} g(E) f(E)\, dE$$

where $f(E)$ is the Fermi–Dirac distribution. Unfortunately, if we substitute for $f(E)$ from eqn (5.1) and $g(E)$ from eqn (5.24) into eqn (5.25), the integral cannot be solved analytically. However, we can solve this problem by using the approximate form of the Fermi–Dirac distribution, i.e.

$$f(E) \approx e^{(E_F - E)/k_B T} \tag{5.26}$$

This expression is valid so long as the difference in energy between the conduction band edge and the Fermi energy is significantly larger than the thermal energy $k_B T$ (see Example 5.4).

Substituting for eqns (5.24) and (5.26) in eqn (5.25) we can then show (see Example 5.9) that the number of conduction electrons per unit volume is given by

$$n = 2\left(\frac{m_e^* k_B T}{2\pi\hbar^2}\right)^{3/2} e^{(E_F - E_g)/k_B T} \tag{5.27}$$

Note that this expression is valid for intrinsic or doped semiconductors. Comparing this expression with our previous simple approximation in eqn (5.13) we can see that the exponential term is identical, but the constant C is replaced by a term which is dependent on T to the power 3/2. Therefore C is not constant with temperature as we assumed previously. However, it is fair to say that the temperature dependence of C is small in comparison with the temperature dependence of the exponential term.

If we substitute the appropriate values for intrinsic silicon into eqn (5.27) and assume that the Fermi energy lies at the centre of the band gap, then we obtain $n_i = 3.5 \times 10^{15}$ m^{-3} at a temperature of 300 K (see Question 5.20). The experimental value is about 1.0×10^{16} m^{-3}, so our modified free electron model appears to be reasonably successful.

Example 5.9 By using the approximate form of the Fermi–Dirac distribution from eqn (5.26) and the density of states function as given by eqn (5.24), show that the number of conduction electrons per unit volume is given by eqn (5.27).

Solution Substituting from eqns (5.26) and (5.24) in eqn (5.25) we have

$$n = C_e \int_{E_g}^{\infty} (E - E_g)^{1/2} e^{(E_F - E)/k_B T} \, dE$$

where

$$C_e = \frac{1}{2\pi^2} \left(\frac{2m_e^*}{\hbar^2} \right)^{3/2}$$

If we make a change of variable

$$x = (E - E_g)/k_B T$$

we have

$$\frac{dx}{dE} = \frac{1}{k_B T} \quad \text{and} \quad x = 0 \text{ at } E = E_g$$

Substituting into the above integral we obtain

$$n = C_e (k_B T)^{3/2} e^{(E_F - E_g)/k_B T} \int_0^{\infty} x^{1/2} e^{-x} \, dx$$

This integral can be found in tables of standard integrals:

$$\int_0^{\infty} x^{1/2} e^{-x} \, dx = \frac{1}{2}\sqrt{\pi}$$

So

$$n = 2 \left(\frac{m_e^* k_B T}{2\pi\hbar^2} \right)^{3/2} e^{(E_F - E_g)/k_B T}$$

We can perform a similar analysis to determine the concentration of holes in the valence band, but there are a few points that we need to treat with some caution. Firstly, we can see from Fig. 5.22(b) that the value of d^2E/dk^2 is less than zero at the top of the valence band—which means that the electrons have a negative effective mass! This rather puzzling result can be resolved if we consider the properties of the holes. Unlike electrons, holes display a tendency to occupy the states of higher energy, so an energy diagram for the holes can be obtained by inverting the energy diagram for electrons in Fig. 5.22(b). Consequently, the negative effective mass of an electron corresponds to a positive effective mass for a hole.

Since a hole is a state which is not occupied by an electron, the concentration of holes, p, is obtained by integrating over all of the vacant states in the valence band. The probability that a state is not occupied is given by $[1 - f(E)]$, so we have

$$p = \frac{1}{V}\int_{-\infty}^{0} g(E)[1 - f(E)]\,dE \tag{5.28}$$

One further point to note is that the number of states near the top of the valence band increases as the energy decreases. Consequently, we can obtain an expression for the density of states by using eqn (4.19) but with E replaced by $-E$ and m by m_h^*, i.e.

$$g(E) = \frac{V}{2\pi^2}\left(\frac{2m_h^*}{\hbar^2}\right)^{3/2}(-E)^{1/2} \tag{5.29}$$

By making a suitable approximation for the term $[1 - f(E)]$ and solving the integral in eqn (5.28) we obtain an expression for the concentration of holes (see Question 5.22):

$$p = 2\left(\frac{m_h^* k_B T}{2\pi\hbar^2}\right)^{3/2} e^{-E_F/k_B T} \tag{5.30}$$

Again, we note that this expression is valid for doped and intrinsic semiconductors.

Although we have obtained separate expressions for the concentrations of conduction electrons and holes, we know that in an intrinsic semiconductor these quantities are the same (see eqn (5.6)). Consequently, if we make these two equations equal to one another then we can determine an expression for the Fermi energy in the intrinsic case. I will leave it as an exercise for the reader (see Question 5.24) to show that in this case the Fermi energy is given by

$$E_F = \frac{E_g}{2} + \frac{3}{4}k_B T \ln\left(\frac{m_h^*}{m_e^*}\right) \tag{5.31}$$

From this equation we can see that at a temperature of absolute zero the Fermi energy is at the exact centre of the band gap, as we assumed in Section 5.3. Although the same statement is not necessarily true at finite temperatures, in most cases the correction to the position of the Fermi energy at room temperature is relatively small. Therefore it is generally a good approximation to assume that the Fermi energy is at the centre of the band gap.

5.12 Summary

In this chapter we have looked at the properties of both intrinsic (i.e. pure) and extrinsic (i.e. doped) semiconductors.

We have seen that the concentration of carriers, and therefore the conductivity, of an intrinsic semiconductor varies exponentially with the magnitude of the band gap. This leads to a huge difference in conductivity between materials which have only a small difference in band gap. The conductivity of a semiconductor is also shown to vary exponentially with temperature, and therefore a material which acts as an insulator at low temperature may be a good conductor at high temperatures.

Aside from thermal excitation, a photon can also be used to excite electrons into the conduction band, and so the conductivity of a semiconductor increases when it is subject to illumination by light of the appropriate frequency. The reverse effect, in which the recombination of a conduction electron and a hole gives rise to a photon, occurs in some materials. These are known as direct gap semiconductors.

Another way to increase the conductivity is to introduce impurities. In particular, impurities which produce a donor or acceptor state close to the relevant band edge are easily ionized, and so at room temperature we can assume that each such impurity generates a carrier. Consequently, only a very small concentration of impurities (less than 10^{-4}%) is required to alter the conductivity of the sample dramatically.

Doping with impurities also affects the ratio of conduction electrons to holes. In an intrinsic semiconductor, the carriers are always created as an electron–hole pair, and so the number of conduction electrons is equal to the number of holes. However, by doping the material with suitable impurities we can create an n- or a p-type semiconductor in which the majority of the carriers are electrons or holes, respectively. This is the basis for the p–n junction and most of the other microelectronic devices that we will discuss in the next chapter.

Questions

*Questions marked * indicate that they are based on the optional sections.*

5.1 Insulators and semiconductors. Show that the approximation in eqn (5.3) is valid for silicon (which has a band gap of 1.11 eV) at 300 K.

5.2 Insulators and semiconductors. The intrinsic carrier concentrations in various semiconductors at 300 K are:

$Ge: n = 3 \times 10^{19}\, m^{-3}$

$Si: n = 1 \times 10^{16}\, m^{-3}$

$GaAs: n = 2 \times 10^{13}\, m^{-3}$

Using the band gap data in Table 5.1, calculate the value of the constant C in eqn (5.4) for each of these materials.

5.3 Insulators and semiconductors. Use eqn (5.4) and the data in Table 5.1 to determine the number of conduction electrons in a 1 cm^3 diamond at 300 K. (Assume that the constant C has a value of $10^{25}\, cm^{-3}$.)

5.4 Insulators and semiconductors. Determine the conductivity of silicon at temperatures of 30 K, 300 K and 1000 K, given that the value of C in eqn (5.4) is $2 \times 10^{25}\, m^{-3}$ and the mobility is $0.135\, m^2\, V^{-1}\, s^{-1}$. Assume that C and the mobility are constant with temperature and consider only the conductivity due to the conduction electrons. Comment on your findings.

5.5 Insulators and semiconductors. The electrical conductivity of a material is given by the equation

$$\sigma = n\mu e$$

where n is the concentration of electrons and μ is the mobility. By considering how n and μ vary as a function of temperature, compare the temperature dependence of conductivity of a semiconductor with that of a metal.

5.6 Holes. Determine the conductivity of intrinsic silicon at 300 K given that the electron and hole mobilities are 0.135 $m^2 V^{-1} s^{-1}$ and 0.048 $m^2 V^{-1} s^{-1}$, respectively, and that the concentration of conduction electrons is 1×10^{16} m^{-3}.

5.7 Holes. Assuming that the electron and hole mobilities are constant with temperature, determine the temperature at which the conductivity of intrinsic germanium is double the value at 300 K.

5.8 Optical properties. Gallium arsenide has a band gap of 1.43 eV. If all of the gallium atoms are replaced with indium, the resulting crystal of indium arsenide has a band gap of 0.36 eV. Assuming that the band gap varies linearly between these two extremes, calculate the percentage of indium that must be used in an alloy of indium gallium arsenide in order to emit light with a wavelength of 1 μm.

5.9 Optical properties. Explain why silicon is opaque, whereas diamond is transparent. (Hint: consider the minimum frequency of light that is absorbed in each case.)

5.10 Optical properties. For an indirect gap semiconductor the difference in wavevector between an electron at the bottom of the conduction band and a hole at the top of the valence band is typically of the order of π/a where a is the length of one side of the unit cell (known as the lattice constant). Determine this quantity for silicon for which the lattice constant is 0.543 nm and compare with the wavevector of a photon corresponding to the indirect band gap energy, E_g = 1.11 eV.

5.11 Effective mass. Determine the cyclotron frequency for a free electron and for a conduction electron in a crystal of GaAs in a magnetic field of 0.05 T. (Use the data in Table 5.2.)

5.12 n-type semiconductors. (i) Use Bohr's model to determine the radius of the donor electron orbit for a phosphorus atom in a silicon crystal. (ii) Estimate the number of silicon atoms in a sphere of this radius. (The nearest neighbour separation in silicon is 0.234 nm.)

5.13 n-type semiconductors. Use Bohr's model to determine the energy of the donor state for a phosphorus impurity in germanium. (The dielectric constant for germanium is 15.8.)

5.14 n-type semiconductors. Determine the concentration of conduction electrons in a sample of silicon if one in every million silicon atoms is replaced by a phosphorus atom. Assume that every phosphorus atom is singly ionized. (Silicon has a molar mass of 0.028 kg mol^{-1} and a density of 2300 kg m^{-3}.)

5.15 n-type semiconductors. Determine the conductivity at 300 K of a sample of n-type silicon with a donor concentration N_d of $5 \times 10^{22} m^{-3}$. Ignore the conductivity of the holes and assume that all of the donors are ionized. Compare your result with the conductivity of intrinsic silicon calculated in Question 5.4. (Use the mobility data from Question 5.4.)

5.16 p-type semiconductors. By considering the probability of an electron occupying a state at the valence band edge, show that the concentration of holes in a doped semiconductor is given by eqn (5.15).

5.17 Majority and minority carriers. A semiconductor has a donor concentration of 6×10^{22} m^{-3} and an acceptor concentration of $1 \times 10^{22} m^{-3}$. Assuming that all of these impurities are ionized and that the intrinsic carrier concentration is $5 \times 10^{15} m^{-3}$, calculate the concentrations of conduction electrons and holes and identify which are the majority carriers and which are the minority carriers.

5.18 Hall effect. A sample of n-type silicon 1 cm wide (i.e. in the *y* direction in Fig. 5.19) and 2 mm thick (in the *z* direction) is subject to a magnetic field $B_z = 0.2$ T. If the current in the sample is 1 mA and the Hall voltage (i.e. the voltage difference across the sample in the *y* direction) is 3×10^{-4} V, determine the concentration of conduction electrons.

5.19 * **Free electron model—semiconductors.** Show that for the free electron model, the effective mass is equal to the free electron mass.

5.20 * **Free electron model—semiconductors.** Use eqn (5.27) to determine the concentration of conduction electrons in a sample of intrinsic silicon at 300 K. (Use the data from Tables 5.1 and 5.2 and assume that the Fermi energy is at the centre of the band gap.)

5.21 * **Free electron model—semiconductors.** A sample of n-type silicon has a doping concentration of 10^{22} m^{-3}. Determine how far the Fermi energy is below the conduction band edge at 300 K assuming that all of the donors are ionized.

5.22 * **Free electron model—semiconductors.** By making a suitable approximation for the term $[1 - f(E)]$ in eqn (5.28), show that the concentration of holes in the valence band is given by eqn (5.30)

5.23 * **Free electron model—semiconductors.** Use eqn (5.30) to determine the concentration of holes in the valence band in a sample of intrinsic silicon at 300 K. (Use the data from Tables 5.1 and 5.2 and assume that the Fermi energy is at the centre of the band gap.) Compare your result with the concentration of conduction electrons calculated in Question 5.20. Can you explain why the values are slightly different?

5.24 * **Free electron model—semiconductors.** Using eqns (5.27) and (5.30), show that the Fermi energy in an intrinsic semiconductor is given by

$$E_F = \frac{E_g}{2} + \frac{3}{4} k_B T \ln\left(\frac{m_h^*}{m_e^*}\right)$$

Use this expression to show that the Fermi energy is at the centre of the band gap if the effective masses of the electrons and holes are equal. Use the data in Table 5.2 to determine to what extent the Fermi energy is shifted from the centre of the band gap in indium antimonide and silicon at a temperature of 300 K.

Chapter 6
Semiconductor devices

6.1 Introduction

The use of semiconductors in electronic and optoelectronic devices is by far the most commercially important application of solid state physics. In particular, the sales of integrated circuits (or silicon chips as they are often called) amount to tens of billions of US dollars each year. In fact, integrated circuits have become so commonplace that it is easy to forget that each one is an incredibly complex circuit, often containing several million transistors. Optical semiconductor devices, such as light-emitting diodes (LEDs) and semiconductor lasers, are also becoming increasingly important with the growing trend to use optical fibres for communication.

In this chapter we will take a look at the underlying physics behind these devices. We have already covered a lot of the groundwork in the previous chapter, but there is one essential topic that we have not yet considered—the p–n junction. A p–n junction is a crystal of semiconductor which is doped p-type in one region and n-type in another. The unique properties of this structure can be understood by examining what happens at the interface between these two regions. Since the p–n junction is the vital component of the majority of semiconductor devices, we will examine it in detail from both a descriptive and a mathematical point of view.

We will also take a look at heterojunctions, which are formed from two different types of semiconductor. Heterojunctions offer many advantages over p–n junctions and we will take a brief look at how they can be used in transistors and optoelectronic devices.

However, we will begin by considering the properties of a junction between two metals. Although this is not directly related to the behaviour of semiconductor devices, it provides some useful clues to help us understand the behaviour of the electrons at a semiconductor junction.

6.2 Junctions between two metals—the contact potential

What happens if we bring two dissimilar metals in contact with one another? For example, consider the metals A and B shown in Fig. 6.1(a). Metal A has a work function

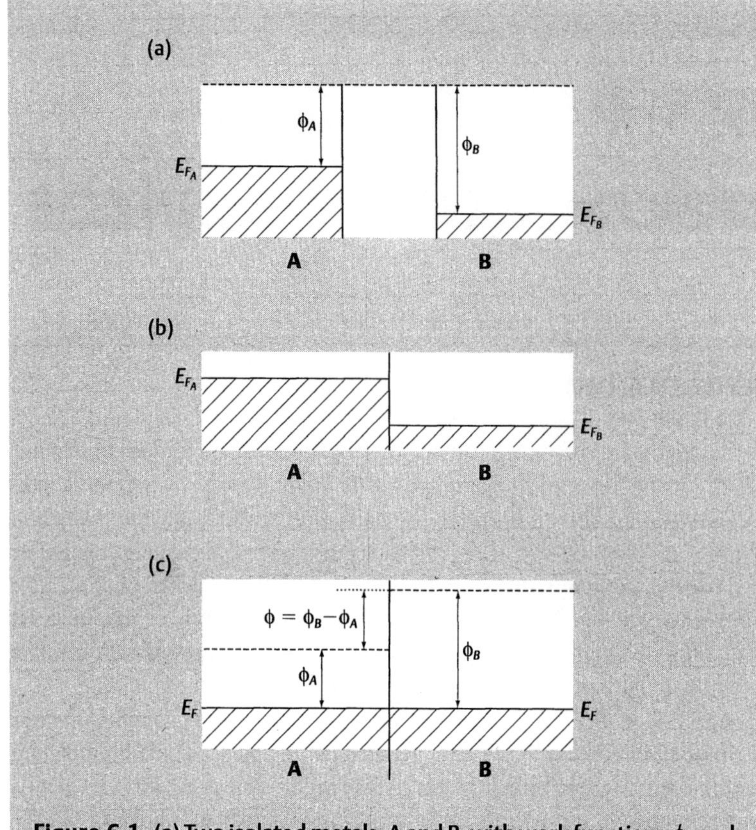

Figure 6.1 (a) Two isolated metals, A and B, with work functions ϕ_A and ϕ_B. (b) When the metals come into contact the electrons flow from metal A to metal B in order to take advantage of the vacant lower energy states in metal B. (c) This continues until the Fermi energies become the same on both sides of the junction. The difference in potential between the two metals is given by $\phi = \phi_B - \phi_A$.

ϕ_A which is smaller than the work function of metal B, ϕ_B. Consequently, when the metals are placed in contact, the electrons in the highest energy states in metal A can move into the lower energy vacant states in metal B, as shown in Fig. 6.1(b). This process continues until the highest occupied energy states on either side of the junction are the same, i.e. the Fermi energies in the two metals are equal (see Fig. 6.1(c)).

However, this is not the end of the story. If the two metals are initially electrically neutral, then the transfer of electrons from A to B means that metal B is now negatively charged whilst metal A has a positive charge. Therefore, there is a potential difference—known as the contact potential—between the two types of metal. From Fig. 6.1(c) we can see that the magnitude of the contact potential, ϕ, is equal to the difference between the work functions of the two metals, i.e.

$$\phi = \phi_B - \phi_A \tag{6.1}$$

6.3 The p–n junction—a qualitative description

Let us now consider a junction between an n-type and a p-type semiconductor. From Chapter 5 we know that the Fermi energies are different in the two samples. In an n-type semiconductor the Fermi energy is near to the conduction band edge, whereas in a p-type material it is close to the valence band edge, as shown in Fig. 6.2(a). Do the Fermi energies become equal in this case?

To answer this question, let us consider what would happen if we put these two materials in contact. (At this stage I should point out that a p–n junction is not made by joining a p-type and an n-type sample together. It is fabricated by taking a single crystal and introducing donor impurities into one region and acceptors into another. However, the idea of placing the two materials in contact is useful because it enables us to see how the equilibrium situation is achieved.) Initially, virtually all of the conduction electrons are on one side of the junction and most of the holes are on the other side, but such an imbalance is unlikely to last for long. As the conduction electrons move about at random, some of them find their way into the p-type material. This process is known as diffusion. Similarly, some of the holes diffuse into the n-type region. However, the process is complicated by the fact that the electrons and holes can recombine with one another.

Let us consider the scenario as a single conduction electron diffuses across the junction from the n-type layer to the p-type side. As it crosses the junction the electron sees a large concentration of holes, and so there is a high probability of the electron recombining with a hole before it has chance to travel far into the p-type region. As more electrons cross the junction and recombine with holes, the number of carriers in the junction region is reduced. We therefore end up with a layer either side of the junction which has a much lower carrier concentration than in the rest of the crystal. This is known as the depletion region. In the above description we have only considered the diffusion of the conduction electrons, but if we consider the diffusion of the holes we end up with exactly the same results (see Question 6.1).

We can see from Fig. 6.2(b) that the above process results in an accumulation of negative charge on the n-type side of the junction and positive charge on the p-type side. How do we represent this build-up of charge on an energy band diagram? Since it is more difficult for the conduction electrons to move from the n-type to the p-type side we can represent this by saying that the energy of the conduction band edge is higher on the p-type side than on the n-type side, as shown in Fig. 6.2(c). We therefore have a contact potential, ϕ, as in the metal junction discussed in the previous section. Since the energy needed to excite an electron across the potential ϕ is equal to $e\phi$, we can see that the difference in energy between the two sides of the junction is $e\phi$. It turns out that this energy difference $e\phi$ is exactly equal to the difference between the Fermi energies in the two materials, and therefore the Fermi energies are aligned, as in the metal–metal junction.

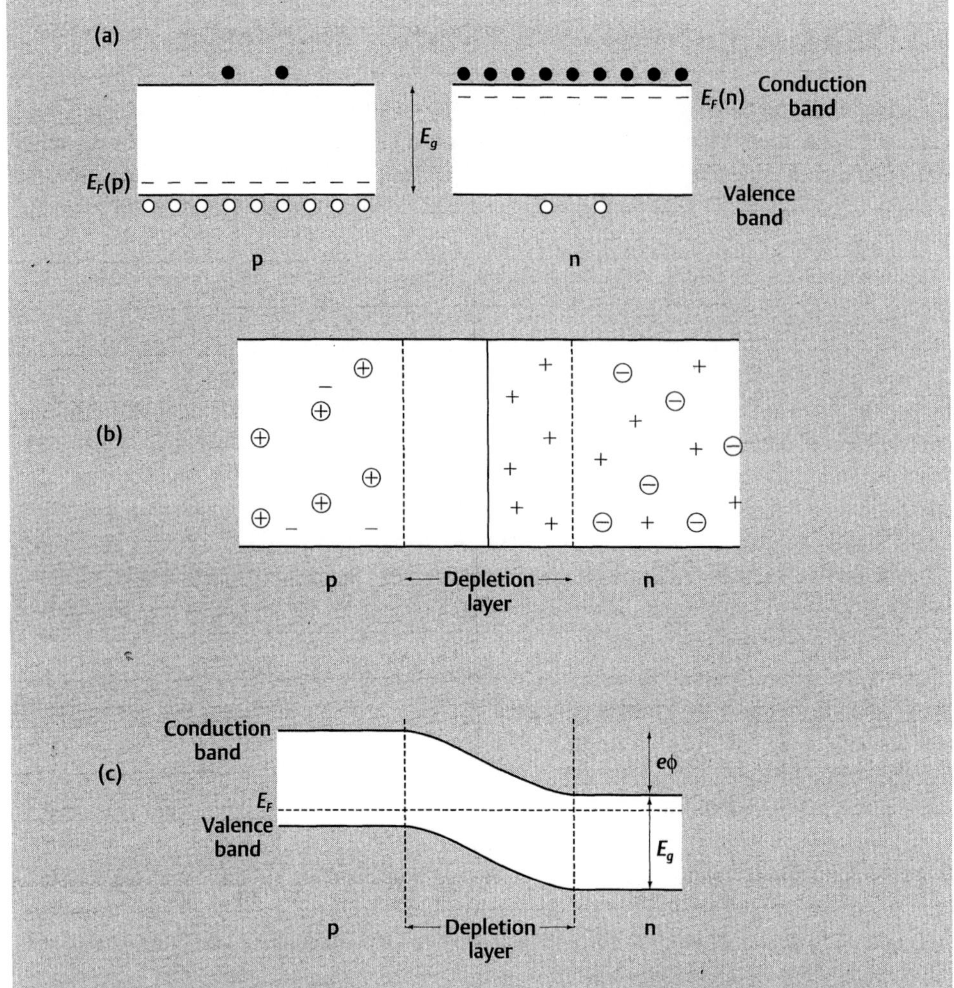

Figure 6.2 (a) The Fermi energy is closer to the valence band edge in a p-type semiconductor and to the conduction band edge in an n-type semiconductor. Conduction electrons and holes are represented by solid and open circles, respectively. (b) When the p- and n-type regions are in contact, diffusion of the carriers results in a depletion region in which there are virtually no carriers. (The acceptor and donor ions are represented by '−' and '+', respectively, and the electrons and holes are represented by circles with '−' and '+' signs, respectively.) (c) The distribution of charge at a p–n junction produces a contact potential ϕ across the junction so that the conduction band edge on the p-type side of the junction is at an energy $e\phi$ higher than that on the n-type side.

6.4 The p-n junction—a quantitative analysis*

Now that we have some idea about the behaviour of electrons and holes at a p–n junction, in this section we will deal with a more mathematical description of the junction. In particular, we will look at ways of estimating the magnitude of the contact potential and the widths of the depletion layers on either side of the junction. It is assumed that the reader is familiar with the free electron treatment of semiconductors discussed in Section 5.11.

Let us begin by considering the variation in potential at the junction. We can do this by solving Poisson's equation which describes how the voltage across the junction, U, varies with the charge density, ρ. (Note that we will use U to represent the intrinsic voltage at any point in the junction, as distinct from V which we use to describe an external potential.) Poisson's equation is used in many instances in electricity and magnetism (e.g. Grant and Phillips—see Further reading) and is given by

$$\frac{d^2U}{dx^2} = -\frac{\rho}{\varepsilon\varepsilon_0} \tag{6.2}$$

where ε is the dielectric constant of the material. In this context x is the direction perpendicular to the junction.

Poisson's equation is not easy to solve for a general case, but we can simplify matters greatly if we assume that there are no free carriers in the depletion region which extends from $x = -l_p$ to $x = +l_n$ (see Fig. 6.3(a)). On the n-type side of the junction the charge density in the depletion layer, ρ_n, is therefore determined by the concentration of donor impurities. If we assume that all of these impurities are ionized, then

$$\rho_n = -eN_d \tag{6.3}$$

Note that we have included a minus sign because the charge on the electron, e, is a negative quantity, so the minus sign ensures that the charge density ρ_n is a positive quantity. Using a similar argument we can see that the charge density on the p-type

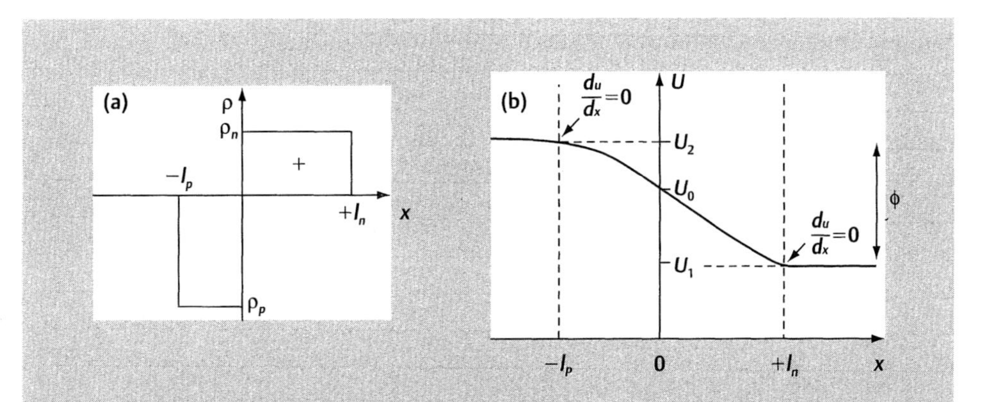

Figure 6.3 (a) The assumed distribution of charge at a p–n junction. The depletion layer is the region between $x = -l_p$ and $x = +l_n$. (b) The variation of the potential U across the junction when no external voltage is applied. (Note that the y axis plots the voltage rather than the energy, so the difference between the two sides of the junction is ϕ rather than $e\phi$.)

$$\rho_p = +eN_a \tag{6.4}$$

Outside the depletion region, i.e. for $x < -l_p$ and $x > l_n$, we assume that the net charge density is zero, as in a bulk-doped semiconductor. This means that the electric field is zero (i.e. $dU/dx = 0$) at $x = -l_p$ and $x = +l_n$, as shown in Fig. 6.3(b).

We can now integrate Poisson's equation. From Example 6.1, we find that the depletion layer widths, l_p and l_n, are related by the equation

$$N_d l_n = N_a l_p \tag{6.5}$$

Note that if we multiply the left hand side of this equation by the electron charge e, then this quantity represents the area of the rectangle marked '+' in Fig. 6.3(a). Similarly, the right hand side of the equation corresponds to the area marked '–' in the diagram. Therefore eqn (6.5) states that the areas of these two rectangles are equal. Of course we could have arrived at this same result by considering the fact that charge must be conserved, i.e. the net amount of negative charge on the p-type side must be exactly equal to the net amount of positive charge on the n-type side.

Example 6.1 By integrating Poisson's equation for the charge density shown in Fig. 6.3(a) and using the boundary conditions

$$\frac{dU}{dx} = 0 \text{ at } x = -l_p \text{ and } x = +l_n$$

show that

$$N_d l_n = N_a l_p$$

Solution Integrating eqn (6.2) for the n-type side gives

$$\frac{dU}{dx} = -\frac{\rho_n x}{\varepsilon\varepsilon_0} + C_1$$

where C_1 is the constant of integration.

Applying the boundary condition $dU/dx = 0$ at $x = l_n$ gives

$$C_1 = \frac{\rho_n l_n}{\varepsilon\varepsilon_0}$$

Therefore

$$\frac{dU}{dx} = \frac{\rho_n}{\varepsilon\varepsilon_0}(l_n - x) \tag{1}$$

Similarly, for the p-type side

$$\frac{dU}{dx} = \frac{\rho_n}{\varepsilon\varepsilon_0}(l_n + x) \tag{2}$$

If we evaluate these two expressions at $x = 0$ we obtain

$$\frac{dU}{dx} = \frac{\rho_n l_n}{\varepsilon\varepsilon_0}$$

and

$$\frac{dU}{dx} = \frac{\rho_p l_p}{\varepsilon\varepsilon_0}$$

Assuming that dV/dx is continuous at $x = 0$, then

$$\rho_n l_n = -\rho_p l_p$$

Substituting for ρ_p and ρ_n from eqns (6.3) and (6.4) gives

$$N_d l_n = N_a l_p$$

By further integration of Poisson's equation we can show (see Question 6.3) that the contact potential ϕ is given by

$$\phi = \frac{e}{2\varepsilon\varepsilon_0}\left(N_d l_n^2 + N_a l_p^2\right) \tag{6.6}$$

Rearranging this expression and using the relationship in eqn (6.5) we can also show (see Question 6.4) that the depletion layer width on the p-type side is

$$l_p = \left(\frac{\phi 2\varepsilon\varepsilon_0}{eN_a}\frac{N_d}{N_a + N_d}\right)^{1/2} \tag{6.7}$$

and that on the n-type side is

$$l_n = \left(\frac{\phi 2\varepsilon\varepsilon_0}{eN_d}\frac{N_a}{N_a + N_d}\right)^{1/2} \tag{6.8}$$

Having obtained expressions for the contact potential and the depletion layer widths, it is also useful to look at the majority and minority carrier concentrations on either side of the junction. We introduce the following notation to denote the concentrations of the carriers at a point *outside* the depletion layers:

n_n is the concentration of conduction electrons in the n-type region;
n_p is the concentration of conduction electrons in the p-type region;
p_n is the concentration of holes in the n-type region;
p_p is the concentration of holes in the p-type region.

This is shown in Fig. 6.4.

If we take the zero of energy to correspond to the valence band edge on the n-type side of the junction then we can see from Example 6.2 that the concentration of conduction electrons on the p-type side is given by

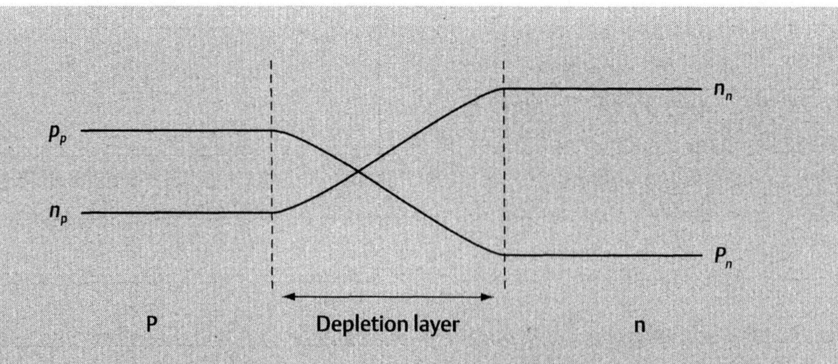

Figure 6.4 Concentrations of electrons, *n*, and holes, *p*, across the p–n junction. The subscripts *n* and *p* denote the values in the n- and p-type regions, respectively, outside the depletion layer.

$$n_p = N_c e^{(E_F - E_g - e\phi)/k_B T} \tag{6.9}$$

where N_c is a constant. Similarly, we can show (see Question 6.5) that the other carrier concentrations are given by

$$n_n = N_c e^{(E_F - E_g)/k_B T} \tag{6.10}$$

$$p_n = N_v e^{-E_F/k_B T} \tag{6.11}$$

$$p_p = N_v e^{(e\phi - E_F)/k_B T} \tag{6.12}$$

where N_v is a constant which in general is different to N_c.

Note that by rearranging eqns (6.9) to (6.12) we can show (see Question 6.6) that the contact potential can also be determined from the equation

$$\phi = \frac{k_B T}{e} \ln\left(\frac{N_a N_d}{n_i^2}\right) \tag{6.13}$$

This is a useful expression because it relates the contact potential to the concentration of acceptors and donors, N_a and N_d, and the intrinsic carrier concentration, n_i, all of which are usually known quantities, and is independent of the depletion layer widths. By inserting typical values for the doping concentrations used in microelectronic devices we find that for a silicon p–n junction the contact potential is typically about 0.7 eV and the depletion layer widths are typically of the order of 0.1 μm to 1.0 μm (see Question 6.7).

Example 6.2 Show that the concentration of conduction electrons on the p-type side of a p–n junction at a point outside the depletion layer is given by

$$n_p = N_c e^{(E_F - E_g - e\phi)/k_B T}$$

where N_c is a constant and the zero of voltage is at the valence band edge on the n-type side of the junction.

Solution We can approach this in a similar way to Example 5.9.

The concentration of carriers is given by

$$n_p = \int g(E) f(E)\, dE$$

where $g(E)$ is the density of states, $f(E)$ is the Fermi–Dirac distribution, and the integral is evaluated from the conduction band edge in the p-type region. From Fig. 6.2(c), we can see that the conduction band edge in the p-type region is at an energy $E_g + e\phi$ above the valence band edge in the n-type region (i.e. the point we have defined as $E = 0$).

As in Example 5.9, the difference between this energy and the Fermi energy, $E - E_F$, is much larger than the thermal energy, $k_B T$, so we can write the Fermi–Dirac distribution in the approximate form

$$f(E) \approx e^{(E_F - E)/k_B T}$$

Also, by comparison with eqn (5.24) we can see that the density of states function is given by

$$g(E) = C_e [E - (E_g + e\phi)]^{1/2}$$

where C_e is a constant (which is given in Example 5.9).

So the value of n_p is

$$n_p = C_e \int_{E_g + e\phi}^{\infty} [E - (E_g + e\phi)]^{1/2} e^{(E_F - E)/k_B T}\, dE$$

Again, following Example 5.9 we make a change of variable

$$x = (E - E_g - e\phi)/k_B T$$

and use the standard integral

$$\int_0^{\infty} x^{1/2} e^{-x}\, dx = \frac{1}{2}\sqrt{\pi}$$

so

$$\begin{aligned} n_p &= C_e (k_B T)^{3/2} e^{(E_F - E_g - e\phi)/k_B T} \int_0^{\infty} x^{1/2} e^{-x}\, dx \\ &= N_c e^{(E_F - E_g - e\phi)/k_B T} \end{aligned}$$

6.5 The p–n junction with an applied voltage—qualitative

In the previous two sections we have considered the behaviour of the electrons and holes at an isolated p–n junction. However, our main interest in p–n junctions is to determine what happens when an external voltage is applied to the system. In particular, we want to examine the current flowing through the junction. Let us begin by looking again at the movement of the electrons and holes when there is no external voltage present.

A popular misconception regarding p–n junctions is that in the equilibrium state the energy barrier $e\phi$ becomes so large that it is no longer possible for any conduction electrons to diffuse across the junction into the p-type layer. This is quite wrong. No matter how large the contact potential, there is always a finite probability that some of the conduction electrons on the n-type side have an energy greater than that of the conduction band edge on the p-type side. These electrons can therefore diffuse into the p-type side and so give rise to a diffusion current across the junction, as shown in Fig. 6.5(a). However, there is also a finite probability of finding some thermally created electrons in the conduction band on the p-type side of the depletion layer. These electrons are attracted to the n-type side of the junction because of the built-in electric field across the junction, and so they give rise to what is called the drift current. Since there is no net flow of charge when the system is in equilibrium, the flow of electrons from n to p must be exactly balanced by the flow of electrons in the opposite direction, i.e. the drift current must be equal in magnitude to the diffusion current. We therefore say that the p–n junction is in a condition of dynamic equilibrium.

Let us now look at what happens when an external voltage, V, is applied across the junction. We will assume that the negative terminal is attached to the n-type side of the junction and the positive terminal to the p-type side. This means that the external voltage is in the opposite sense to the built-in voltage, and so the voltage difference across the junction is reduced, as shown in Fig. 6.5(b). In fact, we can see that the potential across the junction is now $\phi - V$. This means that more of the conduction electrons on the n-type side are able to diffuse across the junction, and so the diffusion current is increased in comparison with the case with zero voltage. Since there is a plentiful supply of conduction electrons on the n-type side, the diffusion current can become very large. In fact, we shall see in the next section that the diffusion current increases exponentially with the magnitude of the applied voltage. On the other hand, the drift current is due to the minority carriers on the p-type side of the junction. Since the concentration of minority carriers depends only on temperature and is not affected by the applied voltage, the drift current does not change from the zero-voltage case. Consequently, we can see that in this instance the diffusion current is larger than the drift current, and so there is a net flow of electrons from the n-type to the p-type side.

What happens if we connect the voltage the opposite way round with the negative and positive terminals on the p- and n-type sides, respectively? Now the voltage is in

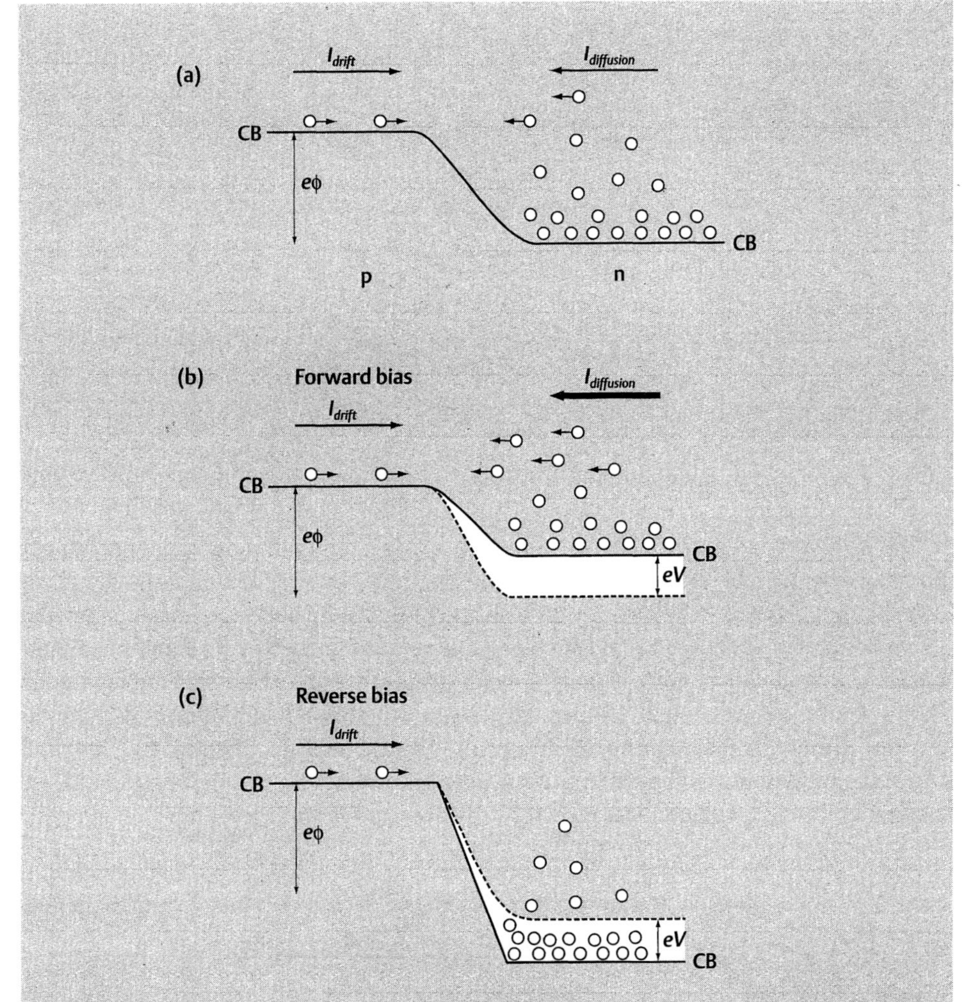

Figure 6.5 Schematic diagrams showing the distribution of electrons in the conduction band on either side of a p–n junction. (a) When there is no applied voltage the diffusion current, $I_{diffusion}$, due to electrons moving from the n-type to the p-type region, is equal to the drift current, I_{drift}, due to electrons moving from p to n. (b) When a forward bias is applied to the p–n junction the potential difference across the junction is reduced and the diffusion current is increased. (c) When a reverse bias is applied, the potential difference across the junction increases. As a result, the diffusion current decreases, and so the net current flow is due to the drift current.

the same sense as the built-in voltage and the potential difference across the junction is increased (Fig. 6.5(c)). Therefore the diffusion current is smaller than in the case with zero voltage, and so the drift current is dominant. However, as we mentioned in the previous paragraph, the drift current is not affected by the external voltage, so although there is a net flow of electrons from the p-type to the n-type side, the magnitude of the current is very small.

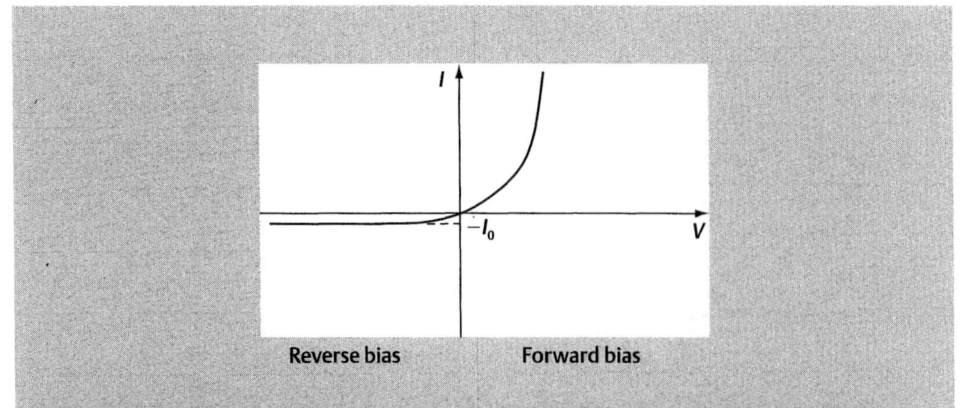

Figure 6.6 The current-voltage characteristic for a p–n junction is typified by a current which increases exponentially with a forward bias, but remains very small for a reverse bias.

Consequently, we can see that the characteristics of the junction vary enormously depending upon the polarity of the applied voltage. In the former case—which is called a forward bias—the current increases exponentially with the voltage, whereas in the latter case—known as reverse bias—the current remains very small regardless of the magnitude of the voltage. The typical variation of current with voltage is shown in Fig. 6.6. To a good approximation we can assume that a p–n junction only allows current to flow one way, and so one use of a p–n junction is as a rectifier to change alternating current into direct current. However, the properties of p–n junctions are exploited in a wide range of other electronic devices, as we shall see later.

6.6 The p–n junction with an applied voltage—quantitative*

In order to obtain expressions for the current flow across a p–n junction we will concentrate on the movement of electrons across the junction, but note that the same results would be obtained if we considered the movement of the holes.

We will begin by considering the case when no voltage is applied to the junction. We know that the drift current is due to the flow of minority carriers across the junction, and in this analysis we will concentrate on the movement of conduction electrons from the p-type to the n-type side (although the same results are obtained if we instead look at the movement of holes into the p-type side). The concentration of conduction electrons on the p-type side of the junction outside the depletion region is denoted by n_p. We have already derived an expression for this term in Example 6.2 and found that it is proportional to $e^{(E_F-E_g-e\phi)/k_BT}$. If we assume that the magnitude of the drift current is proportional to the electron concentration then we predict that the drift current is also proportional to $e^{(E_F-E_g-e\phi)/k_BT}$, i.e.

$$I_{drift} = Ce^{(E_F - E_g - e\phi)/k_B T} \tag{6.14}$$

where C is a constant. (Strictly speaking, this assumption is incorrect. Diffusion is a complicated process which depends on several factors. However, a more in-depth analysis predicts a current with the same exponential dependence as in eqn (6.14). More details are given by Rosenberg—see Further reading.)

Now let us consider the diffusion of conduction electrons from the n-type side to the p-type side. We can assume that the only electrons capable of diffusing across the junction are those on the n-type side which have an energy greater than or equal to the conduction band edge on the p-type side. When there is no external voltage applied to the junction, the conduction band edge on the p-type side is at an energy $e\phi$ greater than the band edge on the n-type side, so following the argument in Example 6.2 we can see that the concentration of electrons on the n-type side which satisfy this criterion is proportional to $e^{(E_F-E_g-e\phi)/k_BT}$. This is the same exponential factor as in eqn (6.14), which of course is just what we would expect because we know that when there is no applied voltage the diffusion current must equal the drift current. Therefore, we can write

$$I_{diffusion} = I_{drift} = I_0 = Ce^{(E_F - E_g - e\phi)/k_B T} \tag{6.15}$$

What happens if we now apply a forward bias, V, to the junction? The concentration of minority carriers is unchanged and so we expect that drift current is still given by eqn (6.14). However, the diffusion current is altered because there are now more electrons on the n-type side which have an energy greater than the conduction band edge on the p-type side, as we can see from Fig. 6.5(b). Since the effect of the external voltage is to raise the conduction band edge on the n-type side by an energy eV in comparison with the conduction band edge on the p-type side, the electrons with an energy $e(\phi - V)$ above the conduction band edge on the n-type side are now able to diffuse across the junction. By following an argument similar to that used in Example 6.2 we can see that the concentration of electrons in this situation is proportional to $e^{(E_F-E_g-e\phi+eV)/k_BT}$. Therefore the diffusion current is given by

$$I_{diffusion} = Ce^{(E_F - E_g - e\phi + eV)/k_B T} \tag{6.16}$$

We therefore find that the net current flow across the junction is given by (see Example 6.3)

$$I = I_{diffusion} - I_{drift} = I_0\left(e^{eV/k_B T} - 1\right) \tag{6.17}$$

If the potential is reversed then the drift current is still unaffected, but the diffusion current is smaller than in the zero-bias case because the potential difference between the two sides of the junction is greater (see Fig. 6.5(c)). The conduction band on the p-type side is now at an energy $e(\phi + V)$ above the conduction band edge on the n-type side, so from the previous discussion we can see that the diffusion current is given by

$$I_{diffusion} = Ce^{(E_F - E_g - e\phi - eV)/k_B T} \tag{6.18}$$

The net current is therefore

$$I = I_{diffusion} - I_{drift} = I_0\left(e^{-eV/k_B T} - 1\right) \tag{6.19}$$

(see Question 6.8).

If we sketch these equations we obtain the current–voltage curve shown in Fig. 6.6 (see Question 6.9). We can see that the reverse bias current has a maximum value of $-I_0$. This is referred to as the reverse saturation current. In a silicon p–n junction the value of I_0 is typically less than 100 nA; therefore it is a good approximation to assume that current flows through the diode in one direction only. So by just considering the concentrations of the carriers on either side of the junction we have been able to demonstrate the rectifying properties of a p–n junction.

Before leaving this subject I should mention that it is possible for large reverse currents to flow if a very large reverse voltage is applied to the junction. This is known as breakdown. However, an understanding of this effect relies on quite different principles to those discussed here and so we will not describe this phenomenon in any more detail.

Example 6.3 Show that when a forward bias V is applied to a p–n junction the net current is given by

$$I_{net} = I_0\left(e^{eV/k_B T} - 1\right)$$

where I_0 is given by eqn (6.15).

Solution From the text we know that the drift current is the same as in the zero-bias case. Therefore the drift current is given by eqn (6.15), i.e.

$$I_{drift} = Ce^{(E_F - E_g - e\phi)/k_B T} = I_0$$

The diffusion current is given by eqn (6.16), i.e.

$$I_{diffusion} = Ce^{(E_F - E_g - e\phi + eV)/k_B T}$$

Since $I_0 = Ce^{(E_F - E_g - e\phi)/k_B T}$, we can write

$$I_{diffusion} = I_0\, e^{eV/k_B T}$$

Therefore the net current is

$$I = I_{diffusion} - I_{drift} = I_0\, e^{eV/k_B T} - I_0 = I_0\left(e^{eV/k_B T} - 1\right)$$

6.7 Transistors—an introduction

The transistor must surely rate as one of the most important inventions of the twentieth century. However, whilst some transistors are implemented as individual discrete devices, the enormous commercial impact of microelectronics is due to the fact that a circuit consisting of a large number of transistors can be fabricated at low cost on a single small slice of semiconductor.

There are two basic types of transistor: the bipolar junction transistor, and the field-effect transistor (or FET). Bipolar transistors are used both as discrete devices and in integrated circuits, but FETs appear only in integrated circuits. In fact, FETs now almost completely dominate the integrated circuit market as they are considerably cheaper to manufacture than bipolar integrated circuits.

The transistor is capable of performing two functions. Firstly, it can amplify an electrical signal, and is used in this context as a discrete device in a wide range of electrical circuits. Secondly, it can be used as a switch, i.e. it has two distinct stable states which correspond to the device being 'ON' or 'OFF'. It is this behaviour which allows integrated circuits to process and store digital (binary) information.

Although both types of transistor can be used in amplifying or switching modes, we will concentrate on the amplification properties of the bipolar transistor (in Section 6.8) and the switching properties of FETs (in Section 6.9).

6.8 Bipolar transistors

The bipolar transistor consists of a sandwich of three alternately doped layers of semiconductor, the middle layer of which is referred to as the base and the other two layers as the emitter and the collector. The base region is very narrow, typically much less than 1 μm for a microelectronic device. Obviously, there are two different ways of arranging the layers, either n–p–n or p–n–p, as shown in Fig. 6.7. In this section we will concentrate on the n–p–n transistor, but the operation of the p–n–p transistor can be explained along similar lines simply by interchanging the words 'electron' and 'hole'.

Let us first of all construct the energy band diagram for the n–p–n transistor. We can do this quite easily because the transistor is essentially just two p–n junctions placed back to back, so from Fig. 6.2(c), we can see that the band structure of the n–p–n transistor with no external voltages applied is as shown in Fig. 6.8(a). As the emitter is more heavily doped than the base and collector, we note that the contact potential between the emitter and base is greater than that between the base and collector.

Now suppose that we apply a voltage across the structure so that the emitter is made negative with respect to the collector. We can see from Fig. 6.8(b) that the junction between the emitter and the base is now forward biased, whereas the junction between the base and collector is reverse biased. This means that a significant number of the conduction electrons in the emitter can flow into the base region. Since the base region is very narrow and the doping concentration is much lower

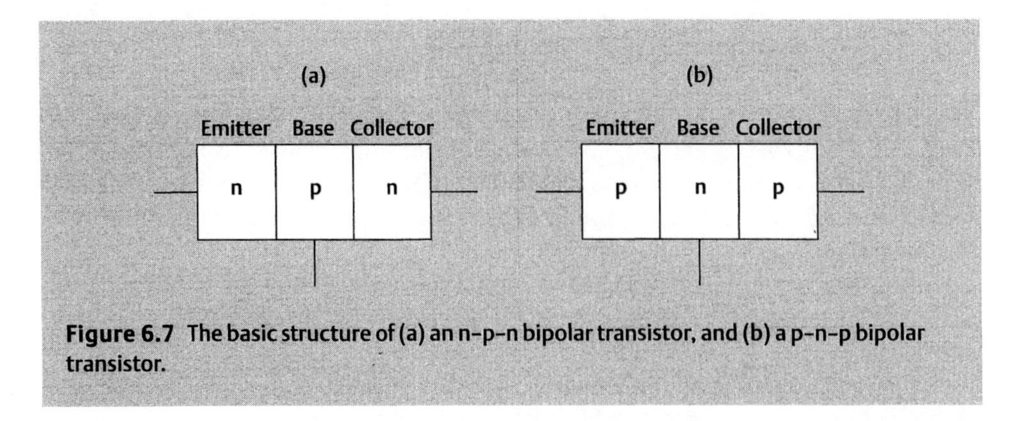

Figure 6.7 The basic structure of (a) an n–p–n bipolar transistor, and (b) a p–n–p bipolar transistor.

than that of the emitter (i.e. there are relatively few holes in the base region), only a small proportion of these electrons recombine with the holes in the base. The rest are swept down into the lower energy states in the collector. So a substantial current of electrons flows from the emitter to the base.

From our discussion of the p–n junction we know that each time an electron recombines with a hole in the base region it leaves behind an uncompensated negative acceptor ion. In a p–n junction this negative charge accumulates and so inhibits the flow of electrons, but in the transistor an electrical contact is made to the base region.

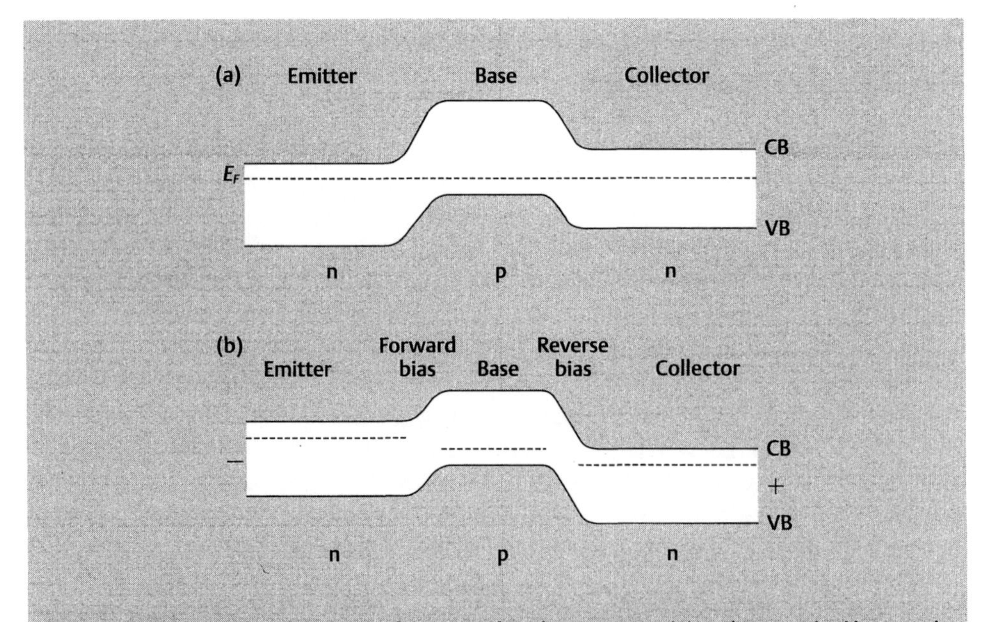

Figure 6.8 Energy band diagram of an n–p–n bipolar transistor (a) with no applied bias, and (b) when a voltage is applied such that the emitter is made negative with respect to the collector. In the latter case the emitter–base junction is forward biased and the base–collector junction is reverse biased.

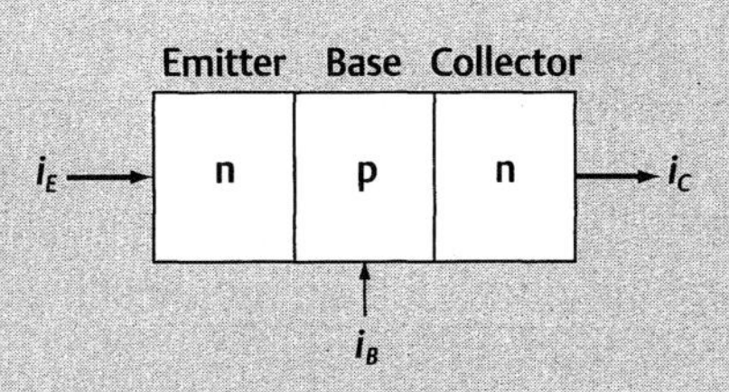

Figure 6.9 Current amplification in a bipolar transistor. A small input current at the base, i_B, causes a much larger current, i_E, to flow from the emitter into the base. As the majority of the conduction electrons that enter the base from the emitter flow straight through to the collector, the output current at the collector, i_C, is approximately equal to i_E. Therefore, current amplification occurs because the output current, i_C, is much larger than the input current, i_B.

This allows a current of holes to flow into the base so that the base remains electrically neutral, i.e. the number of holes drawn into the base exactly compensates those that are lost due to recombination.

To understand how a transistor acts as an amplifier we can look at the above scenario from a slightly different viewpoint. Let us suppose that we supply a hole current i_B to the base of the transistor, as shown in Fig. 6.9. This makes the base positively charged, and so lowers the potential barrier between the emitter and the base, which in turn causes an increased current of electrons to flow from the emitter into the base. If we assume that only 1% of the conduction electrons recombine in the base (and the remaining 99% flow into the collector), then the electron current from the emitter into the base, i_E, must be 100 times larger than i_B in order that the base remains neutral. Consequently, the output current flowing from the collector, i_C, is about 99 times larger than the input current to the base, i_B. The current gain of the amplifier, β, is defined as the ratio of the change in the output current, Δi_C, to the change in the input current Δi_B, i.e.

$$\beta = \frac{\Delta i_B}{\Delta i_C} \tag{6.20}$$

The bipolar transistor can therefore be used to amplify an analogue signal, since a small change in the base current produces a corresponding, but much larger, change in the collector current.

The other mode of operation of a transistor is as a switch. Although a bipolar transistor can be used to perform a switching function, we will study this property in an FET.

6.9 The field-effect transistor

The field-effect transistor, or FET, is similar to the bipolar transistor in that it consists of three distinct regions with alternate doping. However, the principle of operation is quite different, as is the terminology. In an FET these three regions are called the source, gate and drain, as shown in Fig. 6.10.

There are various types of FET, but by far the most common is the MOSFET, where MOS stands for metal-oxide-semiconductor. The name is derived from the fact that the metal contact above the gate region is separated from the semiconductor substrate by a thin oxide layer, usually silicon dioxide, which acts as an electrical insulator. The position of the oxide layer is shown in Fig. 6.10.

Let us begin by considering what happens when a voltage is applied to the metal contact above the gate. If the gate voltage is positive, then the holes in the p-type semiconductor directly below the gate are repelled from the surface, and at the same time the minority carrier conduction electrons in the p-type material are attracted to the surface (see Fig. 6.11(a)). When the gate voltage exceeds a certain level, known as the threshold voltage, we find that there is a region close to the surface in which the concentration of conduction electrons exceeds the concentration of holes. In other words, although the material is doped with acceptors, it behaves as though it is an n-type semiconductor. This region is known as an inversion layer.

We can also see how the inversion layer is created by studying the band diagram in Fig. 6.12. In the p-type semiconductor at a distance well below the interface with the oxide layer, the Fermi energy is close to the valence band edge, as we would expect. However, near to the surface of the semiconductor the electron energies are lowered by the presence of the positive charge on the other side of the oxide layer. This is known as band bending. The semiconductor in this region displays n-type behaviour if the Fermi energy is closer to the conduction band than to the valence band, and therefore the degree of band bending determines whether or not inversion takes place.

We can now see how a MOSFET operates as a switch. If we apply a voltage to the device such that the drain is made positive with respect to the source, then the inversion layer provides a channel for the conduction electrons to flow from the source to the drain, as shown in Fig. 6.11(b). So if the gate voltage is greater than the threshold voltage, a current flows and the device is ON. Conversely, if the gate voltage is less than the threshold voltage, then there is no inversion layer, the current does not flow, and the device is OFF.

The structure we have described here is an n-channel MOSFET, but we can make a

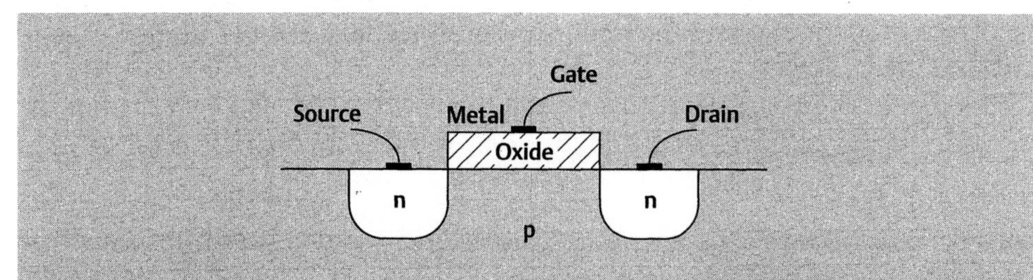

Figure 6.10 The structure of an n-channel MOSFET showing the source, gate and drain region. The electrical contact to the gate is separated from the semiconductor by a thin layer of insulator, typically silicon dioxide.

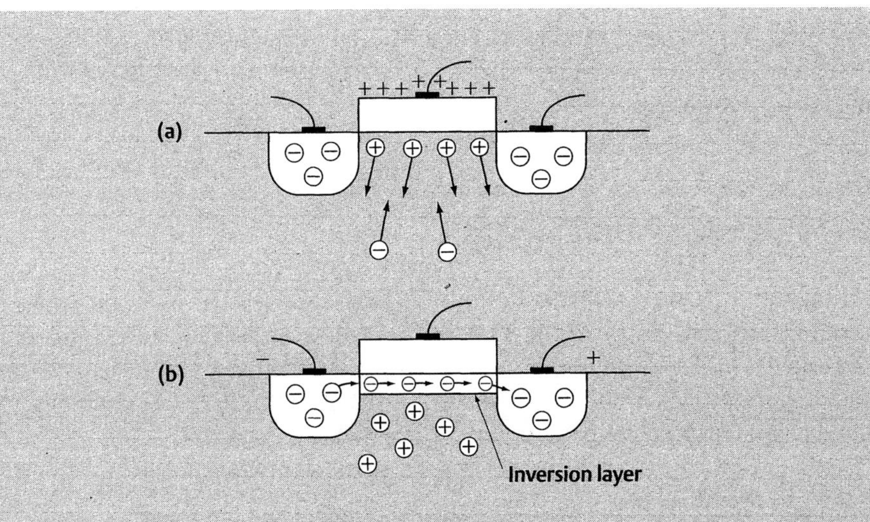

Figure 6.11 (a) When a positive voltage is applied to the gate the holes in the p-type semiconductor are repelled from the surface, and the minority carrier conduction electrons are attracted to the surface. (b) If the gate voltage exceeds the threshold value then an inversion layer is created near the surface. In this layer the material behaves as an n-type semiconductor and so provides a conducting channel between the source and the drain.

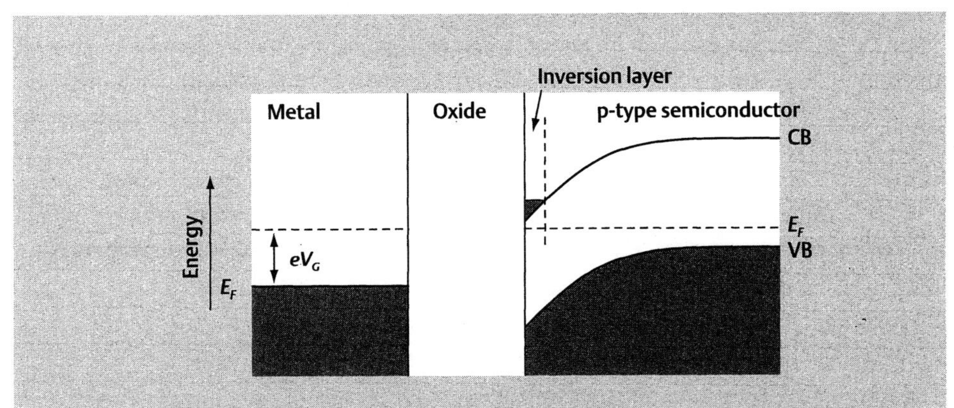

Figure 6.12 The energy band structure at the metal-oxide-semiconductor interface in a MOSFET. A gate voltage V_G raises the Fermi energy by eV_G. The inversion layer corresponds to the region in which the Fermi energy is closer to the conduction band than the valence band.

p-type MOSFET in which the current is carried by holes if we make the gate n-type and source and drain p-type. In general, n-type MOSFETs are preferred because the mobility of the conduction electrons in a semiconductor is much higher than the mobility of the holes; therefore n-channel devices can switch faster than p-channel ones. However, an interesting possibility occurs if we use a pair of transistors consisting of one n-channel and one p-channel MOSFET. This is referred to as complementary MOS or CMOS and has the advantage that the power consumption is extremely low, typically about 100 000 times smaller than that of a conventional MOSFET. CMOS circuits have therefore found uses in many battery-powered applications from portable computers to digital watches.

6.10 The integrated circuit

An integrated circuit is a small slice of semiconductor which contains transistors, capacitors, resistors and all of the metallic interconnects needed to form a complete electrical circuit. Over the last few decades the complexity of these circuits has increased enormously—from four transistors in 1961 to 64 million transistors by 1995. This extraordinary exponential growth rate is shown in Fig. 6.13.

Since the early 1970s there have been two main factors responsible for this exponential growth rate:

(1) the decrease in the size of the individual devices; and
(2) the increase in the area of the circuit.

The increase in the area of the circuit has been achieved by better technology, most noticeably by improving the quality of the semiconductor wafers that form the raw material on which the circuit is produced, and by improving the clean-room environment in which the circuits are manufactured.

From a physics point of view, the more interesting aspect is the decrease in size of the individual components, and so we will concentrate on this area. In particular, we are concerned with one question. How does the behaviour of a transistor alter as we make it smaller?

For MOSFET circuits, which are now by far the most common type of integrated circuit, the reduction in size can be dealt with by defining a scaling parameter, S. The law of constant field scaling, which is generally considered to be the ideal method of scaling, involves:

(1) reducing all linear dimensions of the devices by a factor $1/S$;
(2) reducing the voltages by a factor $1/S$; and
(3) increasing the doping concentrations by a factor S.

These scaling laws are illustrated in Fig. 6.14.

Let us investigate some of the consequences of applying these laws. Rule (1) suggests that the number of devices per unit area increases by a factor of S^2, so if $S = 2$ then we can fit four scaled devices in the area occupied by one unscaled device.

Since both the linear dimensions and the voltages are reduced by $1/S$, the electric field strength (which is simply the change in voltage per unit length) remains constant, as shown in Fig. 6.14(b).

Another consideration is the switching time. We can define the switching time of a transistor to be the minimum time period over which a transistor can be switched from OFF to ON and back to OFF again. The duration of time for which the transistor

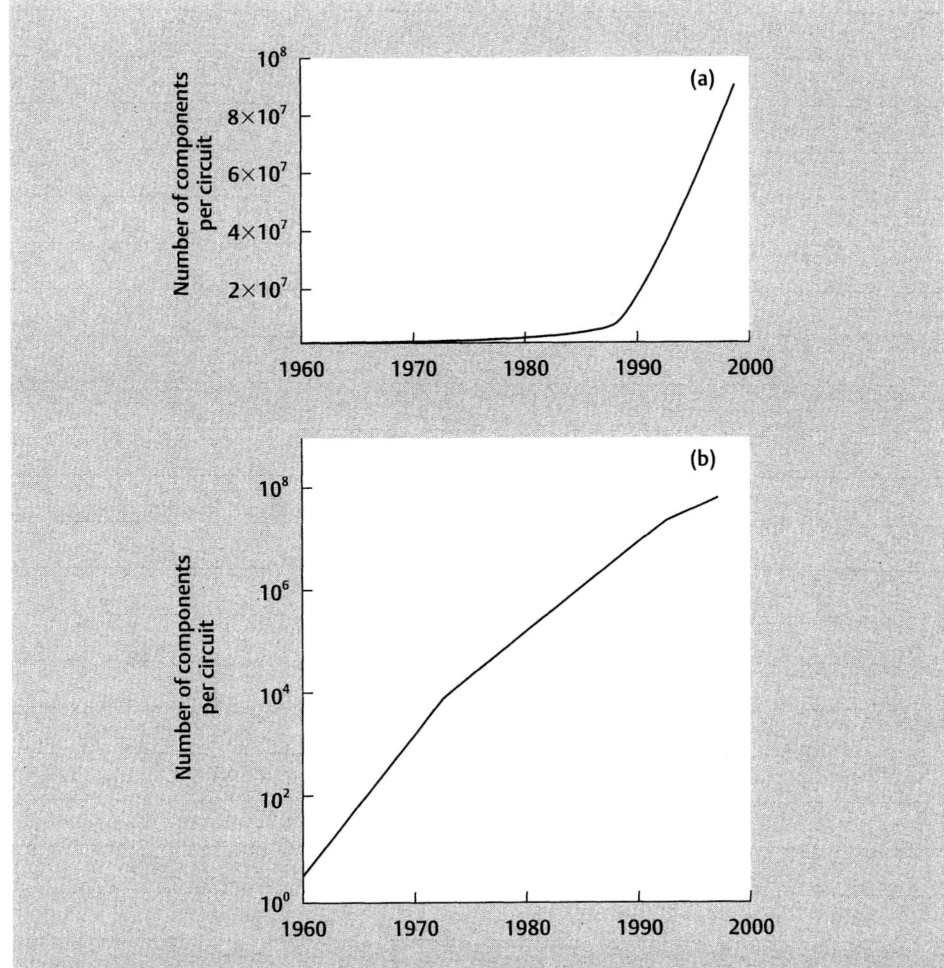

Figure 6.13 (a) Graph showing how the number of transistors on an integrated circuit has increased with time. (b) The same data plotted using a log scale on the *y* axis to show that the trend is exponential.

is in the ON state must be at least long enough to allow an electron (or hole) to pass from the source to the drain, otherwise no current will reach the drain and so we will never know that the device has been switched on momentarily. From Example 6.4 we can see that the switching time decreases as $1/S$, and so we have an added bonus that if the devices are smaller then the transistors also switch faster. This suggests that by making the individual transistors smaller we can build computers which operate at higher speeds.

The power dissipated by a transistor is also of great importance as at present the amount of heat produced per unit area by an integrated circuit is comparable with an element on an electric cooker. Any further increase in the amount of power produced

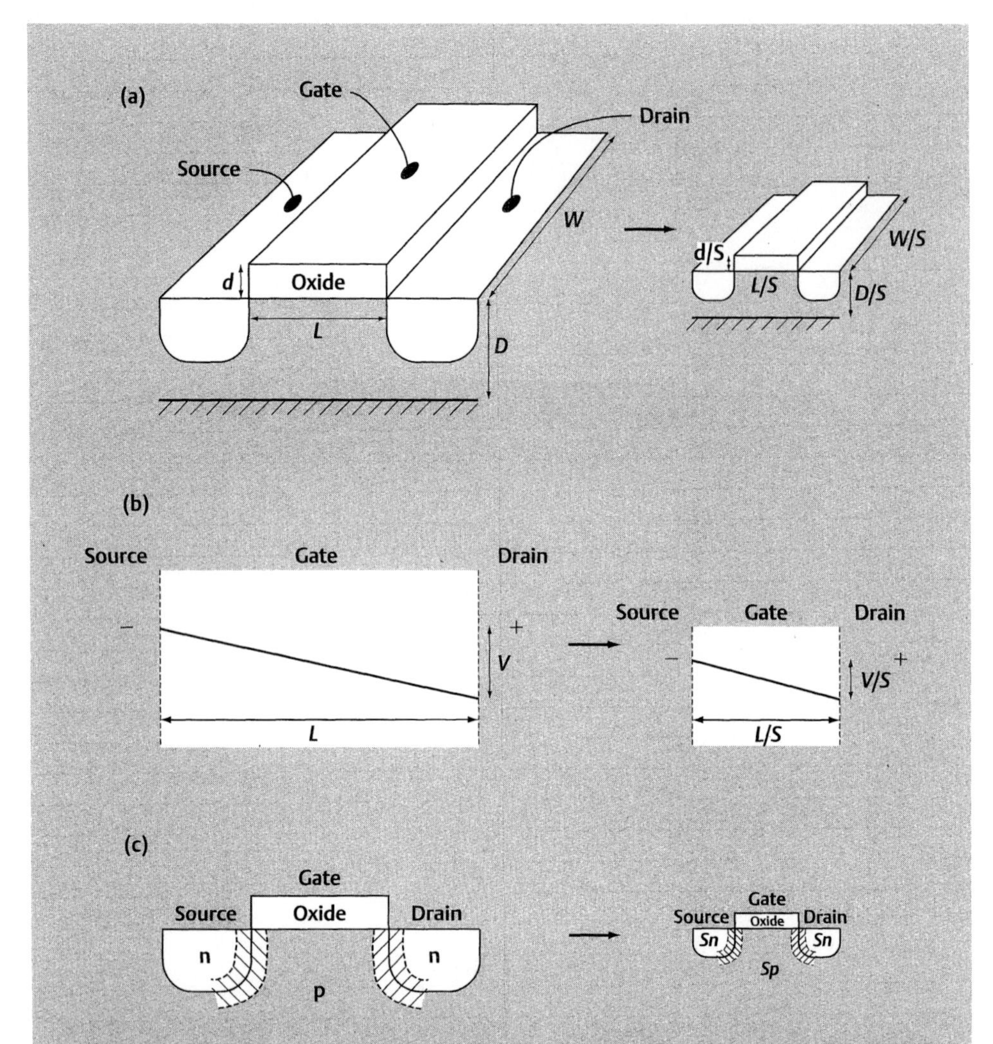

Figure 6.14 The effects of constant field scaling on a MOSFET using a scaling parameter S. (a) Scaling of linear dimensions, (b) scaling of the voltage so that the electric field is constant across the gate region, and (c) scaling of the doping concentration. (The shaded region indicates the depletion layers.)

Example 6.4 Determine how the switching time and power consumption per unit area of a MOSFET circuit are affected by applying the rules of constant field scaling listed above.

Solution The switching time, t, is determined by the average time taken for an electron to cross the gate region. Therefore, if the gate length is L and the drift velocity of the electrons in the gate is $\overline{\Delta v}$ then

$$t \propto \frac{L}{\overline{\Delta v}}$$

We know that scaling changes the linear dimensions by $1/S$, i.e.

$$L \Rightarrow \frac{L}{S}$$

To determine how the drift velocity scales we can use eqn (4.5), i.e.

$$\overline{\Delta v} = \frac{e\mathcal{E}\tau}{m}$$

Since the constant field scaling rules obviously ensure that the electric field, $\mathcal{E}$, remains constant with scaling, and if we assume that the relaxation time, τ, also remains constant with scaling, then we conclude that $\overline{\Delta v}$ is constant.

Therefore the switching time scales as

$$t \Rightarrow \frac{t}{S}$$

The power dissipated by an electrical device is given by

$$P = IV$$

We already know that the voltage scales by $1/S$, but we need to determine how the current scales.

From eqn (4.8) we can write

$$I = n\overline{\Delta v}eA$$

From the scaling laws we know that the doping concentration, n, goes to Sn.

We have already seen that the drift velocity, $\overline{\Delta v}$, is constant.

And since the linear dimensions scale by $1/S$, the cross-sectional area, A, must scale by $1/S^2$.

Therefore

$$I \Rightarrow Sn.\overline{\Delta v}.e.\frac{A}{S^2} \Rightarrow \frac{I}{S}$$

Therefore, the power dissipated per device is

$$P \Rightarrow \frac{V}{S}.\frac{I}{S} \Rightarrow \frac{P}{S^2}$$

However, since the number of devices per unit area increases by S^2, the power per unit area therefore remains constant.

would require even more elaborate cooling systems to be employed. Again we find that constant field scaling provides a satisfactory solution: the power per unit area remains constant as the devices are scaled down in size (see Example 6.4), so at least the problem does not get any worse. Unfortunately, it is not always feasible to scale the voltages at the same rate as the linear dimensions, and so in this case the problem of heat dissipation gets worse as the devices are scaled down in size (see Question 6.14).

Having considered the benefits of scaling, an important question is, how far can this scaling continue? In other words, what is the smallest possible size of a working transistor? Let us tackle this from a physics point of view and assume that any technological problems with actually fabricating such structures can be overcome. One of the main problems appears to be the depletion layers at the p–n junctions in the transistor. In particular, it is found that if we apply the above scaling rules, then the depletion layers do not scale down at the same rate as the other linear dimensions (see Question 6.16). Consequently, as the devices are made smaller, we reach a point at which the depletion regions from the source and drain extend across the whole of the gate region, as shown in Fig. 6.15. Once this situation is reached it becomes relatively easy for a conduction electron to pass from the source to the drain even when there is no gate voltage. Therefore as the devices are scaled down in size it becomes increasingly difficult to switch them off.

How close are we to this limit? Most estimates suggest that the minimum feasible gate length is about 0.07 μm. The microelectronics industries predict that they will be able to build such devices on a commercial scale by the year 2010. Does this mean that the exponential trend in Fig. 6.13 will end in 2010? Not necessarily, but it may be necessary to dispense with the p–n junction if we want to build devices smaller than this. We will consider an alternative possibility in the next section.

6.11 Heterojunctions

So far in this chapter we have discussed devices which rely on the properties of the interface between two regions with different doping. In contrast, a heterojunction is an interface between two different types of semiconductor.

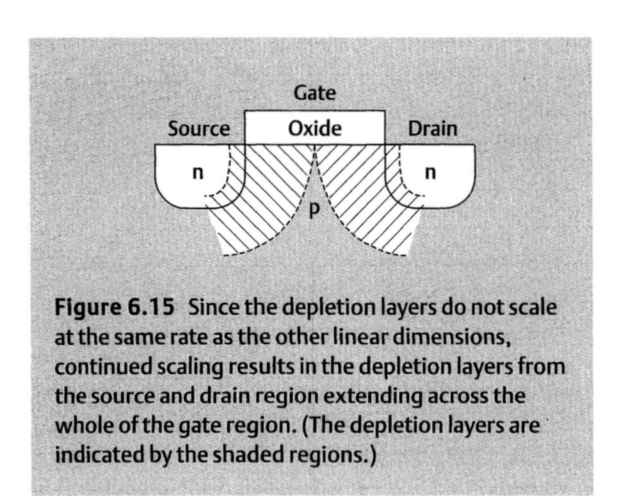

Figure 6.15 Since the depletion layers do not scale at the same rate as the other linear dimensions, continued scaling results in the depletion layers from the source and drain region extending across the whole of the gate region. (The depletion layers are indicated by the shaded regions.)

Let us take a look at a specific example of a heterojunction. A common choice of materials is gallium arsenide (GaAs) and aluminium gallium arsenide (AlGaAs). A crystal of aluminium gallium arsenide is very similar in structure to one of gallium arsenide except that some of the gallium sites are occupied by aluminium atoms. However, certain other properties of the material, such as the band gap, are dependent on the concentration of aluminium. So at a GaAs–AlGaAs heterojunction there is a change in the band gap over a distance of a few atomic spacings. For most practical purposes we can assume that this change in band gap is discontinuous and we refer to this discontinuity as the band offset. In the case of the GaAs–AlGaAs heterojunction, the bands line up so that the smaller band gap of GaAs lies entirely within the larger band gap of AlGaAs, as shown in Fig. 6.16(a). This is called a type I alignment. From the diagram we can see that the band gaps and band offsets are related by the equation

$$E_g(\text{AlGaAs}) = E_g(\text{GaAs}) + \Delta E_v + \Delta E_c \tag{6.21}$$

Unfortunately, there is no simple way of determining how the difference between the band gaps is divided between the valence and conduction band offsets. In fact, in some instances the bands are 'staggered' as shown in Fig. 6.16(b). This is known as a type II alignment. We can still apply eqn (6.21), but in this case the valence band offset is negative.

How can we exploit the properties of a heterojunction in a device? One application is the modulation-doped field-effect transistor, or MODFET. To see why the heterojunction is important in this instance, let us consider the factors that affect the switching speed of a transistor.

We have seen in the previous section that the switching speed is determined by the time taken for an electron or hole to cross the gate region of the transistor, so one way to make the devices operate faster is to reduce the length of the gate. However, an alternative approach is to increase the drift velocity of the carriers.

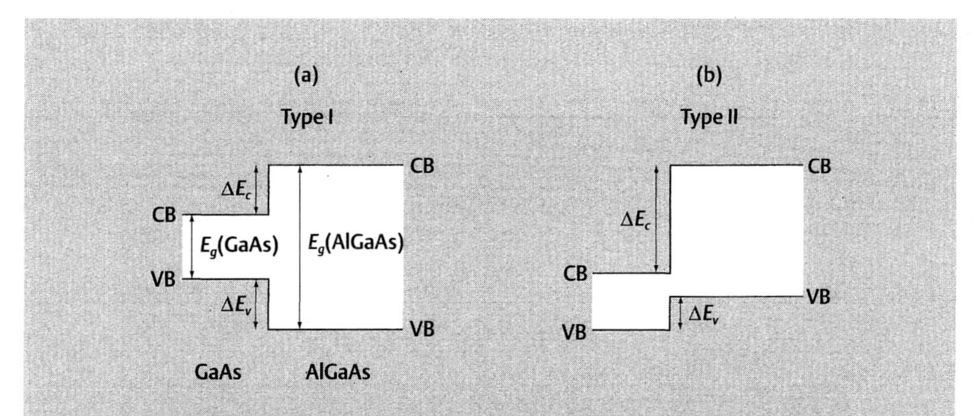

Figure 6.16 The band offsets (a) at a GaAs-AlGaAs heterojunction and (b) at a type II heterojunction. The conduction and valence band offsets are represented by ΔE_c and ΔE_v, respectively.

The drift velocity depends on the mean free path of the carriers, which in a semiconductor is determined by the same factors as in a metal (as discussed in Section 4.4). In other words, the main factors are the thermal vibrations of the atoms and the presence of impurities and imperfections in the crystal. At room temperature the thermal vibrations of the atoms are usually the dominant factor, and so the drift velocity can be increased by reducing the temperature. The limiting factor then becomes the impurities, particularly charged impurities such as the ionized donor and acceptor impurities.

It seems rather ironic that the impurities which provide the conduction electrons and holes are also the main culprit in reducing the drift velocity of the carriers. However, there is a way around this problem. The method uses a heterojunction and a technique known as modulation doping. To see how modulation doping works, let us again consider the example of a GaAs–AlGaAs heterojunction. In this instance we leave the GaAs layer undoped and place the donor impurities only in the AlGaAs layer. Although all of the conduction electrons start off initially in the AlGaAs region, some of them soon diffuse into the GaAs region. Once the electrons enter the lower energy conduction states in the GaAs, they no longer have sufficient energy to get back into the AlGaAs layer, as we can see from Fig. 6.17. Consequently, we end up with a high concentration of conduction electrons in the GaAs layer. Since the GaAs layer is undoped, there are no ionized impurities to affect the motion of the electrons, and so at low temperatures the drift velocity is much higher than in a doped semiconductor.

A MODFET is made by incorporating this modulation-doped heterojunction into the gate region of an FET, as shown in Fig. 6.18. Note that the principle of operation of this device differs slightly from that of a MOSFET because of the absence of the oxide layer. However, the important point is that a high-mobility channel in the undoped GaAs links the source and drain, and so the device can operate much faster than a conventional FET. In fact, one commercial application of MODFETs is as high-frequency amplifiers in satellite receivers.

It should be pointed out that the MODFET incorporates both p–n junctions and heterojunctions, and as in a conventional FET, the size is limited by the depletion layers

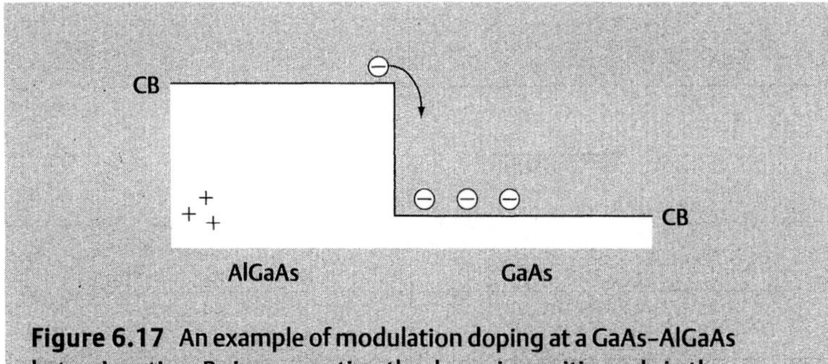

Figure 6.17 An example of modulation doping at a GaAs–AlGaAs heterojunction. By incorporating the donor impurities only in the AlGaAs layer, the conduction electrons accumulate in the high-mobility, lower energy levels in the GaAs region.

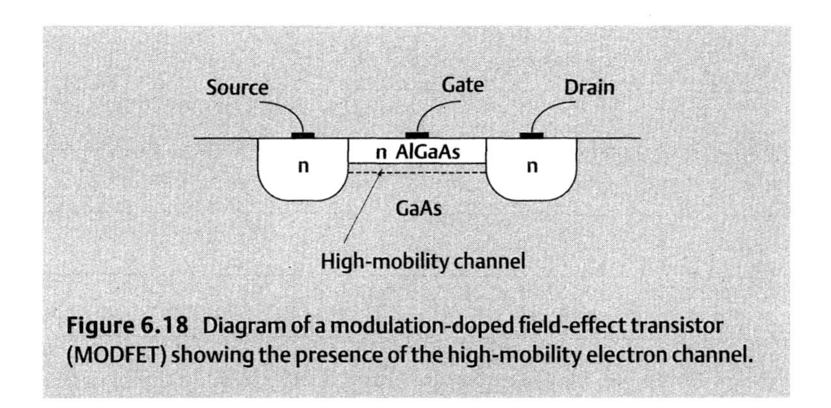

Figure 6.18 Diagram of a modulation-doped field-effect transistor (MODFET) showing the presence of the high-mobility electron channel.

of the p–n junctions. However, since the heterojunction extends over only a few atomic layers, in principle it is possible to make heterojunction devices with dimensions which are a hundred times smaller than devices based on p–n junctions. Such devices operate on very different principles to conventional transistors and there is not the space to consider them here. Instead I will refer you to the articles listed in Further reading.

6.12 Optoelectronic devices

Although we have concentrated in this chapter on electronic devices, we cannot leave this subject without saying a few words about the way in which the optical properties of semiconductors can be utilized.

As we have seen in Chapter 5, the energy of a photon can be used to excite an electron from the valence band into the conduction band, leading to an increase in conductivity, known as photoconductivity. We can therefore use this property to detect light. In practice a photodiode usually incorporates a reverse-biased p–n junction, as shown in Fig. 6.19. When a valence electron in the depletion region on the p-type side of the junction absorbs a photon of the appropriate frequency, the electron is excited into the conduction band where it becomes a minority carrier. It is then swept across into the n-type region by the strong electric field in the depletion layer and so contributes to the drift current. The magnitude of this current is therefore proportional to the intensity of the light. Since this arrangement converts light energy into an electrical current, it can also be used as the basis for a solar cell.

We know that another property of semiconductors is that they can emit light when a conduction electron recombines with a hole, and this process provides the basic mechanism for light-emitting diodes (LEDs) and semiconductor lasers. As we discussed in Section 5.5, it is only direct gap semiconductors, such as GaAs which are suitable for these purposes.

An LED consists simply of a forward-biased p–n junction made from a direct gap material. The wavelength of the emitted light depends on the band gap of the

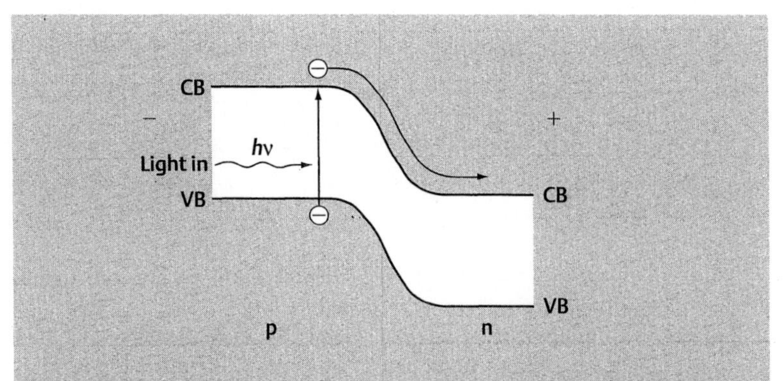

Figure 6.19 The principle of operation of a reverse-biased p–n junction as a photodetector. A valence electron on the p-type side of the junction (and in close proximity to the junction) is photoexcited into the conduction band. It can then take advantage of the lower energy conduction states on the n-type side of the junction and so contributes to the drift current across the junction.

semiconductor, so by using different materials it is possible to produce light with a wide range of colours. In fact, the alloy gallium arsenide phosphide (GaAsP) can emit light across most of the visible spectrum—from red to green—simply by varying the phosphorus content of the material (see Question 6.19). Producing LEDs which emit blue light has proved to be more of a problem until fairly recently, but such devices are now commercially available.

The choice of material not only affects the wavelength of the emitted light, but also determines the efficiency of the recombination process. We can quantify the efficiency as the amount of optical energy produced as a proportion of the amount of electrical energy that is input to the device, i.e.

$$\text{Efficiency} = \frac{\text{optical energy (output)}}{\text{electrical energy (input)}} \times 100\% \tag{6.22}$$

In some materials the efficiency is as high as 30%.

In an LED the photons are produced by spontaneous emission, i.e. the recombination of an electron and a hole is a random process which is independent of the recombination of other electrons and holes. In order to make a semiconductor laser we need to produce coherent light, which requires stimulated emission. This means that the recombination of one electron–hole pair sets off other similar events. As a result, all of the photons that are produced are in phase with one another. A primary requisite of achieving stimulated emission is that we have a population inversion. In other words, we need to have more electrons at the bottom edge of the conduction band than there are electrons at the top edge of the valence band. This situation can be achieved at a p–n junction, as shown in Fig. 6.20, but only if there are very high levels of doping. Unfortunately, the recombination events are spread out over a compara-

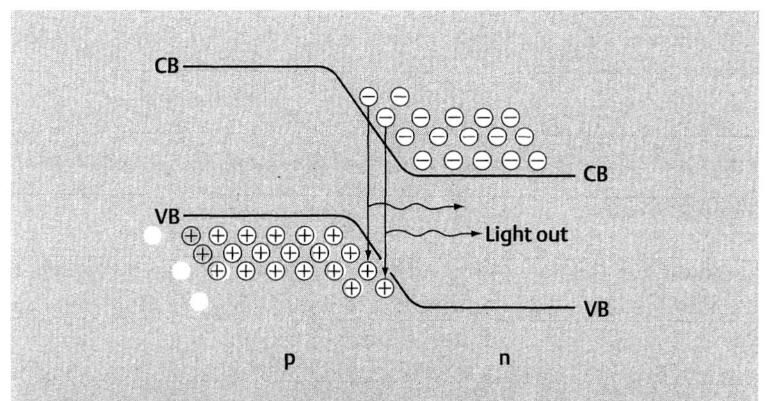

Figure 6.20 The energy band picture for a diode laser. A population inversion can be achieved in a narrow region at a heavily doped p–n junction, leading to the possibility of stimulated emission, as shown. However, the efficiency of such a system is very low because many of the other electron–hole pairs recombine spontaneously.

tively wide region, but the population inversion exists only near the middle of the depletion region. So a large proportion of these events involve spontaneous, rather than stimulated, emission. This means that the lasing process has a low efficiency, and a very large current must flow through the junction in order to replenish continually the supply of electrons and holes and so maintain the population inversion. In fact, the heat produced by this current is so large that it is usually only possible to operate the laser for short bursts or pulses.

A vast improvement in efficiency can be achieved by replacing the p–n junction with a heterostructure, or, to be precise, two heterostructures, as shown in Fig. 6.21. Since both the conduction electrons and holes collect in the same region, the

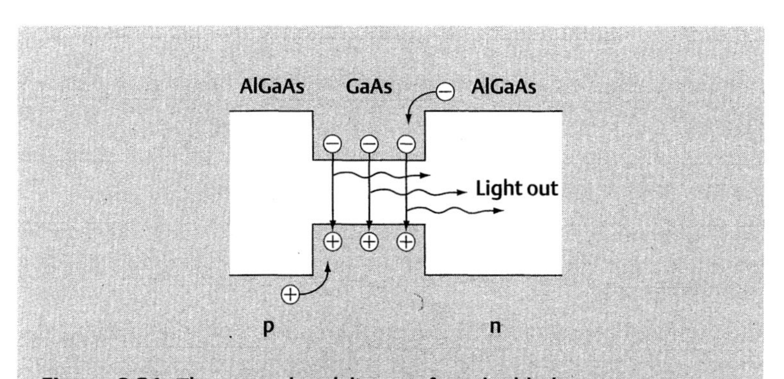

Figure 6.21 The energy band diagram for a double-heterostructure laser. The efficiency is much higher than in the diode laser because the conduction electrons and holes are confined in the GaAs layer and so stimulated emission occurs across all of this region.

proportion of events involving stimulated emission is greatly increased. Double-heterostructure lasers are already in use in compact disc players and in optical fibre telecommunication systems.

6.13 Summary

The basic electronic and optoelectronic properties of p–n junctions can be understood, on both a qualitative and a quantitative level, using the simple models of semiconductors that we introduced in Chapter 5.

On the other hand, transistors are far more complicated because they have three electrical connections rather than two. Consequently, a detailed treatment of the operation of a transistor depends on the circuit in which it is placed. A full mathematical treatment is therefore far beyond the scope of this book. Nevertheless, using our knowledge of doped semiconductors, we can at least understand how a transistor can be used as an amplifier or as a switch.

We have studied some of the factors which affect integrated circuits. Again we are dealing with very complex systems and have had to use simplified models, but our results demonstrate most of the important consequences of integration, and also suggest limits on how far the process of miniaturization can be continued.

Heterojunctions have been the subject of much study in research laboratories around the world for several decades, but until recently only a few commercial devices based on these structures have come to fruition. However, as the demand continues for ever smaller and faster devices, heterojunctions will become increasingly important in a wide range of electronic and optoelectronic applications. An understanding of the operation of these devices often relies more on quantum mechanical concepts—such as electron tunnelling and the wave properties of electrons—than on conventional semiconductor physics.

Questions

*Questions marked * indicate that they are based on the optional sections.*

6.1 p–n junctions—qualitative. By considering the diffusion of holes across a p–n junction, show that the result is the same as that obtained by considering the diffusion of electrons (as in Section 6.3).

6.2 p–n junctions—qualitative. Sketch the energy band profile and calculate the contact potential for a silicon p–n junction if the Fermi energy on the n-type side is 150 meV below the conduction band edge and the Fermi energy on the n-type side is 80 meV above the valence band edge. (The band gap of silicon is 1.11 eV.)

6.3 * p–n junctions—quantitative. (This question is a continuation of Example 6.1.) By further integration of Poisson's equation, show that the contact potential for the p–n junction is given by

$$\phi = \frac{e}{2\varepsilon\varepsilon_0}\left(N_d l_n^2 + N_a l_p^2\right)$$

6.4 ***p–n junctions—quantitative.** Using eqns (6.5) and (6.6) show that the widths of the depletion layer on the p- and n-type sides of a p–n junction are given by eqns (6.7) and (6.8), respectively.

6.5 ***p–n junctions—quantitative.** Show that the carrier concentrations at a p–n junction outside the depletion regions are given by

$$n_n = N_c e^{(E_F - E_g)/k_B T}$$

$$p_n = N_v e^{-E_F/k_B T}$$

$$p_p = N_v e^{(e\phi - E_F)/k_B T}$$

where N_v and N_c are constants and the quantities n_n, p_n and p_p are defined in Fig. 6.4.

6.6 *** p–n junctions—quantitative.** By manipulating eqns (6.9) to (6.12), show that the contact potential ϕ is given by

$$\phi = \frac{k_B T}{e}\ln\left(\frac{N_a N_d}{n_i^2}\right)$$

where N_a and N_d are the concentrations of acceptor and donor impurities, respectively, and n_i is the intrinsic carrier concentration. (Note: assume that all of the impurities are ionized, i.e. $n_n = N_d$ and $p_p = N_a$.)

6.7 ***p–n junctions—quantitative.** A silicon p–n junction has doping levels of $N_d = 2 \times 10^{22}$ m^{-3} and $N_a = 4 \times 10^{21}$ m^{-3}. Determine the values of the contact potential, ϕ, and the depletion layer widths, l_n and l_p, at 300 K. (For silicon the dielectric constant, ε, is 11.7 and the intrinsic carrier concentration, n_i, is 1.0×10^{16} m^{-3} at 300 K.)

6.8 ***p–n junction with an applied voltage.** Show that when a reverse bias V is applied to a p–n junction the net current is given by

$$I_{diffusion} - I_{drift} = I_0\left(e^{-eV/k_B T} - 1\right)$$

where I_0 is given by eqn. (6.15).

6.9 ***p–n junction with an applied voltage.** Use eqns (6.17) and (6.19) to plot the variation of current with applied voltage for a p–n junction.

6.10 ***p–n junction with an applied voltage.** When a reverse bias of 0.3 V is applied to a silicon p–n junction at 300 K the current through the junction is found to be 50×10^{-9} A. Use this information to determine the current when a forward bias of 0.3 V is applied.

6.11 *** p–n junction with an applied voltage.** For the same diode as in Question 6.10, estimate the current for a reverse bias of 0.3 V and for a forward bias of 0.3 V when the temperature is increased to 310 K. (Hint: you may assume that the value of $E_g + e\phi - E_F$ is approximately equal to 1 eV.)

6.12 **Bipolar transistors.** Estimate the current gain of an n–p–n bipolar transistor if only 0.3% of the conduction electrons that enter the base from the emitter recombine in the base region.

6.13 **Integrated circuits.** In 1995 MOSFET circuits containing 64 million devices with a minimum gate length of 0.35 μm came into production. Assuming that the minimum feature size decreases by 11% per year, determine when circuits with a gate length of 0.1 μm will become available. Estimate the corresponding number of devices per circuit at this date assuming that the area of the circuit increases by 9% per year.

6.14 Integrated circuits. If a MOSFET circuit is scaled so that the linear dimensions are changed by $1/S$, the doping concentrations are changed by S, but the voltages remain constant, determine how this affects the switching time of the devices and power produced per unit area. (Note that the electric field does not remain constant in this case.)

6.15 Integrated circuits. If the dimensions of a MOSFET are scaled down by $1/S$, the doping concentration should be increased by S^3 in order to provide the same number of carriers in the source, gate and drain regions. Consider the disadvantages of applying this type of scaling. (Assume that the voltages are scaled by $1/S$.)

6.16 *Integrated circuits. A typical n-channel MOSFET with a gate length of 1.0 μm has acceptor and donor doping concentrations of $2.5 \times 10^{22}\ \text{m}^{-3}$ and $5.0 \times 10^{23}\ \text{m}^{-3}$ in the gate and source/drain regions, respectively. Determine the widths of the depletion layers in the gate region at 300 K. Assuming that the doping concentration is scaled by S as the device dimensions are scaled by $1/S$ (and assuming that the contact potential is approximately constant with scaling), determine the minimum gate width that can be achieved, i.e. such that the depletion layers from the source and drain extend across the whole of the gate region. (Use a dielectric constant, ε, of 11.7 and assume that the intrinsic carrier concentration, n_i, is $1.0 \times 10^{16}\ \text{m}^{-3}$ at 300 K.)

6.17 Heterojunctions. (i) If the band gap of GaAs is 1.42 eV, the band gap of AlGaAs containing 30% Al is 1.80 eV, and the valence band offset is 0.14 eV, determine the magnitude of the conduction band offset. (ii) The band gaps of InAs and AlSb are 0.36 eV and 1.58 eV, respectively. If the conduction band offset is 1.35 eV (i.e. the conduction band edge in AlSb is 1.35 eV higher than in InAs), determine the magnitude of the valence band offset. In each case sketch the alignment of the bands at the heterojunction and identify whether the junction is type I or type II.

6.18 Optoelectronic devices. Using the results obtained from Questions 6.10 and 6.11 explain why photodiodes use a reverse-biased p–n junction rather than a forward-biased junction. Assume that a typical photocurrent in a semiconductor photodiode is about 0.1 mA.

6.19 Optoelectronic devices. The band gap of the alloy $GaAs_{1-x}P_x$ (where x is the proportion of P ions) is given approximately by

$$E_g = (1.42 + 1.3x)\ eV$$

(i) Calculate the proportion of P required to produce a material which emits red light with a wavelength of 680 nm.
(ii) Determine the minimum wavelength of light that can be produced with this alloy given that $GaAs_{1-x}P_x$ is an indirect semiconductor for $x > 0.44$.

Chapter 7
Thermal properties

7.1 Introduction

What happens on an atomic scale when we change the temperature of a solid?

The temperature is a measure of the internal energy of the solid, i.e. the total kinetic and potential energy of the constituent ions and electrons. When the temperature is raised, the internal energy can change in two ways.

Firstly, we know that the ions in a solid vibrate about their equilibrium positions. Raising the temperature causes an increase in the amplitude of these vibrations, which therefore increases the internal energy of the solid.

Secondly, the internal energy can be increased if the electrons move to higher energy states. The probability of finding an electron in a state with energy E at temperature T is given by the Fermi–Dirac distribution, and is shown graphically in Fig. 4.18. In a metal the electrons near the Fermi energy can easily move to higher energy vacant states, and so the number of electrons with energies greater than the Fermi energy increases with temperature, leading to an increase in the total kinetic energy of the electrons. In contrast, in semiconductors and insulators only a very small number of electrons are excited to higher energy states because of the presence of the band gap, so the change in internal energy due to the electrons in these materials is negligible.

We can compare the capabilities of different materials to absorb heat by measuring the heat capacity, the amount of energy required to raise the temperature of the sample by 1 degree. From the above discussion we can see that in semiconductors and insulators the heat capacity is almost entirely due to the vibrations of the ions, whereas in metals we expect that the electrons also contribute. (However, we shall see in Section 7.7 that even in a metal the contribution to the heat capacity from the electrons is small in comparison with that due to the vibrating ions.)

Other thermal properties that we will look at in this chapter are thermal conductivity and thermal expansion. As we shall see, the vibrational motion of the atoms or ions is important in understanding all of these properties, so we will begin by taking a quantitative look at these thermal vibrations.

7.2 Thermal vibrations of the atoms

We know that the atoms in a solid vibrate as a result of their thermal energy, but what are the frequency and amplitude of these vibrations?

To answer this question, let us consider a one-dimensional chain of atoms in which the spacing between adjacent atoms is equal to a_0, as shown in Fig. 7.1(a). The atoms are connected by chemical bonds, as we have seen in Chapter 1, but in this case we are not interested in the details of this bond—all that we need to know is that the bond produces a net force between the atoms which varies as shown in Fig. 7.2. We can see from this diagram that if we decrease the separation between the atoms so that it is less than the equilibrium separation a_0, then we encounter a repulsive force which tends to push the atoms apart. Similarly, if we pull the atoms apart, there is a net attractive force which pushes them back together again. So if we displace atom B by

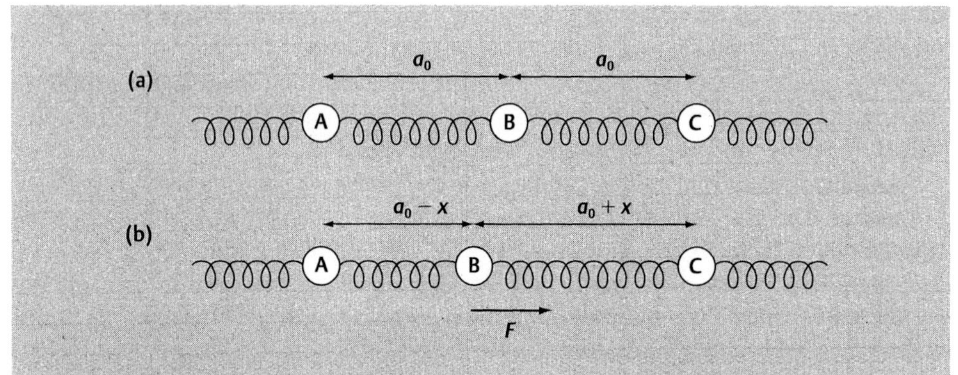

Figure 7.1 (a) A one-dimensional chain of atoms. The separation between each pair of atoms is equal to a_0. (b) If atom B is displaced, there is a net force F pushing it back towards the equilibrium position.

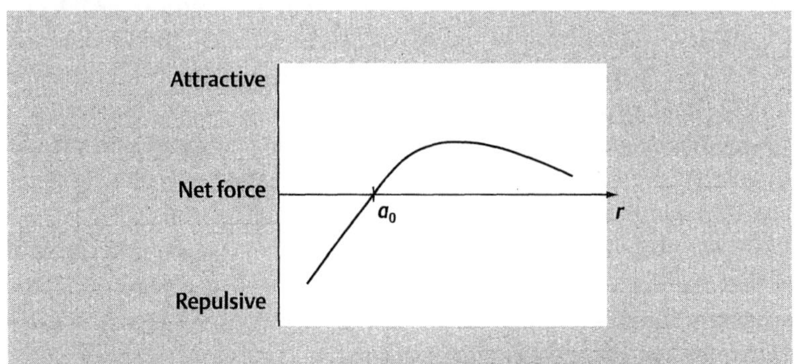

Figure 7.2 A typical plot of the net force between two atoms as a function of the separation, r, between the atoms. The equilibrium separation, a_0, corresponds to the point at which the net force is zero.

moving it a distance x towards atom A, as in Fig. 7.1(b), then we find that there is a repulsive force between A and B and an attractive force between A and C. Therefore, these forces tend to cause atom B to move back towards the equilibrium position.

In order to develop a mathematical treatment of this system, let us first consider a similar scenario on a macroscopic scale. For example, suppose we have a block of mass m held by two identical stretched springs of length l, as shown in Fig. 7.3. When we displace the block from the equilibrium position it executes simple harmonic motion, i.e. the acceleration of the block at any instant is proportional to the magnitude of the displacement, x, and is directed towards the equilibrium position. So we can write the equation of motion of the block as

$$m\frac{d^2x}{dt^2} = -\gamma x \tag{7.1}$$

where γ is the spring constant. The solution of this equation is straightforward and we can show (see Question 7.1) that the angular frequency of the oscillation is

$$\omega = \sqrt{\frac{\gamma}{m}} \tag{7.2}$$

We can apply the same model to the vibrating atoms if we assume that the chemical bonds behave rather like the springs in the above example. The main problem is, how do we determine a suitable value for the force constant, γ, in the atomic model? If we consider the strain that is produced when a force F is applied to a crystal, we find (see Example 7.1) that the force constant is approximately equal to the product of Young's modulus, Y, and the atomic spacing a_0, i.e.

$$\gamma \approx Ya_0 \tag{7.3}$$

Using this equation we obtain values of γ ranging from about 5 N m^{-1} for lead to about 140 N m^{-1} for diamond (see Question 7.2). The corresponding angular frequencies are about 4×10^{12} rad s^{-1} for lead and 8×10^{13} rad s^{-1} for diamond. These figures are a little on the low side compared with experimental measurements, but they demonstrate an important qualitative result: atoms of low mass which are connected by strong

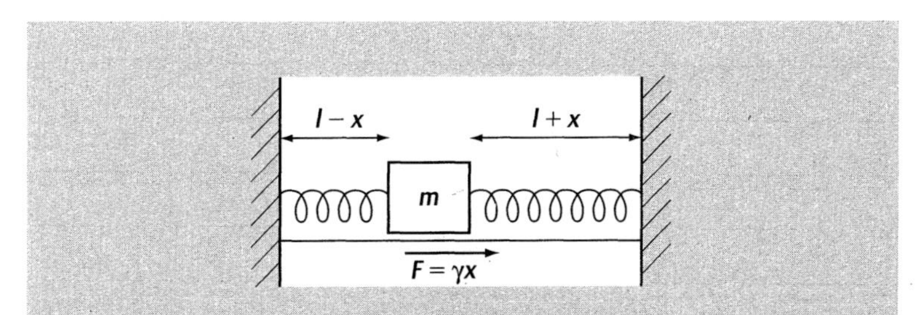

Figure 7.3 A block of mass *m* held by two identical stretched springs of length *l* executes simple harmonic motion when the position of the block is initially displaced from the equilibrium position.

bonds vibrate rapidly, whereas atoms of high mass connected by weak bonds vibrate comparatively slowly.

Example 7.1 By considering the effect of a force F on a one-dimensional chain of atoms, as shown in Fig. 7.4, show that the force constant is given by

$$\gamma = Ya_0$$

where Y is Young's modulus of the material.

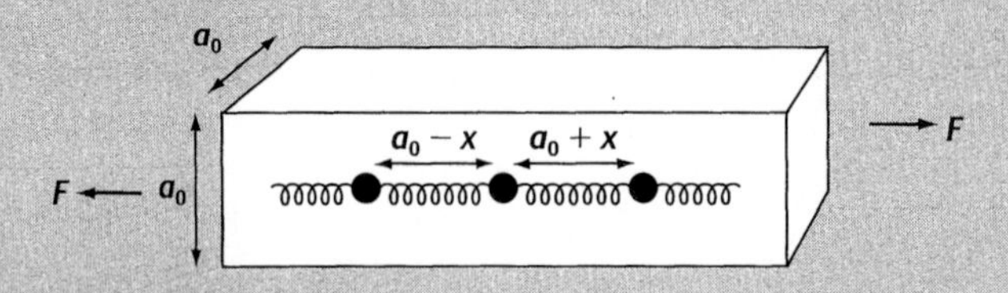

Figure 7.4 Diagram indicating the effect of applying a force F to a one-dimensional chain of atoms.

Solution If the extension produced between each pair of atoms is x, then the strain is x/a_0.

If we assume that the chain of atoms represents a crystal with cross-sectional area a_0^2, then the stress is

$$\sigma = \frac{F}{a_0^2}$$

Since we can express the force as (mass) × (acceleration), from eqn (7.1) we can write $F = \gamma x$, so

$$\sigma = \frac{\gamma x}{a_0^2}$$

Young's modulus is given by

$$Y = \frac{\text{stress}}{\text{strain}} = \frac{\gamma x}{a_0^2}\frac{a_0}{x} = \frac{\gamma}{a_0}$$

Therefore

$$\gamma = Ya_0$$

We can also use the simple harmonic oscillator model to obtain an expression for the amplitude, x_{max}, of the vibration of the atoms. Since the average thermal energy of an atom at temperature T is equal to k_BT, we find (see Question 7.3) that the amplitude of vibration is

$$x_{max} = \left(\frac{2k_B T}{\gamma}\right)^{1/2} \tag{7.4}$$

Consequently, for most metals at room temperature we find that the amplitude of the thermal vibration of the ions is of the order of a few per cent of the equilibrium separation of the ions (see Question 7.4).

7.3 Thermal expansion

It is a common experience that most solids expand when heated. If we are dealing with a rod or bar then the magnitude of this effect is measured by the coefficient of linear expansion, α. This coefficient is defined as the fractional change in length for unit change in temperature. In other words, if a bar which is initially of length L_0 changes in length by ΔL as the temperature is increased by ΔT, then α is given by

$$\alpha = \frac{\Delta L}{L_0}\frac{1}{\Delta T} \tag{7.5}$$

(We can also define a coefficient of volume expansion, β, in a similar way, but in this discussion we will limit our attention to a one-dimensional case, so the linear coefficient is all that is required.)

Although the values of α are very small—typically corresponding to a fractional change in length of about one-hundred-thousandth per kelvin for most solids (see Table 7.1)—if a metal bar is held firmly in place at both ends the tensile stress produced by a temperature change of a few tens of degrees can be sufficient to distort the

Table 7.1 Values of the coefficient of linear expansion, α, for a selection of solids. Note that invar is an alloy of iron and nickel especially designed to produce a small coefficient of linear expansion.

	α ($\times 10^{-5}$ K^{-1})
Invar	0.09
Glass	0.9
Steel	1.1
Copper	1.7
Brass	1.9
Aluminium	2.4
Lead	2.9

bar (see Question 7.5). This means that expansion joints are necessary in bridges, railway lines and roads to stop the structure from buckling on hot days. Thermal expansion can also be used to perform useful functions—for example, in the bimetallic strip the different thermal expansion coefficients of two metals cause the strip to change shape as the temperature is altered (see Fig. 7.5). This arrangement is often used to make a temperature-sensitive switch.

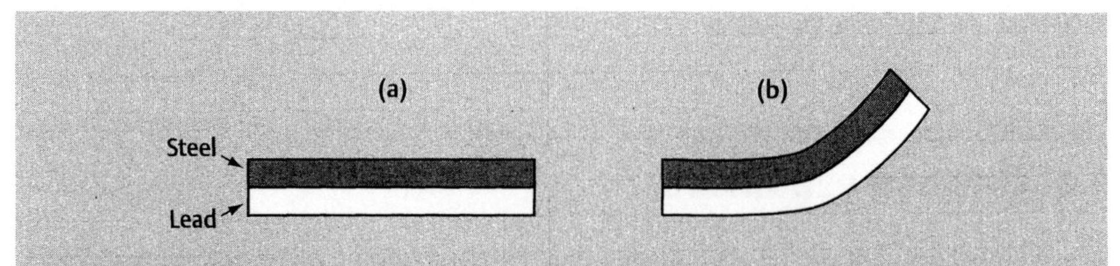

Figure 7.5 A bimetallic strip is formed from two metals with different linear coefficients of expansion. (a) At a particular temperature the two metal strips are of equal length. (b) At a higher temperature the lead strip expands more than the steel strip and so the strips are distorted.

How do we explain thermal expansion on an atomic scale? It seems obvious that a change in the dimensions on a macroscopic scale must be directly related to a change in the interatomic spacing, and this is confirmed by X-ray measurements (see Question 7.7). To see the origin of this change on an atomic scale we need to examine the potential energy curve introduced in Chapter 1. At absolute zero we can assume that all of the atoms have the lowest possible energy, E_{min}, corresponding to the equilibrium separation a_0, as shown in Fig. 7.6(a). At a temperature T, the average energy of the atoms is increased by k_BT and so we can see from the figure that the separation of the atoms varies between a_1 and a_2—in other words, the atoms oscillate between these two extremes. Since the potential energy curve is asymmetric (i.e. it is harder to push the atoms closer together than it is to pull them further apart), we find that the centre of oscillation is at a separation which is greater than a_0. Therefore as the temperature increases, the equilibrium separation a_T increases, as shown in Fig. 7.6(b). (In fact, by using a suitable form for the potential energy curve we can obtain an expression for α from first principles which is in good agreement with experimental values. A derivation is given in Flowers and Mendoza—see Further reading.)

It is interesting to note that although the ions in a crystal are often treated as simple harmonic oscillators (e.g. see Section 7.2), if we consider the thermal expansion of a harmonic oscillator we can see from Fig. 7.6(c) that the distance $a_2 - a_0$ is exactly the same as $a_0 - a_1$. This means that the equilibrium separation remains at a_0 for all temperatures, i.e. the coefficient of linear expansion of a harmonic oscillator is zero! The inclusion of anharmonic terms in the potential is therefore very important when calculating the thermal expansion of a material.

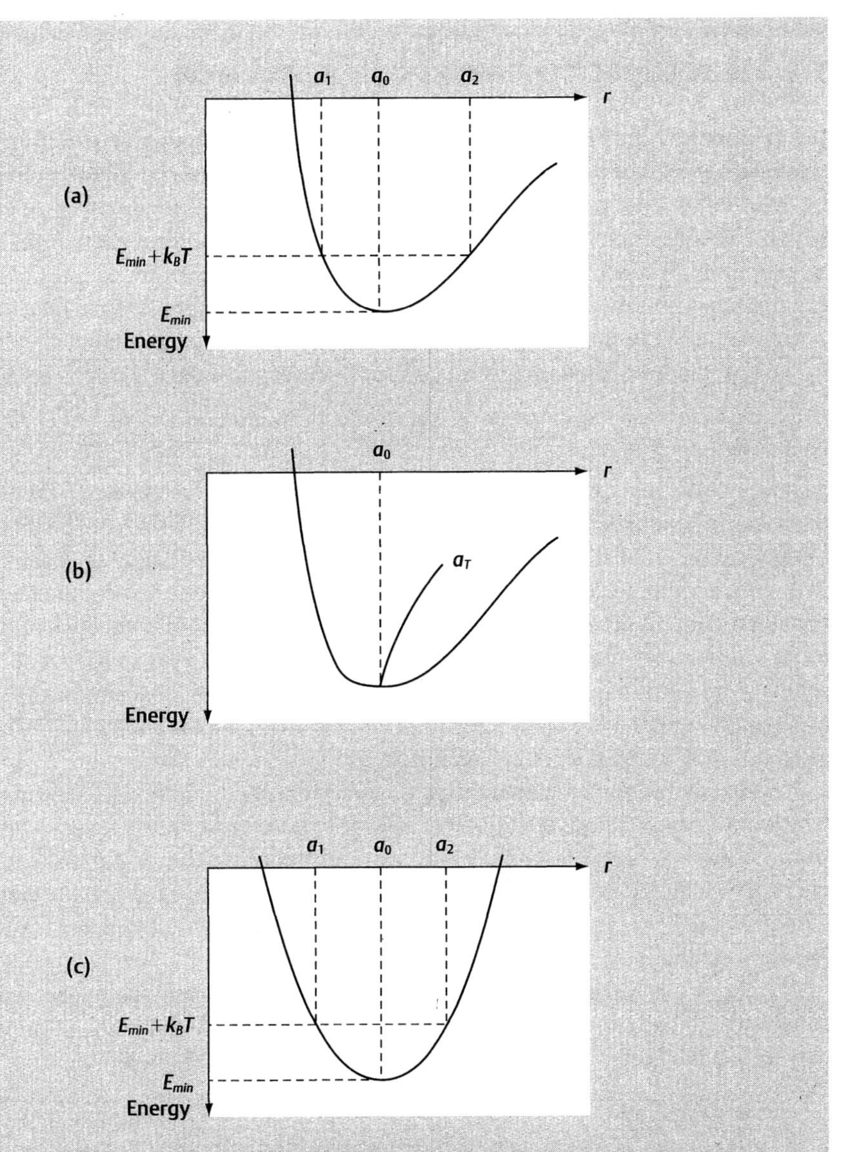

Figure 7.6 Thermal expansion can be explained using a typical potential energy curve. (a) At $T = 0$ K, the equilibrium separation, a_0, corresponds to the minimum energy, E_{min}. At finite temperatures the average energy is $E_{min} + k_BT$, corresponding to an atomic separation of between a_1 and a_2. (b) Since $a_2 - a_0$ is greater than $a_0 - a_1$, the equilibrium separation a_T increases as the temperature increases. (c) For a harmonic potential the value of $a_2 - a_0$ is exactly equal to $a_0 - a_1$, so the equilibrium separation is equal to a_0 for all temperatures.

7.4 Heat capacity—a classical approach

Let us now turn our attention to the specific heat capacity, the amount of energy required to raise the temperature of a given amount of material by 1 K. In this chapter we will define the 'given amount' to be 1 mole, or rather a quantity of substance which contains Avogadro's number (N_A) atoms. Of course, for a monatomic material this is precisely equal to 1 mole, but for a compound the distinction is important. For example, in a crystal of NaCl it is usual to quote the heat capacity for a quantity which contains $N_A/2$ Na ions and $N_A/2$ Cl ions. (This corresponds to 0.5 moles of NaCl.)

The specific heat capacity C is related to the internal energy U by the equation

$$C = \frac{\partial U}{\partial T} \tag{7.6}$$

So if we obtain an expression for the internal energy of a crystal we can then calculate the specific heat capacity using the above equation. We have seen in the introduction to this chapter that in semiconductors and insulators the change in internal energy with temperature is due entirely to the vibrations of the ions. In metals the energy of the electrons also increases with temperature, but from Example 4.6 we know that at room temperature only a few per cent of the valence electrons are thermally excited, so this contribution is likely to be small. Therefore for the time being we will concentrate solely on the heat capacity due to the vibrating ions. (The contribution of the electrons will be considered in Section 7.7.)

How do we determine the vibrational energy of the ions? We will use an argument which you may be familiar with from the kinetic theory of ideal gases—the equipartition theorem. According to the equipartition theorem the amount of energy associated with each degree of freedom of an atom is equal to $\frac{1}{2}k_BT$. (If you are not familiar with this result you can consult one of the introductory general physics texts listed in Further reading.)

How do we apply the equipartition theorem to the vibrating atoms? Let us begin by considering a classical one-dimensional simple harmonic oscillator, such as the system shown in Fig. 7.3. The energy of this system can be written as

$$E = \frac{1}{2} m v_x^2 + \frac{1}{2} \gamma x^2 \tag{7.7}$$

where γ is the force constant (see Section 7.2). The first term on the right hand side of this equation is the kinetic energy and the second term is the potential energy. Each of these quantities represents a degree of freedom of the system, so a one-dimensional oscillator has two degrees of freedom. If we extend this discussion to a three-dimensional oscillator we can see that there are six degrees of freedom in this case (corresponding to kinetic energies $\frac{1}{2}mv_y^2$ and $\frac{1}{2}mv_z^2$ and potential energies $\frac{1}{2}\gamma y^2$ and $\frac{1}{2}\gamma z^2$). Therefore from Example 7.2 we can see that the specific heat capacity of 1 mole (or rather N_A ions) is

$$C = 3R \tag{7.8}$$

where R is the molar gas constant.

Example 7.2 Assuming that each ion in a crystal has six degrees of freedom, each with an average energy of $\frac{1}{2}k_BT$, show that the specific heat capacity for 1 mole of a monovalent solid is $C = 3R$.

Solution The average energy per ion is $6 \cdot \frac{1}{2}k_BT = 3\,k_BT$.
For N_A ions the internal energy is therefore

$$U = N_A . 3 k_B T = 3RT$$

where R is the molar gas constant (which is the product of Avogadro's constant with Boltzmann's constant).
The molar specific heat capacity is therefore

$$C = \frac{\partial U}{\partial T} = 3R$$

So our classical model suggests that for all solids the specific heat capacity of a quantity of material containing N_A atoms is equal to a constant. This is known as the Dulong–Petit law. How does this result compare with experiment? By examining the values in Table 7.2 we can see that the Dulong–Petit law appears to be approximately true for a number of materials at room temperature. However, at 77 K there is a much larger discrepancy. In order to understand why the specific heat capacity is reduced at low temperatures we need to use a quantum theory approach.

Table 7.2 Values of the specific heat capacity at 77 K and 273 K for a quantity of material containing N_A ions. (Note that this corresponds to 1 mole of substance for all of the materials except glass and sodium chloride.) The classical value of the specific heat capacity, $3R$, is equal to 24.9 J K^{-1}.

	77 K (J K^{-1})	273 K (J K^{-1})
Copper	12.5	24.3
Aluminium	9.1	23.8
Gold	19.1	25.2
Lead	23.6	26.7
Iron	8.1	24.8
Sodium	20.4	27.6
Silicon	5.8	21.8
Glass	~4.0	~15.0
Sodium chloride	14.0	24.6
Diamond	0.1	5.0

7.5 Heat capacity—the Einstein model

The first person to apply quantum theory to determine the specific heat capacity of a solid was Albert Einstein. Although this is not one of his better known contributions to physics, it was considered sufficiently important to be cited—along with the photoelectric effect—in his Nobel prize award in 1921. (His theories of relativity were judged to be too controversial to be included!)

As in the classical approach Einstein treated the vibrating atoms as simple harmonic oscillators. In order to simplify matters we will begin by considering an oscillator which vibrates in just one dimension

First of all we need to determine the energies of these oscillators.

According to quantum theory, the energy of a one-dimensional simple harmonic oscillator can only take certain discrete values. These energies are given by

$$E_n = \left(n + \frac{1}{2}\right)\hbar\omega \tag{7.9}$$

where ω is the angular frequency of vibration and n is an integer which takes values $n = 0,1,2,3,\ldots$. (This is a result which is proved in virtually all introductory quantum theory textbooks. However, the derivation requires a considerable amount of mathematical dexterity, so if you are not familiar with this result I have to ask you simply to accept it.)

Next we need to know how many of the oscillators have a particular energy E_n.

At a temperature of 0 K we expect all of the atoms to be in the lowest energy state with $n = 0$. Note that even in this state the energy of the oscillator is not zero but $\frac{1}{2}\hbar\omega$; therefore quantum theory shows that the atoms are not completely stationary even at absolute zero. This term is known as the zero-point energy. To make matters simpler in the following discussion we will measure the energies relative to the zero-point energy. The energies are then simply

$$E_n = n\hbar\omega \tag{7.10}$$

(Note that this is not an approximation—we are simply changing the origin of the energy scale. We can convert our answers back to the original energy scale at any stage simply by adding the term $\frac{1}{2}\hbar\omega$).

At temperatures greater than 0 K some of the atoms occupy higher energy states (with $n = 1, 2, 3$, etc.). According to Boltzmann's distribution, the probability of a particle being in a state with energy E at temperature T is proportional to e^{-E/k_BT}, so the probability, p_n, of an atom occupying the state with energy E_n at temperature T is given by

$$p_n = Ae^{-n\hbar\omega/k_BT} \tag{7.11}$$

where A is a constant.

By using the above equations we can show that the mean energy of these one-dimensional oscillators at temperature T is given by

$$\langle E \rangle = \frac{\hbar\omega}{e^{\hbar\omega/k_B T} - 1} \tag{7.12}$$

A proof of this statement is given in Example 7.3, but it is not particularly straightforward. If you prefer you can simply accept that eqn (7.12) is correct without going through the details.

Following our earlier discussion, we could now add the zero-point energy, $\frac{1}{2}\hbar\omega$, back into this expression. However, since the zero-point energy is not dependent on temperature, it will go to zero when it is differentiated. So whether or not we include the zero-point energy is rather academic as it does not affect the heat capacity.

Having considered the energy of a one-dimensional oscillator, the three-dimensional problem may seem rather daunting. However, since vibrations in perpendicular directions are independent of one another, we can treat the three-dimensional oscillations of an atom by simply considering three separate one-dimensional oscillators. So the internal energy of 1 mole of a monatomic crystal is equal to the energy of $3N_A$ one-dimensional oscillators, i.e.

$$U = 3N_A < E > = 3N_A \frac{\hbar\omega}{e^{\hbar\omega/k_B T} - 1} \tag{7.13}$$

Example 7.3 Show that the mean energy of a system of one-dimensional simple harmonic oscillators at temperature T is given by

$$\langle E \rangle = \frac{\hbar\omega}{e^{\hbar\omega/k_B T} - 1}$$

Solution The mean energy is obtained by summing over all the possible energies E_n multiplied by the probability of an atom occupying that particular state, p_n. Therefore from eqn (7.11),

$$\langle E \rangle = \sum_n E_n p_n = \sum_n E_n A e^{-n\hbar\omega/k_B T}$$

Substituting from eqn (7.10) for E_n gives

$$\langle E \rangle = A \sum_n n\hbar\omega e^{-n\hbar\omega/k_B T}$$

In order to evaluate the expression for the mean energy we need to know the value of the constant A. We can do this because we know that the sum of all of the probabilities must be equal to 1.0 (i.e. the atom must be in one or other of these states), so

$$A \sum_n e^{-n\hbar\omega/k_B T} = 1.0$$

Rearranging this expression we get

$$A = \frac{1.0}{\sum_n e^{-n\hbar\omega/k_B T}}$$

and substituting into the expression for the mean energy gives

$$\langle E \rangle = \frac{\sum_n n\hbar\omega e^{-n\hbar\omega/k_B T}}{\sum_n e^{-n\hbar\omega/k_B T}}$$

To proceed we need to use the mathematical identity

$$\frac{\sum_n n e^{-nx}}{\sum_n e^{-nx}} = -\frac{d}{dx}\left[\log\left(\sum_n e^{-nx}\right)\right]$$

This gives

$$\langle E \rangle = -\hbar\omega \frac{d}{dx}\left[\log\left(\sum_n e^{-nx}\right)\right]$$

where $x = \hbar\omega/k_B T$. This is still not a particularly friendly looking equation, so let us first of all concentrate on simplifying the summation.

The sum of e^{-nx} is a geometric series with common ratio e^{-x}, so we can write

$$\sum_n e^{-nx} = \frac{1-e^{-nx}}{1-e^{-x}} \approx \frac{1}{1-e^{-x}}$$

To obtain the second expression we have assumed that n is very large and so the sum approximates to that for an infinite series. We can now deal with the derivative:

$$\frac{d}{dx}\left[\log\left(\sum_n e^{-nx}\right)\right] = \frac{d}{dx}\left[\log\left(\frac{1}{1-e^{-x}}\right)\right] = -\frac{e^{-x}}{1-e^{-x}}$$

$$= -\frac{1}{e^{x}-1}$$

Substituting back into the expression for the mean energy we finally obtain the result

$$\langle E \rangle = \frac{\hbar\omega}{e^{x}-1} = \frac{\hbar\omega}{e^{\hbar\omega/k_B T}-1}$$

By differentiating this expression we find that the specific heat capacity is given by (see Question 7.8)

$$C = 3R\left(\frac{\hbar\omega}{k_B T}\right)^2 \frac{e^{\hbar\omega/k_B T}}{(e^{\hbar\omega/k_B T}-1)^2} \tag{7.14}$$

If we plot this function (see Fig. 7.7) we can see that it gives very good agreement with experiment at virtually all temperatures. Note also that at high temperatures the result converges to the classical value of $3R$. We can demonstrate that this should occur because when the thermal energy, $k_B T$, is much greater than the energy spacing between the discrete vibrational states, $\hbar\omega$, the expression for the internal energy in eqn (7.13) reduces to the classical result (see Question 7.9). The physical reason for this

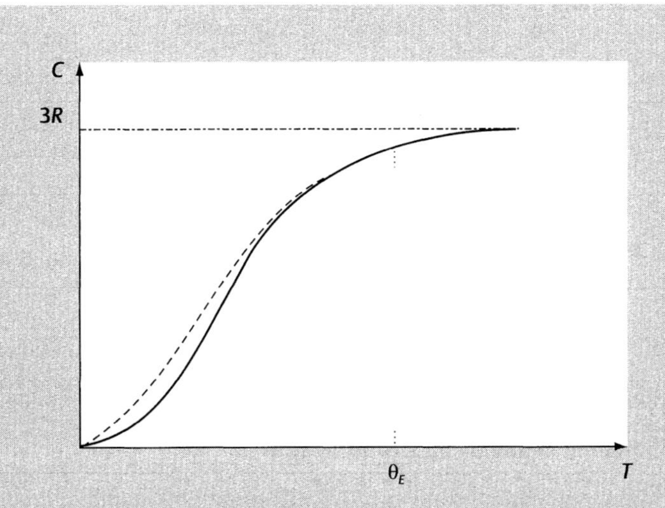

Figure 7.7 A graph of the specific heat capacity, *C*, as a function of temperature, *T*, for a typical solid. The solid line shows the curve predicted by the Einstein model (eqn (7.14)), the dashed line shows the experimental results and the dash-dot line shows the classical value. The Einstein temperature, θ_E, is defined by eqn (7.15).

is that as the thermal energy becomes larger, the distribution of the electrons between the discrete states becomes approximately the same as in the classical case. This is illustrated schematically in Fig. 7.8.

In order to quantify the transition from a classical to a quantum system it is usual to define a characteristic temperature, called the Einstein temperature θ_E, such that

$$\theta_E = \frac{\hbar\omega}{k_B} \tag{7.15}$$

It can easily be shown (see Question 7.10) that the heat capacity at the Einstein temperature is equal to 0.92 of the classical value. Therefore we can assume that above this temperature the heat capacity is given approximately by the classical value, and below this temperature the full quantum formula is required.

In practice, the Einstein temperature is rarely used. Instead values are quoted for the Debye temperature (see Table 7.3). We will explain the meaning of the Debye temperature in the following (optional) section. However, since the Debye and Einstein temperatures are approximately equal to one another, the distinction is not important.

From the data in Table 7.3, we can see that the Debye temperature for many materials is near or below room temperature. Since the heat capacity is approximately equal to the classical value at temperatures above the Debye temperature, this explains why the classical model tends to give good agreement with experimental values of the heat capacity at room temperature. However, it is noticeable that some materials, such as diamond, have Debye temperatures which are well above room temperature. We can explain, at least qualitatively, how the Debye temperature

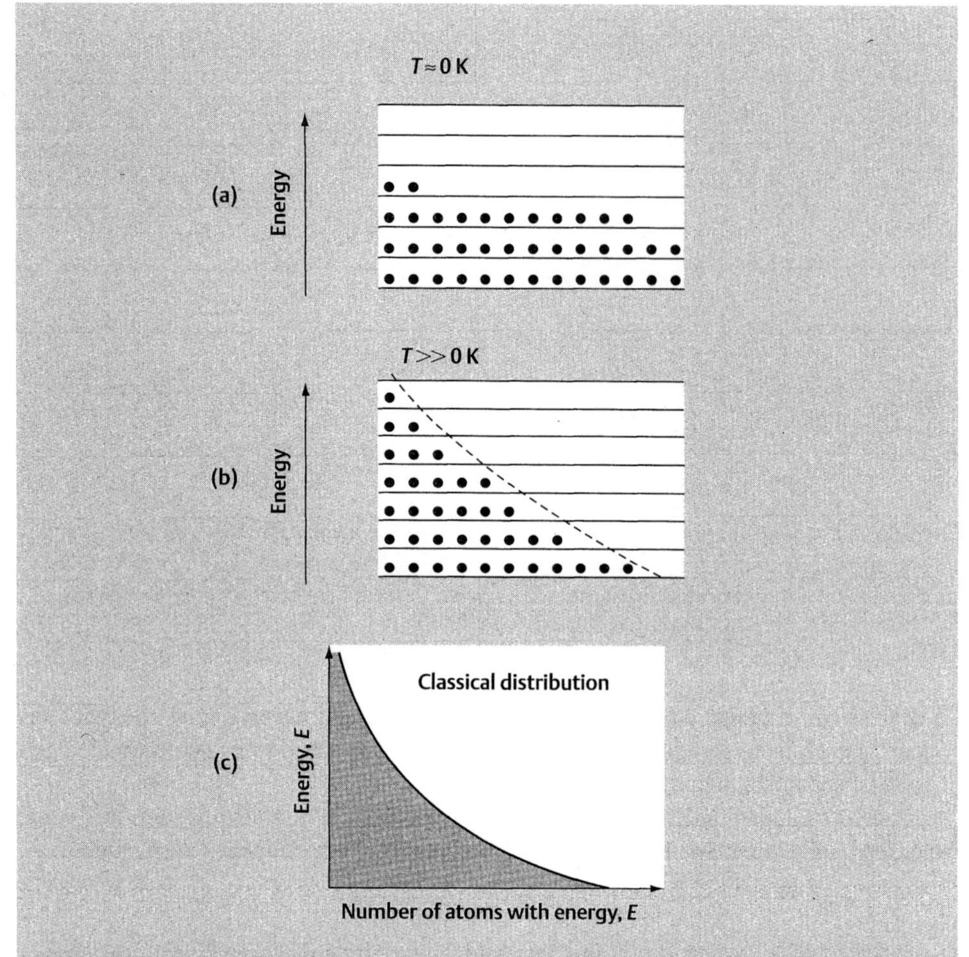

Figure 7.8 Schematic diagrams showing the number of atoms in each vibrational energy state. At low temperatures (a) the discrete nature of the energy states is strongly apparent, but at high temperatures (b) the distribution becomes similar to the classical distribution (c) in which it is assumed that there is a continuous range of energies available to the vibrating atoms.

varies from one material to another by considering the strength of the interatomic bonds and the mass of the atoms. In Section 7.2 we showed that a crystal with light atoms and strong bonds, such as diamond, has a high frequency of oscillation, and so from eqn (7.15) we can see that this leads to a high Einstein (or Debye) temperature. Conversely, materials which have heavy atoms and weak bonds, such as lead and gold, have a low frequency of oscillation, and so a low Einstein (or Debye) temperature. Therefore, we would expect materials which satisfy the latter condition to obey the classical theory of heat capacity at room temperature.

Table 7.3 Values of the Debye temperature, θ_D for a selection of materials.

	θ_D (K)
Aluminium	428
Silver	225
Copper	345
Sodium	160
Gold	165
Lead	95
Silicon	650
Diamond	1850

7.6 Heat capacity—the Debye model*

In the previous section we showed that the Einstein model gives generally good agreement with experimental values of the heat capacity. However, there is a noticeable discrepancy at low temperatures. Experimental results show that for temperatures up to about one-tenth of the Debye temperature the specific heat capacity is proportional to T^3, whereas the Einstein model predicts a value that changes exponentially with temperature (see Question 7.13). So how do we improve on the Einstein model?

One of the main assumptions that we made in the Einstein model is that the atoms behave as independent oscillators, i.e. the vibrations of one atom do not depend on those of its neighbouring atoms. In contrast, if we assume that the vibrating atoms interact strongly with one another then we should study the vibrations of the crystal as a whole.

So let us begin by considering the standing waves that are allowed in a one-dimensional crystal of length L. The first few such states starting with the longest wavelength (or lowest energy) are shown in Fig. 7.9. The amplitude of each of these waves can be described by a function of the form $\psi = \sin k_x x$, with the possible values of k_x equal to $n_x \pi / L$, where n_x is an integer (see Question 7.14). If we extend this argument to a three-dimensional cube of side L then we obtain similar expressions for k_y and k_z.

This gives a macroscopic description of the allowed standing waves in the crystal. To determine the frequency and amplitude of vibration of any particular atom we would need to determine the combined effect of all of these different vibrational modes—quite a formidable task. Fortunately, we do not need to determine the actual frequency of vibration of any particular atom. All we need to know in order to determine the internal energy of the crystal is how many atoms are vibrating at any particular frequency. This information is determined by the density of states, $g(\omega)$, which gives the number of atoms with frequencies between ω and $\omega + d\omega$. By following an argument very similar to that used in Example 4.5 to determine the density of elec-

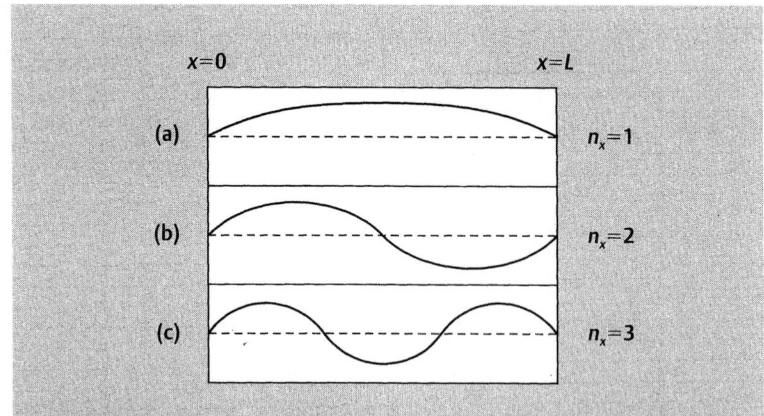

Figure 7.9 The amplitudes of the standing waves in a one-dimensional crystal of length L are given by $\psi = \sin k_x x$ where $k_x = n_x \pi / L$ and n_x is an integer. The waves shown correspond to $n_x = 1, 2$ and 3.

tronic states, $g(E)$, we can show (see Example 7.4) that the density of vibrational states is given by

$$g(\omega) = \frac{\omega^2 V}{2\pi^2 v^3} \tag{7.16}$$

where V is the volume of the crystal and v is the velocity of the wave (which is equal to ω/k_B). As the velocity of the wave is a measure of the speed at which the vibrations pass from one atom to another, and since a sound wave is also due to the vibrations of the atoms, we can assume that the value of v is equal to the speed of sound in the material.

Example 7.4 By considering the allowed standing waves in a crystal of dimensions $L_x = L_y = L_z = L$, show that the density of vibrational states, $g(\omega)$, is given by

$$g(\omega) = \frac{V\omega^2}{2\pi^2 v^3}$$

where v is the velocity of the wave and V is the volume of the crystal.

Solution From Fig. 7.9 we have seen that the allowed values of k_x for a one-dimensional oscillator are

$$k_x = \frac{n_x \pi}{L}$$

Writing similar expressions for k_y and k_z, we find that the allowed values of the wavevector k for a three-dimensional oscillator are given by

$$k^2 = k_x^2 + k_y^2 + k_z^2 = \left(\frac{\pi}{L}\right)^2 \left(n_x^2 + n_y^2 + n_z^2\right) = \left(\frac{n\pi}{L}\right)^2$$

from which we obtain the relationship

$$n = \frac{kL}{\pi}$$

We will use this expression in a moment.

Now each combination of n_x, n_y and n_z corresponds to a different oscillator. From Fig. 4.22 and Example 4.5 we can see that the number of oscillators with a value of n less than a particular value n_{max} is

$$N = \frac{1}{8} \cdot \frac{4}{3} \pi n_{max}^3 = \frac{\pi n_{max}^3}{6} = \frac{\pi}{6} \left(\frac{kL}{\pi} \right)^3 = \frac{k^3 V}{6\pi^2}$$

where V is the volume of the crystal.

Since we want to obtain the density of states in terms of the angular frequency, ω, rather than the wavevector, k, we can use the relationship

$$k = \frac{\omega}{v}$$

where v is the velocity of the wave. Therefore

$$N = \frac{\omega^3 V}{6\pi^2 v^3} \tag{1}$$

Finally, to obtain the density of states, $g(\omega)$, we need to differentiate with respect to ω, i.e.

$$g(\omega) = \frac{dN}{d\omega} = \frac{\omega^2 V}{2\pi^2 v^3}$$

We are now in a position to obtain an expression for the internal energy of the crystal. From eqn (7.12) we already have an expression for the average energy of an oscillator with frequency ω. To obtain the total internal energy of N_A oscillators with a range of different frequencies we therefore need to multiply the expression for the average energy by the density of states, $g(\omega)$, and integrate over all of the possible values of ω. We should also point out that each oscillator has a longitudinal mode of vibration (i.e. along the direction of k) and two transverse modes (which are perpendicular to the direction of k), so a system of N_A atoms has $3N_A$ modes of vibration. Therefore, the internal energy is given by

$$U = 3 \int_0^{\omega_{max}} \frac{g(\omega)\hbar\omega}{e^{\hbar\omega/k_B T} - 1} d\omega \tag{7.17}$$

where the factor of 3 comes from the three vibrational modes of each oscillator and ω_{max} is simply the maximum frequency of the oscillators, i.e. if we notionally fill up the vibrational states starting with those of the lowest frequency (and therefore lowest energy), then ω_{max} is the value of ω corresponding to the highest occupied state. In other words, we can think of ω_{max} as being equivalent to the Fermi energy in the electronic density of states. If we consider the waves from a macroscopic viewpoint, as

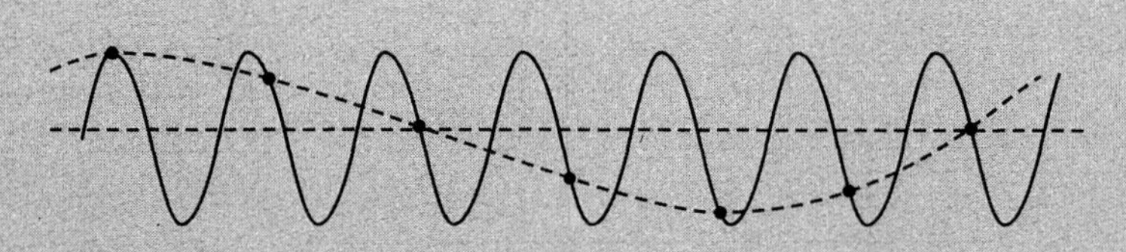

Figure 7.10 Since the waves in a crystal only have any physical existence at the atomic positions, the behaviour of the atoms when the wavelength is less than one atomic spacing (solid line) can always be described by a wave of longer wavelength (as shown by the dashed line).

depicted in Fig. 7.9, then we can see that the value of ω_{max} corresponds to the waves of shortest wavelength that exist in the crystal. For many materials we find that the shortest wavelength corresponds to about one or two atomic spacings (see Question 7.15). In fact, since the wave only physically exists at the atomic positions, it is meaningless to consider the behaviour of waves of shorter wavelength, as shown in Fig. 7.10.

In order to evaluate the specific heat capacity we need to differentiate the expression for the internal energy. Unfortunately, the integral in eqn (7.17) cannot be solved analytically. However, at low temperatures we can obtain an approximate value for the integral (see Question 7.16), and so we find that the specific heat capacity in this regime is given by

$$C = \frac{12R\pi^4}{5}\left(\frac{T}{\theta_D}\right)^3 \tag{7.18}$$

Therefore at low temperatures we predict a dependence on T^3 as observed in experiments. The Debye temperature, θ_D, is given by

$$\theta_D = \frac{\hbar\omega_{max}}{k_B} \tag{7.19}$$

We can see that the form of this equation is very similar to the definition of the Einstein temperature in eqn (7.15). In fact, if we assume that the value of ω in the Einstein model corresponds to the maximum vibrational frequency, ω_{max}, in the Debye model, then we can see that the Einstein and Debye temperatures are identical.

7.7 Heat capacity of the electrons*

In the previous sections we have concentrated on the specific heat capacity of the vibrating ions, but let us now turn our attention to the effect that the valence electrons in a metal have on the heat capacity.

According to classical theory, each valence electron in a metal has an energy of $\frac{3}{2}k_BT$, but we have seen (in Example 4.2) that this assumption leads to a heat capacity that is far larger than observed experimentally. The problem is resolved if we take into account the Fermi–Dirac distribution. We then find that only a few per cent of the valence electrons can be thermally excited into vacant states (see Section 4.10), and so the contribution of the electrons to the heat capacity is negligible in comparison with that due to the vibrating ions at room temperature.

We might therefore be tempted to neglect the contribution of the electrons entirely. This certainly seems reasonable at room temperature. However, at very low temperatures we know that the specific heat capacity of the vibrating ions also becomes very small, as can be seen from Fig. 7.7, so under these conditions the effect of the electrons can be important.

Let us then try to estimate the heat capacity of the valence electrons in a metal. We begin by considering the density of occupied states, as shown in Fig. 7.11. If we assume that only those electrons which have energies within $k_B T$ of the Fermi energy can be thermally excited, then the number of electrons involved is approximately given by the area of the shaded rectangle in Fig. 7.11. If we further assume that each of these electrons has the classical energy $\frac{3}{2}k_B T$, then we can see from Example 7.5 that the electronic contribution to the heat capacity is

$$C_e = \frac{9}{2}\frac{k_B^2 T N}{E_F} \tag{7.20}$$

where N is total the number of valence electrons. Note that in order to compare this quantity with the specific heat capacity of the vibrating ions we need to calculate the electronic heat capacity for a quantity of material containing N_A ions. So for a monovalent metal $N = N_A$, for a divalent metal $N = 2N_A$, etc.

Although the above result is obtained by rather a hotchpotch of classical and quantum methods, it turns out to be almost identical to the result obtained using a more rigorous quantum theory approach.

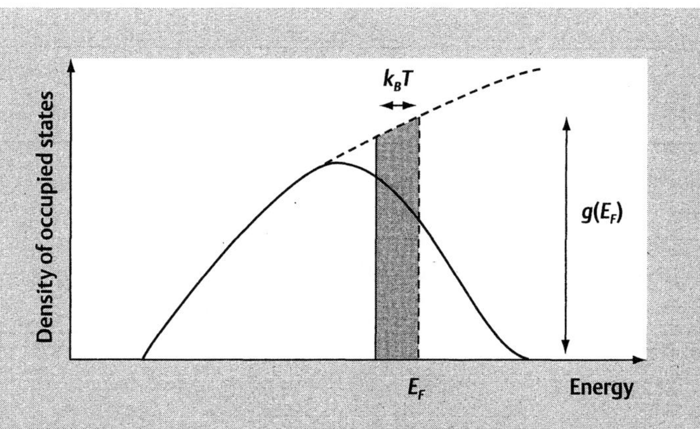

Figure 7.11 The density of occupied electron states for a metal. The shaded region shows the band of energies that lie within an energy $k_B T$ below the Fermi energy, E_F.

Example 7.5

Show that the heat capacity of the valence electrons in a metal is given by

$$C_e = \frac{9}{2}\frac{k_B^2 T N}{E_F}$$

Solution The number of valence electrons which can be thermally excited at temperature T is approximately given by the area of the rectangle in Fig. 7.11, i.e. $g(E_F) \cdot k_B T$, where $g(E_F)$ is the density of states at the Fermi energy.

If we assume that each of these electrons has an energy $\frac{3}{2}k_B T$ then the total energy of these electrons is

$$U = \frac{3}{2}k_B T g(E_F) k_B T = \frac{3}{2}k_B^2 T^2 g(E_F)$$

(Note that this expression represents the temperature-dependent part of the internal energy of the electrons. We have neglected the energies of the other electrons which do not change with temperature, and so do not contribute to the heat capacity.)

From Question 7.17 we know that

$$g(E_F) = \frac{3N}{2E_F}$$

where N is the total number of valence electrons in the metal. Substituting for $g(E_F)$ in our expression for the internal energy U gives

$$U = \frac{3}{2}k_B^2 T^2 \frac{3N}{2E_F} = \frac{9}{4}k_B^2 T^2 \frac{N}{E_F}$$

Therefore the electronic contribution to the heat capacity is

$$C_e = \frac{dU}{dT} = \frac{9}{2}k_B^2 T \frac{N}{E_F}$$

Substituting values in eqn (7.20) we find that most metals have an electronic specific heat capacity of about 0.2 J mol^{-1} K^{-1} at room temperature (see Question 7.18), which is about a hundred times smaller than the heat capacity of the vibrating ions. So at room temperature we are justified in neglecting the contribution of the electrons to the heat capacity. However, it is interesting to note that the electronic heat capacity is proportional to T, whereas from the previous section we know that the ionic heat capacity at low temperatures is proportional to T^3. Therefore at very low temperatures the electronic heat capacity becomes larger than the ionic contribution, as we can see from Fig. 7.12. By equating eqns (7.18) and (7.20) we find that the temperature at which the electronic heat capacity becomes comparable with the ionic heat capacity is of the order of a few kelvin (see Question 7.20).

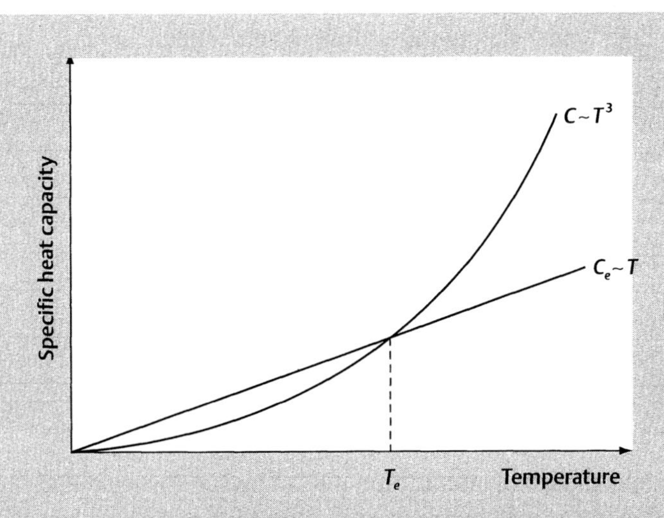

Figure 7.12 At low temperatures the heat capacity of the vibrating atoms, C, varies in proportion to T^3, whereas the electronic heat capacity, C_e, is proportional to T. Consequently, there is a temperature T_e below which the electronic contribution to the heat capacity is dominant. T_e typically is of the order of a few kelvin.

7.8 Thermal conduction

Which materials are good thermal conductors? It is well known that metals are good conductors of heat—as most people have painfully discovered at one time or another, e.g. when a metal spoon is left in a heated pan. In contrast, most non-metals, such as plastics and glass, are poor thermal conductors. This suggests that there is a similarity between electrical and thermal conduction. In fact, an empirical relationship discovered by Wiedemann and Franz in 1853 states that the thermal conductivity, κ, and the electrical conductivity, σ, of a metal at temperature T are related by the equation

$$\frac{\kappa}{\sigma T} = L \tag{7.21}$$

where L is a constant.

The above presents strong evidence to suggest that the mechanism of thermal conduction is similar to that of electrical conduction, i.e. it is due to movement of the delocalized valence electrons. However, there are two noticeable differences between thermal and electrical conduction:

1. The ratio of the conductivity of a good electrical conductor (i.e. a material with a plentiful supply of delocalized electrons) to that of a poor electrical conductor (with virtually no delocalized electrons) is of the order of 10^{24} (see Fig. 5.1), whereas the ratio of thermal conductivities is only about 10^5 (see Table 7.4).

Table 7.4 Values of the thermal conductivity, κ, at 295 K for a selection of materials.

	κ ($\mathrm{W\,m^{-1}\,K^{-1}}$)
Diamond	~2000
Copper	400
Gold	310
Aluminium	230
Silicon	160
Sodium	140
Glass	~1.0
Polystyrene	0.02

2. A few materials, such as diamond, are found to be excellent thermal conductors, despite the fact that they are electrical insulators.

This suggests that there must be some process other than the transport of electrons which contributes to thermal conduction. An obvious possibility is that thermal conduction is related to the vibrational motion of the atoms.

We can see how this can occur by considering a one-dimensional chain of atoms, as shown in Fig. 7.1. If one end of the chain is heated, the atoms at that end begin to vibrate more than those at the other end. Since the bonds between the atoms behave rather like springs (see Section 7.2), the increase in the amplitude of vibration of one atom affects the vibration of the neighbouring atom, and so the disturbance travels along the chain rather like a Mexican wave in a football stadium. Therefore, heat energy is transferred by thermal waves travelling through the material. In some cases it is useful to think of these thermal waves as particles, called phonons. This is analogous to the description of light waves in terms of photons. In fact, phonons are similar to photons in many respects. They are simply packets of energy, and so the movement of a phonon through a material transfers heat energy from one region to another.

Unlike electrical conduction, which is due solely to the movement of electrons, we therefore find that thermal conduction involves two types of particles, electrons and phonons. We can write the total thermal conductivity as

$$\kappa = \kappa_e + \kappa_p \qquad (7.22)$$

where κ_e and κ_p are the electron and phonon contributions, respectively.

Let us take a look first at the effect of the phonons. We will assume that the thermal conductivity due to the phonons is given by

$$\kappa_p = \frac{1}{3} n C \lambda_{ph} v_{ph} \qquad (7.23)$$

where n is the number of moles per unit volume, C is the molar specific heat capacity of the vibrating atoms, λ_{ph} is the mean free path of the phonons and v_{ph} is the velocity of the phonons. This formula is obtained by treating the phonons in the same way as the molecules in an ideal gas. (A derivation can be found in Flowers and Mendoza—see Further reading). Since we have already discussed the value of the specific heat capacity in the previous sections, we will concentrate here on the mean free path and velocity of the phonons.

To get some idea of the velocity of the phonons we should remember that sound is also transmitted by the vibrations of the atoms. Therefore, we can assume that the velocity of the phonons is approximately equal to the velocity of sound in the material.

The mean free path of the phonons is more difficult to estimate, but it is affected by the same factors that determine the mean free path of the conduction electrons in a metal (as discussed in Section 4.4), namely:

- interactions with other phonons (i.e. the vibrating atoms);
- interactions with impurities; and
- interactions with imperfections in the crystal structure.

We can use this information to make some qualitative statements about the temperature dependence of the thermal conductivity. At low temperatures there is not much thermal energy, and so there are relatively few phonons, which means that the chances of phonons interacting strongly with one another is low. We also note that the energy of the phonons is small, and therefore (since $E = hc/\lambda$) the wavelength associated with the phonons is large. As a consequence, the phonons are not easily scattered by impurities and imperfections, as shown in Fig. 7.13(a). In contrast, at high temperatures there are a large number of phonons and so the probability of interactions occurring with other phonons is much higher. Also, the wavelength is short, which means that the phonons interact strongly with impurities (see Fig. 7.13(b)). We can therefore predict that the mean free path of the phonons increases as the temperature decreases. However, we also know that the heat capacity decreases for temperatures below the Einstein (or Debye) temperature, and goes to zero as the

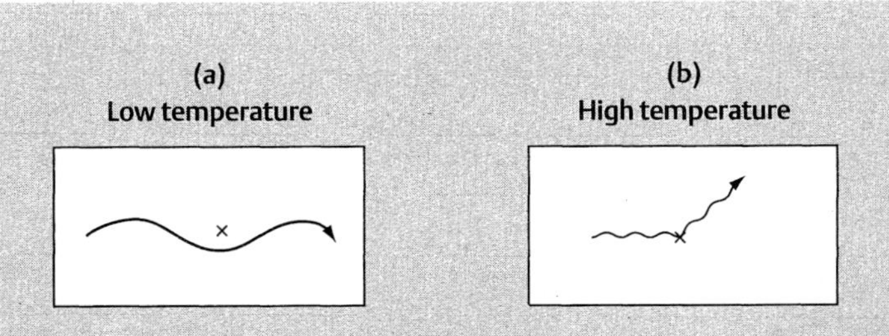

Figure 7.13 Schematic diagrams showing the interaction of a phonon with an impurity (indicated '×'). (a) At low temperatures, the wavelength of the phonon is large and so does not tend to interact strongly with impurities. (b) At high temperatures, the phonon wavelength is much shorter and so the phonon interacts strongly with impurities.

temperature drops to absolute zero (see Fig. 7.7). Since the thermal conductivity is determined by the product of the mean free path and the heat capacity, we can conclude that as the temperature is lowered the thermal conductivity first increases, and then drops to zero, as shown in Fig. 7.14. For most materials it is found that the maximum thermal conductivity, κ_{max}, corresponds to a temperature of approximately $\theta_D/10$.

Let us now turn our attention to the thermal conductivity of the electrons. We will assume that the electron contribution to the thermal conductivity is given by an equation similar to eqn (7.23) except that:

- the specific heat capacity of the ions, C, is replaced by the electronic specific heat capacity, C_e;
- the mean free path of the phonons, λ_{ph}, is replaced by the mean free path of the electrons, λ_e; and
- the velocity of the phonons, v_{ph}, is replaced by the electron velocity, v_e.

Therefore, the electron contribution to the thermal conductivity is given by

$$\kappa_e = \frac{1}{3} n C_e \lambda_e v_e \tag{7.24}$$

We know that the specific heat capacity of the electrons is only about 1% of that due to the vibrating ions, but this is more than compensated by the fact that the mean velocity of the electrons at the Fermi energy is about 10^6 m s^{-1} (see Section 4.10), i.e. about 1000 times larger than the velocity of the phonons. If we assume that the mean free path is approximately the same for both phonons and electrons, then we can see that the electronic contribution to the heat capacity of a metal is about 10 times larger than the phonon contribution (see Question 7.23). In other words, we expect the thermal conductivity of metals to be dominated by the behaviour of the free electrons, whereas non-metals (which have no free electrons) should have a thermal conductivity about an order of magnitude smaller than that of a metal.

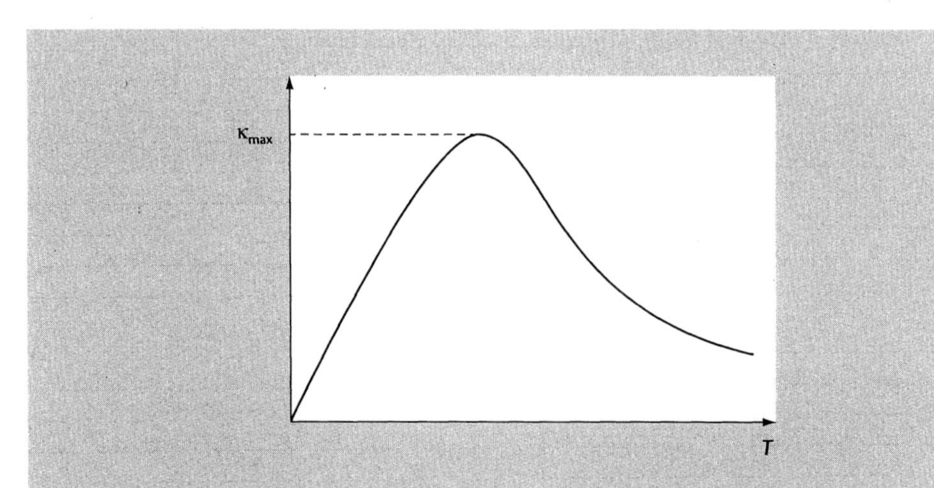

Figure 7.14 Typical variation of thermal conductivity, κ, with temperature, T.

Although this result appears to be correct for the majority of non-metals, one noticeable exception is diamond. As we can see from Table 7.4, the thermal conductivity of diamond at room temperature is actually much larger than that of a metal. The reasons for this can be found by applying the comments that we have already made about the behaviour of the phonons. Because a diamond is very pure and has an almost perfect crystal structure, the mean free path of the phonons in a diamond is exceedingly long (see Question 7.24). We should also note that since the Debye temperature of diamond is at nearly 2000 K, the peak of the thermal conductivity curve in Fig. 7.14 corresponds to about 200 K, which is not that far below room temperature. (In comparison, most other materials have a Debye temperature of less than 600 K, and therefore the peak thermal conductivity is less than 60 K.)

An interesting consequence of the high thermal conductivity of diamond is that if you place a diamond on your tongue (which is very sensitive to temperature) it always feels cold, because the diamond conducts heat away from the tongue very effectively. In comparison, if you try this with a small piece of glass, the thermal conductivity of the material is much lower and so the region of the glass that is in contact with your tongue soon becomes the same temperature as your tongue. So if you are ever tempted to buy a diamond from a dubious source, just remember this simple way of using thermal conductivity to test the authenticity of a diamond.

7.9 Summary

In this chapter we have seen that in order to understand the thermal properties of solids we need to consider both the movement of the valence electrons and the thermal vibrations of the atoms. The latter can also be treated as particles called phonons.

The electrons are important in determining the thermal conductivity of a metal, but their contribution to the heat capacity is negligible except at very low temperatures. In contrast, phonons provide the dominant contribution to the heat capacity in virtually all materials, and also account for the thermal conductivity of electrical insulators.

We have also seen that many thermal properties can be described quite adequately using classical models. For example, we obtain good results for the thermal conductivity using a classical model in which the phonons and delocalized valence electrons are treated in the same way as the molecules in an ideal gas. However, although the classical model of heat capacity generally agrees with experiment at room temperature, we need a quantum approach to explain the behaviour at lower temperatures.

Questions

Questions marked * indicate that they are based on the optional sections.

7.1 **Thermal vibration of the atoms.** For a system executing simple harmonic motion with an equation of motion given by eqn (7.1), show that the angular frequency of the oscillation is $\omega = \sqrt{\gamma/m}$.

7.2 **Thermal vibration of the atoms.** Obtain values of the force constant, γ, and the angular frequency of vibration of the atoms, ω, for diamond, copper and lead using the following data:

For diamond: $Y = 9.5 \times 10^{11}$ N m^{-2}, $a_0 = 0.154$ nm, $m = 12$ u;
For copper: $Y = 1.3 \times 10^{11}$ N m^{-2}, $a_0 = 0.256$ nm, $m = 64$ u;
For lead: $Y = 1.5 \times 10^{11}$ N m^{-2}, $a_0 = 0.350$ nm, $m = 207$ u.
(The atomic mass unit, u equals 1.66×10^{-27} kg.)

7.3 **Thermal vibration of the atoms.** By treating the atoms as simple harmonic oscillators and assuming that the average thermal energy of an atom at temperature T is k_BT, show that the amplitude of the atomic vibrations is given by

$$x_{max} = \left(\frac{2k_BT}{\gamma}\right)^{1/2}$$

7.4 **Thermal vibration of the atoms.** Given that the actual value of γ for copper is about 100 N m^{-1} and the atomic spacing is 0.256 nm, estimate the amplitude of vibration of the atoms at 300 K as a percentage of the equilibrium separation.

7.5 **Thermal expansion.** Both ends of a steel girder of length 20 m are clamped in position at a temperature of 5°C. Calculate the change in length that occurs when the temperature is raised to 35°C and determine the stress that is produced in the girder as a result. (The linear coefficient of expansion for steel is 1.1×10^{-5} K^{-1} and Young's modulus is 2.1×10^{11} N m^{-2}.) If the yield stress of the girder is 5.0×10^{7} N m^{-2}, is the thermal stress sufficient to cause a permanent deformation of the girder?

7.6 **Thermal expansion.** A bimetallic strip is 20 cm long and consists of a layer of steel and a layer of aluminium, each 1mm thick. The strip is clamped at one end. If the strip is flat (as in Fig. 7.5(a)) at 10°C, calculate the radius of curvature, r, and the deflection of the free end, x (as shown in Fig. 7.15), at 20°C. Assume that each layer expands such that the middle of the layer (as shown by the dashed line) has the equilibrium lattice spacing at any given temperature. The value of Δr in the figure is therefore 1 mm.

7.7 **Thermal expansion.** In a Bragg reflection experiment using aluminium at 300 K a sharp peak is observed at an angle of 15.46°. At 800 K the same peak has a Bragg angle of 15.27°. Use this information to calculate the coefficient of linear expansion of aluminium.

7.8 **Heat capacity—Einstin model.** Using the expression for the internal energy U given in eqn (7.13), show that the specific heat capacity is given by eqn (7.14).

7.9 **Heat capacity—Einstein model.** Show that the expression for the internal energy U in eqn (7.13) reduces to the classical value (i.e. $U = 3RT$) when $k_BT \gg \hbar\omega$. Explain why we obtain the classical result even though we have treated the system using quantum theory.

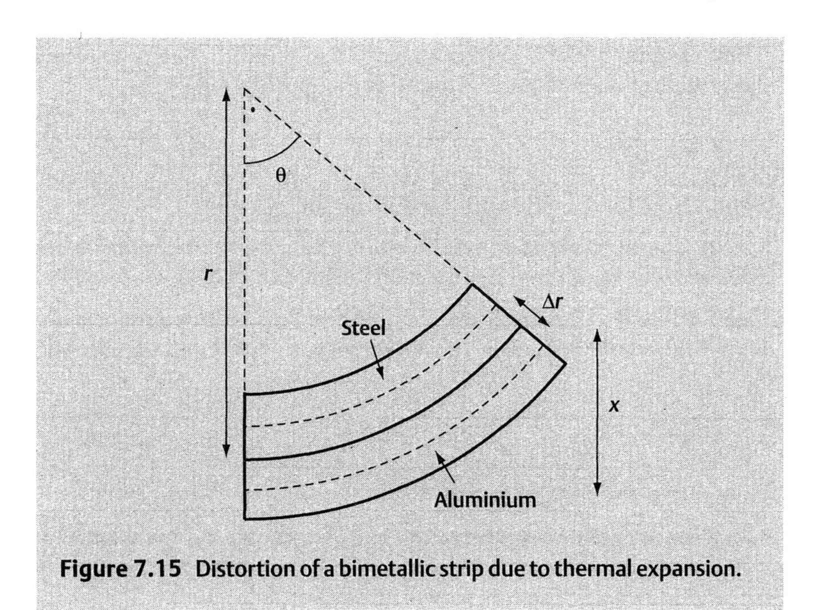

Figure 7.15 Distortion of a bimetallic strip due to thermal expansion.

7.10 **Heat capacity—Einstein model.** Show that at the Einstein temperature, θ_E, the specific heat capacity given by eqn (7.14) is equal to $0.921 \times 3R$.

7.11 **Heat capacity—Einstein model.** A particular solid consists of N_A atoms which behave as simple harmonic oscillators each of which has an angular frequency of 5×10^{13} rad s^{-1}. By calculating the specific heat capacity at a number of temperatures (e.g. 50 K, 100 K, 200 K, etc.), plot a graph of the specific heat capacity as a function of temperature.

7.12 **Heat capacity—Einstein model.** Using eqn (7.15) and the data in Table 7.3, calculate the force constant γ for diamond, copper and lead. Compare your values with those obtained in Question 7.2. Which set of values do you think are most accurate?

7.13 **Heat capacity—Einstein model.** Show that at very low temperatures the Einstein specific heat capacity, as given by eqn (7.14), is proportional to $e^{-\hbar\omega/k_BT}$.

7.14 * **Heat capacity—Debye model.** Determine the nodes of the equation $\psi = \sin(n_x \pi/L)$ in the range $x = 0$ to $x = L$ for the cases $n_x = 1, 2$ and 3.

7.15 * **Heat capacity—Debye model.** Use the data in Table 7.3 to estimate the minimum vibrational wavelength in copper. (The speed of sound in copper is about 4000 m s^{-1}.) Compare your result with the atomic spacing, which is about 0.256 nm.

7.16 ***Heat capacity—Debye model.** By making a suitable substitution in eqn (7.17) and using the standard integral

$$\int_0^\infty \frac{x^3}{e^x - 1} dx = \frac{\pi^4}{15}$$

show that the specific heat capacity at low temperatures in the Debye model is given by

$$C = \frac{12R\pi^4}{5}\left(\frac{T}{\theta_D}\right)^3 \text{ J mol}^{-1}\text{ K}^{-1}$$

where $\theta_D = \hbar\omega_{max}/k_B$.

7.17 * **Heat capacity of the electrons.** Using the free electron model (see Section 4.8), show that the density of states at the Fermi energy, $g(E_F)$, is given by

$$g(E_F) = \frac{3N}{2E_F}$$

where N is the number of valence electrons per unit volume.

7.18 * **Heat capacity of the electrons.** Use the data in Table 4.2 to determine the molar electronic heat capacity of copper, zinc and aluminium at 295 K.

7.19 * **Heat capacity of the electrons.** Show that the ratio of the electronic and ionic contributions to the heat capacity of a monovalent metal at room temperature is given by

$$\frac{3}{2}\frac{k_B T}{E_F}$$

(Assume that the ionic contribution is given by the classical value of $3R$.)

7.20 * **Heat capacity of the electrons.** Determine the temperature at which the electronic and ionic heat capacities of copper are equal to one another. (The Debye temperature of copper is 345 K and the Fermi energy is 7.00 eV.)

7.21 **Thermal conduction.** Explain why the range of thermal conductivities of solids is much smaller than the range of electrical conductivities.

7.22 **Thermal conduction.** Using the values of thermal conductivity from Table 7.4 and electrical conductivity from Table 4.1, determine the value of L in eqn (7.21) for copper, gold, aluminium and sodium at 273 K.

7.23 * **Thermal conduction.** Compare the contribution of electrons and phonons to the thermal conductivity of a monovalent metal (i.e. one valence electron per atom) at 300 K. Assume that the ionic specific heat capacity is given by the classical value, the Fermi energy is 5 eV, the speed of sound in the metal is 4000 m s^{-1}, and that the mean free paths of the electrons and phonons are approximately the same.

7.24 **Thermal conduction.** Using the data in Tables 7.2 and 7.4, estimate the mean free path of phonons in diamond at 295 K. Diamond has a density of 3.52×10^3 kg m^{-3}, a molar mass of 12 g mol^{-1} and the speed of sound in diamond is about 5000 m s^{-1}.

Chapter 8
Magnetic properties

8.1 Introduction

Mention magnetic materials and most people think of the small pieces of metal that you find in various household gadgets, travel games, children's toys, fridge magnets, and so on. These applications exploit the well-known property that like poles repel and unlike poles attract. Indeed, this property of magnets was known to the Ancient Greeks and Chinese over 2000 years ago. However, magnets display a much wider range of properties, which are exploited in numerous different applications. For example, the magnetic iron core is a vital part of electrical motors and transformers. Magnets also play a very important role in many modern technologies. Indeed the revolution in information technology that has taken place over the last few decades owes as much to the developments in magnetic storage of information as it does to the ubiquitous silicon chip.

In this chapter we will concentrate not so much on the applications of magnets, but rather on understanding why some materials are magnetic, and—perhaps equally puzzling—why most materials are not magnetic. So which materials are magnetic? We tend to think of magnetism as being the exclusive property of iron-based materials, although a few other metals, such as cobalt and nickel, could also be added to this list. The majority of solids are generally considered to be non-magnetic, at least at room temperature, but this is not quite correct. Strictly speaking all materials are magnetic to some degree in that they become magnetized in the presence of a magnetic field. However, in most cases the effects are very weak and they remain magnetized only for so long as the external magnetic field is applied. The materials which are of interest from a technological point of view are usually those which not only have a large magnetization, but also retain their magnetization even when the external field is removed. Such materials are called permanent magnets.

In order to understand the properties of permanent magnets—and to explain why so few materials are permanent magnets—we need to begin by looking at the magnetic properties of individual atoms. First, though, we will define some macroscopic quantities associated with magnetism.

8.2 Macroscopic magnetic quantities

There are several macroscopic quantities that are used in describing magnetic properties of materials, and different textbooks tend to use different terminologies (or, worse, use the same term to mean something slightly different!). So we will begin by defining exactly what is meant by the notation and terminology used in this chapter.

All magnets (with the possible exception of some exotic particles which we will not consider in this book) exist as a magnetic dipole, i.e. they have a north pole and a south pole. If the moment of the magnetic dipole is $\mathbf{m}_m$, then the dipole moment per unit volume is referred to as the magnetization **M**, and the energy of the dipole in a magnetic field $\mathbf{B}_0$ is $-\mathbf{m}_m \cdot \mathbf{B}_0$. These are vector quantities because they are associated with a particular direction, but in this chapter we will generally deal only with the component of these quantities measured along the direction of the magnetic field. In this case we can treat all of the variables as scalar quantities.

The magnetization is related to the magnetic field by the equation

$$\chi_m = \frac{\mu_0 \mathbf{M}}{\mathbf{B}_0} \tag{8.1}$$

where χ_m is the magnetic susceptibility and μ_0 is the permeability of a vacuum. χ_m is a dimensionless quantity—the values of χ_m quoted in this chapter are referred to a unit volume.

Confusion tends to arise in the distinction (or lack of it!) between the magnetic field $\mathbf{B}_0$ and the magnetic induction **B**. In this chapter we can think of $\mathbf{B}_0$ as the external, or applied, magnetic field, whereas **B** is the total field in the presence of a magnetic material with magnetization **M**. The two are related by the equation

$$\mathbf{B} = \mathbf{B}_0 + \mu_0 \mathbf{M} \tag{8.2}$$

For most materials the quantity $\mu_0\mathbf{M}$ is very small and so $\mathbf{B} \approx \mathbf{B}_0$. It is therefore very tempting to treat these quantities as though they are the same. However, for permanent magnets the quantity $\mu_0\mathbf{M}$ is not small and in this case it is important to make the distinction between **B** and $\mathbf{B}_0$.

Finally, a note on units. Both **B** and $\mathbf{B}_0$ are measured in tesla (T). The tesla is a very large unit. For example, the Earth's magnetic field is less than 10^{-4} T, and even powerful electromagnets rarely produce a field of more than a few tesla. The magnetization is measured in units of $\mathrm{J\,T^{-1}\,m^{-3}}$. (In some textbooks the magnetization is quoted in units of $\mathrm{A\,m^{-1}}$. These units are equivalent, i.e. $1\ \mathrm{J\,T^{-1}\,m^{-3}} = 1\ \mathrm{A\,m^{-1}}$.) The permeability of a vacuum, μ_0, is usually quoted in units of $\mathrm{H\,m^{-1}}$, but for the purposes of this chapter it may be useful to think of μ_0 in terms of the equivalent units $\mathrm{T^2\,m^3\,J^{-1}}$. Although these units look rather clumsy, they are useful for checking that equations are dimensionally correct (see Question 8.1).

8.3 Atomic magnets

In the classical picture of an atom we imagine that the electrons describe a circular orbit around the nucleus, as shown in Fig. 8.1. Each orbit can therefore be thought of as a loop of electrical current. From electromagnetic theory we know that a loop of current produces a magnetic field, and so we should not be surprised that the electrons in an atom also generate a magnetic field.

Of course, we now know that this picture of the atom is rather naive. According to quantum theory the electrons do not really describe circular orbits. Nevertheless, quantum theory also predicts that the electrons in an atom produce a magnetic field. In Chapter 1 we introduced the quantum numbers n, l, m_l and m_s to label the electrons in an atom. In this chapter we will be concerned mainly with the orbital magnetic quantum number, m_l, which can take values between $-l$ and $+l$, and the spin magnetic quantum number, m_s, which takes values of $\pm\frac{1}{2}$. It turns out that the orbital magnetic quantum number produces a magnetic field which is the same as the classical result obtained for an orbiting electron. However, the spin magnetic quantum number of the electron also produces a contribution to the magnetic field. This term has no classical equivalent, so any attempt to calculate the magnetic field of an atom based on classical mechanics is doomed to failure. From the outset we must therefore consider the magnetic moment of an atom from a quantum viewpoint.

For an atom containing a single electron the component of the spin magnetic moment, μ_s, in the direction parallel to the magnetic field is given by

$$\mu_s = -\frac{e m_s \hbar}{m_e} = \mp \frac{e\hbar}{2m_e} \tag{8.3}$$

where m_e is the mass of the electron and m_s is the spin magnetic quantum number. The quantity $e\hbar/2m_e$ is known as the Bohr magneton. It is denoted by μ_B and has a value of 9.27×10^{-24} J T^{-1}. The component of orbital magnetic moment is given by

$$\mu_l = -\frac{e m_l \hbar}{2m_e} = -m_l \mu_B \tag{8.4}$$

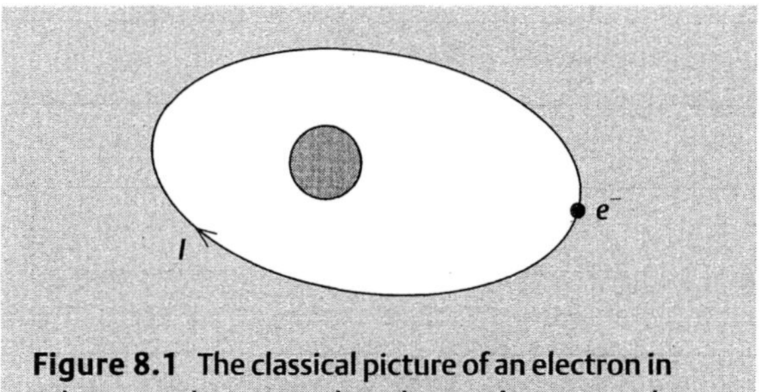

Figure 8.1 The classical picture of an electron in orbit around at atom. The orbiting electron can be thought of as a current I flowing round a loop.

When an atom has more than one electron the situation is rather more complicated. For a particular subshell (i.e. for a given combination of n and l), we can define the total spin angular momentum as

$$S = \sum m_s \tag{8.5}$$

and the total orbital angular momentum as

$$L = \sum m_l \tag{8.6}$$

If a subshell is filled, the values of S and L are both zero (see Question 8.2), so we conclude that filled subshells make no contribution to the total angular momentum of an atom. If a subshell is partially filled, then we need to know how the electrons are distributed between the different m_l and m_s states in order to determine the values of S and L. Fortunately, there are two simple rules (known as Hund's rules) which determine how these states are occupied in a partially filled subband:

1. The states are occupied so that as many electrons as possible (within the limitations of the exclusion principle) have their spins aligned parallel to one another, i.e. so that the value of S is as large as possible.
2. When it is determined how the spins are assigned, the electrons occupy states such that the value of L is a maximum.

The total angular momentum, J, is obtained by combining L and S as follows:

- if the subshell is less than half filled then $J = L - S$;
- if the subshell is more than half filled then $J = L + S$;
- if the subshell is exactly half filled then $L = 0$ (see Question 8.3) and so $J = S$.

By applying these rules we can determine the values of S, L and J for any atom, as demonstrated in Example 8.1.

Example 8.1 Determine the values of S, L and J for Cr^{3+} which has three electrons in the 3d subshell. (All lower energy subshells are filled.)

Solution The 3d subshell corresponds to $l = 2$ (see Section 1.3), and therefore m_l takes integer values between −2 and +2, i.e. there are five allowed values of m_l, so the subband has a capacity to hold 10 electrons.

In Cr^{3+}, the three 3d electrons can therefore all occupy states with $m_s = +\frac{1}{2}$, and therefore $S = \frac{3}{2}$.

The maximum value of L is obtained if two electrons occupy the state with $m_l = +2$ and one occupies the $m_l = +1$ state. However, the exclusion principle forbids two electrons with the same value of m_s from occupying the same m_l state. Consequently, the largest allowed value of L corresponds to having one electron in each of the states $m_l = +2$, $+1$ and 0, and therefore $L = 3$.

Since the subband is less than half filled, the value of J is

$$J = L - S = 3 - \frac{3}{2} = \frac{3}{2}$$

Once a value is determined for the total angular momentum quantum number J, the maximum component of the magnetic moment of the atom in the direction parallel to the magnetic field is given by

$$m_J = -g\mu_B J \tag{8.7}$$

where g is the Landé splitting factor which is given by the equation

$$g = \frac{3J(J+1) + S(S+1) - L(L+1)}{2J(J+1)} \tag{8.8}$$

(A derivation of this expression for g is given by Enge, Wehr and Richards—see Further reading.)

From the above discussion we can conclude that most atoms exhibit a magnetic moment. What we now need to consider is how this magnetic moment is affected when we form a solid.

8.4 Which materials have a magnetic moment?

In the previous section we have seen that the filled electron subshells in an atom do not affect the magnetic momentum of the atom because they have a net angular momentum of zero, i.e. $S = L = 0$ and so $J = 0$. This means that inert atoms have a magnetic moment of zero because they have only filled electron subshells. Similarly, ionic materials also have a magnetic moment of zero because in this case the electrons are transferred from one atom type to another so that the resulting ions have only filled subshells.

In covalent materials the outer subshell is only partially filled, and so we would expect these materials to have a finite magnetic moment. However, each covalent bond is formed by a pair of electrons with opposite spin and with a net orbital angular momentum of zero. So although a hydrogen atom has a finite magnetic moment, a hydrogen molecule does not. Similarly, covalent solids also have a net magnetic moment of zero.

We therefore reach a rather surprising conclusion: although most atoms have a non-zero magnetic moment, it appears that in the majority of solids the effects cancel and the resultant magnetization is zero.

Actually, this is not quite true. The presence of a magnetic field also affects the orbital motion of the electrons in an atom in such a way that the atom generates a magnetic field which opposes the external field. This is referred to as diamagnetism and occurs in all types of atom. It produces a finite magnetic susceptibility in all solids, but the effect is very weak (except in the case of superconductors, which we will discuss in Chapter 9).

Since it appears that most non-metals display only a very small magnetic susceptibility, let us now turn our attention to metallic materials. In a simple metal, such as sodium or aluminium, the valence electrons are assumed to be delocalized, in other words they are no longer attached to any particular atom. As a result the metal ions contain only filled electron subshells, and so each has a total angular momentum of zero. However, although the ions do not contribute to the magnetic moment, the delocalized electrons do produce a non-zero magnetic moment. This is known as Pauli (or free electron) paramagnetism and we will consider this subject in more detail in the next section.

Like diamagnetism, free electron paramagnetism is a weak effect, but the two effects act in the opposite sense: paramagnetism is caused by the magnetic dipole moments becoming aligned with the magnetic field, whereas diamagnetism results from an induced magnetic field which opposes the applied field. Consequently, paramagnetism gives rise to a positive magnetic susceptibility, whilst diamagnetism produces a negative susceptibility. So in a simple metal the effect of diamagnetism partially cancels the free electron paramagnetism. It also follows that a paramagnetic material (i.e. a material in which the paramagnetic response is stronger than the diamagnetic response) is attracted by either pole of a magnet, whereas a diamagnet (in which the effects of diamagnetism are stronger than those of paramagnetism) repels both poles of a magnet.

We still have not found any materials which display strong magnetic behaviour, and so we can begin to appreciate why only a few solids are usually considered to be magnetic. However, a group of metallic elements known as the transition metals and the so-called rare earth and actinide elements have more promising characteristics. These materials occupy the central columns of the periodic table and have an electronic structure which is very different to that of the simple metals. For example, the arrangement of the outer electrons in the iron group of transition metals are shown is Table 8.1. If we look down the table we can see that in addition to the one or two 4s electrons in the outer shell there is also an incomplete subshell of 3d electrons. When

Table 8.1 Electronic configurations of the iron group of transition metals showing the number of 3d and 4s electrons in each atom. In each case the 3d subshell, which has a capacity of 10 electrons, is incomplete. (All of the elements shown here also have filled shells of 1s, 2s, 2p, 3s and 3p electrons.)

	3d	4s
Scandium	1	2
Titanium	2	2
Vanadium	3	2
Chromium	5	1
Manganese	5	2
Iron	6	2
Cobalt	7	2
Nickel	8	2

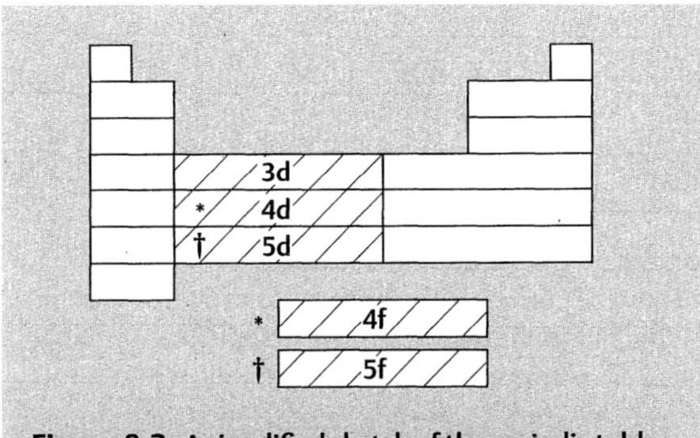

Figure 8.2 A simplified sketch of the periodic table indicating those elements which have incomplete subshells of electrons even when incorporated in a solid. These are the elements with the greatest potential for having strong magnetic properties.

these elements take part in metallic bonding it is only the 4s electrons which become delocalized. To a good approximation we can assume that the 3d electrons remain localized on each atom, and since this subshell is not filled, it produces a non-zero magnetic moment. So at last we have found a group of materials which are capable of producing a significant magnetic susceptibility in solid form. Similar considerations apply to the other rows of transition metals where it is the 4d and 5d subshells which are incomplete, and to the rare earth and actinide groups where the 4f and 5f subshells are not filled (see Fig. 8.2). These elements give rise to what we will refer to as Curie paramagnetism, and under certain circumstances they can also form permanent magnets. We will consider these possibilities later in this chapter.

8.5 Pauli paramagnetism*

Pauli paramagnetism is due to the magnetic moment associated with the spin angular momentum of the delocalized electrons in a metal, as described in the previous section. In this section we will use a simple argument to obtain an expression for the magnetic susceptibility in this case. (Note that it is helpful, but not essential, to have read Section 4.8 before reading this section.)

Let us begin by considering the orientation of the spins of these free electrons in the metal. When there is no external magnetic field present, the spins of the electrons are randomly orientated, as shown schematically in Fig. 8.3(a), and so the magnetic moment in any given direction averages to zero. However, when a magnetic field $\mathbf{B}_0$ is applied to the metal, the electron spins line up so that they are either parallel or antiparallel to the field (see Fig. 8.3(b)). From the definition in Section 8.2 we know that the magnetic energy of those electrons with spins aligned parallel to the field is $-m_m B_0$, whereas those which are antiparallel to the field have an energy $+m_m B_0$.

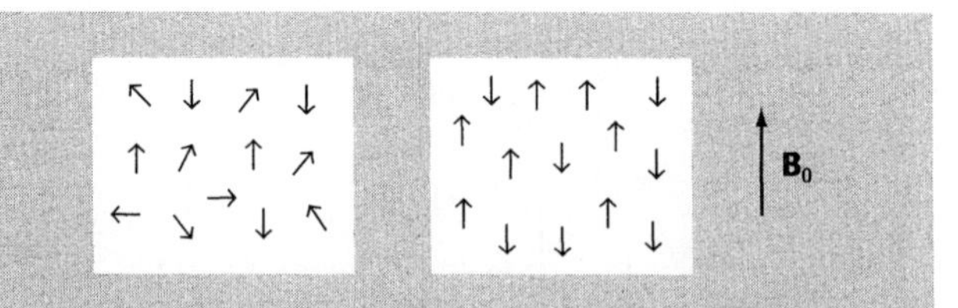

Figure 8.3 The delocalized electrons in a metal have magnetic dipole moments due to the spin of the electrons. (a) When there is no magnetic field the spins are randomly orientated. (b) When a magnetic field $\mathbf{B}_0$ is applied, the spins are aligned parallel or antiparallel to the field.

Therefore it is energetically favourable for the spins to be aligned parallel to the field. If we denote the number of electrons per unit volume with spins aligned parallel to the field by $N\uparrow$ and the number with spins antiparallel by $N\downarrow$, then we can see that the magnetization is determined by $N\uparrow - N\downarrow$.

If the electrons were free electrons, then we would expect them all to occupy the lowest energy state, and have their spins aligned parallel to the magnetic field. But in a metal the electrons must obey the exclusion principle, so the problem becomes more complicated. What proportion of the electrons have their spins parallel to the applied magnetic field in this case? To answer this question we need to take into account the density of electronic states (see Section 4.7). In fact, we will consider the density of states separately for electrons with spins parallel to the field and those with spins antiparallel to the field. For zero field the energies of the electrons are unaffected by the direction of spin, and so there are equal numbers of electrons with opposing spins (see Fig. 8.4(a)). When a magnetic field $\mathbf{B}_0$ is applied the energies of the electrons with spins parallel to the field are reduced by $m_m B_0$, and those with spins antiparallel to the field are increased by $m_m B_0$. Consequently, the density of states functions are shifted with respect to one another, as shown in Fig. 8.4(b). The highest energy electrons with spins antiparallel to the field, which occupy the shaded region in Fig. 8.4(c), can therefore move into lower energy states by changing their direction of spin. As a result we find that the number of electrons with spins parallel to the field, $N\uparrow$, is greater than the number with spins antiparallel $N\downarrow$, and so the magnetization is non-zero.

If we denote the density of states at the Fermi energy by $g(E_F)$, then we can show (see Example 8.2) that the magnetic susceptibility of the free electrons is

$$\chi_m = \mu_0 g(E_F) m_m^2 \tag{8.9}$$

Since the magnetic dipole moment m_m is due only to the spin magnetic moment of the electron, we can replace m_m by the Bohr magneton μ_B. Also, the density of states at the Fermi energy, $g(E_F)$, is given by (see Question 7.17)

$$g(E_F) = \frac{3N}{2E_F} \tag{8.10}$$

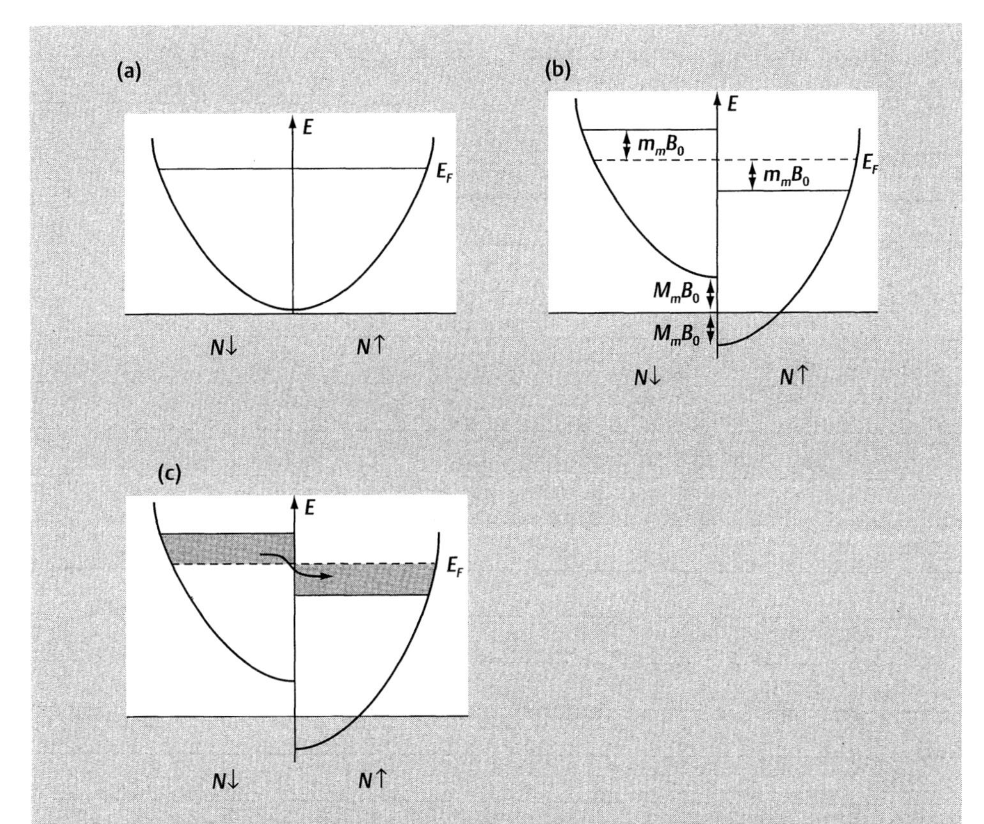

Figure 8.4 The density of states for electrons in a metal. *N*↑ corresponds to the number of electrons with spins aligned parallel to the *z* direction and *N*↓ represents the number with spins antiparallel to the *z* direction. (a) When no magnetic field is applied, there are equal numbers of electrons with spins parallel and antiparallel to any given direction. (b) When a magnetic field $\mathbf{B}_0$ is applied along the *z* direction the energies of electrons with spins parallel to the magnetic field are lowered by $m_m B_0$ and the energies of electrons with spins antiparallel to the field are raised by $m_m B_0$. (c) As a result, some of the electrons with spins antiparallel to the field (as shown by the shaded region on the left) change their spin and move into available lower energy states with spins aligned parallel to the field. (Note that the density of states curves are the same as shown in Fig. 4.23, but the axes have been swapped so that the energy, *E*, is plotted along the *y* axis and the corresponding number of states with energy *E* is plotted along the *x* axis.)

where *N* is the number of valence electrons per unit volume. So we can write the expression for the magnetic susceptibility as

$$\chi_m = \frac{3N\mu_B^2\mu_0}{2E_F} \tag{8.11}$$

It should be pointed out that the energy of the dipole associated with an electron spin, $m_m B_0$, is very small in comparison with the Fermi energy E_F. Consequently, only a tiny fraction of a per cent of the total number of valence electrons in a metal are able

Example 8.2 Show that the magnetic susceptibility due to the delocalized electrons in a metal is given by

$$\chi_m = g(E_F) m_m^2 \mu_0$$

Solution The number of electrons in the shaded region on the left hand side of Fig. 8.4(c) is

$$\frac{1}{2} g(E_F) m_m B_0$$

where $g(E_F)$ is the density of states at the Fermi energy and the factor of $\frac{1}{2}$ is because we are only using one-half of the electron distribution (i.e. those with spins aligned antiparallel to the field).

If the total number of valence electrons per unit volume is N, then the number aligned parallel to the field is

$$N\uparrow = \frac{N}{2} + \frac{1}{2} g(E_F) m_m B_0$$

and the number aligned antiparallel to the field is

$$N\downarrow = \frac{N}{2} - \frac{1}{2} g(E_F) m_m B_0$$

The difference is therefore

$$N\uparrow - N\downarrow = g(E_F) m_m B_0$$

Since each electron contributes a magnetic moment m_m, the magnetization is

$$M = (N\uparrow - N\downarrow) m_m = g(E_F) m_m^2 B_0$$

From eqn (8.1) we have

$$\chi_m = \frac{\mu_0 M}{B_0} = \mu_0 g(E_F) m_m^2$$

to contribute to the magnetic susceptibility (see Question 8.7); therefore Pauli paramagnetism is a very weak effect.

Pauli paramagnetism produces a positive contribution to χ_m, but the diamagnetic contribution to χ_m is negative, and so partially cancels the Pauli paramagnetism. In fact, for many metals such as copper, silver and gold, the diamagnetic contribution is larger in magnitude than the paramagnetic contribution from the delocalized electrons (see Table 8.2), and so these materials display diamagnetic properties.

Table 8.2 Measured values of the magnetic susceptibility χ_m for a unit volume (in SI units).

	χ_m
Lithium	2.1×10^{-5}
Sodium	7.2×10^{-6}
Copper	-9.8×10^{-6}
Silver	-2.6×10^{-5}
Gold	-3.6×10^{-5}
Aluminium	2.2×10^{-5}
Tin	0.2×10^{-6}
Lead	-1.7×10^{-5}

8.6 Curie paramagnetism

Curie paramagnetism is exhibited by materials which incorporate atoms of the type indicated in Fig. 8.2. In a solid these atoms have incomplete electron subshells, and therefore a non-zero dipole moment.

When there is no external magnetic field we can assume that these atomic magnetic dipoles are randomly orientated (similar to the electron spins shown in Fig. 8.3(a)), but when a magnetic field is applied we expect the atomic dipoles to become aligned with the field. However, this does not happen. Quantum theory states that a magnetic dipole cannot be aligned parallel to a magnetic field, rather it precesses around the direction of the field so that the dipole makes a constant angle θ with the direction of the field, as shown in Fig. 8.5. The angle θ is not arbitrary. In fact, if the total angular momentum of the atom is J, then it is found that there are only $2J + 1$ possible orientations of the dipole with respect to the field. The corresponding components of the magnetic moment, m_J, in the direction of the field are given by

$$m_J = +g\mu_B J, +g\mu_B(J-1), \ldots, -g\mu_B(J-1), -g\mu_B J \tag{8.12}$$

as shown in Fig. 8.6.

From the definition in Section 8.2 we know that the energies of these dipoles in a field B_0 are given by

$$-m_J B_0 = -g\mu_B J B_0, -g\mu_B(J-1)B_0, \ldots, +g\mu_B(J-1)B_0, +g\mu_B J B_0 \tag{8.13}$$

Unlike electrons, atoms are not governed by the exclusion principle, and so at zero kelvin we would expect all of the dipoles to occupy the lowest energy state. But, at a

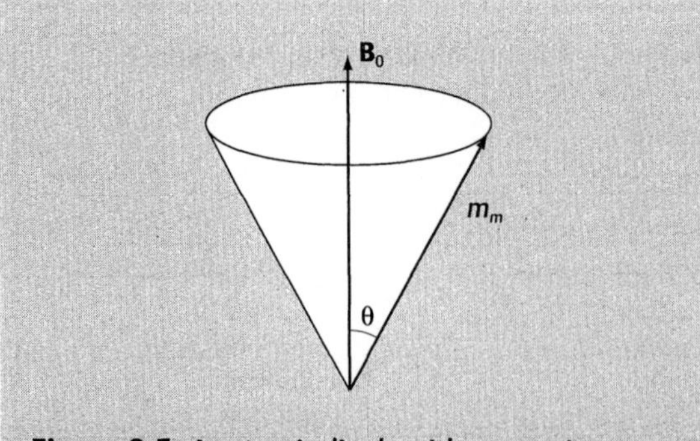

Figure 8.5 An atomic dipole with magnetic moment m_m precesses around a magnetic field $\mathbf{B}_0$ so that the axis of the dipole makes a constant angle θ with the direction of the field.

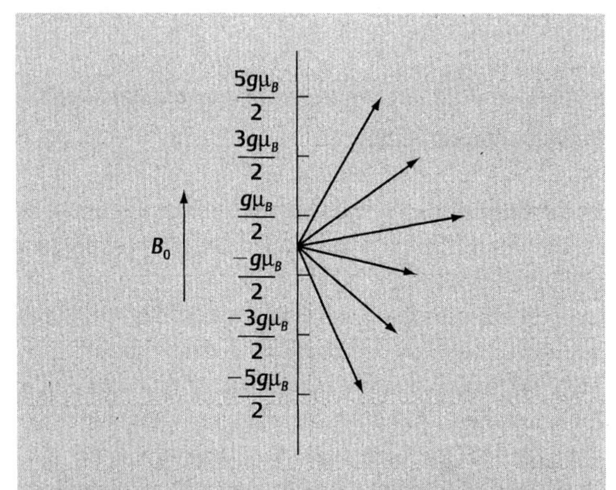

Figure 8.6 The allowed values of magnetic moment m_J along the direction of the magnetic field B_0 for an atom with total angular momentum $J = \frac{5}{2}$.

finite temperature, T, classical statistical arguments show that the probability of finding an atomic dipole in a state with energy E is proportional to

$$e^{-E/k_B T}$$

where the values of E are equal to $-m_J B_0$, as given by eqn (8.13).

Consequently, at very low temperatures we predict that most of the dipoles are in the lowest energy state with $m_J = g\mu_B J$. This means that the majority of the dipoles have their largest possible component of the magnetic moment aligned along the direction of the field, so the magnetic susceptibility is relatively large. In contrast, at high temperatures there are a substantial number of dipoles in the higher energy states in which the value of m_J is smaller or may even be negative (i.e. the dipole is aligned in the opposite direction to the applied field). In fact at room temperature the dipoles become almost evenly distributed between the available states (see Question 8.8) so that the magnetic susceptibility goes to zero.

By using the above expressions we can show (see Example 8.3) that the susceptibility is actually inversely proportional to the temperature, i.e.

$$\chi_m = \frac{C}{T} \tag{8.14}$$

where C is the Curie constant. This relationship is known as Curie's law. A detailed calculation (see Question 8.9) shows that the Curie constant is given by

$$C = \frac{\mu_0 N g^2 \mu_B^2 J(J+1)}{3k_B} \tag{8.15}$$

where N is the number of magnetic ions per unit volume. This equation is sometimes written as

$$C = \frac{\mu_0 N \mu_B^2 p^2}{3k_B} \tag{8.16}$$

where p is known as the effective Bohr magneton number and is given by

$$p^2 = g^2 J(J+1) \tag{8.17}$$

This quantity is often used for making comparisons between theory and experiment.

Example 8.3 Using the above expressions for the energies of the magnetic dipoles and the probabilities of a dipole possessing a given energy, show that the temperature dependence of the magnetic susceptibility can be written as

$$\chi_m = \frac{C}{T}$$

where C is a constant.

Solution The probability that an atomic dipole has a dipole moment m_J is given by

$$Ae^{m_J B_0/k_B T}$$

where A is assumed to be constant as a function of temperature.

If there are N magnetic ions per unit volume, then the total number of dipoles in this state is

$$NAe^{m_J B_0/k_B T}$$

and the magnetic moment of these atoms is

$$m_J NAe^{m_J B_0/k_B T}$$

To obtain an expression for the magnetization M (i.e. the total magnetic moment per unit volume) of the crystal we sum over all possible values of m_J, i.e.

$$M=\sum_{-J}^{J} m_J N A e^{m_J B_0 / k_B T}$$

Since at room temperature $m_J B_0$ is very much smaller than the thermal energy $k_B T$ (see Question 8.7), we can write the exponential term as

$$e^{m_J B_0 / k_B T} \approx 1+\frac{m_J B_0}{k_B T}$$

Therefore the magnetization becomes

$$\begin{aligned} M &= \sum_{-J}^{J} N A m_J\left(1+\frac{m_J B_0}{k_B T}\right) \\ &= N A \sum_{-J}^{J}\left(m_J+\frac{m_J^2 B_0}{k_B T}\right) \end{aligned}$$

Since the sum over m_J from +J to -J is obviously zero, we can write

$$M=N A \sum_{-J}^{J} \frac{m_J^2 B_0}{k_B T}=\frac{N A B_0}{k_B T} \sum_{-J}^{J} m_J^2$$

Therefore

$$M=\frac{C' B_0}{T}$$

where C' is a constant. Using eqn (8.1) we have

$$\chi_m=\frac{\mu_0 M}{B_0}=\frac{\mu_0 C'}{T}=\frac{C}{T}$$

where C (= $\mu_0 C'$) is known as the Curie constant.

The Curie law holds true for many paramagnetic materials containing rare earth group metals (the elements from lanthanum to lutetium), but the agreement is not so good for the iron group of transition metals (see Question 8.11). We can explain this as follows. In the rare earth group elements, the incomplete 4f subshell of electrons which produces the magnetic dipole moment lies inside the orbits of the filled 5s and 5p subshells, as shown in Fig. 8.7(a). Consequently, the 4f electrons are relatively unaffected by the presence of other neighbouring atoms. In contrast, in the iron group of metals the incomplete 3d subshell of electrons is the outermost electrons on the ion (see Fig. 8.7(b)). As a result, they interact significantly with other ions, and so the dipole moments of the individual ions are altered.

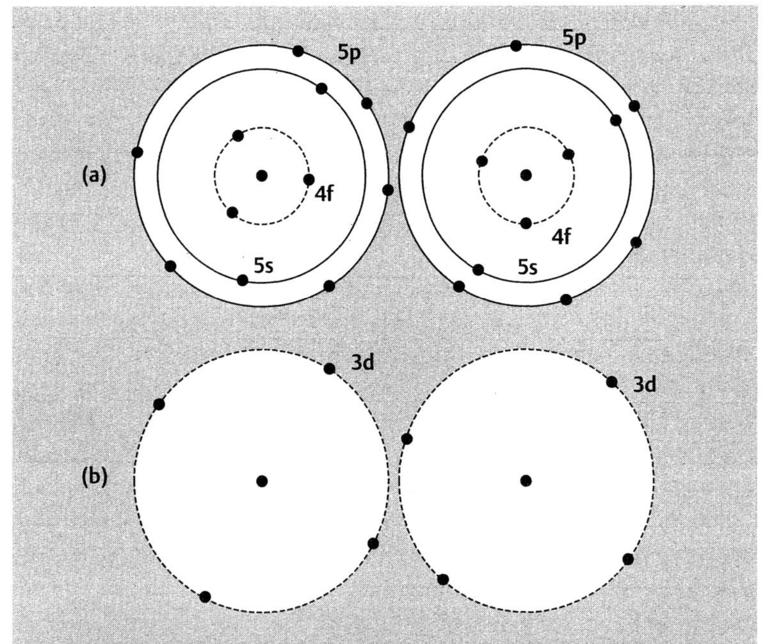

Figure 8.7 (a) In the rare earth metals the incomplete 4f subshell of electrons is within the 5s and 5p subshells and so the 4f electrons are not strongly affected by any neighbouring atoms. (b) In the iron group of metals the 3d electrons are the outer electrons, and so interact strongly with other nearby atoms. (In each case the incomplete subshell is indicated by a dashed circle.)

8.7 Ordered magnetic materials

So far we have seen that although individual isolated atoms behave as small atomic magnets, the magnetic properties of these atoms are either lost, or at best randomly orientated in a solid. In the former case we get only a very weak magnetic response due to diamagnetism or Pauli paramagnetism. In the latter case the magnetic susceptibility is larger, but an external magnetic field is still required in order to align the magnetic dipoles and produce a polarization.

However, we know that some materials retain their magnetization even in the absence of a magnetic field. These are referred to as permanent magnets. To explain this type of behaviour we require some mechanism for causing the magnetic dipoles on neighbouring ions to become mutually aligned without the aid of an external magnetic field. Our understanding of this effect relies on the interactions between the electrons on adjacent ions. From our comments at the end of the previous section we know that the 3d electrons in the iron group of transition metals interact strongly with similar electrons on neighbouring ions, so these materials would appear to be prime candidates for exhibiting such alignment. In fact, we find that the interactions between the 3d electrons in these materials have two effects on the magnetic moments of the ions. Firstly, the interaction affects the orbital angular momentum of the 3d electrons in such a way that the average orbital angular momentum on neighbouring ions cancels. In other words the magnetic moment due to the orbital angular momentum becomes zero (see Question 8.12). Secondly, the spins of the electrons interact in such a way that there is a correlation between the spins of the 3d electrons on neighbouring ions. This is called the exchange interaction and was proposed by Werner Heisenberg in 1928.

To illustrate the exchange interaction, let us consider the behaviour of two adjacent ions, each of which possesses a single 3d electron. If the interaction between the electrons is sufficiently strong, we find that in some materials it is energetically favourable for the electrons in the two ions to have the same spin, as shown in Fig. 8.8(a); therefore the spin magnetic moments of the ions are parallel. Although the exchange interaction is effective over a very short range—it is usual to assume that only the spins of electrons on adjacent ions can be altered by this effect—we can see that if each ion affects the dipole moment of each of its neighbouring ions, then all of the atomic dipoles in the crystal will be aligned in a common direction, as shown in Fig. 8.9(a). Such a material is called a ferromagnet.

The exchange interaction therefore provides a mechanism for aligning the atomic dipoles without the need for an external magnetic field, and so a ferromagnet can have a large magnetization even when no external field is present. However, we should point out that the above scenario does not occur in very many materials. A more common result is that the exchange interaction causes the electrons on adjacent ions to have opposite spins, as shown in Fig. 8.8(b); therefore the spin magnetic

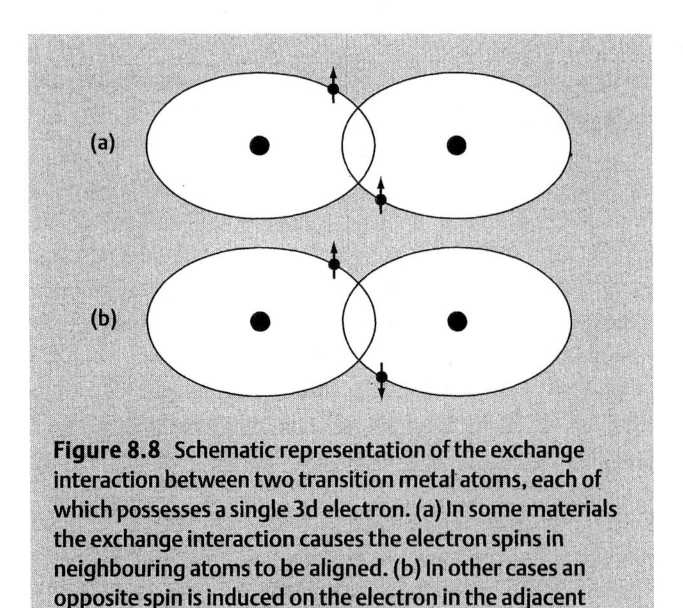

Figure 8.8 Schematic representation of the exchange interaction between two transition metal atoms, each of which possesses a single 3d electron. (a) In some materials the exchange interaction causes the electron spins in neighbouring atoms to be aligned. (b) In other cases an opposite spin is induced on the electron in the adjacent atom.

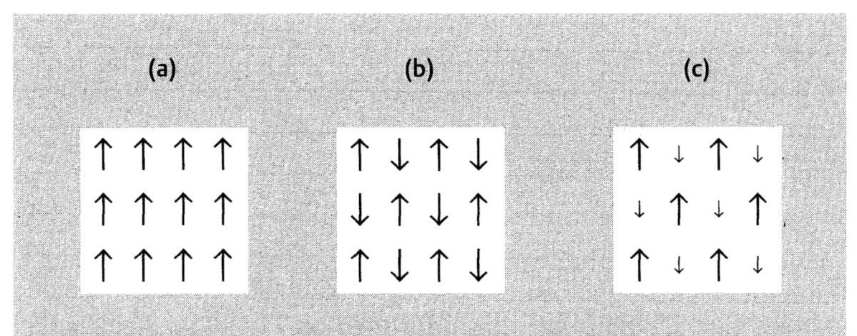

Figure 8.9 The three simplest types of ordering of atomic magnetic moments in a magnetic material are (a) ferromagnetic (adjacent magnetic moments are aligned), (b) antiferromagnetic (adjacent magnetic moments are antiparallel), (c) ferrimagnetic (adjacent magnetic moments are antiparallel but of unequal magnitude).

moments are antiparallel. If we extend this line of reasoning to a crystal, this suggests that the magnetic dipole moments are arranged as shown in Fig. 8.9(b). In this case the neighbouring dipoles cancel each other so that the net magnetization of the crystal is zero! Such materials are described as antiferromagnets.

Unfortunately, there is no simple method of prescribing whether any particular combination of ions will produce parallel alignment of spins (ferromagnetism) or antiparallel alignment (antiferromagnetism).

We should also note that these are not the only possible ways of ordering the atomic dipoles. For example, there are many materials where the magnetic dipoles are arranged in an antiparallel formation, but the magnitudes of the dipoles pointing in one direction are larger than those aligned in the opposite direction. The simplest such arrangement is shown in Fig. 8.9(c). Since the cancellation of the dipole moments is not complete, these materials, which are known as ferrimagnets, exhibit a finite magnetization. In some cases the magnetization of ferrimagnetic materials can be quite large, as for example in the naturally occurring material magnetite (also known as lodestone).

Magnetite is an example of a group of materials called ferrites and it is worth looking at these materials in more detail. Ferrites have the generic chemical formula MFe_2O_4 where M is a doubly ionized metal ion. The two iron ions are both Fe^{3+} and are arranged in an antiferromagnetic arrangement with their dipoles aligned in opposite directions. This means that there is a net magnetic moment due to the M^{2+} ions, as shown in Fig. 8.10. There is a ferromagnetic interaction between the M^{2+} ions so that all of the magnetic dipoles of these ions are aligned in a common direction. In magnetite, M is a doubly charged iron ion (Fe^{2+}), so we can write the chemical formula as Fe_3O_4, but it is important to remember that not all of the iron ions are identical in this case. It is quite simple to calculate the magnetization produced by these M^{2+} ions (see Example 8.4) and the result is in good agreement with the experimental data in Table 8.3.

Another possible way of ordering the dipoles is if the magnetic dipoles in adjacent

Example 8.4 Calculate the magnetization of magnetite, Fe_3O_4, assuming that the magnetization is due only to the Fe^{2+} ions (which have six 3d electrons) and that only the spin angular momentum of the electrons contributes to the magnetic moment of the ions. (The molar volume of Fe_3O_4 is $4.40 \times 10^{-5}\ m^{-3}$.)

Solution According to Hund's rules, the six 3d electrons in each Fe^{2+} ion are arranged so that the spins of five of the electrons are parallel to one another and the sixth electron is antiparallel. We can represent this as

$$\uparrow\uparrow\uparrow\uparrow\uparrow\downarrow$$

Since each electron has a spin magnetic moment of μ_B, the net magnetic dipole moment per ion is $4\mu_B$, and so the dipole moment per mole is $4\mu_B N_A$ where N_A is Avogadro's number (because in 1 mole of Fe_3O_4 there are N_A Fe^{2+} ions).

Consequently, the magnetization (i.e. the dipole moment per unit volume) is

$$M = \frac{4\mu_B N_A}{\text{molar volume}} = \frac{(4)(9.27\times 10^{-24}\ \mathrm{J\,T^{-1}})(6.02\times 10^{23}\ \mathrm{mol^{-1}})}{4.40\times 10^{-5}\ \mathrm{m^{-3}\,mol^{-1}}}$$
$$= 5.07\times 10^{5}\ \mathrm{J\,T^{-1}\,m^{-3}}$$

This is in good agreement with the value of $4.80 \times 10^5\ J\,T^{-1}\ m^{-3}$ in Table 8.3.

Table 8.3 Values of the saturation magnetization at 300 K, shown as M_s in units of $\mathrm{J\,T^{-1}\,m^{-3}}$ and as $\mu_0 M_s$ in units of T, and the Curie temperature, θ_C, for a selection of ferromagnetic and ferrimagnetic materials.

	M_s ($\times 10^5\ \mathrm{J\,T^{-1}\,m^{-3}}$)	$\mu_0 M_s$ (T)	θ_C (K)
Iron	17.1	2.15	1043
Cobalt	14.0	1.76	1388
Nickel	4.85	0.61	627
Gadolinium*	20.6	2.60	292
CrO_2	5.18	0.65	386
Fe_3O_4	4.80	0.60	858
$MnFe_2O_4$	4.10	0.52	573
$NiFe_2O_4$	2.70	0.34	858

*Values of M_s and $\mu_0 M_s$ for gadolinium are at $T = 0$ K because $\theta_C < 300$ K.

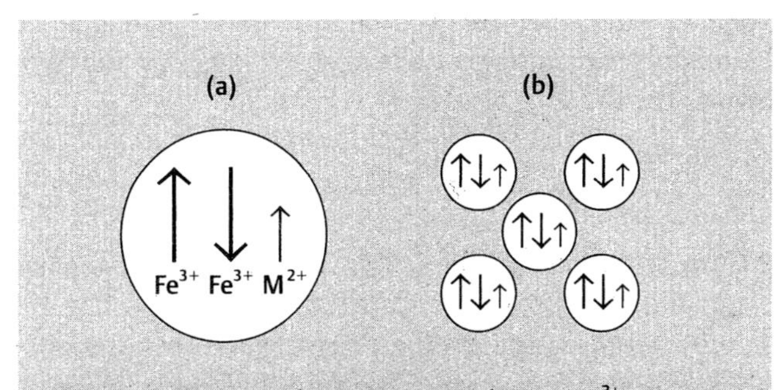

Figure 8.10 (a) In a ferrite, MFe_2O_4, the two Fe^{3+} ions are aligned antiparallel to one another, but there is a net magnetic moment due to the M^{2+} ion. (The oxygen ions do not make any significant contribution.) (b) In a crystal the magnetic moments of the M^{2+} ions are aligned in parallel with one another.

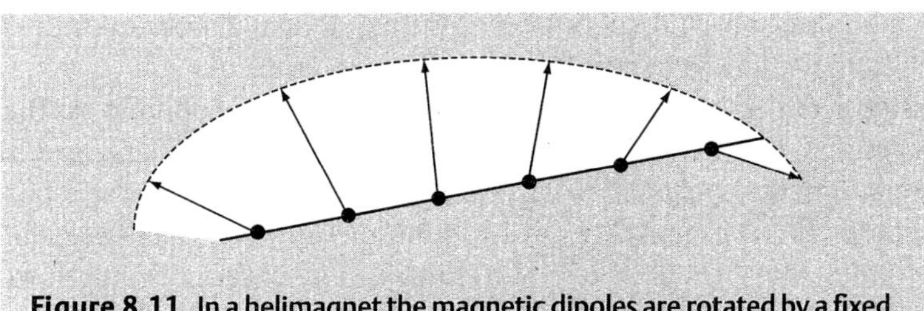

Figure 8.11 In a helimagnet the magnetic dipoles are rotated by a fixed angle relative to the adjacent dipole so that they produce a helix.

planes are misaligned in such a way that the dipoles form a helix, as shown in Fig. 8.11. For example, in magnesium dioxide the angle between the dipoles is about 129°. These materials are called helimagnets. (In a sense we can think of antiferromagnetism as a special case of helimagnetism in which the angle between the dipoles in adjacent planes is 180°.)

8.8 Temperature dependence of permanent magnets

When a permanent magnet is heated it is found that there is a temperature above which the magnetization of the material disappears. The transition point is known as the Curie temperature after Pierre Curie who investigated this phenomenon in the late nineteenth century, although the effect appears to have been discovered by William Gilbert, a scientist in the court of Queen Elizabeth I, almost 300 years earlier.

Above the Curie temperature the material behaves like a paramagnet—the dipoles are randomly aligned unless they are exposed to an external magnetic field. The magnetic susceptibility of this paramagnetic material is given by a modified form of the Curie law, known as the Curie–Weiss law,

$$\chi_m = \frac{C}{T - \theta_C} \tag{8.18}$$

where θ_C is the Curie temperature. The Curie–Weiss law is therefore simply the Curie law with the origin of the temperature scale shifted from absolute zero to the Curie temperature. In other words, as the temperature of a ferromagnet is lowered towards the Curie point, all of the atomic dipoles move to the lowest energy state and therefore become aligned along a common direction. The ordering of the atomic dipoles at temperatures below the Curie point therefore seems to follow naturally from this law.

In iron, cobalt and nickel the Curie temperature is well above room temperature (see Table 8.3), but in some instances the Curie point is below room temperature. For obvious reasons, such materials are not normally considered for use as permanent magnets.

Antiferromagnets also undergo a transition from an ordered phase to a paramagnetic phase at a temperature called the Néel temperature, θ_N. In fact, a large proportion of compounds which display Curie paramagnetism at room temperature are found to become antiferromagnetic at lower temperatures.

The existence of the Curie temperature can be explained as follows. As the temperature is increased, the thermal vibrations of the ions become so large that the alignment of the magnetic dipoles is destroyed. We can assume that this happens when the thermal energy kT is comparable with the interaction energy between the neighbouring dipoles. Since thermal effects are random in nature, we might expect the transition to occur gradually over a temperature range of several degrees, but in fact the Curie temperature is remarkably well defined. To understand why this is so, let us consider what happens if we have a ferromagnetic material in the absence of any external magnetic field. Suppose that initially the temperature is higher than the Curie temperature. In this case the magnetic dipoles are orientated in random directions. If the temperature is now lowered, then as we reach the Curie temperature we find that the thermal vibrations are weak enough to allow a few neighbouring magnetic dipoles to become aligned. As a result, these aligned dipoles create a local magnetic field within the sample, and so other neighbouring dipoles tend to become aligned in the same direction, which in turn further increases the strength of the internal magnetic field. So we have a runaway process which does not stop until all of the magnetic dipoles in the crystal are aligned in the same direction. Such a process is known as a cooperative transition. Consequently, the change from a disordered paramagnet to an ordered magnetic material takes place over a very narrow temperature range.

8.9 Band theory of ferromagnetism*

We have described how ferromagnetism (and ferrimagnetism) can be understood by assuming that the electron spins on neighbouring ions are aligned, and for magnetite we have seen that we obtain excellent agreement with experiment if we assume that the magnetization is due to the alignment of the 3d electrons on the Fe^{2+} ions (see Example 8.4). If we apply a similar argument to iron and assume that all of the ions are Fe^{2+} ions, then we should have a net magnetic moment equivalent to four electrons per ion, giving a total magnetization of about 31×10^5 J T^{-1} m^{-3} (see Question 8.14), but this is considerably larger than the value in Table 8.3. If we instead work backwards from the experimental result, as in Example 8.5, then we find that this value corresponds to a net magnetic moment of 2.18 electrons per ion.

Example 8.5 Determine the magnetic moment per ion for a sample of iron given that the magnetization of iron is 17.1×10^5 J T^{-1} m^{-3}, the density of iron is 7.87×10^3 kg m^{-3} and the molar mass of iron is 0.056 kg mol^{-1}. (Assume that iron consists entirely of Fe^{2+} ions which each have six 3d electrons.)

Solution The volume of 1 mole of iron, V_m, is

$$V_m = \frac{\text{molar mass}}{\text{density}} = \frac{0.056 \text{ kg mol}^{-1}}{7.87 \times 10^3 \text{ kg m}^{-3}} = 7.12 \times 10^{-6} \text{ m}^3 \text{ mol}^{-1}$$

Therefore the volume per ion is

$$V_i = \frac{V_m}{N_A} = \frac{7.12 \times 10^{-6} \text{ m}^3 \text{ mol}^{-1}}{6.02 \times 10^{23} \text{ mol}^{-1}} = 1.18 \times 10^{-29} \text{ m}^3$$

Since the magnetization M is the dipole moment per unit volume, the dipole moment of a single ion is

$$\mu = M.V_i = (17.1 \times 10^5 \text{ J T}^{-1} \text{m}^{-3})(1.18 \times 10^{-29} \text{m}^3) = 2.02 \times 10^{-23} \text{J T}^{-1}$$

We can express this result in terms of the effective number of electrons aligned with the field, n_{eff}, by dividing by the spin magnetic moment of a single electron, μ_B, i.e.

$$n_{eff} = \frac{\mu}{\mu_B} = \frac{2.02 \times 10^{-23} \text{ J T}^{-1}}{9.27 \times 10^{-24} \text{ J T}^{-1}} = 2.18$$

This appears to be a ludicrous result. How can we have 0.18 of an electron aligned along a particular direction? The problem is resolved if we think of the 3d electrons as forming an energy band, rather than identifying them with an individual ion. If we assume that the electron spins are aligned either parallel (↑) or antiparallel (↓) with a given direction, then the magnetization is determined by $N\uparrow - N\downarrow$, as in the case of Pauli paramagnetism (see Section 8.5). However, in a ferromagnet the proportion of electrons that flip their spins is much greater than in Pauli paramagnetism for two reasons:

1. The density of states at the Fermi energy in an iron group transition metal is very large because the 3d energy band has a capacity to hold 10 electrons per atom and

covers only a small range of energies (i.e. the interaction between the 3d electrons is considerably weaker than that between the outer valence electrons). A comparison of the density of states in the 3d and 4s bands is shown in Fig. 8.12. If we draw separate densities of states for N↑ and N↓ and sketch them as in Fig. 8.4, we can see that a small change in energy ΔE results in a comparatively large number of electrons changing state (see Fig. 8.13).

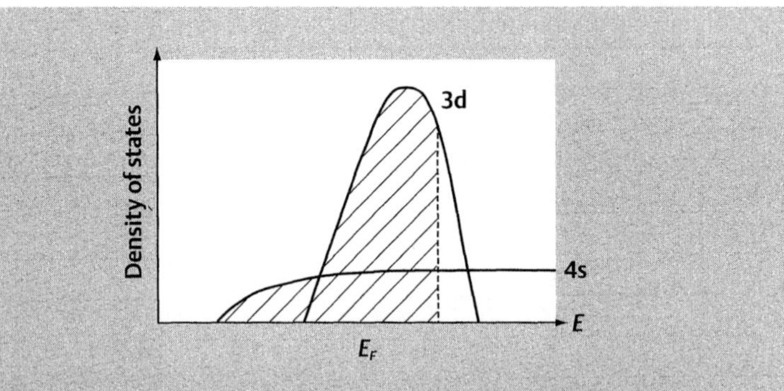

Figure 8.12 The density of states in the 3d band is much higher than that in the 4s band (with which it overlaps) because it covers a narrower range of energies and accommodates a larger number of electrons. For iron (and other transition metals in the same group) the Fermi energy lies within the 3d band and so the density of states at the Fermi energy is very high.

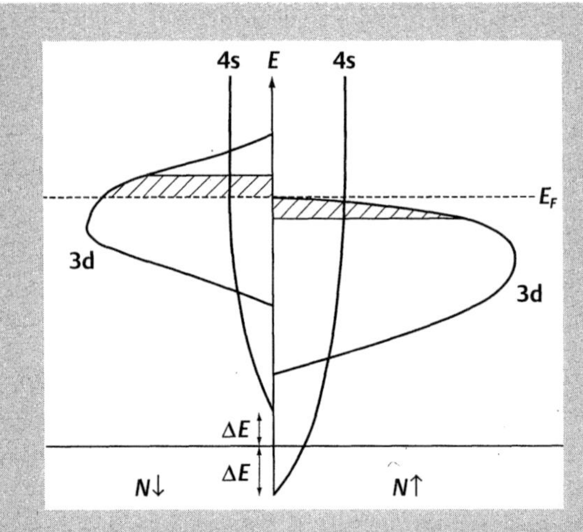

Figure 8.13 In a ferromagnetic material the energies of electrons with spins parallel to a particular direction (↑) are decreased by the interaction energy ΔE and the energies of electrons with spins antiparallel to this direction (↓) are increased by ΔE. The diagram is similar to Fig. 8.4(c), but the density of states at the Fermi energy is much larger because of the 3d band and the energy ΔE is much larger than the energy $m_m B_0$ in Fig. 8.4.

2. The energy shift ΔE in Fig. 8.13 is very much greater than the corresponding energy change in Fig. 8.4. For a ferromagnet, ΔE is a measure of the interaction energy between the 3d electrons, which we can assume is roughly comparable in magnitude with the thermal energy at the Curie temperature, θ_C. For iron this energy is about 90 meV (see Question 8.16). In comparison, the energy change that occurs in Pauli paramagnetism is $m_m B_0$, which even in a large magnetic field of 1 T is only about 0.06 meV (see Question 8.7).

Therefore the alignment of electron spins in a ferromagnet results in a much larger magnetization than occurs in Pauli paramagnetism.

8.10 Ferromagnetic domains

One of the apparently contradictory features of magnetic materials is that a ferromagnet is not always magnetized even when the temperature is lower than the Curie point. This appears to be at odds with our explanation of ferromagnetism on an atomic scale. How can the atomic magnetic dipoles be aligned and yet produce a magnetization of zero?

A solution to this puzzle was suggested by the French physicist Pierre Weiss. He proposed that a ferromagnet is divided into several domains. The dipoles within each domain are mutually aligned in a common direction, but the direction of magnetization is different in neighbouring domains. This is shown schematically in Fig. 8.14.

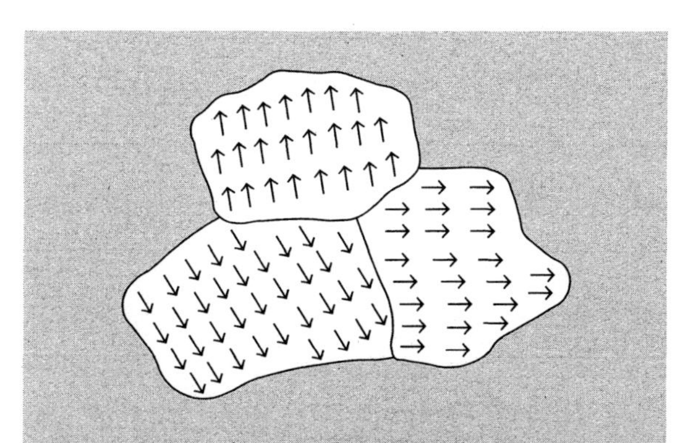

Figure 8.14 Schematic diagram of the domains in a ferromagnet showing how the atomic dipoles are aligned in the same direction within a domain, but are in different directions in neighbouring domains. The dimensions of the grains are typically 10 μm to 100 μm. (Note that although the diagram may appear to be similar to that of the crystal grains in Fig. 2.3, it is important to realize that domains and crystal grains are generally quite different entities.)

Between adjacent domains there is a narrow region, called a Bloch wall, in which each dipole is slightly misaligned with the neighbouring dipoles. As a result, over a distance of typically a few hundred atomic spacings the orientation of the dipoles changes from the direction of one domain to that of the adjoining domain, as shown in Fig. 8.15.

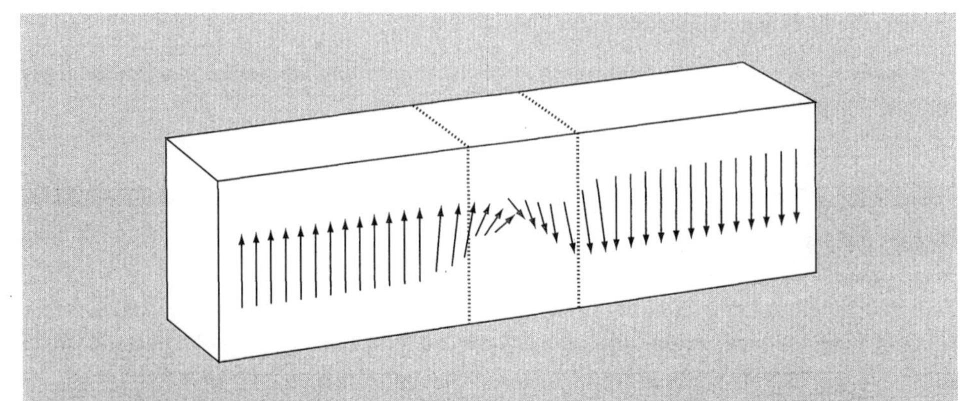

Figure 8.15 Illustration of how the magnetic dipoles change their orientation from one domain to the next. The width of this region is typically a few hundred atomic spacings.

Figure 8.16 The domain walls in a single crystal of nickel obtained by the Bitter method. Tiny grains of magnetic iron oxide collect along the lines where the Bloch walls meet the surface.

Although the concept of domains was introduced in 1907, it was not until 1931 that domains were actually observed. The original technique for making domains visible, known as the Bitter method, consists of dusting the surface of a ferromagnet with fine particles of magnetic iron oxide. The particles are attracted to the domain walls and so produce an outline map of the domains on the surface of the crystal as shown in Fig. 8.16. An alternative technique is to shine polarized light onto the surface of the ferromagnet. The different directions of magnetization in each domain cause the polarization of the reflected light to be altered slightly, producing a pattern as shown in Fig. 8.17.

Why do domains form? To answer this question let us consider how the magnetic dipoles are arranged in a sample of iron which has just been smelted from the ore and allowed to cool in the absence of any external magnetic field. We can consider two possibilities. Firstly, if all of the magnetic dipoles in the sample are aligned then the material is magnetized and produces a magnetic field which extends beyond the sample, as shown in Fig. 8.18(a). However, a large amount of energy is stored in this external magnetic field. Alternatively, the existence of domains means that the magnetic dipoles could be arranged as in Fig. 8.18(b). In this case the magnetization is zero

Figure 8.17 Polarized light reflected from a magnetic surface is rotated by slightly different amounts according to the magnetization of the atomic dipoles in each domain. This picture shows the magnetic domains in a sample of cobalt samarium.

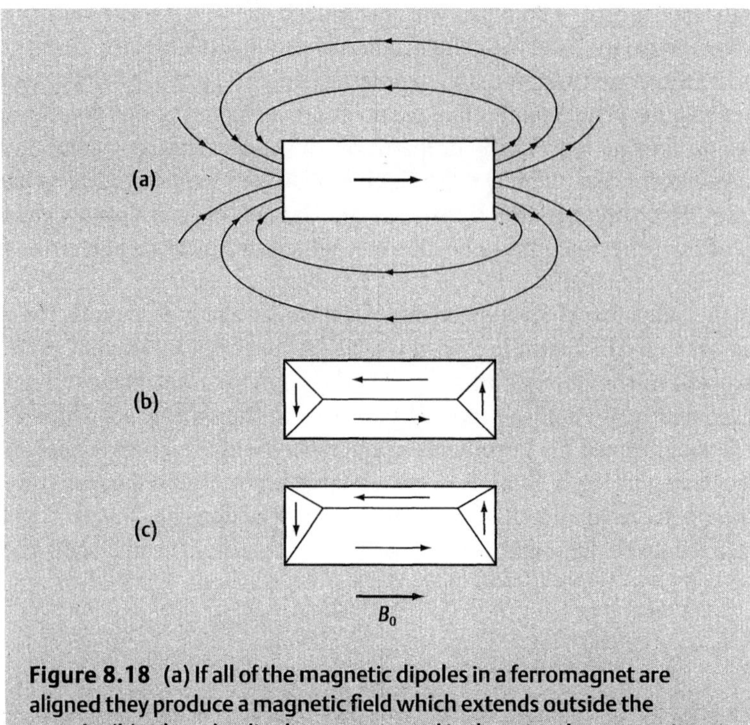

Figure 8.18 (a) If all of the magnetic dipoles in a ferromagnet are aligned they produce a magnetic field which extends outside the sample. (b) When the dipoles are arranged in domains the magnetization of the crystal can be zero, as in this instance. (c) When an external magnetic field B_0 is applied, the domain walls move so that a greater number of the magnetic dipoles are aligned parallel to the field than antiparallel. As a result the magnetization of the sample is non-zero.

and there is no magnetic field outside the sample, but energy is involved in creating the domain walls. Which is the most likely outcome? The answer is the configuration which has the lowest energy. From experience we know that a freshly smelted piece of iron is not usually magnetized, and so we can assume that the energy contained in the domain walls in the latter case is less than that in the magnetic field in the former case.

Let us then consider what happens when an external magnetic field is applied to the sample. It is now energetically favourable for more of the magnetic dipoles to be aligned with the external field, and so the domain walls move accordingly, as shown in Fig. 8.18(c), and the material becomes magnetized.

The most interesting part of this process is that when the external field is removed, the domain walls do not necessarily return to their original positions (i.e. as in Fig. 8.18(b)) and so the material remains magnetized. The residual magnetization when the external field has been reduced to zero is called the remanent magnetization, M_R.

We can understand why this occurs by considering the analogous situation of pushing a crate up a slight incline. If the frictional forces between the crate and the incline

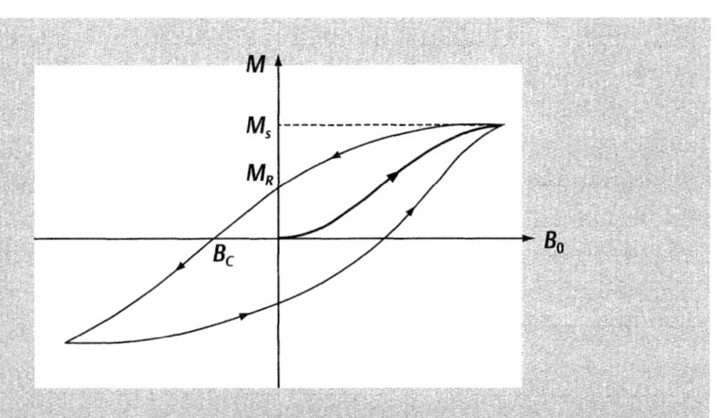

Figure 8.19 A typical hysteresis curve for a ferromagnet showing the magnetization M as a function of the external magnetic field B_0. The heavy curve shows the response when an external magnetic field is applied to an initially unmagnetized sample. Once the sample has become magnetized it follows the lighter curve. Consequently, when the magnetic field is removed the magnetization decreases to a value M_R when the external field is zero. A reverse field of magnitude B_C directed in the opposite sense to the magnetization of the sample is needed to return the magnetization to zero. The saturation magnetization, M_s, corresponds to the situation in which all of the magnetic dipoles in the sample are aligned with the external field.

are sufficiently large, then the crate will not slide back to the bottom of the slope even when the external force is removed. In a ferromagnet the friction is provided by the presence of imperfections in the crystal structure such as grain boundaries. Energy is dissipated in moving the domain walls past these obstacles, and when the external field is removed the domain walls are pinned by these imperfections and so the sample remains magnetized. To return the sample to its original state (i.e. to demagnetize it) requires that a magnetic field be applied in the opposite direction to the magnetization of the sample. This is known as the coercive field, B_C.

From the above discussion we can see that if the magnetization of the sample is plotted against the external magnetic field strength, then the magnetization and demagnetization curves are different. This produces an effect called hysteresis. A typical hysteresis curve for a ferromagnet is shown in Fig. 8.19. We will consider these hysteresis curves in more detail in the following section.

8.11 Soft and hard magnets

What are the ideal characteristics of a permanent magnet? It should have a large remanent magnetization (i.e. the magnetization should be large even when there is no external field), and the coercive field should be large so that the sample is not easily demagnetized. However, perhaps the most important feature is that it should have a large energy product. The energy product is given by the maximum value of the expression $B\,B_0/\mu_0$ and represents the energy stored in the magnet per unit volume. We can estimate this value graphically by plotting the magnetic induction B as a function of the external magnetic field B_0 (as shown in Fig. 8.20). (Note that this curve is very similar in shape to the plot of magnetization against magnetic field shown in Fig. 8.19. This is because the magnetic induction is related to the magnetization by eqn (8.2).) From the top left quadrant of this curve we can determine the maximum value of the product $B\,B_0$. Dividing by the permeability μ_0 we obtain a quantity with units of energy per unit volume (see Question 8.17). The energy product is important from the technological viewpoint because by increasing the energy product we can achieve a given magnetic field with a smaller magnet.

Materials which display the above properties are known as hard magnets and a typical hysteresis curve is shown in Fig. 8.21(a). However, these characteristics are not

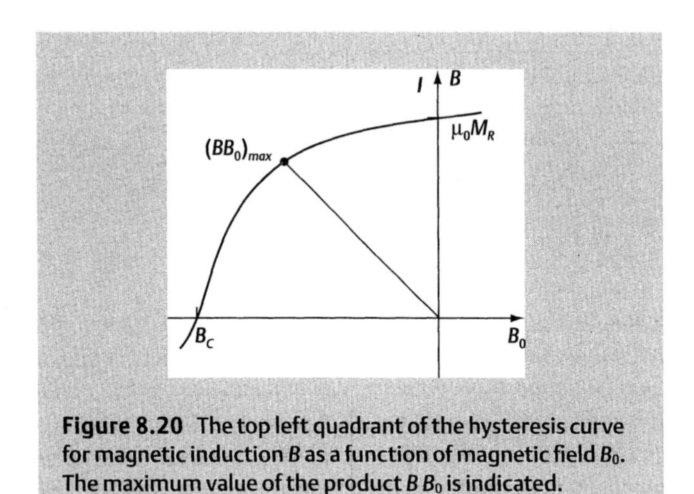

Figure 8.20 The top left quadrant of the hysteresis curve for magnetic induction B as a function of magnetic field B_0. The maximum value of the product $B\,B_0$ is indicated.

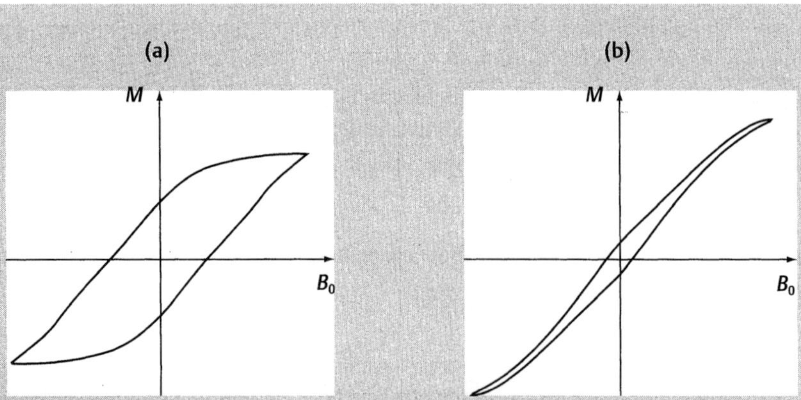

Figure 8.21 Typical hysteresis curves for (a) a hard magnet and (b) a soft magnet.

ideal for all applications. For example, the magnets used in motors and transformers may change their direction of magnetization many times each second. In this case we require the time lag between the changing magnetic field and the magnetization of the magnet to be as small as possible. This is achieved by making a soft magnet which has a low coercive field (see Fig. 8.21(b)). (Note that the magnitude of the remanent magnetization is not particularly important in this case, but the saturation magnetization—corresponding to the situation in which all of the magnetic dipoles are aligned with the field—should ideally be as large as possible.)

We can understand the difference between hard and soft magnets on a microscopic scale by again considering the movement of the domain walls. To create a soft magnet we need the domain walls to move as easily as possible, or using the analogy from the previous section, we want the friction to be as low as possible. This can be achieved if we have a material with large crystal grains to minimize the interaction between the domain walls and the grain boundaries. Conversely, hard magnets are usually made from fine-grained samples. In fact, the best hard magnets consist of magnetic particles which are so small (typically 10 nm to 100 nm across) that each particle is a single domain. In this case the only way to demagnetize the material is to cause all of the dipoles within the domain to change direction simultaneously, which means that the coercive field is very large.

Coarse-grained pure iron is a rather good soft magnet, but small amounts of silicon are usually added in order to increase the electrical resistivity of the material. This is important for applications in motors and transformers because the increased resistivity reduces losses due to eddy currents. Even lower losses are required for magnetic components in electronic circuits. For these applications special alloys with tradenames such as Permalloy and Supermalloy have been developed. These materials are roughly 80% nickel and 20% iron. As they are far more expensive than iron–silicon alloys, their use is confined to small-scale applications.

A large variety of different materials and techniques have been used to create better hard magnets, and the properties of a selection of these are compared in Table 8.4. Alnico was introduced in the 1930s and is a fine-grained alloy consisting of roughly 50% iron with various concentrations of aluminium, nickel and cobalt (hence the

Table 8.4 Values of the coercive field, Bc, the remanent magnetization $\mu_0 M_R$ and the energy product $(BB_0/\mu_0)_{max}$ for a selection of permanent magnets.

	Bc (T)	$\mu_0 M_R$ (T)	$(BB_0/\mu_0)_{max}$ (kJ m^{-3})
Pure iron	7×10^{-5}	1.2	~0.05
Steel	5×10^{-3}	0.9	~1
Alnico V	0.07	1.3	55
Co_5Sm	1.0	0.8	160
$Nd_2Fe_{14}B$	1.2	1.2	300

name, which is simply derived from the chemical symbols AlNiCo). The remarkable magnetic properties of cobalt samarium alloy Co_5Sm were discovered in the late 1960s and enabled very small, powerful magnets to be produced, but the high cost of this material has restricted its use in commercial applications. More recently, the compound $Nd_2Fe_{14}B$ has been shown to have properties superior to Co_5Sm and is considerably cheaper.

It is interesting to note that whilst these new materials have substantially increased the coercive field and the energy product, the remanent magnetization has changed very little, and is generally comparable with that of pure iron.

8.12 Applications of magnetic materials for information storage

We mentioned in the introduction to this chapter that an important application of magnetic materials is in information storage. This includes audio and video tapes, floppy disks and hard disks for computers. In this section we will take a brief look at how magnetic materials are used in this context.

The basic principle of magnetic storage of information is broadly similar for the above-mentioned systems regardless of whether the data stored is digital (as on computer disks) or analogue (as on most audio and video tapes). In each case the magnetic medium—the tape or disk—is covered by many small magnetic particles, each of which behaves as a single domain. A magnet—known as the write head—is used to record information onto this medium by orientating the magnetic particles along particular directions as shown in Fig. 8.22. Reading the information from the tape or disk is simply the reverse of this process—the magnetic field produced by the magnetic particles induces a magnetization in the read head.

The choice of material for the magnetic recording medium depends on several factors. For example, the coercivity must be low enough so that the orientation of the particles can be altered by the write head, but high enough so that the orientation is not accidentally affected by other external magnetic fields. The Curie temperature should also be well above the temperatures to which the material will normally be exposed, otherwise thermal effects can alter the orientation of the particles. This latter consideration causes potential problems with chromium dioxide (CrO_2), which for some time was a popular material for audio tapes. Although this material has many excellent qualities as a recording medium, problems were experienced because the Curie temperature is only 128°C.

Magnetic hard disks are used in computers because they are considerably cheaper than semiconductor memory and also have the advantage that they are non-volatile, i.e. unlike most semiconductor circuits, a magnetic system does not require a constant electrical power supply in order to remain in the same memory state. In order to keep pace with the advances in semiconductor technology it has been necessary to increase continually the storage density of these magnetic hard disks, from about 2

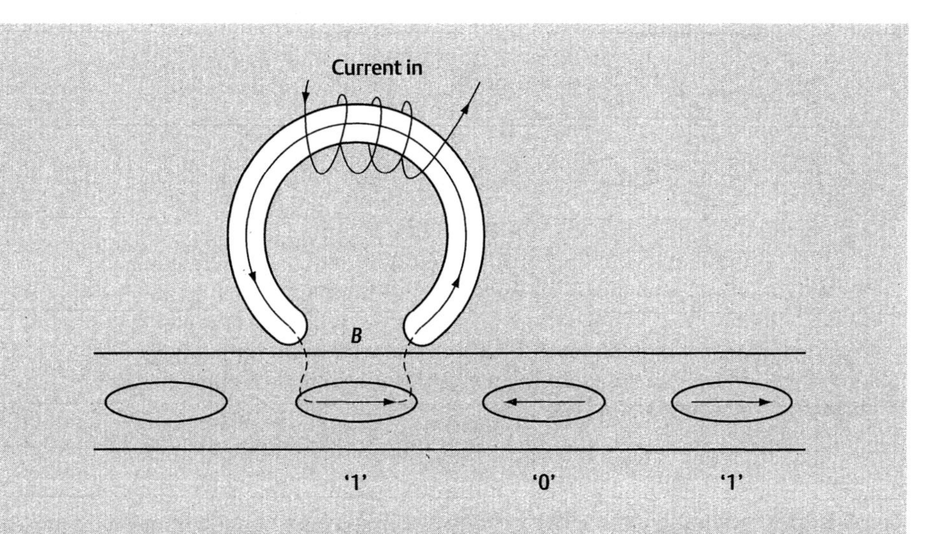

Figure 8.22 Schematic diagram showing the process of recording information onto a magnetic medium. A coil is used to convert an electrical signal into a magnetic field in the write head, which in turn affects the orientation of the magnetic particles on the magnetic tape or disk. In this instance there are only two possible directions of magnetization of the particles, so the data is in binary form and the two states can be identified as '0' and '1'.

MB cm^{-2} (million bits per square centimetre) in the early 1980s to about 75 MB cm^{-2} in 1996. However, increasing the storage density leads to several problems.

Firstly, as the size of the magnetic particles is reduced, the magnetic field that they produce becomes smaller, typically of the order of 0.001 T in a 75 MB cm^{-2} hard disk. The type of read head described above, which depends on magnetic inductance, cannot function with such small fields, and so the read heads in hard disks are now made using magnetoresistive materials. In magnetoresistive materials the electrical resistivity of the material changes in the presence of a magnetic field, and so it is possible to convert the magnetic field directly into an electrical response. In most materials this effect produces only a very small change in resistivity, but in 1988 it was discovered that a structure fabricated from extremely thin (approximately three atoms thick!) alternate layers of iron and chromium has a magnetoresistance about a hundred times larger than that of any natural material. This effect is known as giant magnetoresistance (or GMR) and is discussed in more detail in articles listed in Further reading. Such extraordinary sensitivity can be used to detect even smaller magnetic fields, meaning that much higher storage densities may be possible in the near future.

A second problem with increasing the storage density is that as the size of the magnetic particles is reduced and the particles are packed more closely it becomes increasingly likely for the particles to demagnetize each other. A solution to this is to align the particles so that they are perpendicular to the surface of the disk, as shown in Fig. 8.23.

An entirely different approach to magnetic data storage is to use magneto-optic

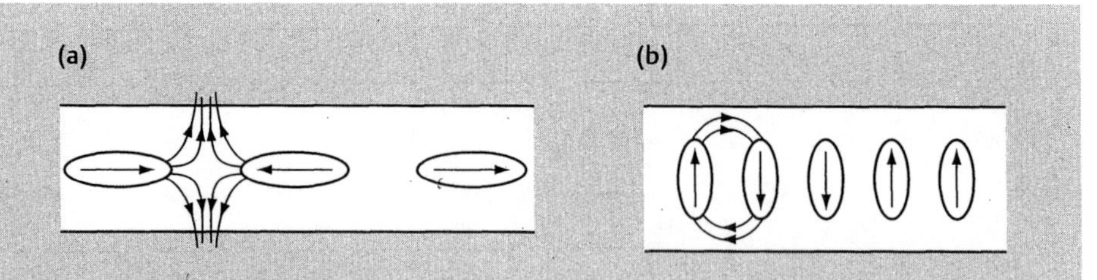

Figure 8.23 (a) In a conventional magnetic medium the magnetic particles are orientated parallel to the recording surface. As a result, particles magnetized in opposing directions tend to demagnetize one another. (b) By orienting the particles perpendicular to the surface the density of particles is increased and the magnetic fields reinforce the magnetization of neighbouring particles.

disks. In this technology the write process is achieved by using a laser beam to heat a magnetic particle above the Curie temperature as a magnetic field is applied. As it cools down the particle retains the magnetization of the applied field. To read the information another (weaker) polarized laser beam is directed onto the surface. The polarization of the reflected beam is determined by the direction of magnetization of each particle, as mentioned in Section 8.10.

8.13 Summary

Magnetism is such a common everyday phenomenon that we tend to take it for granted, but hopefully by reading this chapter you now have a better appreciation of what is involved in making a material magnetic.

We have seen that although most atoms are magnetic, these effects tend to cancel when we form a solid. One exception is transition metals which retain a non-zero dipole moment even when they are ionized in a solid. For most such materials at room temperature the dipoles are randomly aligned and the material displays only a very weak response to an external magnetic field, but in a few cases the electrons on neighbouring dipoles interact to produce a spontaneous ordering of the dipoles. In many of these instances the alignment is such that neighbouring dipoles cancel one another (antiferromagnetism), and so we are left with only a very small number of materials which behave as permanent magnets.

We have also seen that the existence of magnetic domains can explain the fact that ferromagnets (and ferrimagnets) are often in an unmagnetized state. In fact, the only reason why a magnetic material can have a finite magnetization in the absence of an external magnetic field is because the movement of the domain walls is not a reversible process.

Despite the fact that magnetic materials have been used in a wide range of applications for many years, there are still many properties of magnetic materials which are not completely understood. The design of improved soft and hard magnets is still something of a black art. The use of multilayered magnetic materials in which the layers are only a few atoms thick has revealed many new possibilities for magnetic applications and is a subject of huge research interest. In particular, the discovery of giant magnetoresistance has already found some commercial applications, and could even be used to make tiny magnetic transistors.

Questions

Questions marked * are more challenging or are based on the optional sections.

8.1 Macroscopic magnetic quantities. Use the definitions in Section 8.2 to check that eqns (8.1) and (8.2) are dimensionally correct.

8.2 Atomic magnets. Show that for a filled electron subshell in an atom the total spin and orbital angular momenta, S and L, are zero.

8.3 Atomic magnets. Show that if an electron subshell is half filled then $L = 0$.

8.4 Atomic magnets. Determine the values of S, L and J for (a) Ni^{2+}, which has eight electrons in the 3d subshell, and (b) Gd^{3+}, which has seven electrons in the 4f subshell. (Note: the f subshell corresponds to $l = 3$.)

8.5 Atomic magnets. Determine values of the Landé splitting factor g for (a) a system with no spin angular momentum, and (b) a system with no orbital angular momentum.

8.6 Pauli paramagnetism. Using data from Tables 4.1 and 4.2, determine the magnetic susceptibility, χ_m, for sodium and gold. By comparing your result with the measured values of χ_m in Table 8.2, determine the diamagnetic contribution to the magnetic susceptibility in each case.

8.7 * Pauli paramagnetism. Estimate the proportion of valence electrons in a metal that contribute to Pauli paramagnetism in a magnetic field of 1 T. (Note that this is a very large magnetic field.) Assume that the Fermi energy is 3 eV and that the density of states function is constant for all energies above the valence band edge.

8.8 * Curie paramagnetism. In the metallic state the transition metal scandium has a single electron in the 3d subshell. Calculate the values of the total angular momentum J and the Landé splitting factor g, and use these values to determine the energy of the lowest energy dipole moment in a field of 0.5 T. Determine the relative probability of the atomic dipoles occupying the highest and lowest energy levels at temperatures of 1 K and 300 K.

8.9 * Curie paramagnetism. By extending the derivation in Example 8.3, show that the Curie constant is given by

$$C = \frac{\mu_0 N g^2 \mu_B^2 J(J+1)}{3k_B}$$

Note that an expression for the constant A in Example 8.3 can be obtained by using the fact that the total probability of a dipole having one of the $2J+1$ possible orientations is equal to 1.0, i.e.

$$\sum_{-J}^{J} Ae^{m_J B_0/k_B T} = 1.0$$

You will also need to use the mathematical identity

$$\sum_{-J}^{J} J^2 = \frac{1}{3} J\,(J+1)(2J+1)$$

8.10 Curie paramagnetism. A solid contains a dilute concentration of Nd^{3+} ions, each of which possess three 4f electrons. Assuming that there are $10^{25}\ m^{-3}$ of these ions, calculate the magnetic susceptibility of the sample at 1 K and 300 K.

8.11 Curie paramagnetism. Calculate the value of the effective Bohr magneton number p for the following ions and compare your results with the experimental values (shown in parentheses.)

Nd^{3+} (3.5) three 4f electrons

Gd^{3+} (8.0) seven 4f electrons

Cr^{3+} (3.8) three 3d electrons

Cr^{2+} (4.9) four 3d electrons

Fe^{2+} (5.4) six 3d electrons

Ni^{2+} (3.2) eight 3d electrons

8.12 Ordered magnetic materials. Calculate the effective Bohr magneton number p for Cr^{3+}, Cr^{3+}, Fe^{3+} and Ni^{3+} assuming that the total orbital angular momentum L is zero. Compare your results with the experimental data given in Question 8.11.

8.13 Ordered magnetic materials. Calculate the magnetization of the ferrite $NiFe_2O_4$ assuming that the magnetization is due only to the Ni^{2+} ions (which have eight 3d electrons). The molar volume is $4.30 \times 10^{-5}\ m^{-3}$ and you can assume that only the spin angular momentum of the electrons contributes to the magnetic moment of the ions.

8.14 * Band theory of ferromagnetism. Estimate the magnetization of iron given that the density and molar mass of iron are 7.87×10^3 kg m^{-3} and 0.056 kg mol^{-1}, respectively. (Assume that metallic iron consists entirely of Fe^{2+} ions, each of which has six 3d electrons.)

8.15 * Band theory of ferromagnetism. Using the value of magnetization in Table 8.3, determine the magnetic moment per ion for cobalt given that the density and molar mass of cobalt is 8.9×10^3 kg m^{-3} and 0.059 kg mol^{-1}, respectively. (Assume that metallic cobalt consists entirely of Co^{2+} ions, each of which has seven 3d electrons.)

8.16 * Band theory of ferromagnetism. Use the values of the Curie temperature in Table 8.3 to estimate the magnitude of the interaction energy ΔE for iron, nickel and cobalt.

8.17 Soft and hard magnets. Prove that the term $B\,B_0/\mu_0$ has units of [energy]/[volume].

8.18 Soft and hard magnets. Determine expressions for the energy product $(B\,B_0/\mu_0)_{max}$ for the hysteresis curves shown in Figs 8.24(a) and (b). Assume that these curves represent extremes of behaviour, and use these expressions and the values of B_C and $\mu_0 M_R$ in Table 8.4 to place upper and lower limits on the energy products for Alnico V, Co_5Sm and $Nd_2Fe_{14}B$.

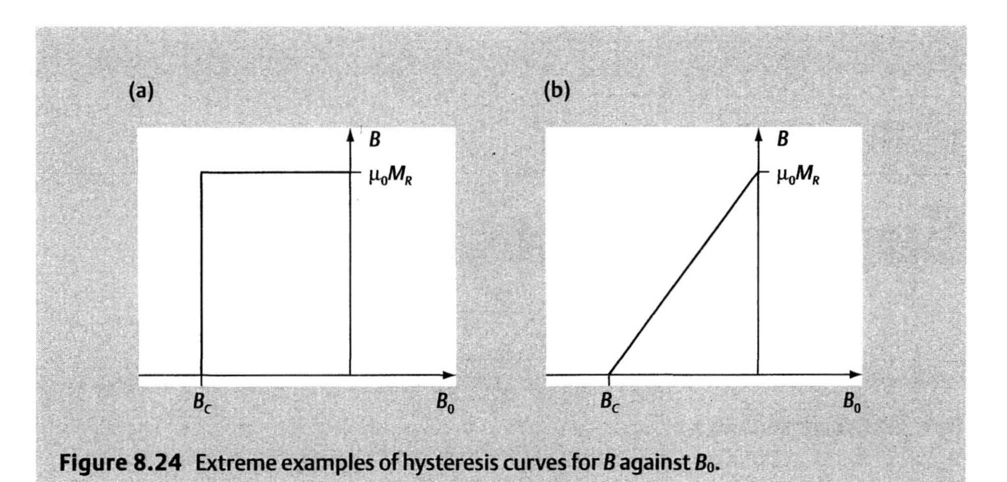

Figure 8.24 Extreme examples of hysteresis curves for B against B_0.

Chapter 9
Superconductivity

9.1 Introduction

When I give a lecture on superconductivity, I usually begin by asking the audience what they already know about the properties of superconductors. Invariably, I obtain the answer that these materials have zero resistivity and that the phenomenon only occurs at very low temperatures. So let us take this as a starting point. Bearing in mind what we know about the electrical properties of metals, do we expect the resistivity of a metal to become zero at very low temperatures?

From our discussion in Chapter 4 we know that one source of electrical resistance is the vibrations of the ions. As the temperature is lowered, the amplitude of the vibrations becomes smaller and so the electrical resistance is reduced. Indeed, experimental measurements on most metals at around about room temperature show that the resistivity is directly proportional to temperature. If we extrapolate this result it suggests that the resistivity should become zero at a temperature of about 0 K.

However, we are forgetting that there are other sources of resistance, such as the presence of impurities and imperfections in the crystal structure. At room temperature these contributions to the resistivity are generally negligible for reasonably pure metals, but they become increasingly important as the temperature is reduced. Since a real sample of metal always contains some impurities and imperfections, we are left with a residual resistivity, ρ_0, at absolute zero. This value is typically about 1% of the resistivity at room temperature for a pure sample. The relationship between resistivity and temperature for a 'normal' metal is shown in Fig. 9.1. (Note that in this chapter we will use the term 'normal' to signify a material which is not in the superconducting state.)

We are left with an interesting conundrum. Our theory of electrical resistance, which seems to have worked very well so far, suggests that all metals should exhibit a finite resistivity even at absolute zero. How then do we explain the fact that superconductors exist? It is obvious that we need either to modify our theory or else to come up with an entirely new explanation for this phenomenon. We will come back to this problem shortly, but first let us take a more detailed look at some of the properties of superconductors.

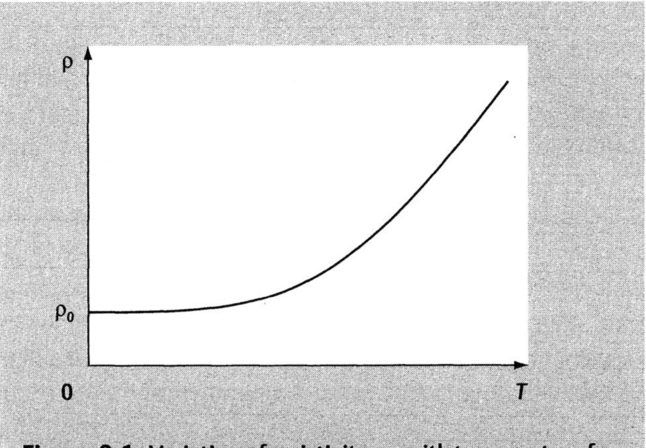

Figure 9.1 Variation of resistivity, ρ, with temperature for a metal which does not become a superconductor. At absolute zero there is a residual resistivity ρ_0.

9.2 The discovery of superconductivity

The phenomenon of superconductivity was discovered by Heike Kamerlingh Onnes in 1911 whilst investigating the electrical properties of metals at very low temperatures. He had previously measured the resistivity of platinum and found that for temperatures down to about 1 K, the resistivity followed a curve similar to that in Fig. 9.1. However, when he performed a similar experiment with a sample of mercury he obtained a quite different result. At a temperature of about 4.2 K the resistance of the sample dropped sharply to a value close to zero, as shown in Fig. 9.2.

Kamerlingh Onnes was unable to measure either the resistivity of the sample in

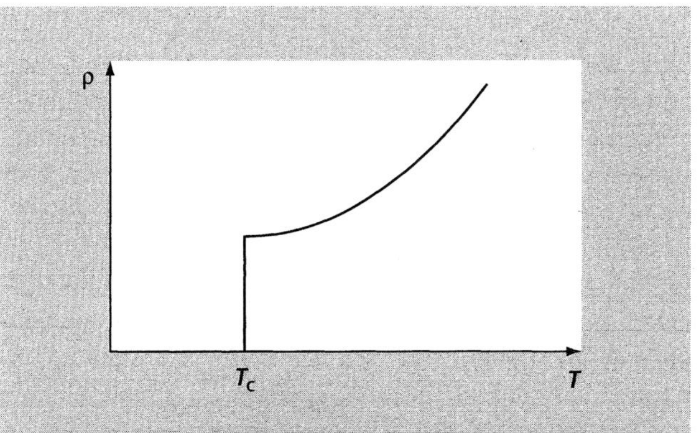

Figure 9.2 Variation of resistivity with temperature for a metal which becomes a superconductor at a critical temperature T_c.

this state, or the temperature range, ΔT, over which the transition from normal metal to superconductor takes place. We now know that ΔT is generally of the order of 10^{-3} K for a pure sample. However, no experiment has yet been able to measure the resistivity of a material in the superconducting state. The best that can be achieved is to place an upper limit on the resistivity. (We will discuss this further in the next section.)

Let us take a closer look at the transition from the normal to the superconducting state. When a physical property of a material changes abruptly at a particular temperature we say that a phase change has taken place. For example, the volume of a fixed quantity of water changes discontinuously at 0°C as it becomes ice. Similarly, the transition from a normal metal to a superconductor is also referred to as a phase change, although in this case the physical appearance of the material is not altered. The temperature at which this phase change occurs is called the critical temperature, T_c. For most superconductors the critical temperature is only a few degrees above absolute zero, but there are a number of materials which have critical temperatures in the region of 100 K. These are known as high-temperature superconductors and we will look at these in more detail in Section 9.8.

Although a superconducting material behaves as a normal metal at temperatures above T_c, Kamerlingh Onnes soon discovered that the converse is not true, i.e. the material does not necessarily behave as a superconductor when the temperature is below T_c. Whether or not a sample of material superconducts is found to depend not only on the temperature, but also on the magnetic field to which the sample is exposed and on the amount of current flowing through the sample.

Let us first of all consider the effects of a magnetic field. The maximum magnetic field that a material can tolerate and still remain superconducting is zero at T_c, but increases to a maximum value B_c as the temperature is reduced to absolute zero. Experimentally, the critical magnetic field, B_T, at temperature T is generally found to obey the relationship

$$B_T = B_c\left[1-\left(\frac{T}{T_c}\right)^2\right] \tag{9.1}$$

We can represent this result graphically by making a plot of the magnetic field strength against temperature. The above equation is represented by the curve shown in Fig. 9.3. What this graph shows is that for combinations of magnetic field strength and temperature corresponding to points under this curve the material behaves as a superconductor, whereas outside this region the material behaves as a normal metal.

We can understand the existence of a critical current density if we remember that a current flowing through a wire produces a magnetic field. It is therefore possible for the current flowing in a superconductor to produce a magnetic field which exceeds the critical magnetic field of the material (seeQuestion 9.1). However, I should point out that there are other factors which also limit the critical current density (e.g. see Section 9.5), so the actual critical current density of a sample may be considerably smaller than the value predicted by the above-mentioned calculation.

Since there are three independent parameters which affect whether or not a material superconducts, the criteria for achieving superconductivity should be represented

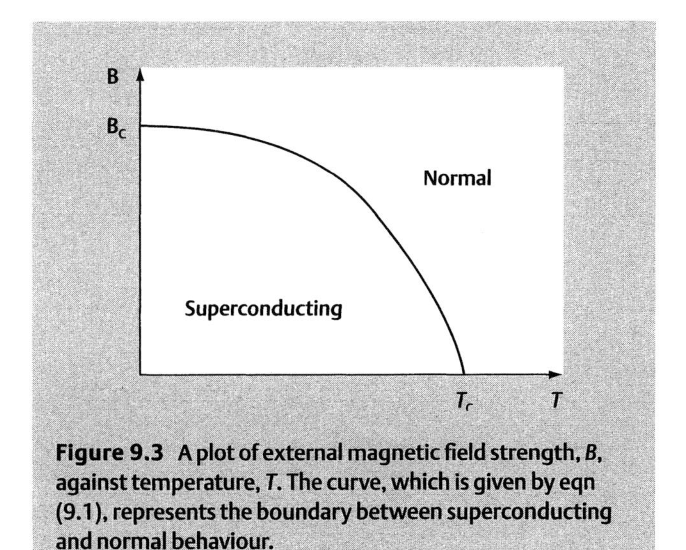

Figure 9.3 A plot of external magnetic field strength, B, against temperature, T. The curve, which is given by eqn (9.1), represents the boundary between superconducting and normal behaviour.

by a three-dimensional graph, as in Fig. 9.4. For all combinations of temperature, magnetic field and current density corresponding to points below the critical surface the material behaves as a superconductor, and for all other combinations the material is a normal metal.

Consequently, in order to use a superconductor in a particular application it is necessary to consider not only the temperature at which the device will operate, but also the maximum magnetic field and current density to which it will be subjected. A good example of this is the development of superconducting magnets. The discovery of materials with zero resistivity lead Kamerlingh Onnes to believe that he could construct powerful electromagnets using a solenoid of superconducting wire. However, he was thwarted by the fact that the magnetic field produced by the current in the solenoid caused the material to revert to a normal metal at very low fields. This dream has now been realized with the discovery of niobium alloys which typically have critical temperatures of around 20 K, critical magnetic fields of up to 50 T at liquid helium temperatures and critical current densities in excess of 10^7 A cm^{-2}.

9.3 Is the resistivity of a superconductor zero, or just very small?

The resistivity of a superconductor is certainly far too small to be able to measure it by any conventional means. Instead, experiments have been performed in which a pulse of current is induced into a superconducting loop and then left for a period of time in order to try and measure the decrease in the magnitude of the current.

To see how this works, consider what happens if we induce a pulse of current into

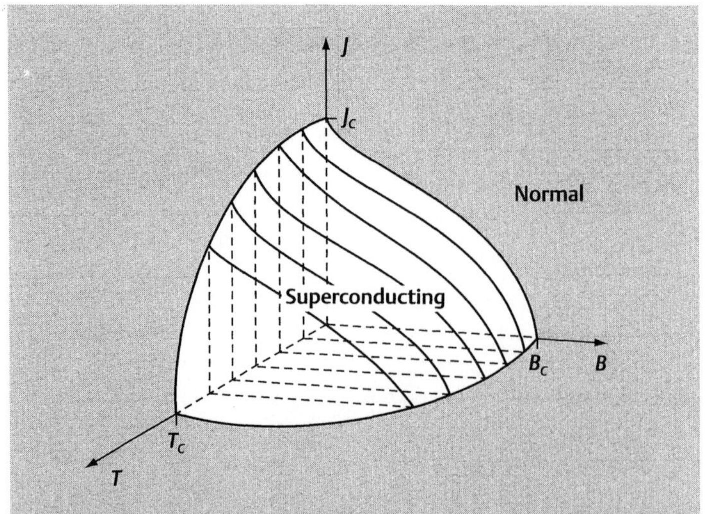

Figure 9.4 The ability of a material to superconduct depends on three factors: the temperature, T, the magnetic field strength, B, and the current density J. T_c is the critical temperature in the absence of an external magnetic field and with no current flowing in the sample, B_c is the critical magnetic field strength with no current flowing at $T = 0$ K, and J_c is the critical current density at $T = 0$ K with no external magnetic field. The critical surface shown in the figure separates the superconducting and normal states.

a loop of wire which behaves as a normal metal. Let us suppose that when the current is induced the electrons are all moving with a small drift velocity in the direction of the current. After a brief time some of the electrons are deflected as a result of interactions with phonons or impurities. Since there is no electric field to keep pushing them in the direction of the current flow, we can assume that the direction of motion of the electrons after the interaction is completely random. This means that these electrons no longer contribute to the current. Consequently, by the time the majority of the electrons have undergone at least one interaction, the current has died away to almost nothing.

We can determine the magnitude of the current at time t after the pulse of current is induced by using the relationship

$$I(t) = I_0 e^{-t/\tau} \tag{9.2}$$

where I_0 is the original magnitude of the induced current and τ is the relaxation time. Since τ is typically about 10^{-13} s for a normal metal, we can see from Example 9.1 that in this case the current dies away to just 1% of its original value in less than a picosecond.

Let us now consider the same experiment performed with a superconducting loop of wire. If a superconductor really has zero resistivity, then the relaxation time would be infinite. This means that the current would continue to flow undiminished for

Example 9.1 Determine the time taken for an induced current in a normal metal loop to decay to 1% of its original value, assuming a relaxation time of 10^{-13} s.

Solution From eqn (9.2)

$$t = \tau \ln\left(\frac{I_0}{I(t)}\right)$$

To determine the time when $I(t) = 0.01\ I_0$, i.e. $I_0 / I(t) = 100$, we have

$$\begin{aligned} t &= (1.0 \times 10^{-13}\,\text{s}) \ln(100.0) \\ &= 4.6 \times 10^{-13}\,\text{s} \end{aligned}$$

ever. Is this really what happens in practice, or is there a small resistivity which ultimately causes the current to deteriorate? In order to try and answer this question experiments have been performed in which a current is induced in a superconducting ring and then left for a long period of time—in some cases several years. (If there was ever a prize for the most boring experiment, this must surely be a strong contender!) Even after a period of several years there has been no measurable decrease in the magnitude of the current.

Obviously, an experiment of this type can never prove that the resistivity of a superconductor is actually zero; all it can provide is an upper limit for the resistivity of these materials. The experiments have shown that the resistivity of a superconductor is certainly less than $10^{-25}\ \Omega$ m. This is a factor of 10^{17} times lower than the resistivity of the best metallic conductors at room temperature (see Table 4.1), so for all practical purposes we can treat the resistivity of a superconductor as zero. Incidentally, theoretical calculations suggest that the resistivity of a superconductor is not quite zero, but that the experiment would need to be left for a time considerably longer than the age of the universe in order to produce a significant change in the current!

9.4 The Meissner effect

One of the most memorable and perplexing physics experiments to witness is a demonstration of the Meissner effect. If you have not seen this before let me describe it to you. All that is needed is a small magnet and a sample of superconductor cooled to below the critical temperature. (With the advent of high-temperature superconductors this demonstration can easily be performed in the lecture theatre or classroom using a shallow tub of liquid nitrogen to act as the refrigerator.) The demonstration consists simply of placing the magnet on top of the superconductor. The perplexing feature is that the magnet does not actually make contact with the superconductor, instead it appears to levitate above the surface (see Fig. 9.5).

This gravity-defying feat is no magician's trick—the magnet is really suspended in mid-air. The explanation of this strange phenomenon is that a superconductor acts as

Figure 9.5 Demonstration of the Meissner effect. A small permanent magnet is levitated above the surface of a high-temperature superconductor (which is cooled by liquid nitrogen).

a perfect diamagnet. This means that it repels both poles of a magnet, and it is these magnetic forces which overcome the gravitational force of attraction. We will try to explain this behaviour based on the premise that a superconductor is simply a metal which has zero resistivity. However, we shall see that this approach cannot account for all of the features of the Meissner effect.

Let us begin by considering what happens when a metal object is placed in a magnetic field. According to the laws of electromagnetic induction, a change in magnetic field induces a current on the surface of the metal which opposes the applied magnetic field. In a normal metal this current dies away very rapidly, but in a superconductor the current persists and produces a magnetic field which continues to oppose the external field. This means that inside the superconductor the magnetic field is zero—the external field is precisely cancelled by the field due to the induced current. Consequently, the lines of magnetic flux are found to bend around a superconductor rather than passing through it (see Fig. 9.6). In other words, the magnetic field is repelled by the superconductor. This repulsion is what explains the behaviour of the 'floating' magnet. The height of the magnet above the surface of the superconductor is determined by the balance point between the gravitational force acting downwards on the magnet and the repulsive force from the superconductor which is pushing it upwards (see Question 9.5).

It appears then that we can explain the Meissner effect using classical electromagnetism and the fact that a superconductor has zero resistivity. However, let us consider a slightly different scenario. Suppose that we have a sample of metal which becomes a superconductor at a critical temperature T_c (in the absence of a magnetic field) and which has a critical magnetic field at B_c of 0 K. Initially the sample is placed in a magnetic field B (which is less than B_c) at a temperature greater than T_c, as indi-

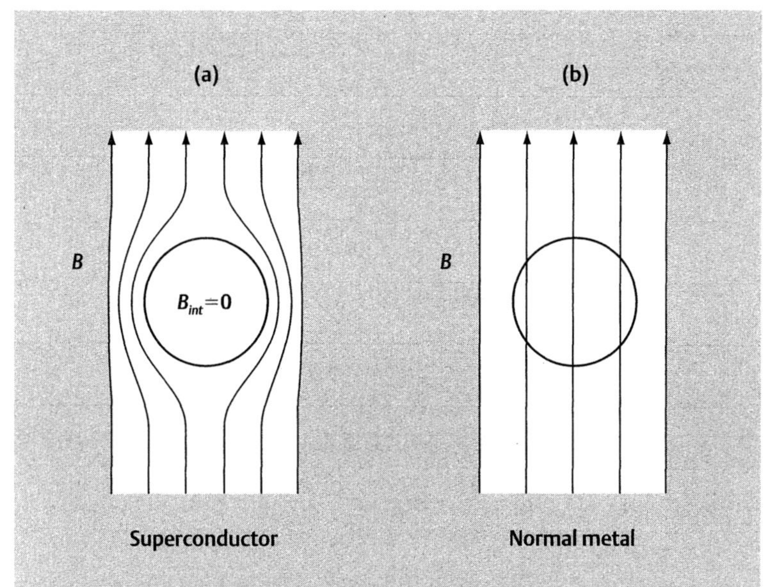

Figure 9.6 (a) When a superconductor is placed in an external magnetic field a supercurrent I_s flows in the surface layer of the superconductor which precisely balances the external field. Consequently, the lines of flux due to the external magnetic field are repelled by the superconductor and the magnetic field inside the sample is zero. (b) When a normal metal is placed in a magnetic field the field penetrates the sample completely.

cated by point A in Fig. 9.7. Obviously, the sample behaves as a normal metal under these conditions and the magnetic field completely permeates the sample, as in Fig. 9.6(b). Now we reduce the temperature, keeping the magnetic field constant, i.e. following the line AB in Fig. 9.7. At point B the sample becomes a superconductor. The question is, to what extent does the magnetic field penetrate the sample at this point? Following our previous discussion we note that there is no change in the external magnetic field, and so there should be no current induced on the sample. Therefore, the magnetic field should pass right through the sample as before. However, we find that even in this scenario, when the sample becomes superconducting the magnetic field is once again expelled from the sample.

As there appears to be no way of explaining this latter result using classical physics, we must conclude that the expulsion of the magnetic field from a superconductor is simply an additional property of superconductors. Therefore, we can say that a superconductor has two independent properties: it has zero (or extremely small) resistivity and it displays the Meissner effect.

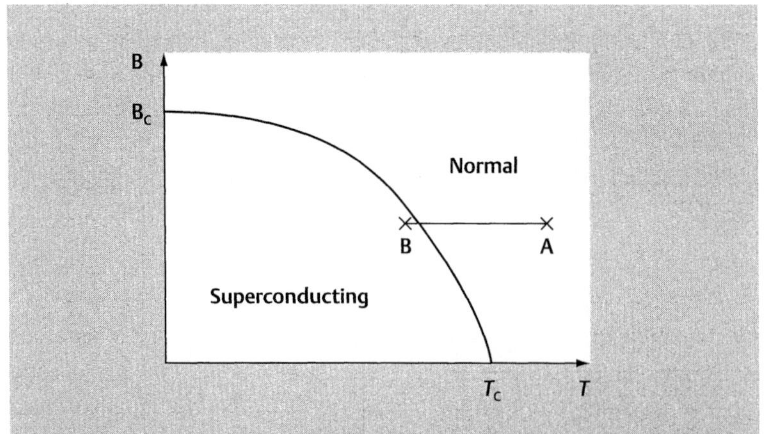

Figure 9.7 A demonstration of the Meissner effect. The magnetic field is expelled by the superconductor even when there is no change in the magnetic field (see text for details).

9.5 Type II superconductors

The properties of an idealized superconductor in a magnetic field B are summarized in Fig. 9.8. We have seen that the resistivity and the internal magnetic field are zero for a small external field, and that both change discontinuously at the critical field B_c. The magnetization also changes discontinuously at B_c; it is large and negative for a superconductor because the induced magnetic moment opposes the external field, but typically becomes small and positive at the critical field as the sample becomes a normal paramagnetic metal.

This type of behaviour is characteristic of most pure metals (at least those which become superconductors). These materials are known as type I superconductors. However, not all superconductors behave in this manner. Other materials, particularly alloys, behave quite differently, and are referred to, rather predictably, as type II superconductors.

The characteristics of a type II superconductor are shown in Fig. 9.9. We can see that there are two critical fields, B_{c1} and B_{c2}, relevant to this type of superconductor. (The critical field B_c of a similar type I superconductor is also shown for comparison.) It is apparent that at the lower critical field B_{c1} (which is less than B_c) the magnetic flux begins to penetrate the sample, and so the material is no longer a perfect diamagnet. However, the resistivity of the sample remains zero until the upper critical field B_{c2} is reached, and it is only at this point that the magnetic field inside the specimen becomes equal to the external field.

The commercial importance of type II superconductors is due to the fact that the upper critical field B_{c2} may be considerably larger than the critical field in a type I superconductor. Although the largest critical field for a type I superconductor is only about 0.1 T, type II superconductors have been discovered with upper critical fields of up to 50 T

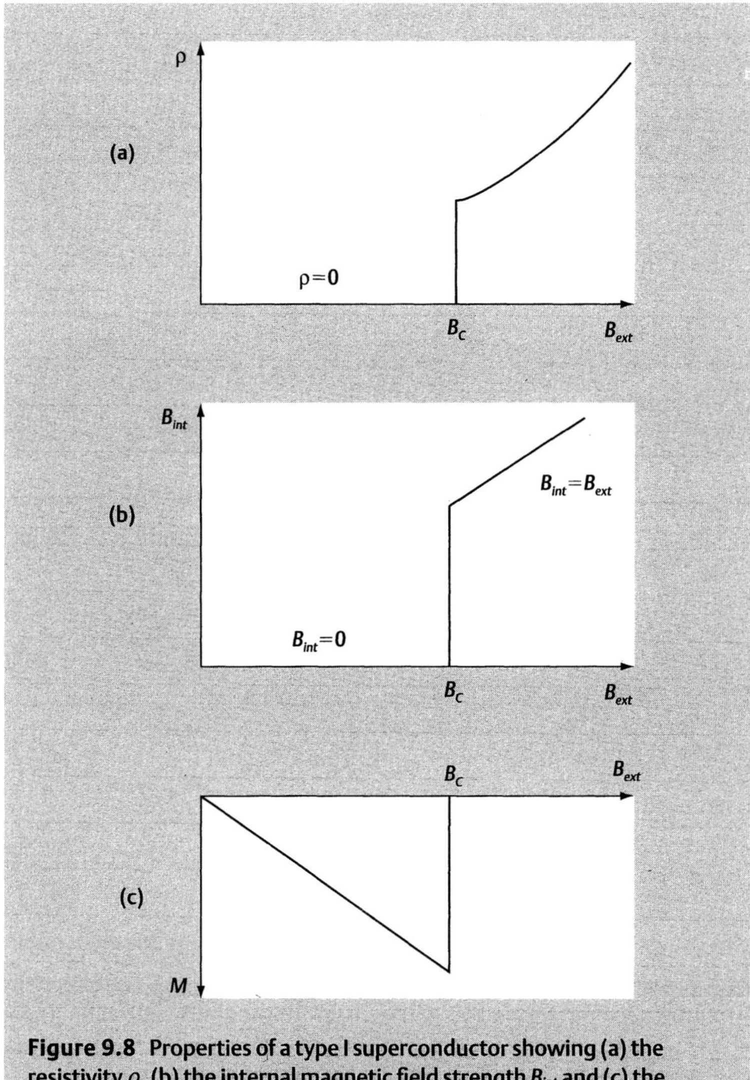

Figure 9.8 Properties of a type I superconductor showing (a) the resistivity ρ, (b) the internal magnetic field strength B_{int} and (c) the magnetization M as a function of the external magnetic field strength B_{ext}.

(and even higher values may be achieved with high-temperature superconductors—see Section 9.8).

How can we explain the difference between the behaviour of a type I and a type II superconductor? Let us first consider how the magnetic flux begins to penetrate the superconductor. Suppose we take a type II superconductor and expose it to a magnetic field greater than B_{c1}. If very tiny crystals of iron are carefully deposited on to the surface of the sample then we find that the particles form clusters, as shown in Fig. 9.10. This behaviour can be explained if we assume that the material underneath

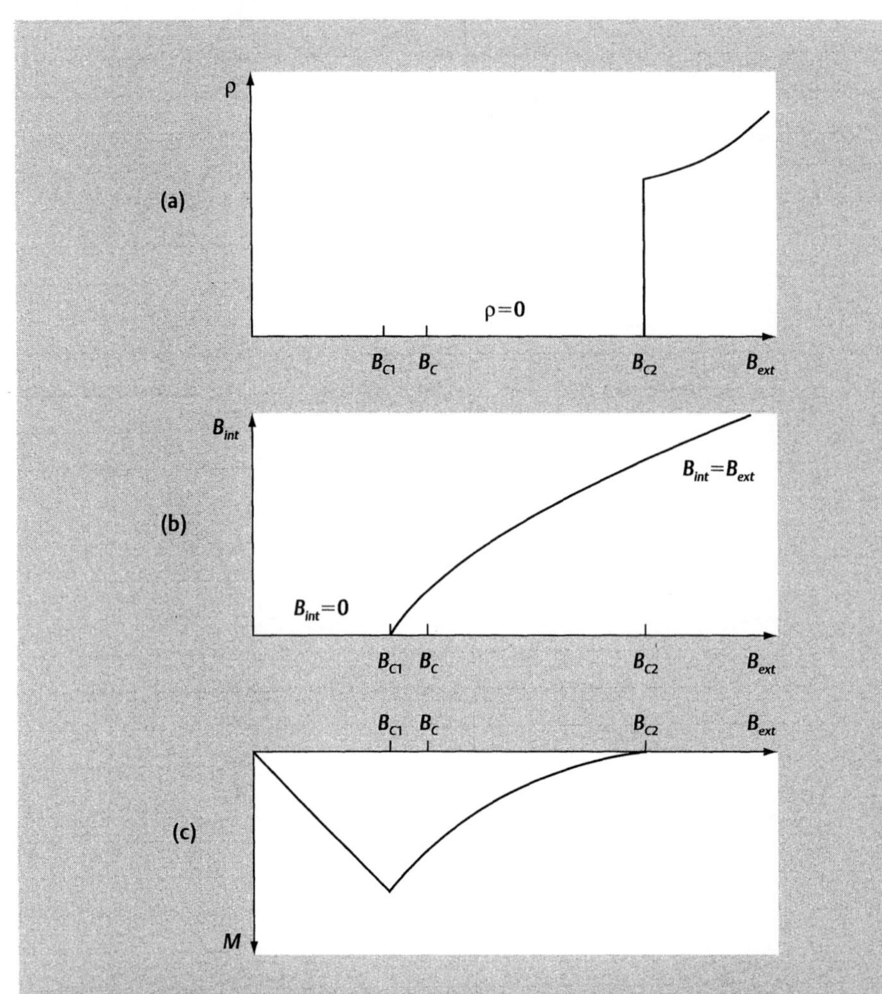

Figure 9.9 Properties of a type II superconductor showing (a) the resistivity ρ, (b) the internal magnetic field strength B_{int}, and (c) the magnetization M as a function of the external magnetic field strength B_{ext}. It can be seen that in this case there are two critical magnetic fields, B_{c1} and B_{c2}. The critical field B_c for a comparable type I superconductor is also indicated.

these clusters corresponds to regions of normal metal which contain some magnetic flux. Consequently, we say that the superconductor is in a mixed state since both normal and superconducting regions are present in the same sample. It turns out that these normal regions form filaments which extend right through the sample like the letters in a stick of rock (see Fig. 9.11). The number of filaments increases with the external magnetic field strength so that at the upper critical field the sample is filled with these filaments and the entire material becomes normal.

What about the resistivity in the mixed state? We expect the resistivity to remain zero so long as there is a continuous superconducting path from one end of the sam-

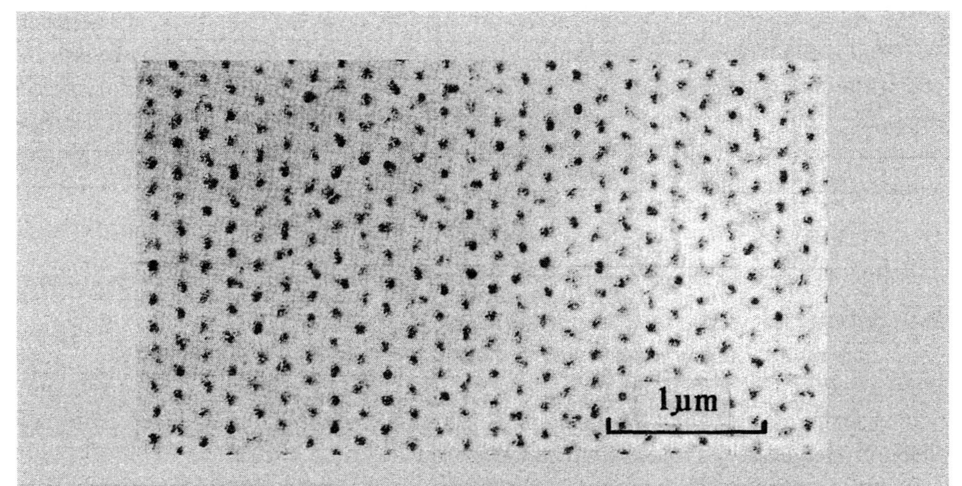

Figure 9.10 Photograph showing the filaments of normal material in the mixed state of a type II superconductor. The filaments are made visible by depositing a fine ferromagnetic powder on the surface.

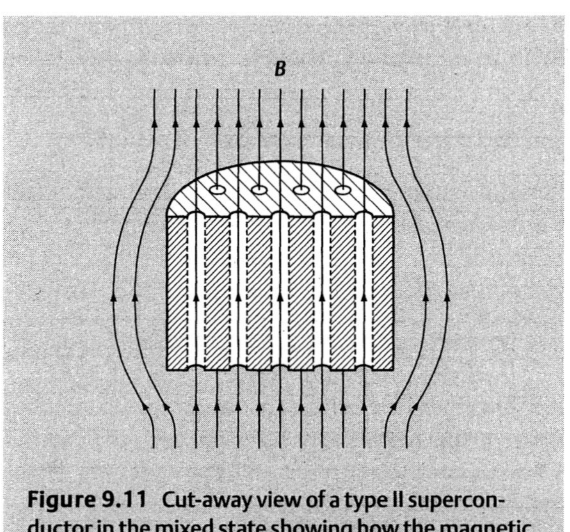

Figure 9.11 Cut-away view of a type II superconductor in the mixed state showing how the magnetic field passes through the filaments of normal material. The shaded regions remain superconducting.

ple to the other. Therefore the resistivity should not return until the upper critical field, B_{c2}, is reached.

There is one main problem with type II superconductors. When a current flows through the superconducting regions of the sample it exerts a Lorentz force on the magnetic flux in the normal filaments. If the filaments are caused to move as a result of this force, then energy is dissipated and this gives rise to a resistance. In general, this means that the critical current density is very low. However, it is possible to increase the critical current density by introducing impurities into the sample. Suitable impurities 'pin' the filaments so that they do not move so easily. This makes it possible to fabricate superconductors which have both a high critical current density and a high critical magnetic field.

9.6 How do superconductors superconduct?

Having examined some of the properties of superconductors, let us now try to explain why these materials behave as they do. In particular, how is it that electrons can move through a superconductor without encountering any resistance?

There is considerable experimental evidence to suggest that superconductivity is not an extension of the model of electrical conductivity that we introduced in Chapter 4. We have already mentioned the following:

1 the resistivity of a normal metal is not expected to go to zero even at absolute zero;
2 the discontinuous change in resistivity at the critical temperature (or critical magnetic field etc.) cannot be explained by the conventional theory; and
3 a normal metal with zero resistivity would not display all facets of the Meissner effect.

In addition, there are two other points which are worth noting:

4 materials which become superconductors are not necessarily good conductors when they are normal metals—in fact, the best conductors at room temperature, copper and silver, do not superconduct even at temperatures of 0.001 K;
5 the critical temperatures of two different isotopes of the same element are found to be proportional to $M^{-1/2}$ where M is the atomic mass. This is referred to as the isotope effect.

These points all suggest that the electrons in a superconductor do not behave in the same way as those in a normal metal. So how do we begin to explain the properties of superconductors? The isotope effect gives us an important clue. If the thermal motion of the ions is assumed to be simple harmonic, we can show that the frequency of the vibrations is also proportional to $M^{-1/2}$ (see Question 9.8). So maybe there is a relationship between superconductivity and the vibrations of the ions.

Further support for this idea is given by point (4). If a metal is a good conductor at room temperature then this suggests that the electrons in this material are not easily deflected by the vibrating ions. So if the mechanism of superconductivity depends on the interaction between the electrons and the vibrating ions, then this would explain why good conductors do not generally become superconductors. In contrast, there is a strong interaction between the electrons and the vibrating ions in those materials which are poor conductors at room temperature, and so these materials are good candidates for becoming superconductors.

These ideas are central to the current understanding of superconductivity which was formulated by John Bardeen, Leon Cooper and Robert Schrieffer in 1957. This model is usually referred to by the acronym BCS theory.

Cooper pairs

The basic assumption in BCS theory is that the electrons team up in pairs, called Cooper pairs. The formation of Cooper pairs can be explained as follows.

Let us suppose that the temperature is close to absolute zero, so that the ions can be considered to be more or less stationary. As an electron moves through the lattice of positive ions, there is an attractive force between the electron and the ions which distorts the lattice by a small amount (Fig. 9.12). This produces a region of the crystal which has a slightly higher concentration of positive charge than the neighbouring region and so it is possible for a second electron to be attracted to this region. We therefore have a pair of electrons which are joined together by their mutual attraction for the ions in the lattice.

However, it is not quite this simple. The electrons will only form a Cooper pair if the second electron has approximately the same speed as the first, but is travelling in the opposite direction. A detailed calculation shows that when the electrons pair up in this fashion they have a lower energy than that of two normal conduction electrons.

How do we know that Cooper pairs really exist?

Fortunately there is a property which we can measure to demonstrate the existence of these pairs. Let us consider a metal at absolute zero. In a normal metal we know that all of the electron states up to the Fermi energy are filled and all those above are empty (Fig. 9.13(a)). Suppose that we excite two electrons from below the Fermi energy into the higher energy vacant states. This would appear to increase the energy of the system. However, if the electrons form a Cooper pair they can reduce their energy to a level which is below that of the two individual electrons, thereby lowering the total energy of the system. If a large number of Cooper pairs form in this way so that

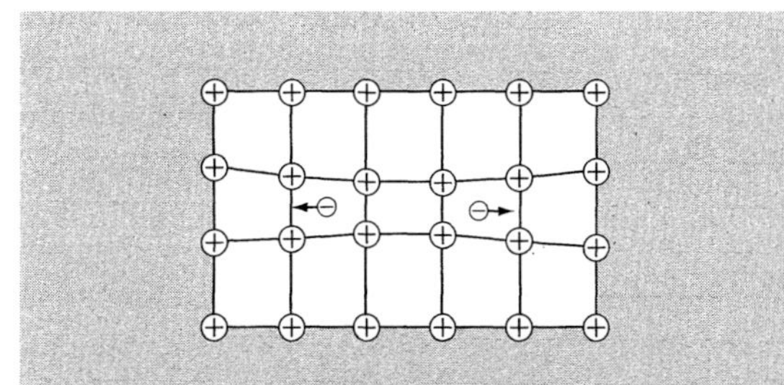

Figure 9.12 Diagram illustrating the formation of a Cooper pair. As an electron moves through the crystal it affects the positions of the ions producing a potential which is attractive to a second electron moving in the opposite direction to the first.

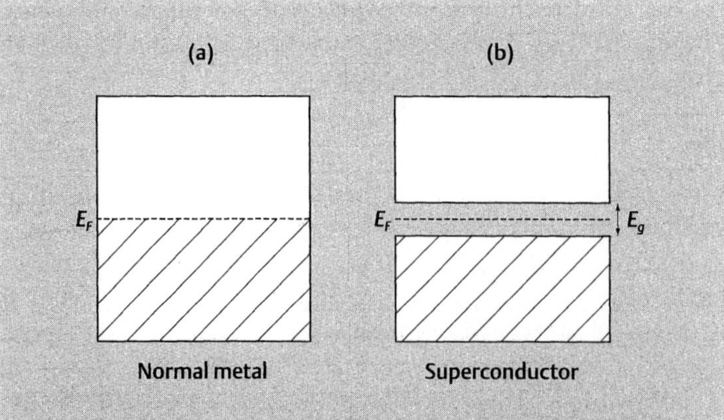

Figure 9.13 Occupation of energy levels at absolute zero in (a) a normal metal and (b) a superconductor. E_F denotes the Fermi energy. Note that in the superconductor there is a gap between the highest filled states and the lowest vacant states.

all of the electrons in the vicinity of the Fermi energy become paired up, then there is an energy gap, E_g, between the highest occupied states (corresponding to the electron pairs) and the lowest unoccupied single electron states (Fig. 9.13(b)).

This picture is reminiscent of a semiconductor, except that the band gap is very much smaller. BCS theory predicts that at absolute zero the band gap is given by

$$E_g = 3.53 k_B T_c \tag{9.3}$$

where k_B is Boltzmann's constant. For a typical superconductor the energy gap is just a few meV (see Example 9.2). This is typically a thousand times smaller than the band gap in a semiconductor. Nevertheless, we can demonstrate the existence of the superconductor energy gap in the same way as in a semiconductor: by shining light onto the sample. If the frequency of the light is such that the energy of the photon is larger than the energy gap E_g, then the photon can be absorbed. The photon therefore supplies the energy required to destroy the electron pair, and so an electron pair becomes two single electron states. If the light is sufficiently intense to destroy all of the Cooper pairs, then the resistivity of the material becomes non-zero.

Example 9.2 Calculate the energy gap for superconducting tin and the minimum frequency of light that can be absorbed given that the critical temperature of tin is 3.72 K.

Solution Using eqn (9.3)

$$\begin{aligned} E_g &= 3.53\, k_B T_c = (3.53)(1.38 \times 10^{-23}\,\mathrm{J\,K^{-1}})(3.72\,\mathrm{K}) \\ &= 1.81 \times 10^{-22}\,\mathrm{J} = 1.13 \times 10^{-3}\,\mathrm{eV} \end{aligned}$$

(For comparison, the experimental value is 1.15×10^{-3} *eV*.) Light is absorbed when the frequency ν is

$$\nu = \frac{E_g}{h} = \frac{(1.81 \times 10^{-22}\,\mathrm{J})}{(6.63 \times 10^{-34}\,\mathrm{J\,s})} = 2.73 \times 10^{11}\,\mathrm{s}^{-1}$$

Note that this frequency is about 1000 times lower than that of visible light and corresponds to the microwave region of the electromagnetic spectrum.

Further evidence for the existence of the band gap can be obtained using the structure shown in Fig. 9.14(a) in which a superconductor is separated from a normal metal by a layer of insulator. The presence of the insulator might suggest that there is no current flow across the junction, but this is not necessarily the case. Quantum theory shows that if the layer of insulator is very narrow (no more than a few nanometres thick), then it is possible for electrons to 'tunnel' through this region and appear on the other side of the junction. (If you are unfamiliar with this idea, I have discussed it in more detail in Chapter 8 of *The Quantum Dot*.) Can the electrons tunnel from the superconductor to the normal metal in this instance? The answer is no, at least if the temperature is close to absolute zero. This is because of the need to conserve energy—tunnelling can only occur if there are vacant states on one side of the junction at the same (or lower) energy as occupied states on the other side. However, if a positive bias is applied to the normal metal, this has the effect of raising the energy of the electrons in the superconductor with respect to those in the normal metal. If the potential V is large enough, the most energetic electrons in the superconductor will be able to tunnel into the vacant states in the metal, as shown in Fig. 9.14(c). We can see from Fig. 9.14(b) that this condition occurs if the change in energy *e*V is greater than half the band gap energy, i.e. there is a critical voltage V_c given by

$$V_c = \frac{E_g}{2\,e} \tag{9.4}$$

at which a current begins to flow. If the current through the junction is plotted as a function of voltage we obtain a curve similar to that shown in Fig. 9.14(d), from which the magnitude of the energy gap can be determined. Measurements on a wide range of superconductors show that in most cases the band gap is in good agreement with eqn (9.3).

How does the formation of Cooper pairs explain the absence of resistance in a superconductor?

One of the simplest and clearest explanations of this phenomenon is due to Little (see Further reading). The argument goes as follows. Let us begin by considering the velocities of the electrons in a normal metal at a temperature close to absolute zero (i.e. so that only the lowest energy states are occupied). Since kinetic energy is proportional to the velocity squared, we expect the velocities of the electrons to be distributed as in Fig. 9.15(a). In this instance there is no flow of current since the effect of

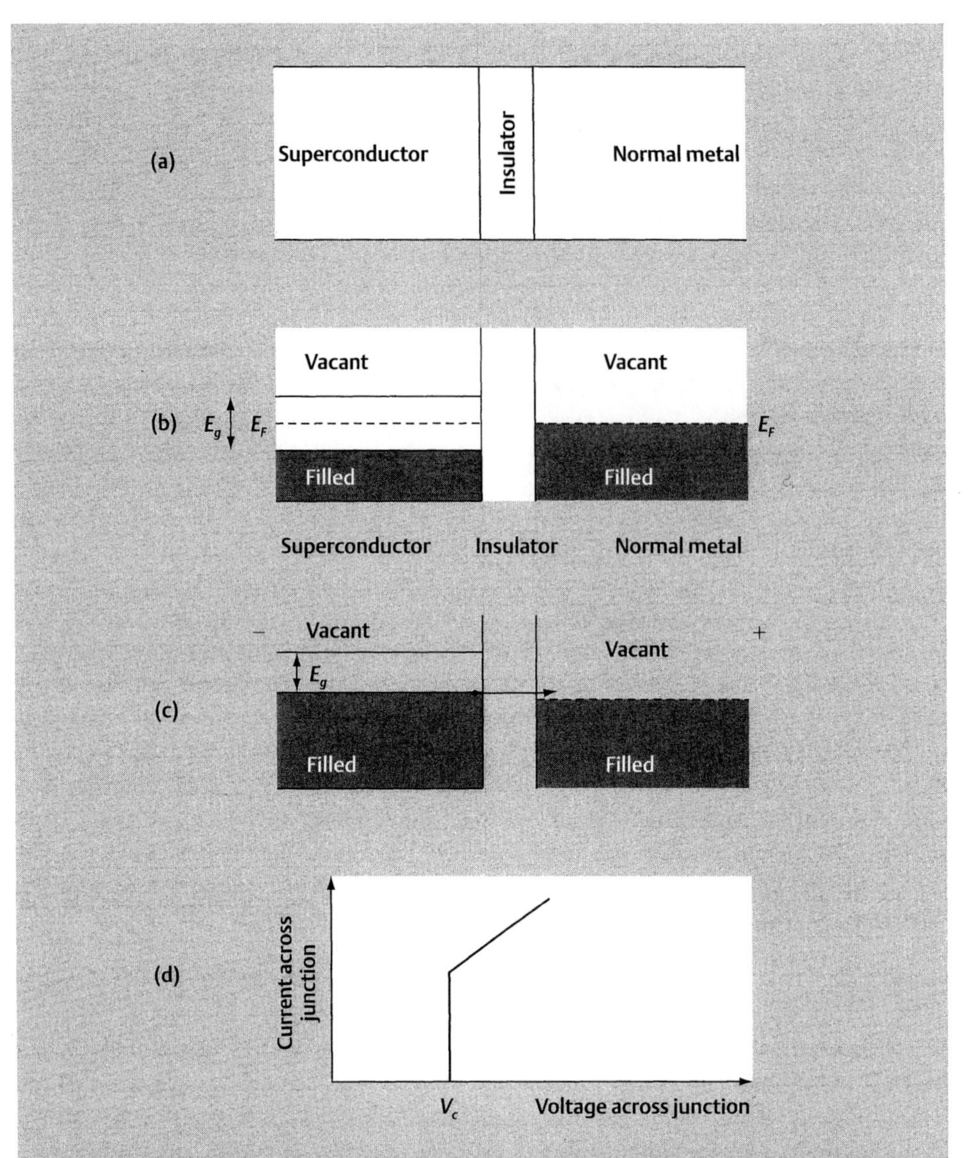

Figure 9.14 (a) A schematic diagram of a tunnel junction between a superconductor and a normal metal. Tunnelling can occur if there are vacant states on one side of the junction at the same (or lower) energy as occupied states on the other side. (b) The electron energy bands at zero bias and at a temperature of 0 K. Note that there are no allowed energy levels in the insulator for the energy range shown. This means that electrons can only pass from one side to the other by tunnelling through the insulator layer. Tunnelling is prohibited in this instance because there are no suitable vacant states on either side of the insulator. (c) When the junction is subject to a bias $V \geq E_g/2e$, electrons from the superconductor can tunnel into vacant states in the normal metal and a current flows across the junction. (d) A plot of current against voltage shows that for a voltage less than the critical value V_c there is no current across the junction, but at the voltage V_c there is a sharp increase in current.

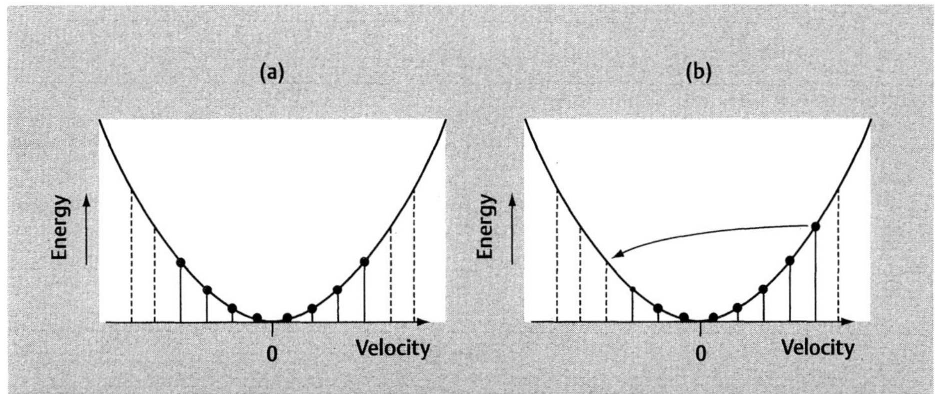

Figure 9.15 A schematic diagram showing the distribution of electron velocities in a one-dimensional normal metal (a) with zero current, and (b) with a net current flow from left to right.

each electron moving to the right is exactly cancelled by an electron of equal velocity moving to the left. When a current flows, there is a shift in the distribution, as shown in Fig. 9.15(b). We now have an unstable configuration because there are vacant states with energies below those of the highest occupied states, but this state of affairs will not persist for long. When the electrons with the highest speeds moving to the right encounter vibrating ions, impurities or imperfections they are deflected into the lower energy states, as shown in the diagram. The excess energy is converted into heat energy and so this process gives rise to a resistance.

In a superconductor the situation is different because the electrons form pairs. In Fig. 9.16(a) there is no flow of current because the electrons are moving at exactly the same speed but in opposite directions, so the velocity of each pair, v_{pair}, is zero. When a current flows there is a shift in the distribution, as shown in Fig. 9.16(b), so that all of the pairs have the same non-zero velocity. This picture looks similar to that for the normal metal shown in Fig. 9.15(b). However, although there appears to be a lower energy state which is vacant, this state cannot be occupied unless we separate one of the pairs of electrons. Consequently, it is only energetically favourable for an electron to be scattered into one of these lower energy states if the energy gained in this process is greater than the binding energy of the Cooper pair. If this is not the case then the paired electron is not deflected. This means that there is no dissipation of energy and so the current flows without resistance.

An alternative explanation for the phenomenon of zero resistance can be obtained by considering the interactions between electron pairs. We used the idea of the electrons deforming the lattice (see Fig. 9.12) to account for the bond between an isolated electron pair, but what happens if there are other nearby electron pairs which cause the lattice to deform along different directions? We would expect the bond between the first pair to be weakened and possibly even destroyed by these other pairs. However, BCS theory predicts that large numbers of electron pairs can coexist in the

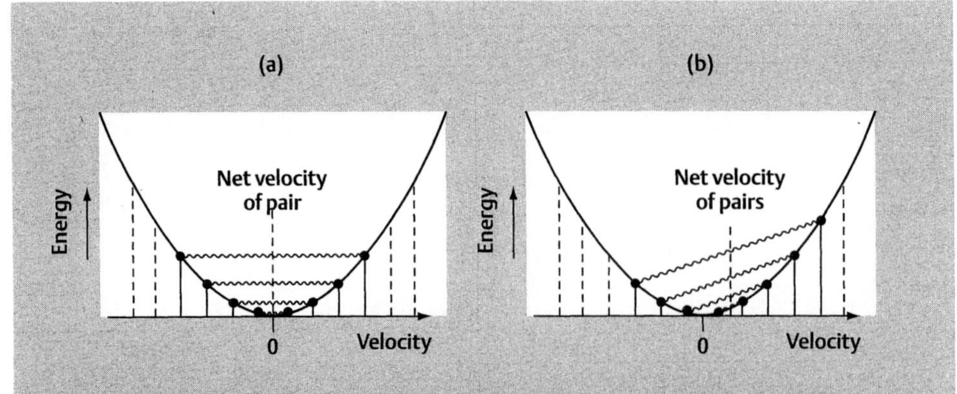

Figure 9.16 A schematic diagram showing the distribution of electron velocities in a one-dimensional superconductor (a) with zero current, and (b) with a net current flow from left to right. The pairing of electrons is shown by the zigzag line. Note that the net velocity of the pairs (i.e. the velocity corresponding to the centre of mass of each pair) is the same for all electron pairs.

same region of the crystal (see Question 9.12). The only way in which this can be achieved is if the motion of the electron pairs is correlated. To see what this means we need to consider a wave representation of the electron pairs. It turns out that the waves corresponding to the electron pairs not only have the same velocity, but also have the same wavelength and phase. When light waves have this property (as in laser light) we say that the light is coherent. Consequently, the motion of the electron pairs in a superconductor can also be described as coherent. We can therefore picture the electron pairs moving in unison through the crystal, with each pair tuning in to the natural frequency of vibration of the ions. Because of this collective behaviour, it is not possible to scatter a single electron pair into a state with a different velocity. The only way in which any energy can be dissipated is if all of the electron pairs are simultaneously scattered into a new state. Since the probability of such an event is exceedingly small, the resistivity of the material is essentially zero.

Why is there a well-defined critical temperature?

The existence of a critical temperature can be explained as follows. The bond between the electrons in a Cooper pair is very weak, and so can easily be destroyed by the thermal motion of the ions. At absolute zero all possible electron pairs are formed, but as the temperature is increased, the number of pairs decreases. The critical temperature is therefore the temperature at which the last electron pair is destroyed.

However, this explanation presents us with a further problem. Measurements on a wide range of samples of a given material show that the critical temperature is remarkably constant—to within 0.01 K in most cases. This seems to be at odds with the fact that Cooper pairs are destroyed by thermal effects. Since thermal processes are statistical in nature, our above explanation seems to suggest that the critical temperature should vary significantly from one sample to another.

We can resolve this problem if we consider the effects not only of the paired electrons but also of the single, unpaired electrons. Each time an electron pair is destroyed we get two more single electrons. Since the motion of the unpaired electrons is not correlated, these electrons produce random fluctuations in the movement of the ions, and so further weaken the bonds between the remaining electron pairs. Consequently, we end up in a vicious circle. As more electron pairs are destroyed it becomes increasingly easier to separate the electron pairs that remain. So rather than a gradual decrease in the number of paired electrons with increasing temperature, what we observe is a catastrophic collapse of the superconducting state at a particular temperature. (For a more quantitative description of this process, see Question 9.11.) The breakdown of the superconducting state is therefore another example of a cooperative transition, as described in Section 8.8.

9.7 Type I or type II?*

In the previous section we described the BCS theory and used it to explain the absence of resistance in a superconductor. This may be all you wish to know about the theory of superconductors, in which case you can safely skip to the next section without affecting your understanding of the remainder of this chapter. However, there are other properties of superconductors which we have not yet explained, such as the Meissner effect and the relationship between type I and type II superconductors.

The Meissner effect and the occurrence of a critical field in type I superconductors can be explained if we assume that Cooper pairs cannot exist in a magnetic field. This means that a material in an external magnetic field can only become a superconductor if the magnetic field is expelled from the sample. We can determine whether or not this is likely to happen by examining the energy of the system. There are two terms to consider, the energy required to expel the magnetic field and the reduction in energy achieved if electron pairs are formed. If the latter term is greater, then it is favourable for electron pairs to form and the magnetic field is expelled. The critical field corresponds to the situation in which these two terms are equal. Consequently, the Meissner effect is simply the result of the system reaching the minimum energy configuration.

In a type II superconductor the situation is not so simple. For magnetic fields up to the lower critical field, both types of superconductor behave in the same way. However, in order to understand the formation of the mixed state we need to introduce two new length scales, the penetration depth and the coherence length. We will first of all describe the physical meaning of these two parameters.

We have seen that the field is expelled from a superconductor by a current flowing around the surface of the superconductor. However, since the current must flow in a layer of finite thickness, this suggests that the magnetic field actually penetrates some distance into the superconductor. This problem was first addressed by Fritz and Heinz London in 1935. They proposed that the magnetic field decays exponentially, so that at a distance z below the surface the strength of the magnetic field inside the sample, B_{int}, is given by

$$B_{int} = B_{ext} e^{-z/\lambda} \tag{9.5}$$

where B_{ext} is the field strength immediately outside the sample and λ is the length over which the field strength decreases by a factor of $1/e$, as shown in Fig. 9.17. To a rough approximation this means that we can consider the magnetic field to be confined to a surface layer of thickness λ. This is the quantity referred to as the penetration depth. A typical value for the penetration depth in a pure metal is a few tens of nanometres. Consequently, in a bulk sample of type I superconductor we can claim that the magnetic field is expelled from virtually all of the sample.

The coherence length, ξ, appears in the somewhat later theory of Landau and Ginzburg. Essentially it can be thought of as a measure of the typical separation of the electrons in a Cooper pair (see Fig. 9.18). For a pure metal the coherence length is of the order of a few hundreds of nanometres. This means that the region occupied by a Cooper pair typically contains several million other Cooper pairs (see Question 9.12).

Let us then take another look at a type II superconductor in the mixed state and consider the effect of these two length scales. We know that in the mixed state the magnetic field enters the sample in the form of filaments of normal material. Since the superconducting regions of the sample must be shielded from the magnetic field in these filaments, each filament must be surrounded by a screening current, as shown schematically in Fig. 9.19. We can therefore picture each filament as a cylinder of material with a radius equal to the penetration depth. What we want to know is why this scenario occurs in some materials and not in others. The answer can be found by considering the ratio of the coherence length and the penetration depth.

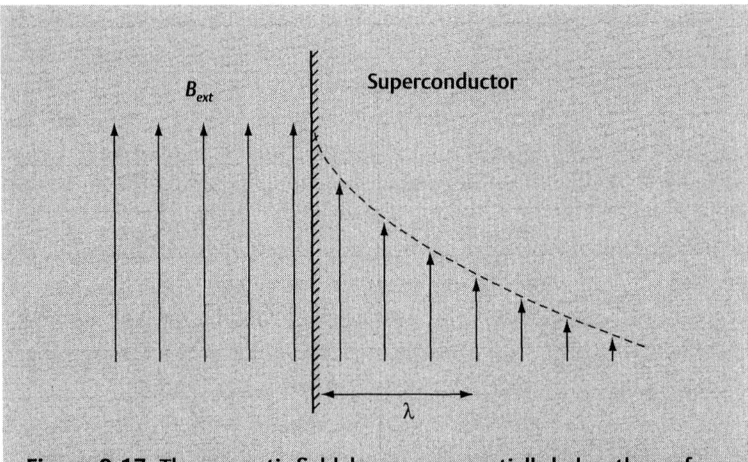

Figure 9.17 The magnetic field decays exponentially below the surface of a superconductor. The penetration depth, λ, is the distance over which the magnetic field strength decreases by a factor of $1/e$.

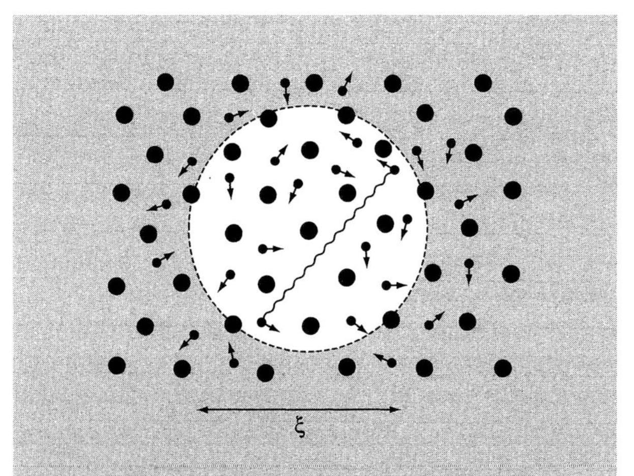

Figure 9.18 The coherence length, ξ, is the effective diameter of a Cooper pair, as shown by the dashed circle. In this diagram the large solid circles represent the ions and the small arrowed circles are the electrons.

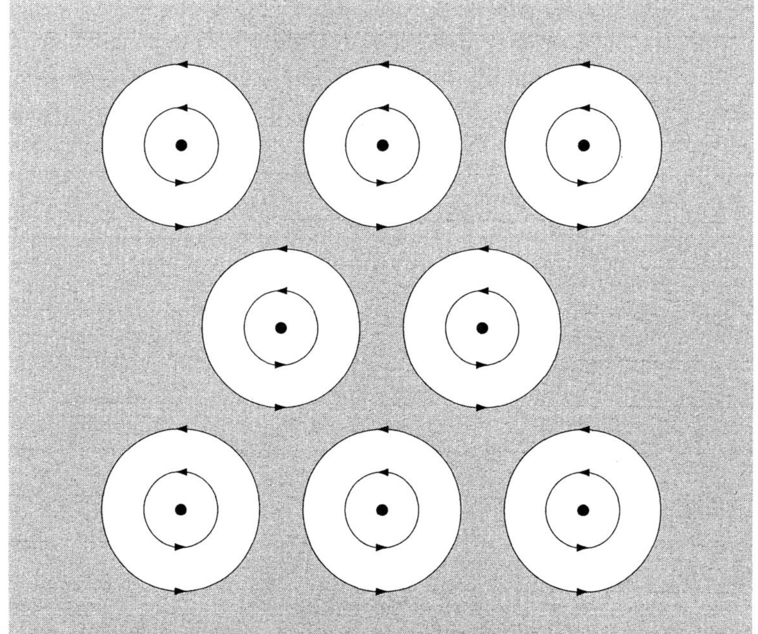

Figure 9.19 Schematic representation of the surface of a type II superconductor in the mixed state, as shown in Fig. 9.10. Each normal filament is screened by a circulating current.

Let us imagine what would happen if we tried to introduce a filament of normal material into a pure superconductor. The above values suggest that the coherence length in a pure metal is approximately 10 times longer than the penetration depth. Consequently, the Cooper pairs would extend right through the cylinder (Fig. 9.20(a)). So from our previous discussion we conclude that there can be no magnetic field in this filament. This is consistent with the behaviour of a type I superconductor. On the other hand, if the coherence length is shorter than the penetration depth, there is a region of normal metal at the centre of the cylinder (Fig. 9.20(b)), and so a magnetic field can pass through this region.

This analysis suggests that the criterion for a material becoming a type II superconductor is that the penetration depth λ is longer than the coherence length ξ. In fact, this is remarkably close to the truth. The solution of the Ginzburg–Landau equations shows that it is energetically favourable for filaments of normal material to form if

$$\lambda > \xi / \sqrt{2} \tag{9.6}$$

The Ginzburg–Landau equations also predict that the upper critical magnetic field of a type II superconductor, B_{c2}, is related to the type I critical field, B_c, by the equation

$$B_{c2} = \sqrt{2} \frac{\lambda}{\xi} B_c \tag{9.7}$$

This suggests that the way to optimize the critical field is to decrease the coherence length whilst increasing the penetration depth. It turns out that this can generally be achieved by adding impurities or by forming an alloy (which in a sense can be considered to be an extremely impure material). Figure 9.21 demonstrates this effect for a selection of lead–indium alloys. We can see that as the impurity concentration—in this case, the proportion of indium atoms—is increased, the upper critical field, B_{c2}, increases accordingly. This reasoning explains why the majority of pure metals behave as type I superconductors, whereas alloys typically exhibit type II behaviour.

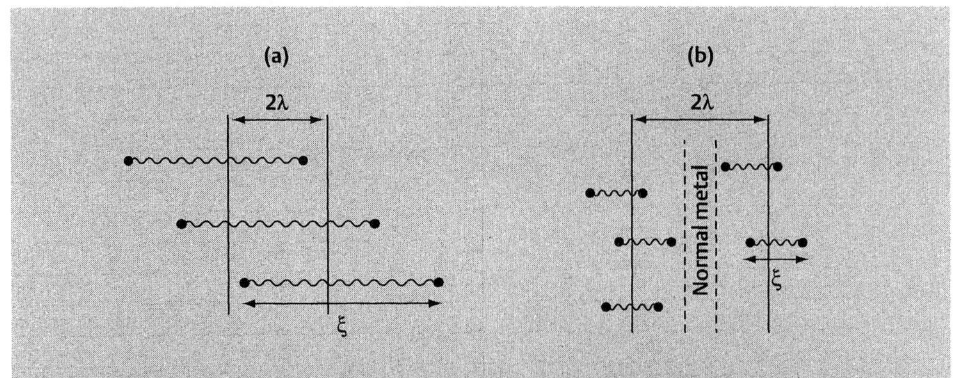

Figure 9.20 Schematic illustration to show that a region of normal metal can only exist within a superconductor if the penetration depth, λ, exceeds the coherence length, ξ. (Details are given in the text.)

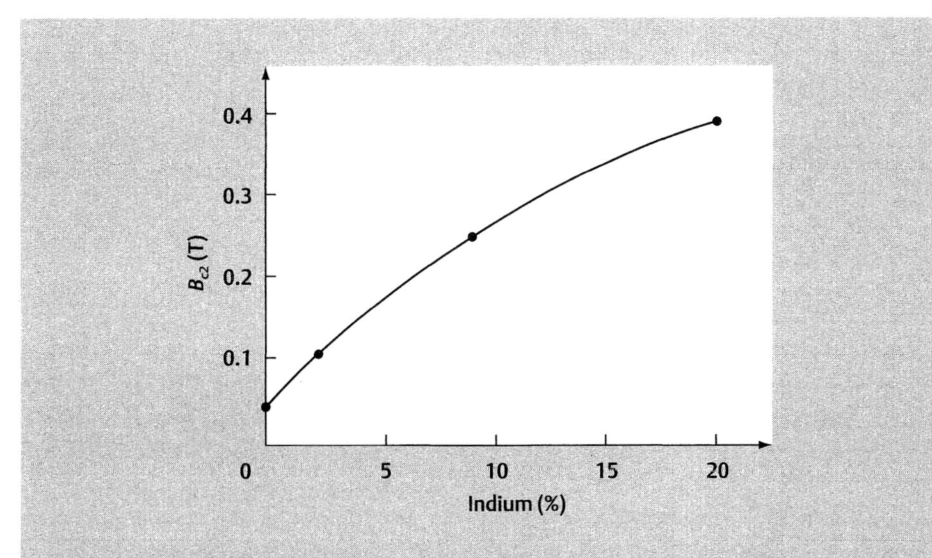

Figure 9.21 Variation of the upper critical field, B_{c2}, for lead–indium alloys as a function of the indium content (defined as a percentage of the number of atoms of lead that are replaced by indium). (Data from Livingston.)

9.8 High-temperature superconductors

Although BCS theory explains many of the phenomena associated with superconductivity, there is still no comprehensive theory which allows us to predict which materials become superconductors and at what temperatures. Consequently, the discovery of new superconductors with increasingly higher critical temperatures has been slow and erratic, as shown in Fig. 9.22(a). By the mid 1980s the highest recorded critical temperature was 23 K for the alloy Nb_3Ge, and it was generally believed that the maximum possible critical temperature for any material was in the region of 40 K.

All this changed in 1986 when Alex Müller and Georg Bednorz at IBM Zurich made a dramatic breakthrough. They discovered a material with a critical temperature of 35 K. The result was important not only because it represented a significant increase in the maximum critical temperature but also because the material, lanthanum–barium–copper oxide (LBCO), belongs to a new class of superconducting materials.

Up to the time of the discovery, most known superconducting materials were metals. Only a few groups in the world were investigating the superconducting properties of oxides and only a handful of superconducting oxides were known. Over the next couple of years there was a dramatic increase in the research into these materials, and the results were stunning. In 1987 another oxide, yttrium–barium–copper oxide or YBCO, was found to superconduct at temperatures up to 90 K, and by the following year the highest critical temperature had been pushed up to 125 K for a related material, thallium–barium–calcium–copper oxide or TBCCO (see Fig. 9.22(b)). In the rush of enthusiasm several groups reported superconductivity at temperatures close

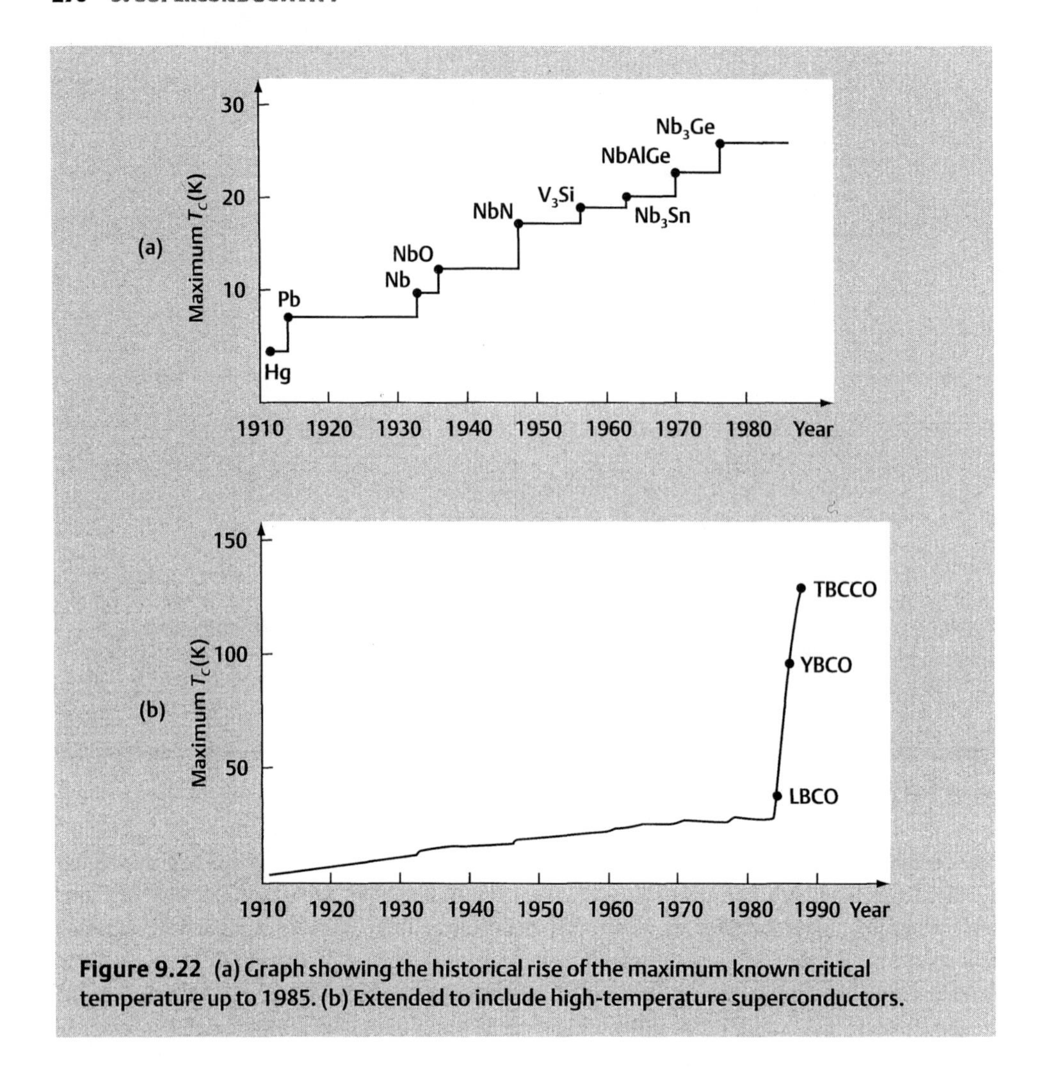

Figure 9.22 (a) Graph showing the historical rise of the maximum known critical temperature up to 1985. (b) Extended to include high-temperature superconductors.

to room temperature, but these results were never substantiated. To date, the highest critical temperature has been increased by only a few degrees over the value for TBCCO.

So room temperature superconductivity is not yet a reality, and is unlikely to be unless some other new family of superconducting materials is discovered. However, the critical temperature of these new oxide superconductors comfortably exceeds another significant temperature—the boiling point of liquid nitrogen (at 77 K). This means that the superconducting state can be achieved using liquid nitrogen, which is far cheaper and easier to handle than liquid helium.

The high critical temperature is not the only unusual property of these oxide superconductors. They also exhibit extreme type II behaviour. Estimates of the upper critical field vary widely, but even the most conservative calculations suggest that the material will remain superconducting in magnetic fields of 250 T. (If you read the previous section, the reason for this behaviour is that the coherence length is very short, of the order of 1 nm, and the penetration depth is large, typically about 700 nm.) However, at present there is no hope of using these materials to generate magnetic fields of this magnitude because the critical current density is far too low.

The low critical current density is due to the fact that the electrical behaviour of these materials is highly anisotropic, i.e. it varies considerably along different crystal directions. This point is worth explaining further. The distinguishing feature of these high-temperature oxide crystals is that the copper and oxygen atoms form a series of parallel planes, as shown in Fig. 9.23. It is found, for example, that the critical current density for a current flowing perpendicular to these planes is typically a hundred times smaller than that for current flowing parallel to the planes. This does not present any problems when we are dealing with a single crystal, but in a polycrystalline wire the critical current density is limited by the lower value.

There are still some practical difficulties to be overcome before high-temperature superconductors can be used commercially. The limitation on the current-carrying capacity is one problem that we have already mentioned. Another is the fact that these materials tend to be brittle and so cannot easily be formed into wires. However, these are not insurmountable problems. Considerable progress has already been made in these areas and high-temperature superconductors will almost certainly become commonplace in the twenty-first century.

A further question concerns the theoretical understanding of these materials. As

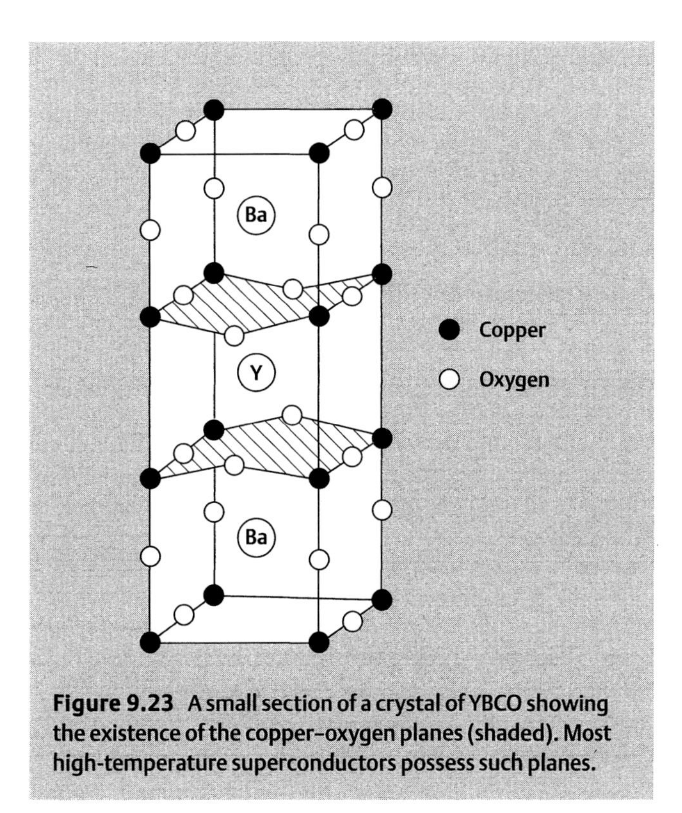

Figure 9.23 A small section of a crystal of YBCO showing the existence of the copper–oxygen planes (shaded). Most high-temperature superconductors possess such planes.

yet we do not understand why the materials have such high critical temperatures, or even how they superconduct. We do know that electron pairs are involved, as in conventional superconductors, but it is not yet clear whether BCS theory can explain the properties of these new superconductors. This problem is currently being addressed by some of the world's top theorists.

9.9 Superconducting magnets

In the two final sections of this chapter we will briefly consider a couple of applications of superconductors. Conceptually the simplest, and perhaps most obvious, application is to use the supercurrent to generate an intense magnetic field. To see the benefits of using a superconductor, let us first of all consider the difficulties associated with an electromagnet made from a normal metal wire.

A conventional electromagnet is made by passing a current through a metal wire wound in the shape of a solenoid. For a long solenoid, electromagnetic theory tells us that the magnetic field produced by this arrangement is given by the equation (see Further reading)

$$B = \mu_0 n I \tag{9.8}$$

where n is the number of turns per unit length and I is the current in the wire.

Example 9.3 An electromagnet is formed by winding a solenoid with 150 turns per metre using copper wire of diameter 3 mm. If the maximum current density for copper is 400 A cm^{-2}, determine the largest magnetic field that can be produced with this solenoid.

Solution The maximum current in wire is determined by the maximum current density, J_{max}, multiplied by the cross-sectional area of the wire, i.e.

$$I_{max} = J_{max}\pi r^2 = (400\times 10^4\ \text{A m}^{-2})(3.14)(1.5\times 10^{-3}\ \text{m})^2 = 28.3\ \text{A}$$

Using eqn (9.8), the maximum field is

$$\begin{aligned} B_{max} &= \mu_0 n I_{max} = (1.26\times 10^{-6}\ \text{H m}^{-1})(150\ \text{m}^{-1})(28.3\ \text{A}) \\ &= 5.34\times 10^{-3}\ \text{T} \end{aligned}$$

What is the largest magnetic field we can generate with this arrangement? The answer is related to the current-carrying capacity of the wire, which in turn is determined by the amount of heat generated by the current. For most practical purposes, the maximum current density in a copper wire is about 400 A cm^{-2}. From Example 9.3, we can see that this corresponds to a maximum magnetic field of a few millitesla. This is about a hundred times larger than the Earth's magnetic field, but is far too small for most practical applications. However, by placing an iron core within the solenoid, this magnetic field can be enhanced by a factor of about a thousand to produce a maximum field of about 2 T. The main disadvantage of this arrangement is that the iron core is extremely heavy and cumbersome.

Let us now consider a similar structure using a superconducting wire. In this case the maximum current is determined by the critical current density. Since it is not uncommon to achieve critical current densities of 10^7 A cm^{-2} it is possible to produce very large magnetic fields in a superconducting solenoid (see Question 9.15). An additional advantage is that no iron core is required in this case. This dramatically reduces the size and weight of the electromagnet and so opens up a whole new range of applications which cannot be performed using large and heavy conventional electromagnets.

Which superconducting materials are appropriate? Obviously the material must have a critical field which is higher than the maximum field we want to generate with the magnet. Type I superconductors are usually unsuitable as they have critical fields of only a fraction of a tesla, but with type II superconductors it is possible to achieve upper critical fields of up to 50 T. Unfortunately, it is not yet possible to make a 50 T superconducting magnet because the critical current density in the mixed state of a type II superconductor is considerably smaller than in a type I superconductor (see Section 9.5). Nevertheless, magnetic fields of up to 20 T can be generated at liquid helium temperatures using superconducting wire made from niobium alloys.

With conventional superconductors, the cost and size of the refrigeration system limits the use of superconducting magnets to large-scale applications. A much larger market would be available for high-temperature superconductors operating at 77 K. In addition, the extremely high upper critical field in these materials suggests that it may be possible to produce magnets of several hundred tesla. However, the current state of the art is a long way from this goal. At 77 K the critical current densities are too small to be of use, but more promising results have been obtained at 4 K. It seems to be only a matter of time before practical systems operating at liquid nitrogen temperatures are achieved.

9.10 SQUID magnetometers*

The charmingly named SQUID (an acronym for Superconducting QUantum Interference Device) magnetometer is capable of measuring exceedingly small magnetic fields. It has already found applications in such diverse areas as medicine (measuring the small magnetic fields produced by activity in the brain), prospecting (detecting changes in the Earth's magnetic field due to the presence of oil or other mineral deposits) and particle physics (searching for quarks and other exotic particles). To understand how a SQUID works we need to describe two further properties of superconductors.

Firstly, let us consider a loop of superconductor in a magnetic field. To simplify the discussion we will assume that it is a type I superconductor. Let us suppose that the material is initially in the normal state, and the temperature is then reduced until it becomes superconducting. At this point the magnetic field is expelled from the material. However, the magnetic field is still present in the hole at the centre of the loop. What happens if the external field is now removed? Since a magnetic field cannot pass through a superconductor, the flux inside the loop is trapped, as shown in Fig. 9.24. The flux is maintained by the presence of a current I_s circulating around the inner surface of the loop.

Let us examine this circulating current in more detail. Since the electron pairs which constitute the current act as a single coherent wave, it is tempting to draw an analogy between the electron pair waves in the superconducting ring and the electron waves in an atom (see Fig. 1.1). Is this really valid? The interference of electron waves in an atomic orbit is strictly a quantum phenomenon which takes place over a range of less than 1 nm. In general, we find that quantum effects are only noticeable on this sort of length scale. However, experimental measurements have confirmed that the same interference effects can be observed in a superconducting ring with a circumference of several metres! It follows that the inner circumference of the ring must correspond to an integral number of wavelengths of the Cooper pairs. What are the consequences of this result? In an atom we found that this leads to a quantization of the energy of the electrons. In the case of a superconducting loop it is the circulating current which can only take certain values. This implies that the amount of magnetic flux contained in the loop is also quantized. We will leave out the gory details (which are given in more advanced texts, such as Rosenberg) and simply quote the result that the magnetic flux in a superconducting loop is quantized in units of $h/2e$. This quantity is known as a flux quantum, ϕ_0, and has a value of 2.07×10^{-15} T m^{-2}.

The second property of superconductors that we are going to describe concerns the behaviour of a Josephson junction. This is a structure which is similar to the one

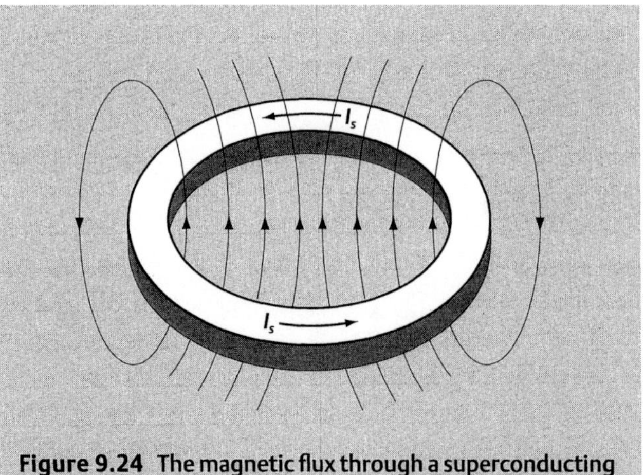

Figure 9.24 The magnetic flux through a superconducting loop remains trapped in the loop even when the external field is removed. The flux is maintained by the presence of a circulating current in the surface layer of the superconductor.

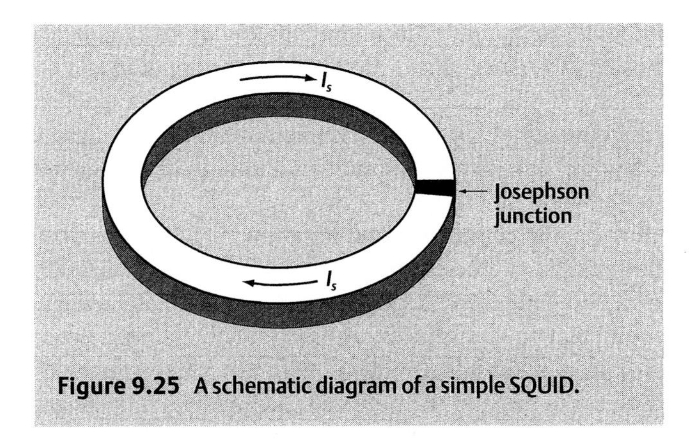

Figure 9.25 A schematic diagram of a simple SQUID.

shown in Fig. 9.14(a), except that the materials on both sides of the junction are superconductors. We have already seen (Section 9.6) that single electrons can tunnel through a layer of insulator, but in 1962 Brian Josephson made the startling prediction that electron pairs could tunnel through a narrow layer of insulator between two superconductors. This means that the layer of insulator not only allows a current to flow, but also behaves as though it is part of the superconductor! However, the superconductivity of the Josephson junction can easily be destroyed—the critical field and current density are much smaller than for the bulk superconductors. Consequently, this structure is often referred to as a weak link.

The essential characteristic of a SQUID is that it is a loop of superconductor which contains one or more Josephson junctions, as shown in Fig. 9.25. Let us see how this structure responds to a magnetic field. If we turn on a weak magnetic field in the vicinity of the structure, a circulating current is induced in the ring and no magnetic flux can enter the middle of the ring. However, if the magnetic field strength is increased so that the circulating current exceeds the critical current density of the Josephson junction, then the junction becomes normal, allowing some of the magnetic flux to enter the loop. As one quantum of magnetic flux enters the loop, the circulating current decreases in magnitude and the junction becomes superconducting again. If the magnetic field strength is increased further, the process is repeated, each time allowing just one flux quantum to enter the loop. Consequently, the magnitude of the circulating current varies periodically according to the strength of the external field. By monitoring the current flowing in the loop it is possible to detect changes in flux of less than one flux quantum, allowing the measurement of exceedingly weak magnetic fields (see Question 9.16).

9.11 Summary

Superconductivity is such a fascinating subject that it is hard to do it justice in just a single chapter. We have seen that superconductors are not simply materials which happen to have zero electrical resistivity. They display a wide range of other unique properties such as the Meissner effect, the Josephson effect and the quantization of magnetic flux.

Although first discovered in 1911, our understanding of these materials has developed quite slowly and there are still some fundamental questions which remain unanswered.

For different reasons, the commercial exploitation of these materials has also been slow. We have described a couple of applications in which superconductors have already had some success, but many other possible applications have been suggested. For example, a novel way to store electrical energy without loss is to keep a circulating current in a giant superconducting ring. A prototype system using a superconducting loop more than 1 km in diameter has already been demonstrated. At the other end of the scale, microelectronic devices based on the Josephson effect have been constructed and several attempts have been made to use these devices to build a superconducting computer.

One of the drawbacks of using superconductors is the need to maintain the system at extremely low temperatures, usually in the region of 4 K. Until recently, the cost and size of the refrigeration plant has meant that only large-scale projects have been viable. However, advances in cooling systems and the discovery of high-temperature superconductors has opened up a whole new range of possibilities. Who knows, in 20 years' time superconductors may be as common in household consumer items as microchips are at present.

Questions

Questions marked * indicate that they are based on the optional sections.

9.1 Critical current density. Determine the maximum current density in a lead wire of diameter 4 mm at a temperature of 4.2 K if the wire is to remain superconducting. The relevant values for lead are $T_c = 7.19$ K and $B_c = 8.03 \times 10^{-2}$ T. (Note that the magnetic field due to a current I in a wire is given by $B = \mu_0 I / 2\pi r$, where r is the distance from the centre of the wire.)

9.2 Critical temperature. A superconductor with a critical temperature of 20 K obeys eqn (9.1). At 4.2 K the critical field is found to be 30 T. Determine the maximum operating temperature if the material is used to make a 10 T superconducting magnet.

9.3 Persistent currents. A current is left to flow in a superconducting loop for a period of 2.5 years. At the end of this time the magnitude of the current is found to be unchanged within the experimental uncertainty of 1%. Use this information to determine a lower limit for the relaxation time. Given that the relaxation time for a normal metal at room temperature is about 10^{-14} s, determine a lower limit for the ratio of the conductivity of a superconductor to that of a normal metal. (You may assume that the number density of electrons is the same in each case.)

9.4 Meissner effect. A small magnet is placed on the surface of a disc of normal material. Explain what happens when the disc is cooled down so that it becomes superconducting.

Can this be explained from a classical viewpoint if we simply treat the superconductor as a material which has zero resistivity?

9.5 Meissner effect. A small magnet of mass 10 g and magnetic moment $m_m = 0.30\ \mathrm{J\,T^{-1}}$ levitates above the surface of a thin disc of type I superconductor. Determine the height of the magnet above the surface given that the magnetic potential energy for a magnet with magnetic moment m_m at distance x from a perfect diamagnet is $\mu_0 m_m^2/32\pi x^3$, where μ_0 is the permeability of free space. How will your answer change if the disc is replaced by a type II superconductor in the mixed state?

9.6 Type II superconductors. Modify Fig. 9.3 to show the behaviour of a type II superconductor as a function of external magnetic field and temperature. Indicate the region in which the material is in the mixed state. Adapt eqn (9.1) to give the lower and upper critical fields, $B_{c1}(T)$ and $B_{c2}(T)$, at temperature T.

9.7 Type II superconductors. Nb_3Sn is a type II superconductor with an upper critical field of 21 T at 4.2 K. Determine the critical current density for a 1.0 mm diameter wire of Nb_3Sn at 4.2 K assuming that the only limitation is the magnetic field produced by the current. The measured critical current density turns out to be $1.0\times10^9\ \mathrm{A\,m^{-2}}$. Explain why there is a discrepancy between these results.

9.8 Isotope effect. The transition temperature for two isotopes of mercury with atomic masses of 199.5 u and 203.4 u are found to be 4.185 K and 4.146 K, respectively. If the critical temperature, T_c, is related to the mass of the ions, M, by

$$T_c = AM^{-\alpha}$$

where A and α are constants, determine the value of α for mercury. Assuming that T_c is proportional to the frequency of vibration of the ions, show that this result is in good agreement with a model in which the ions behave as simple harmonic oscillators.

9.9 Supercurrents. A supercurrent of 1000 A flows in a lead wire of radius 3 mm. Assuming that the number of Cooper pairs is $5.0\times10^{27}\ \mathrm{m^{-3}}$, estimate the average speed of the Cooper pairs. Compare this with the average velocity of the electrons given that the Fermi energy is 9.37 eV. Explain why the values are so different.

9.10 Superconductor energy gap. A superconductor and a semiconductor are illuminated with electromagnetic radiation of frequency $\nu = E_g/h$ (where E_g is the magnitude of the relevant band gap). Describe how the radiation affects the resistivity in each case.

9.11 Cooper pairs. Sketch a graph of the number of Cooper pairs as a function of temperature given that the number of pairs at temperature T, $n_s(T)$, is related to the number at absolute zero, $n_s(0)$, by

$$n_s(T) = n_s(0)\left[1-\left(\frac{T}{T_c}\right)^4\right]$$

At what fraction of the critical temperature is the number of pairs decreased by a half relative to the value at 0 K?

9.12 Cooper pairs. The coherence length represents the average separation of a Cooper pair. Given that the coherence length for a type I superconductor is typically 200 nm and that the number density of Cooper pairs, n_s, at absolute zero is about $10^{28}\ \mathrm{m^{-3}}$, estimate the number of other Cooper pairs that overlap this region.

9.13 **High-temperature superconductors.** A tunnel junction (as shown in Fig. 9.14(a)) is constructed in which the superconductor is YBCO which has a critical temperature of 90 K. Determine the minimum voltage for which a current will flow across the junction at 0 K assuming that eqn (9.3) applies for this material.

9.14 * **High-temperature superconductors.** The penetration depth and coherence length of a high-temperature superconductor are measured as 700 nm and 1 nm, respectively. Estimate the upper critical field for such a material assuming that the critical field for a corresponding type I compound is of the order of 0.1 T.

9.15 **Superconducting magnets.** A magnet is formed by winding a solenoid with 100 turns per metre using a 5 mm diameter superconducting wire with a critical current density, J_c, of 10^6 A cm^{-2}. Determine the largest magnetic field that can be produced with this solenoid.

9.16 * **SQUIDs.** A SQUID is constructed from a superconducting loop with a radius of 5 mm. If the current in the SQUID can be measured to an accuracy of 1%, determine the minimum change in magnetic field that can be detected with this instrument. (Assume that the change in the supercurrent is directly proportional to the change in the magnetic flux.)

Chapter 10
Dielectrics

10.1 Introduction

In previous chapters we have dealt with metals and semiconductors. In this chapter we will turn our attention to insulators, or dielectrics as we shall refer to them here. These are materials for which the electrical conductivity at room temperature is so small that we can assume that they do not conduct electricity at all. Following our discussion in Chapter 5 we will assume that this applies to materials which have a band gap of 3 eV or more.

Although these materials have almost zero conductivity, this does not mean that they are unaffected by an electric field. For example, let us consider a parallel plate capacitor, where the plates are of area A and separation d, as shown in Fig. 10.1(a). When there is a vacuum (or air) between the plates the capacitance is given by

$$C_0 = \frac{\varepsilon_0 A}{d} \tag{10.1}$$

However, if we place a dielectric material between the plates of the capacitor we find that the electric field within the capacitor induces a surface charge on the dielectric,

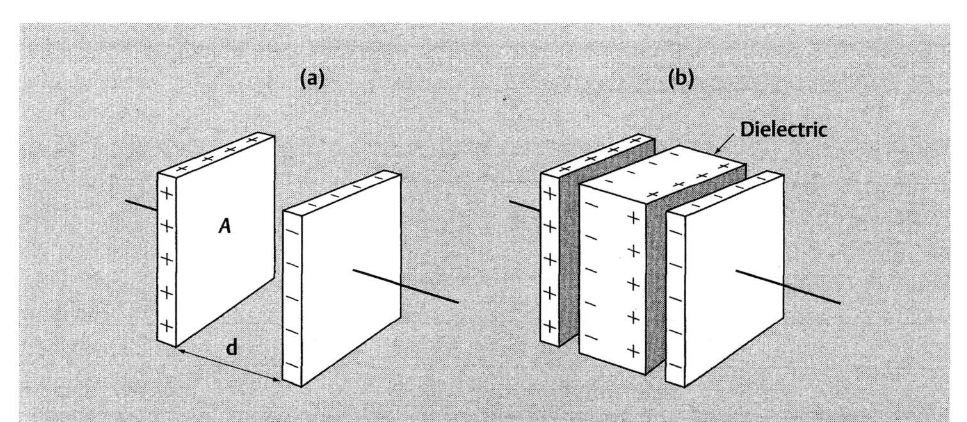

Figure 10.1 (a) A parallel plate capacitor with plates of area A and separation d. When there is no dielectric present the capacitance C_0 is given by eqn (10.1). (b) When a dielectric material is placed between the plates of the capacitor, the dielectric becomes polarized, and so the charge on the plates increases. The capacitance C is given by eqn (10.2).

which therefore increases the charge on the capacitor plates (see Fig. 10.1(b)). The capacitance is now given by

$$C = \frac{\varepsilon \varepsilon_0 A}{d} \tag{10.2}$$

where ε is the dielectric constant of the material. From the above two equations we can see that

$$\varepsilon = \frac{C}{C_0} \tag{10.3}$$

So the dielectric constant can be found by measuring the value of the capacitance with and without the dielectric present (see Question 10.1).

If we assume that the surface charge density, or polarization P, of the dielectric is proportional to the electric field strength $\mathcal{E}$, then we can write

$$P = \varepsilon_0 \chi_e \mathcal{E} \tag{10.4}$$

where χ_e is the electric susceptibility of the material. It can be shown that the electric susceptibility, χ_e, is related to the dielectric constant, ε by the equation

$$\varepsilon = 1 + \chi_e \tag{10.5}$$

(A more detailed discussion of the relationship between the electric susceptibility and the dielectric constant is given in *The Feynman Lectures on Physics, Volume II*—see Further reading.)

Having dealt briefly with the macroscopic properties of dielectrics in a capacitor, let us now look at how we can explain this behaviour from an atomic viewpoint.

10.2 Induced polarization

Let us begin by considering the effect of an electric field on an isolated atom. When there is no field applied we can assume that on average the distribution of the electrons is spherically symmetric, and therefore the centroid of negative charge, G^-, lies at the centre of the atom, as shown in Fig. 10.2. Of course, the positive charges are located in the nucleus, which is also at the centre of the atom, so in this case the centroids of positive and negative charge coincide. However, if we apply an electric field $\mathcal{E}$, the electron orbits are distorted, and so the centroids of charge become separated by a distance δ (see Fig. 10.3). The atom therefore acquires an induced dipole moment p which we can write as

$$p = q\delta \tag{10.6}$$

where q is the total amount of negative (or positive) charge in the atom.

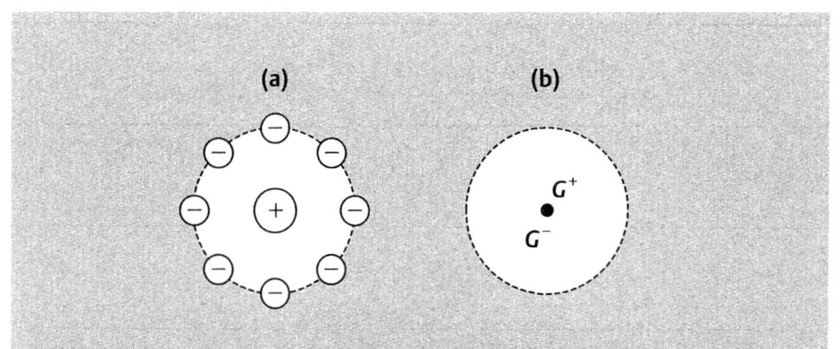

Figure 10.2 (a) For an isolated atom in zero electric field the average distribution of the electrons is spherically symmetric, and so (b) the centroids of positive and negative charge, G^+ and G^-, coincide.

If we extend these ideas to a dielectric crystal, then we again expect an external electric field to produce a distortion of the electron orbits, as in Fig 10.3. Consequently, one end of the crystal acquires a net positive charge and the other has a net negative charge, as shown in Fig. 10.4. We can quantify this effect by determining the polarization of the crystal. Since the polarization is equal to the dipole moment per unit volume, the polarization of the crystal can be written as

$$P = Nq\delta \tag{10.7}$$

where N is the number of atoms per unit volume.

The above analysis gives us an expression for the polarization of the crystal as a function of the average displacement of the electrons, but in order to make a comparison with our macroscopic definition of polarization in eqn (10.4) it would be helpful to determine the microscopic relationship between the polarization and the external electric field. If we assume that the displacement of the centroids of charge is proportional to the electric field strength $\mathcal{E}$, then we can write the dipole moment of a single isolated atom as

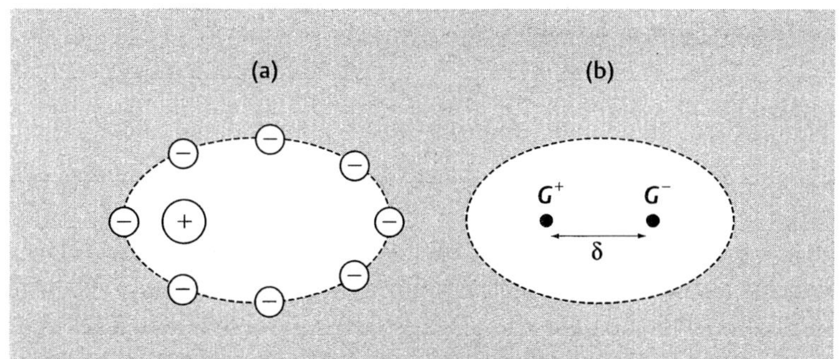

Figure 10.3 (a) The electron orbits of an isolated atom in an electric field $\mathcal{E}$ are distorted, and so (b) the centroids of charge are separated by a distance δ.

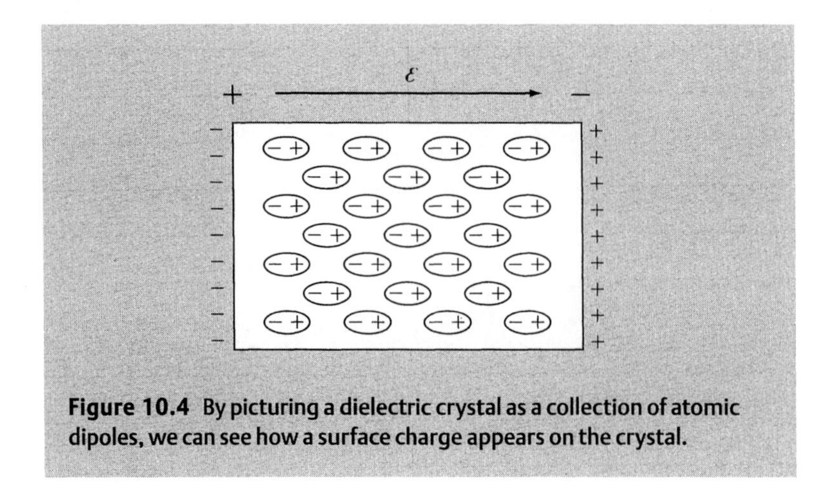

Figure 10.4 By picturing a dielectric crystal as a collection of atomic dipoles, we can see how a surface charge appears on the crystal.

$$p = \alpha \mathcal{E} \tag{10.8}$$

where α is a constant which we will refer to as the atomic polarizability. Extending this idea to a crystal we might expect that the polarization of the crystal is found by multiplying this expression by the number of atoms in the crystal, N. However, this is not quite correct. Although the external electric field is given by $\mathcal{E}$, the actual electric field acting on an atom in the crystal is also affected by the electric field due to the dipole moments of the neighbouring atoms. The dipole moment of a particular atom in the crystal is therefore determined by the net, or local, electric field $\mathcal{E}_{loc}$ in the vicinity of that atom. Therefore, for an atom in the crystal the dipole moment is

$$p = \alpha \mathcal{E}_{loc} \tag{10.9}$$

The magnitude of the local field obviously depends on the position of the atom in the crystal. For instance, the local field for an electron at the centre of the crystal is obviously different from that for an atom at the surface of the crystal. However, for a crystal of macroscopic dimensions, the surface atoms constitute a very small proportion of the crystal. Consequently, if we neglect the contribution from the surface atoms, then the polarization of the crystal is given by

$$P = N \alpha \mathcal{E}_{loc} \tag{10.10}$$

where $\mathcal{E}_{loc}$ is now defined as the average local field experienced by an atom within the crystal.

In order for this expression to be of any use we need to be able to relate the local electric field, $\mathcal{E}_{loc}$, to the external electric field, $\mathcal{E}$. One way to estimate the local field on a particular atom is to consider the electric field that exists when we remove this atom from the crystal. If the electric field due to the surrounding dipoles is denoted by $\mathcal{E}'$, then from Fig. 10.5 we can see that the local field is

$$\mathcal{E}_{loc} = \mathcal{E}' + \mathcal{E} \tag{10.11}$$

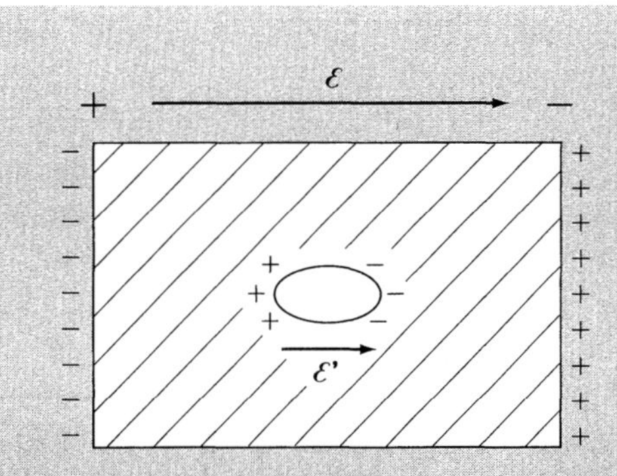

Figure 10.5 An estimate of the local field at a particular atom can be obtained by imagining that we remove the atom in question. The local field $\mathcal{E}_{loc}$ is then given by the combined effect of the external field, $\mathcal{E}$, and the field due to all of the other dipoles, $\mathcal{E}'$.

Since the field $\mathcal{E}'$ acts in the same direction as the external field, we find that the local field is larger than the external field. We will not go through a derivation of the form of $\mathcal{E}'$ as it is rather tedious and is included in most more advanced solid state texts (see Further reading). Instead, we will simply quote the result that

$$\mathcal{E}_{loc} = \frac{1}{3}(\varepsilon + 2)\mathcal{E} \tag{10.12}$$

Example 10.1 By using the microscopic and macroscopic equations for the polarization of a crystal and assuming that the local electric field in a dielectric is given by eqn (10.12), show that the atomic polarizability is given by eqn (10.13).

Solution From eqn (10.12) we have

$$\mathcal{E}_{loc} = \frac{1}{3}(\varepsilon + 2)\mathcal{E}$$

Substituting in eqn (10.10) gives the polarization

$$\begin{aligned} P &= N\alpha\mathcal{E}_{loc} \\ &= \frac{N\alpha}{3}(\varepsilon + 2)\mathcal{E} \end{aligned}$$

From eqns (10.4) and (10.5) we can obtain an expression for the polarization in terms of the dielectric constant, ε,

$$P = \varepsilon_0 \chi \mathcal{E} = \varepsilon_0 (\varepsilon - 1)\mathcal{E}$$

Equating these two expressions gives

$$N\alpha = 3\varepsilon_0 \frac{\varepsilon - 1}{\varepsilon + 2}$$

where ε is the dielectric constant of the material. Note that if we put $\varepsilon = 1$ (as in a vacuum), then we get $\mathcal{E}_{loc} = \mathcal{E}$, as we would expect.

Now that we have both a microscopic and a macroscopic relationship for the polarization of the crystal, we can use these equations to obtain an expression for the atomic polarizability, α, in terms of the macroscopic dielectric constant, ε. From Example 10.1 we find that this relationship can be written as

$$N\alpha = 3\varepsilon_0 \frac{\varepsilon - 1}{\varepsilon + 2} \tag{10.13}$$

This is known as the Clausius–Mossotti formula. A more general form of this equation takes into account the fact that the dielectric may consist of more than one type of atom with different atomic polarizabilities. In this case we obtain

$$\sum_j N_j \alpha_j = 3\varepsilon_0 \frac{\varepsilon - 1}{\varepsilon + 2} \tag{10.14}$$

where N_j is the concentration and α_j is the polarizability of atom type j.

10.3 Other polarization mechanisms

Induced polarization effects occur in all dielectrics, but there are two other types of polarization that occur only in specific types of materials.

In ionic crystals, an applied electric field pulls the positive and negative ions in different directions, as can be seen in Fig. 10.6. This means that there is an additional contribution to the polarization of the crystal. We can therefore write the total polarization as

$$P = P_e + P_i \tag{10.15}$$

where P_e is the electronic contribution (i.e. the induced polarization) and P_i is the ionic contribution. If we assume that the ions have charges of $\pm e$ (i.e. that they are singly ionized), then by analogy with eqn (10.7) we can see that the polarization P_i is given by

$$P_i = NeA \tag{10.16}$$

where A is the amplitude of the displacement and N is the total number of ions per unit volume.

The other type of polarization occurs in materials where the molecules possess a permanent dipole moment. In this case the centroids of positive and negative charge do not coincide even when there is no external electric field present. We have already discussed an example of this behaviour when we considered the water molecule in Chapter 1 (see Fig. 1.22). Since the dipole moment of these polar molecules is typically several orders of magnitude larger than the induced dipole moment in non-polar

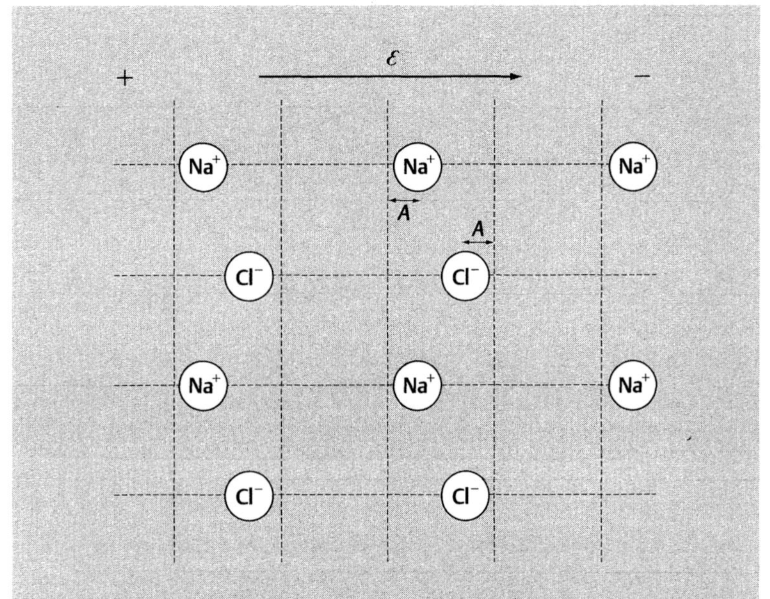

Figure 10.6 The effect of applying an electric field parallel to the [100] direction in a sodium chloride crystal. The deviations of the ions are grossly exaggerated in this diagram.

molecules, this suggests that the polarization, and therefore the dielectric constant in polar materials, should be much greater than that in non-polar materials. Although this is true for liquids—because the molecules can rotate so that the dipoles are aligned with the external electric field—in a solid the molecules are fixed and so cannot usually respond in this way. However, there are some solids in which the permanent dipoles continue to contribute to the polarization at temperatures well below the melting point. An example is hydrogen chloride, the molecules of which form dipoles as depicted in Fig. 10.7. The dielectric constant of hydrogen chloride changes abruptly at a temperature of about 50 K below the melting point, as can be seen from

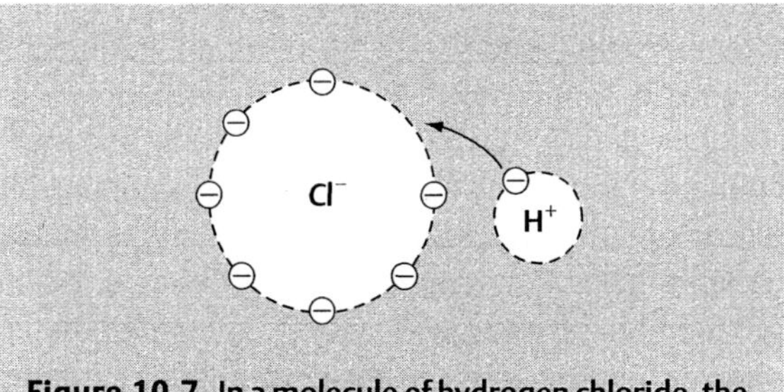

Figure 10.7 In a molecule of hydrogen chloride, the electron from the hydrogen atom is strongly attracted to the chlorine atom, and so effectively we have a Cl^- ion and a H^+ ion.

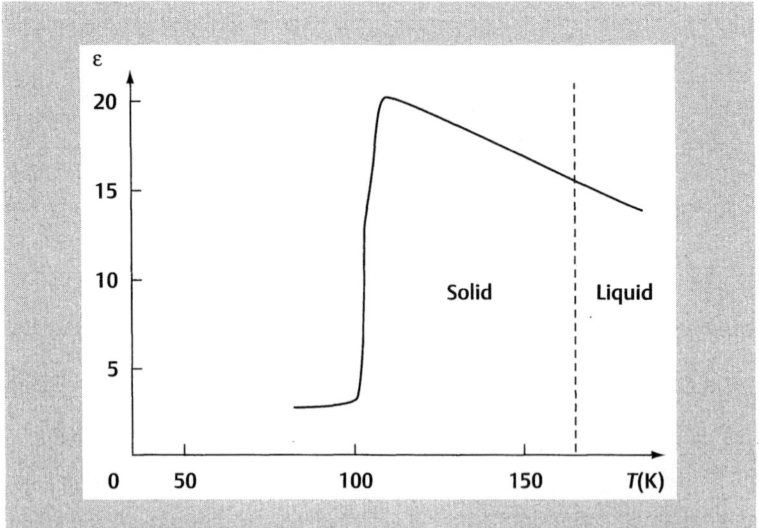

Figure 10.8 Variation of the dielectric constant, ε, with temperature for hydrogen chloride. The abrupt change corresponds to the temperature at which the molecules are no longer able to align themselves with the external electric field.

Fig. 10.8. Below this temperature the dielectric constant is due mainly to ionic and induced polarizations, but above this temperature it appears that the molecules are able to orientate themselves along the direction of the field and so the dielectric constant is much larger.

10.4 The frequency dependence of the dielectric constant

So far we have considered the dielectric constant in the static case where the electric field is assumed to be constant over a long period of time. In this and the following section we will look at what happens if the electric field varies with time.

If the electric field varies slowly, then we would expect the electrons, ions or molecules (depending on the type of polarization) to follow the movements of the field. So the polarization changes periodically in phase with the electric field, as shown in Fig. 10.9, and the dielectric constant is therefore just the same as in the static case.

At higher frequencies this may no longer be true. For example, let us consider the effect of an electric field which alternates at a frequency of about 5×10^{14} Hz. Since light is an electromagnetic wave, and the above frequency corresponds to the visible light range, this can be achieved simply by shining light onto the material. In this case we can obtain an expression for the dielectric constant from electromagnetic theory. Light travelling through a non-magnetic medium (for which the relative permeability $\mu = 1$) has a velocity

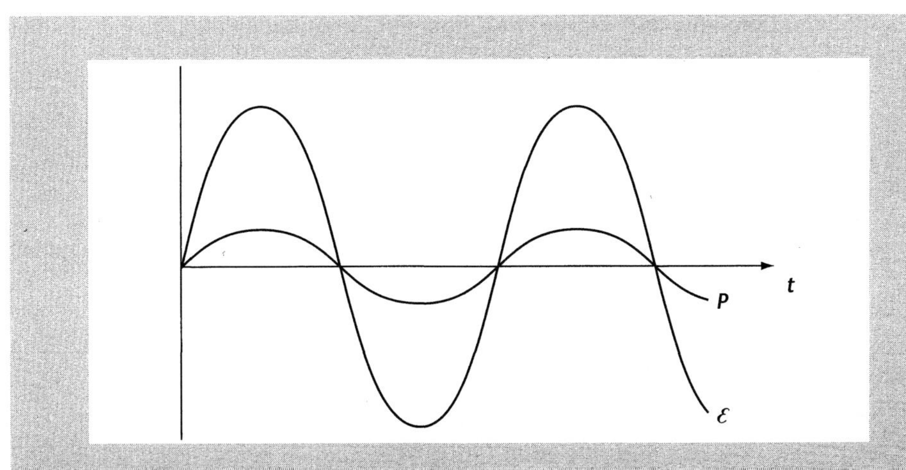

Figure 10.9 For a slowly varying electric field $\mathcal{E}$, the polarization of the dielectric is expected to vary at the same frequency as the applied field.

$$v = \sqrt{\varepsilon_{opt}\varepsilon_0\mu_0} \tag{10.17}$$

(where ε_{opt} is the dielectric constant at optical frequencies), whereas light travelling through a vacuum has a velocity

$$c = \sqrt{\varepsilon_0\mu_0} \tag{10.18}$$

(If you are unfamiliar with these expressions, consult one of the introductory texts mentioned in Further reading.) From these two equations we can see that

$$\varepsilon_{opt} = \left(\frac{v}{c}\right)^2 \tag{10.19}$$

We also know that the refractive index N is equal to the ratio of the speed of light in a medium to the speed of light in a vacuum, i.e. $N = v/c$, so our expression for the dielectric constant can be written as

$$\varepsilon_{opt} = N^2 \tag{10.20}$$

Since we can easily determine the refractive index of a material, we can therefore make experimental measurements of ε_{opt}. Values for a selection of materials are given in Table 10.1.

We can see from the data in Table 10.1 that for most materials the dielectric constant at optical frequencies, ε_{opt}, is significantly lower than that in the static case, ε_s. This result can be explained by considering how the different polarization mechanisms are affected by a rapidly alternating electric field. We have seen that the induced polarization is due to the movement of electrons. Since the electrons are capable of responding quickly to any changes in the external field, we expect this mechanism to continue to operate effectively even at optical frequencies. On the

Table 10.1 Values of the dielectric constant in a constant (static) electric field ε_s and at optical frequencies ε_{opt} for a selection of dielectric materials.

	ε_s	ε_{opt}
Diamond	5.68	5.66
NaCl	5.90	2.34
LiCl	11.95	2.78
TiO_2	94	6.8
Quartz	3.85	2.13

other hand, the types of polarization discussed in Section 10.3 are due to the displacement of the ions or molecules. These particles are far more massive than the electrons, and so are unable to keep pace when the electric field varies rapidly. In practice, we find that the dipolar and ionic contributions to the polarization fall off rapidly for frequencies above about 10^{11} Hz and 10^{13} Hz, respectively, whereas induced polarization effects persist up to about 10^{16} Hz, well above the visible light range.

We can conclude that at optical frequencies the only contribution to the polarization of the crystal is from the induced polarization, whereas at lower frequencies other polarization mechanisms may contribute and so produce larger values of the dielectric function. Therefore, in most cases the dielectric constant is significantly larger in the static case (and at low frequencies) than at optical frequencies. However, in a few materials, such as diamond, the dielectric function is almost the same in the static and optical cases. Why is this? It is because there are no dipole or ionic effects in a diamond crystal, and so even in the static case the only contribution is from the induced polarization. This result also confirms our assumption that the induced polarization effects remain virtually constant even up to optical frequencies.

Consequently, we expect the only contribution to the dielectric constant at optical frequencies to be from the induced polarization, whereas at lower frequencies the other mechanisms also contribute.

We can therefore explain qualitatively why the dielectric constant varies as a function of frequency. In the next section we will develop quantitative models to describe this behaviour. The treatment assumes that you are familiar with forced, damped oscillations (see Further reading). It is an optional section which you can omit without affecting your understanding of the remainder of this chapter.

10.5 Resonant absorption and dipole relaxation*

To obtain a quantitative picture of the polarization effects in an ionic crystal, let us assume that the vibrating ions behave like tiny harmonic oscillators. Their motion is therefore described by equations of the form

$$\frac{d^2x}{dt^2}+\omega_0^2x=0 \tag{10.21}$$

where ω_0 is the natural frequency of vibration of the ions.

When an electric field $\mathcal{E}$ which varies periodically with frequency ω is applied to the crystal, the ions are subject to a driving force given by $e\mathcal{E}e^{i\omega t}$ and the equation of motion becomes

$$\frac{d^2x}{dt^2}+\omega_0^2x=\frac{e\mathcal{E}}{m}e^{i\omega t} \tag{10.22}$$

where m is the mass of the ions.

However, if we solve this equation we find that the amplitude of vibration becomes infinite when the driving frequency ω is equal to the natural frequency of the system ω_0 (see Question 10.6). Since this obviously does not happen in practice we need to introduce a damping term. The damping can be attributed to interactions with phonons (i.e. interactions with the other vibrating ions in the crystal) which cause energy to be dissipated. For harmonic oscillators it is usual to assume that the damping is proportional to the velocity of the particle, and so the equation of motion becomes

$$\frac{d^2x}{dt^2}+\gamma\frac{dx}{dt}+\omega_0^2x=\frac{e\mathcal{E}}{m}e^{i\omega t} \tag{10.23}$$

where γ is the damping coefficient.

By solving eqn (10.23), we obtain an expression for the dielectric constant which is complex, i.e. we can write

$$\varepsilon=\varepsilon'+i\varepsilon'' \tag{10.24}$$

(see Example 10.2). The real part, which is given by

$$\varepsilon'=\frac{Ne^2}{\varepsilon_0 m}\frac{\omega_0^2-\omega^2}{(\omega_0^2-\omega^2)^2+\gamma^2\omega^2}+\varepsilon_{opt} \tag{10.25}$$

is due to vibrations which are in phase with the external field, whilst the imaginary part

$$\varepsilon''=\frac{Ne^2}{\varepsilon_0 m}\frac{\gamma\omega}{(\omega_0^2-\omega^2)^2+\gamma^2\omega^2} \tag{10.26}$$

corresponds to vibrations which are out of phase with the field.

If we plot these functions (see Fig. 10.10), we find that when the frequency of the applied field is very different from the natural frequency, ω_0, the real part of the dielectric function, ε', is approximately constant with frequency and the imaginary part, ε'', is virtually zero. Since the imaginary part represents vibrations which are out of phase with the driving field, this shows that the ions are moving in phase with the external field at these frequencies (as in Fig. 10.9), and so the dielectric constant, ε, is approximately the same as the real part, ε'. In contrast, when the frequency of the applied field is close to the natural frequency of the system there is a sudden change in the real part of the dielectric function. There is also a non-zero imaginary term, ε'',

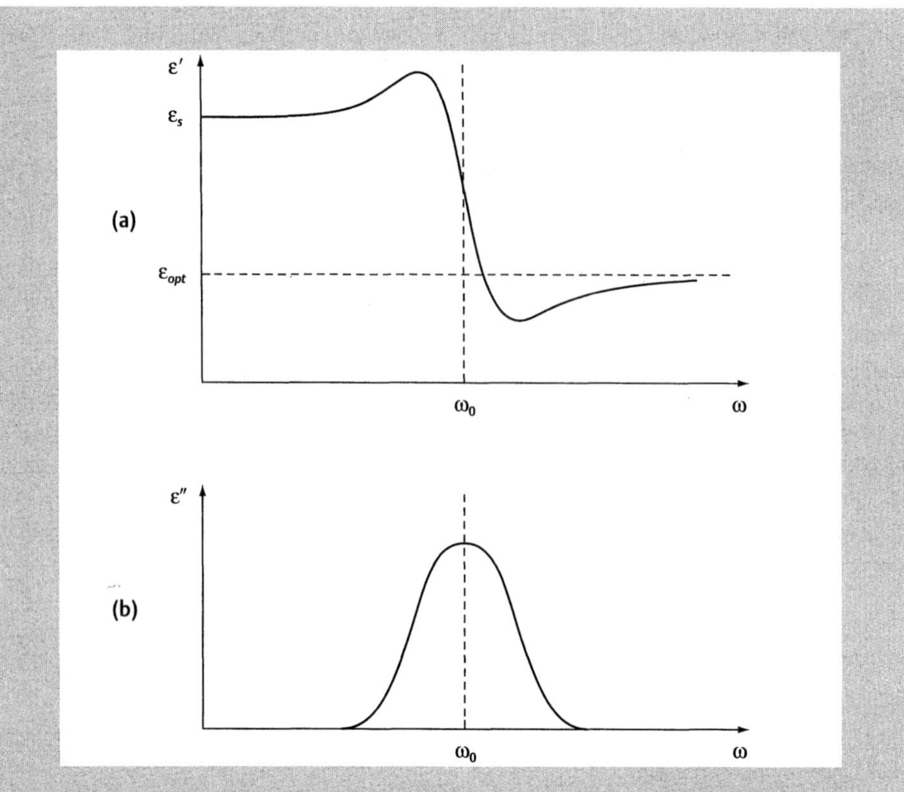

Figure 10.10 (a) The real part of the dielectric constant, ε', and (b) the imaginary part, ε'', as given by eqns (10.25) and (10.26), plotted as a function of the frequency of the applied electric field, ω.

which means that for some of the time the ions are out of phase with the external field. This leads to a dissipation of energy.

We can also see from Fig. 10.10(a) that the dielectric constant is larger at low frequencies (i.e. when $\omega \ll \omega_0$) than at high frequencies (when $\omega \gg \omega_0$), as we would expect from our discussion in the previous section. This is due to the fact that the amplitude of vibration of the ions is much greater at the lower frequencies. It is quite simple to show (see Question 10.7) that the difference between the real parts of the dielectric constant at low and high frequencies is given by

Example 10.2 Determine the amplitude of the vibration of the ions by solving eqn (10.23), and hence obtain an expression for the dielectric constant in terms of the frequency of the applied electric field, ω, and the natural frequency of vibration of the ions, ω_0.

Solution The solution to eqn (10.23) is of the form

$$x = Ae^{i\omega t}$$

Substituting into eqn (10.23) we obtain

$$\omega^2 Ae^{i\omega t} + i\omega\gamma Ae^{i\omega t} + \omega_0^2 Ae^{i\omega t} = \frac{e\mathcal{E}}{m} e^{i\omega t}$$

Cancelling the factor of $e^{i\omega t}$ in each term and rearranging, we obtain an expression for the amplitude A:

$$\begin{aligned} A &= \frac{e\mathcal{E}}{m} \frac{1}{\omega_0^2 - \omega^2 + i\omega\gamma} \\ &= \frac{e\mathcal{E}}{m}\left(\frac{\omega_0^2-\omega^2}{(\omega_0^2-\omega^2)^2+\gamma^2\omega^2} + \frac{i\gamma\omega}{(\omega_0^2-\omega^2)^2+\gamma^2\omega^2}\right) \end{aligned} \quad (1)$$

To obtain the second expression we have multiplied the top and bottom of the function by the complex conjugate of the denominator and separated the terms into real and imaginary parts.

Now that we have obtained an expression for the amplitude of vibration of the ions in the electric field, let us see how this affects the dielectric constant, ε.

From eqns (10.7), (10.15) and (10.16), the total polarization (due to induced and ionic effects) is given by

$$P = P_i + P_e = NeA + Nq\delta$$

Using this expression and eqns (10.4) and (10.5) we can express the dielectric constant as

$$\varepsilon = \frac{P}{\varepsilon_0\mathcal{E}} + 1 = \frac{NeA}{\varepsilon_0\mathcal{E}} + \frac{Nq\delta}{\varepsilon_0\mathcal{E}} + 1 \quad (2)$$

At optical frequencies we can assume that there is no ionic contribution, and so

$$\varepsilon_{opt} = + \frac{Nq\delta}{\varepsilon_0\mathcal{E}} + 1$$

Substituting in eqn (2) we obtain

$$\varepsilon = \frac{NeA}{\varepsilon_0\mathcal{E}} + \varepsilon_{opt}$$

Now substituting for A from eqn (1) we can write

$$\varepsilon = \varepsilon' + i\varepsilon''$$

where the real part is

$$\varepsilon' = \frac{Ne^2}{\varepsilon_0 m} \frac{\omega_0^2-\omega^2}{(\omega_0^2-\omega^2)^2+\gamma^2\omega^2} + \varepsilon_{opt}$$

and the imaginary part is

$$\varepsilon'' = \frac{Ne^2}{\varepsilon_0 m} \frac{\gamma\omega}{(\omega_0^2-\omega^2)^2+\gamma^2\omega^2}$$

$$\varepsilon_s - \varepsilon_{opt} \approx \frac{Ne^2}{\varepsilon_0 m \omega_0^2} \tag{10.27}$$

Substituting in appropriate values for these terms produces results which in many cases are in reasonable agreement with experimental measurements.

To describe how the polarization due to polar molecules varies as a function of frequency we need to employ a different approach. In this case we begin by defining a characteristic response time τ, which represents the time taken for a molecule to relax back to its original position after being disturbed by an external field. We will not go into the details of this approach, but it can be shown (see e.g. Solymar and Walsh) that the real part of the dielectric function in this instance is given by

$$\varepsilon' = \varepsilon_{opt} + \frac{\varepsilon_s - \varepsilon_{opt}}{1 + \omega^2 \tau^2} \tag{10.28}$$

and the imaginary part by

$$\varepsilon'' = \frac{(\varepsilon_s - \varepsilon_{opt})\omega\tau}{1 + \omega^2 \tau^2} \tag{10.29}$$

By studying these expressions we find that the real part of the dielectric function is equal to ε_s in the static case ($\omega = 0$), and ε_{opt} at high frequencies (see Question 10.8), as we would expect. For intermediate values of ω we can see from Fig. 10.11 that there is quite a sharp drop in ε' around the frequency $\omega = 1/\tau$, suggesting that the dipolar contribution to the polarization is 'switched off' at this frequency, and there is a corresponding peak in the imaginary term, ε'', showing that energy is dissipated in this frequency range.

So the dielectric function for polar materials appears to be very similar to the ionic model. However, the value of τ at room temperature is typically about 10^{-11} s, so the

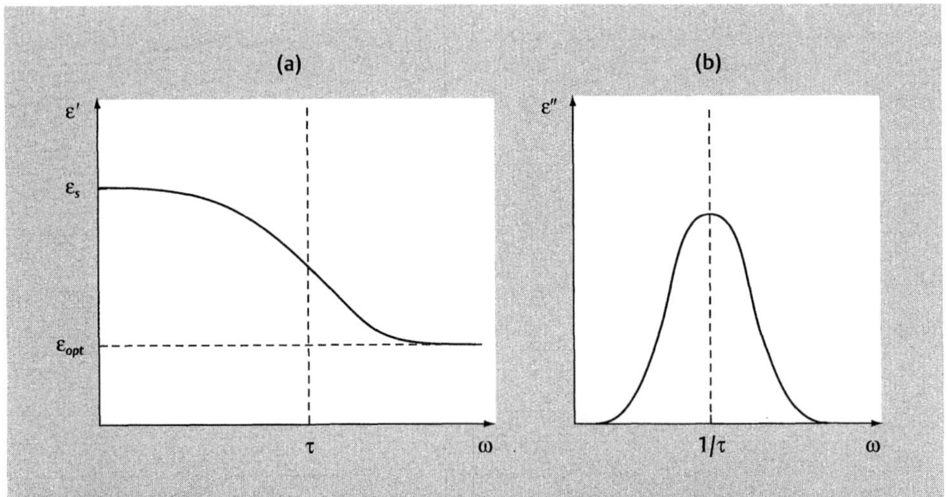

Figure 10.11 (a) The real part of the dielectric function, ε', and (b) the imaginary part, ε'', for polar molecules, as given by eqns (10.28) and (10.29).

frequency at which the dipolar contribution switches off is typically about three orders of magnitude lower than that for the ionic materials. In fact, since most polar materials also have an ionic contribution to the polarization, the dielectric function for a polar material is typically given by a curve such as that in Fig. 10.12. As the frequency of the external field increases there is a sharp drop at the frequency $1/\tau$ and a second sudden change at the frequency ω_0.

10.6 Impurities in dielectrics

If a material has a band gap greater than about 3 eV, then the energy required to excite an electron from the valence band into the conduction band is greater than can be supplied by a photon of visible light (see Question 10.9). Consequently, we would expect these materials to be transparent to visible light.

However, this is not always the case. One reason is that light is diffracted by the grain boundaries in a polycrystalline solid. We will consider this in more detail in Chapter 11. However, even a single crystal of dielectric material may not be completely transparent because the presence of impurity states in the band gap means that the material can absorb photons which have an energy less than the band gap energy. In this case it is only certain specific wavelengths of light which are absorbed. For example, a crystal of alumina (Al_2O_3) is normally transparent and colourless, but the presence of chromium impurities allows green light to be absorbed, and so when viewed in white light the crystal appears red (because this is the complementary colour to green). Such a crystal is called a ruby, and the presence of these impurity states is important in the operation of a ruby crystal laser. The appearance of other transition metal impurities in alumina causes the crystal to appear green (emerald) or blue (sapphire). Similarly, the inclusion of boron in diamond is responsible for the

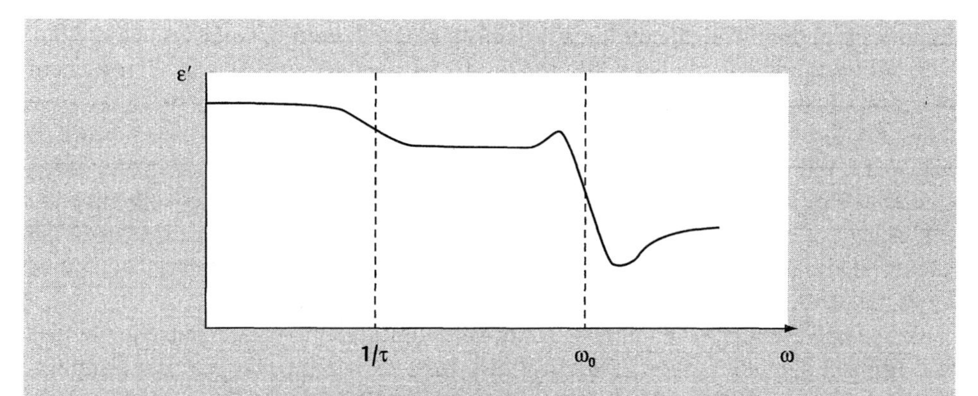

Figure 10.12 The real part of the dielectric function, ε', for a typical polar dielectric material. There are two sharp drops corresponding to the dipolar and the ionic contributions becoming inoperative at frequencies of $1/\tau$ and ω_0, respectively.

deep blue colour of some diamonds such as the famous Hope diamond. (I should, however, point out that not all coloured gemstones are the result of impurities. In some cases a defect in the crystal structure, such as a vacancy, can also produce absorption at visible light frequencies.)

Impurities can also be used to control the conductivity of a dielectric, in much the same way that they are used in semiconductors. However, the binding energies are often much larger. For example, boron can be used as an acceptor impurity in diamond (because it has a valence of three), but the impurity level is 0.36 eV above the valence band edge, so consequently only a very small proportion of these impurities are ionized at room temperature (see Question 10.10). This is not necessarily a problem—it just means that higher doping concentrations are required than are conventionally used in semiconductors if the device is to be operated at room temperature. However, one advantage of using large-band-gap materials is that they are far more suitable for use in higher temperature applications. Semiconductor devices cease to operate correctly at temperatures of a few hundred degrees Celsius because the concentration of intrinsic carriers becomes so large that it is no longer possible to distinguish clearly between the minority and majority carriers, but this is not a problem when the band gap is much larger (see Question 10.11). (More details about the use of diamond in electronic devices can be found in the article by Geis and Angus—see Further reading.)

10.7 Piezoelectricity

In certain types of dielectric material the application of an external stress produces electrical charges on the surface of the material (see Fig. 10.13). In other words, the applied stress gives rise to a polarization, which can be detected as a voltage difference between the ends of the sample. This is known as the piezoelectric effect.

Materials which display the piezoelectric effect also exhibit a converse effect: if the material is placed in an electric field it deforms and becomes strained (see Fig. 10.14).

To understand why this phenomenon occurs we need to consider the arrangement of the atoms in the crystal. A characteristic of piezoelectric crystals is that they have a low degree of symmetry. For example, let us consider the crystal structure shown in Fig. 10.15. When there is no applied stress, the ions are arranged in a pattern which has three-fold symmetry, and the centroids of negative and positive charge coincide for each group of ions (Fig. 10.15(a)). However, when a stress is applied in the direction shown in Fig. 10.15(b), the three-fold symmetry is destroyed and the crystal acquires a net polarization, P.

The piezoelectric effect is of interest not only for its novelty value, but also because it is utilized in a wide range of applications. In fact, there are far too many applications of piezoelectricity to be able to mention them all, so we will consider just a few examples.

One application which has been used for many years is the microphone. In this case a piezoelectric crystal is used to convert a sound wave into an electrical signal. The

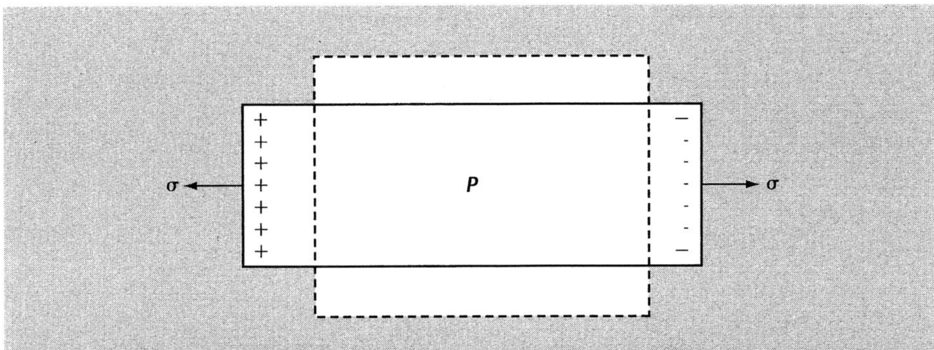

Figure 10.13 Applying a stress σ to a piezoelectric material gives rise to surface charges which produce a polarization P.

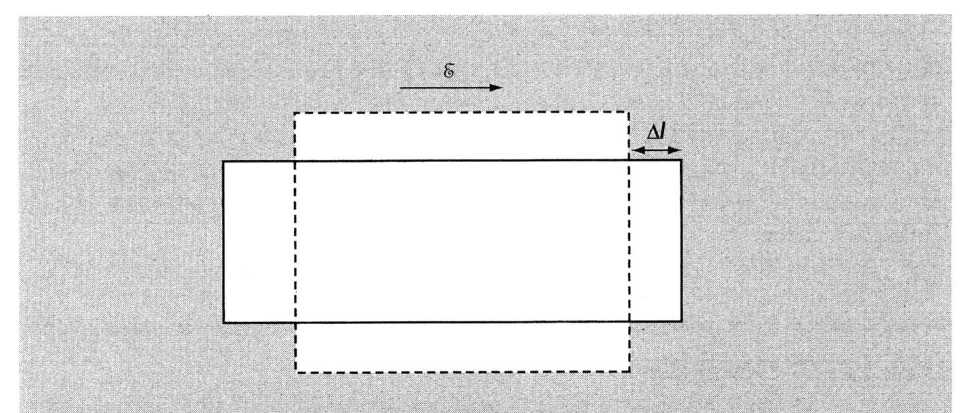

Figure 10.14 The converse piezoelectric effect. A piezoelectric material becomes strained when placed in an electric field.

basic principle is as follows. Since sound propagates as a pressure wave, it causes minute variations in the positions of the atoms in the piezoelectric crystal, which in turn produces a voltage across the crystal. The voltage changes in time with the same frequency as the pressure wave and the magnitude of the voltage corresponds to the amplitude of the pressure wave; therefore we obtain an electrical representation of the sound.

A rather different response is exploited in a quartz oscillator. In a quartz crystal, the displacements of the atoms are negligible unless the driving frequency is such that standing waves are set up in the crystal. This means that the quartz crystal vibrates only at a certain specific frequency. If the crystal is placed in an electrical circuit it can therefore be used to regulate the frequency of the circuit. In fact, the frequency of vibration is constant to within about 10^{-6}%, which is sufficient to keep a quartz watch accurate to within 1 second over a period of a year. Quartz oscillators are also used in a similar manner for maintaining a constant frequency in radio and television transmitters.

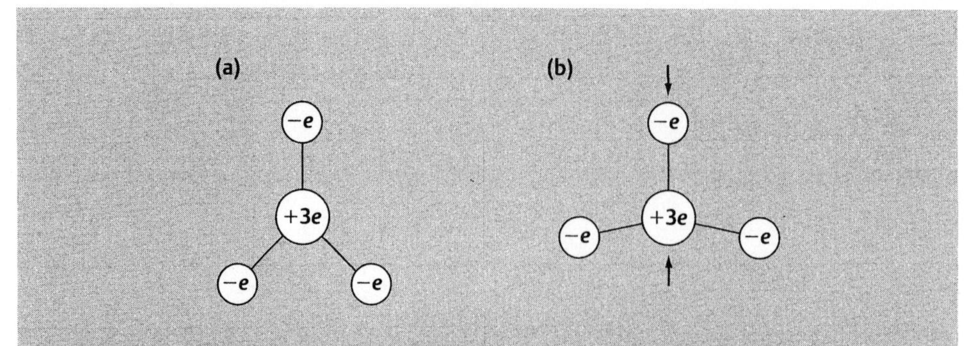

Figure 10.15 (a) A crystal structure with three-fold symmetry and zero polarization. (b) When a stress is applied, the symmetry is altered and the material acquires a non-zero polarization even in the absence of an electric field.

One further application worth mentioning is the role of piezoelectricity in the scanning tunnelling microscope (STM). The STM uses a fine probe to scan the surface of a crystal, as discussed in Chapter 2, and is capable of resolving individual atoms. A vital part of this apparatus is the piezoelectric crystals which are used to support the probe. By applying voltages to these crystals, the dimensions of the crystals are altered and so the probe is moved. In this way the position of the probe can be controlled to an accuracy of better than 0.01 nm.

10.8 Ferroelectrics

Ferroelectrics are another special category of dielectrics which, under certain circumstances, exhibit a finite polarization even in the absence of any applied stress or external electric field.

The name is something of a misnomer as these materials do not usually contain any iron. Rather, the term is derived from the fact that the materials have many properties which are similar to those of ferromagnetic solids (see Chapter 8). For example, if we plot the polarization P as a function of the external electric field strength $\mathcal{E}$ for a ferroelectric material we observe a hysteresis effect, as shown in Fig. 10.16. This is reminiscent of the relationship between the magnetization and external magnetic field in a ferromagnet (see Fig. 8.19). The terminology used to quantify the behaviour of a ferroelectric is also similar to that applied to ferromagnets: the saturation polarization P_S is the maximum value of polarization, the remanent polarization P_R is the net polarization when the electric field is zero, and the coercive field $\mathcal{E}_C$ is the electric field needed to return the polarization to zero after the material has been polarized.

The behaviour of ferroelectric materials can be explained by a theory analogous to that used to explain the properties of ferromagnets (see Chapter 8). It is assumed that the neighbouring dipoles are aligned by some form of mutual interaction, and

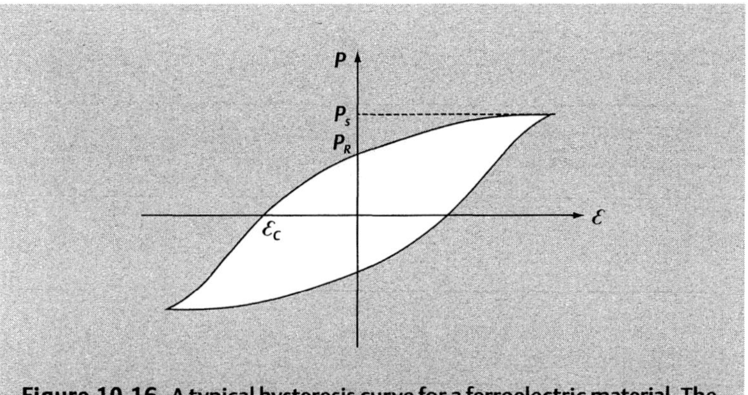

Figure 10.16 A typical hysteresis curve for a ferroelectric material. The saturation polarization, P_S, the remanent polarization, P_R, and the coercive field, $\mathcal{E}_C$, are indicated on the diagram.

that the aligned dipoles form domains, as in a ferromagnet. The existence of the domains in a ferroelectric material can be demonstrated by using polarized light, as shown in Fig. 10.17. A further similarity with ferromagnetic materials is that for many ferroelectric materials there is a characteristic transition temperature, known as the ferroelectric Curie temperature. Above this temperature the thermal jostling of the ions destroys the alignment of the dipoles and so the hysteresis effects disappear.

Ferroelectric materials typically have complicated crystal structures with low symmetry. An example is barium titanate, $BaTiO_3$. On the face of it, barium titanate appears to form a crystal structure with a high degree of symmetry. The Ba^{2+} ions form a simple cubic structure, the Ti^{4+} ions occupy positions at the body centre of this cubic structure and the O^{2-} ions are approximately at the centres of the cubic faces, as shown in Figs 10.18(a) and (b). However, a detailed investigation shows that at room temperature the oxygen ions are not quite at the centres of the cubic faces, but are displaced slightly in a common direction, as shown in Fig. 10.18(c). It is obvious from this diagram that the centroids of positive and negative charge do not coincide, and so the material has a finite polarization even when there is no external electric field present (see Example 10.3).

What happens when a stress is applied to a barium titanate crystal? It is found that the separation of the positive and negative lattices changes, and so the material is also a piezoelectric. In fact, it turns out that all ferroelectric materials display the piezoelectric effect. (It should be noted, however, that the converse is not true, i.e. not all piezoelectric materials are ferroelectrics.)

In the past many ferroelectric materials have been used for their piezoelectric properties, but in recent years other applications have been considered which exploit the unique properties of ferroelectrics. For example, ferroelectric materials have a very high dielectric constant, with values of ε as large as several thousand. This means, for example, that capacitors can be made much smaller than with ordinary dielectrics—a very important consideration in microelectronic circuits

Figure 10.17 The existence of domains in a ferroelectric material shown by imaging with crossed polarizers.

Example 10.3 If the displacement δ between the lattices of positive and negative ions (see Fig. 10.18(c)) in barium titanate is 0.01 nm at room temperature, determine the magnitude of the remanent polarization, P_R. (The molar volume of $BaTiO_3$ is V_m = 3.8×10^{-5} m^3.)

Solution One mole of $BaTiO_3$ consists of

N_A ions of Ba^{2+}
N_A ions of Ti^{4+}
$3N_A$ ions of O^{2-}

So the total charge associated with 1 mole is $\pm$ 6 $N_A e$, and therefore from eqn. (10.16) the dipole moment for 1 mole is

$$p_m = 6 N_A e \delta$$

Since the polarization is equal to the dipole moment per unit volume, the remanent polarization (i.e. the residual polarization when no electric field is applied) is

$$P_R = \frac{p_m}{V_m} = \frac{6 N_A e \delta}{V_m}$$

$$= \frac{(6.0)(6.02 \times 10^{23}\ \text{mol}^{-1})(1.60 \times 10^{-19}\ \text{C})(0.01 \times 10^{-9}\ \text{m})}{3.8 \times 10^{-5}\ \text{m}^3\ \text{mol}^{-1}}$$

$$= 0.15\ \text{C m}^{-2}$$

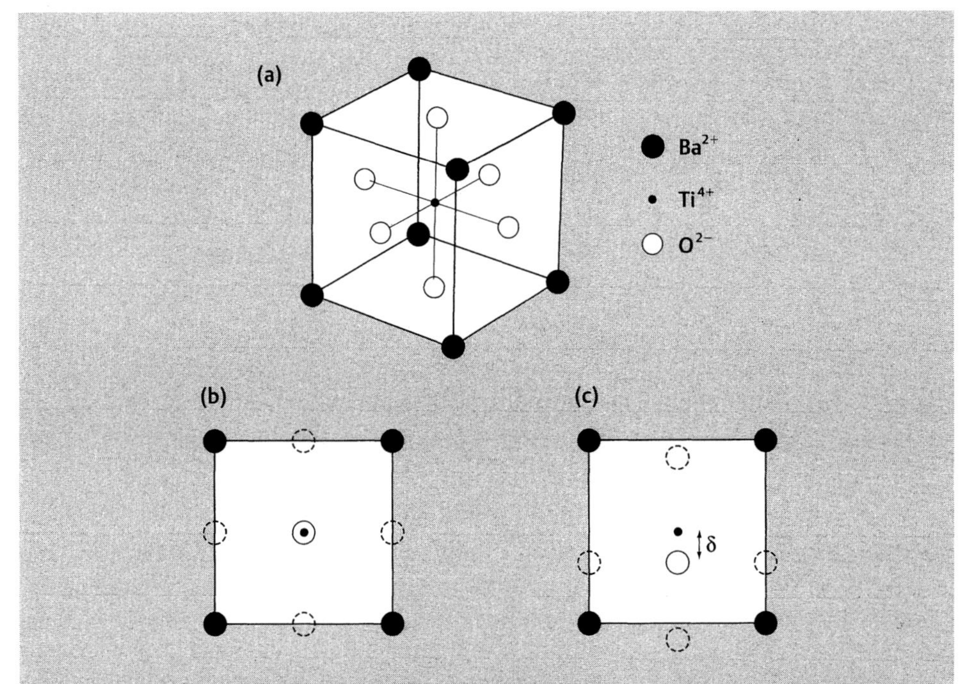

Figure 10.18 (a) A simplified picture of the unit cell of a barium titanate crystal. (b) The same unit cell shown in a plan view. The barium ions (large solid circles) and the oxygen ion at the centre of the cube face (open circle) are in the same plane. The titanium ion (small filled circle) and the other oxygen atoms (dashed circles) lie in a plane parallel to, and half a lattice spacing below, the one containing the barium atoms. (c) The actual structure of barium titanate at room temperature. The lattices of positive and negative ions are displaced by a distance δ relative to one another, where $\delta \approx 10^{-11}$ m.

where space is at a premium. Another application is in microelectronic memories. We can see from Fig. 10.19 that a crystal such as barium titanate has two possible polarization states in the absence of an electric field. The polarization of a ferroelectric crystal can therefore be used to store binary information. An additional feature is that a ferroelectric memory element is non-volatile, unlike a semiconductor memory circuit, because the crystals do not require a continuous supply of power in order to remain in the same state. (For more information about using ferroelectric materials in microelectronic applications, see the article by Scott mentioned in Further reading.)

10.9 Dielectric breakdown

Throughout this chapter we have assumed that the conductivity of a dielectric material is virtually zero. However, above a certain critical field, $\mathcal{E}_B$, breakdown occurs

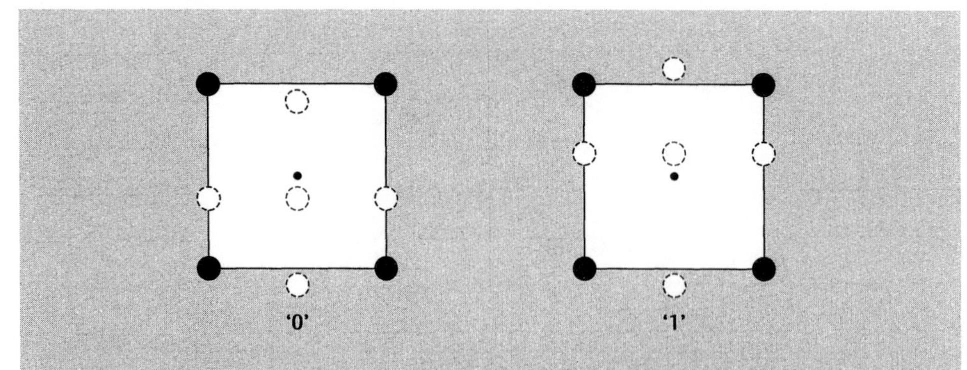

Figure 10.19 Plan view of a barium titanate unit cell showing the two stable polarization states in the absence of an external electric field.

and the material becomes an electrical conductor, as shown in Fig. 10.20. The magnitude of the breakdown field varies considerably between different samples of the same material, but a typical value for the breakdown field for d.c. voltages is about 10^8 V m^{-1}.

How and why does this effect occur? There are in fact several causes of dielectric breakdown, but for d.c. fields it is usually due to a sort of avalanche process, which is shown schematically in Fig. 10.21. To understand this diagram we need to start at the left hand side with a single conduction electron. In very high electric fields this electron is accelerated rapidly so that when it interacts with an atom in the crystal it has sufficient excess energy to liberate an electron from the atom. This means that we now have two conduction electrons. Both electrons gain energy from the applied field, interact with two other atoms and liberate another electron from each atom. And so the process continues with the number of conduction electrons increasing exponentially.

10.10 Summary

In this chapter we have seen that when a dielectric is placed in an electric field it becomes polarized. Using an atomic model of the crystal we have shown that there are three factors which can contribute to the polarization: induced polarization, which occurs in all dielectrics, and ionic and dipole contributions, which occur only in specific materials. By considering how these mechanisms respond at different frequencies, we have also been able to understand how the polarization, and therefore the dielectric function, varies with the frequency of the applied electric field.

Aside from the insulating properties of dielectrics, many useful applications can also be achieved by exploiting the piezoelectric or ferroelectric properties of certain

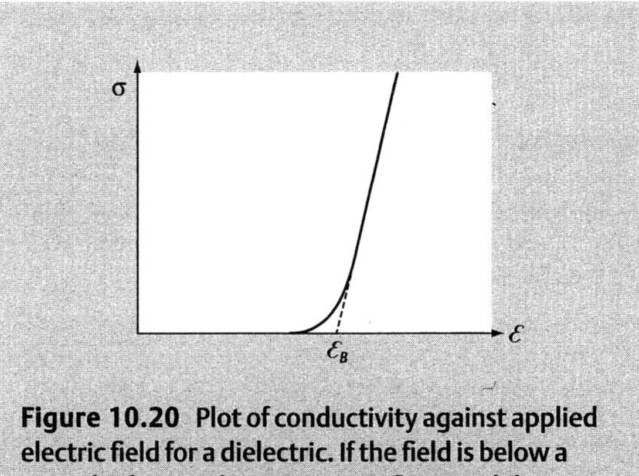

Figure 10.20 Plot of conductivity against applied electric field for a dielectric. If the field is below a critical value, $\mathcal{E}_B$, then no current flows and the material is an electrical insulator, but if the field exceeds $\mathcal{E}_B$ then a current flows and the material behaves as an electrical conductor.

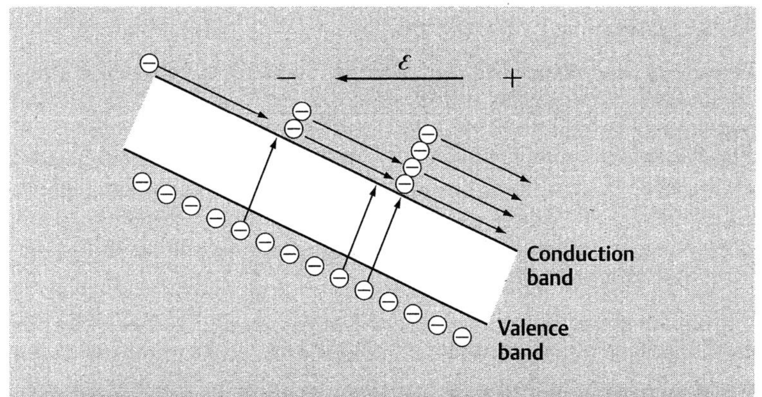

Figure 10.21 Schematic illustration showing the mechanism of avalanche breakdown in a dielectric. The conduction electron on the extreme left is accelerated by the field. When the electron is scattered by the lattice, the excess energy is used to excite another electron into the conduction band. Each of these electrons then excites another electron, and so on.

dielectric materials. In particular, many new applications are currently being explored for ferroelectric materials.

Questions

*Questions marked * indicate that they are based on the optional sections.*

10.1 Capacitors. The capacitance of a parallel plate capacitor increases from 0.085 μF to 0.195 μF when a sheet of polythene is placed between the plates. Use this information to determine the dielectric constant of polythene.

10.2 Induced polarization. Using the data in Table 10.1, determine the local field in a crystal of LiCl when placed between the plates of a parallel plate capacitor. The potential difference between the plates is 12 V and they are 0.1 mm apart.

10.3 Induced polarization. Determine the dipole moment and the displacement of the centroids of positive and negative charge for a neon atom in an electric field of 5×10^4 V m^{-1}. The atomic polarizability of neon is 4.3×10^{-41} F m^2.

10.4 Dipolar polarization. (a) Estimate the dipole moment of a hydrogen chloride molecule assuming that there is a complete transfer of the electron from the hydrogen atom to the chlorine atom. (The separation of the hydrogen and chlorine nuclei is 0.128 nm.) (b) Given that the measured dipole moment of a hydrogen chloride molecule is 3.3×10^{-30} C m, determine the actual amount of charge transferred from the hydrogen to the chlorine atom.

10.5 Frequency dependence of ε. (a) Given that the atomic polarizabilities of K^+ and Cl^- ions measured at optical frequencies are 1.48×10^{-40} F m^2 and 3.29×10^{-40} F m^2, respectively, and that the molar volume of KCl is 3.71×10^{-5} m^3, use the Clausius–Mossotti formula to determine the optical dielectric constant ε_{opt} for KCl. (b) Assuming that the atomic polarizabilities are the same in the static case as at optical frequencies, use the data given in part (a) to determine the amplitude of the displacements of the ions in KCl when placed in a static electric field of 10^5 V m^{-1}. (The static dielectric constant of KCl is 4.84.)

10.6 * Resonant absorption. Solve eqn (10.22) by assuming a solution of the form $x = A\,e^{i\omega t}$. Hence show that the amplitude of vibration becomes infinite when $\omega = \omega_0$.

10.7 * Resonant absorption. (a) Show that the difference in the real part of the dielectric constant (as given by eqn (10.25)) between the static case and the optical case is

$$\varepsilon_s - \varepsilon_{opt} = \frac{Ne^2}{\varepsilon_0 m \omega_0^2}$$

(b) Use the above equation to calculate $\varepsilon_s - \varepsilon_{opt}$ for sodium chloride given that the natural frequency of vibration $\omega_0 = 3 \times 10^{13}$ rad s^{-1}, the density is 2200 kg m^{-3} and the average mass of the Na^+ and Cl^- ions is 29.2 u (where u is the atomic mass unit which is 1.66×10^{-27} kg).

10.8 * Dipole relaxation. (a) Using eqn (10.28), obtain simplified expressions for the real part of the dielectric function ε' in a polar material at frequencies $\omega = 0$, $\omega = 1/\tau$ and $\omega \gg 1/\tau$.
(b) Using eqn (10.29), show that the energy dissipation is greatest (i.e. ε'' has a maximum) at $\omega = 1/\tau$.

10.9 Impurities. Given that the shortest wavelength of light that is visible to the eye is about 380 nm, determine the value of the largest band gap which is able to absorb visible light.

10.10 Impurities. If the binding energy for a boron impurity in diamond is 0.36 eV, determine what proportion of these impurities is ionized at 300 K. (Since the band gap of diamond is much larger than the thermal energy k_BT, the concentration of intrinsic carriers is negligible, and so we can assume that even at a temperature of 300 K the Fermi energy is midway between the valence band edge and the acceptor state.)

10.11 Impurities. Determine the concentration of thermally excited carriers in silicon and diamond at 1200 K using the equation

$$n = Ce^{-E_g/k_BT}$$

assuming that $C \approx 2 \times 10^{26}\ \mathrm{m^{-3}}$ and that the band gaps of silicon and diamond are 0.88 eV and 5.3 eV, respectively at this temperature. Explain why a silicon device based on a p–n junction could not be used at these temperatures.

10.12 Ferroelectrics. Calculate the displacement δ between the lattices of positive and negative ions in barium titanate (see Fig. 10.18(c)) when the polarization is saturated. (The molar volume of $BaTiO_3$ is $V_m = 3.8 \times 10^{-5}\ \mathrm{m^3}$ and the saturation polarization P_S is $0.26\ \mathrm{C\,m^{-2}}$).

10.13 Dielectric breakdown. If the type of plastic used for insulating electric wires has a breakdown field of $5 \times 10^7\ \mathrm{V\,m^{-1}}$, determine the minimum thickness required to insulate a domestic household supply (assuming a voltage of 240 V). Compare this figure with the actual thickness of insulation used in practice, which is typically about 0.5 mm.

Chapter 11

Crystallization and amorphous solids

11.1 Introduction

Although this book is primarily concerned with the physics of solids, in this chapter we will take a brief look at the behaviour of a material near the transition point—as a solid changes to a liquid, or, conversely, as a liquid changes to a solid. In general, this transition is characterized by a distinct change in the atomic structure from the regular ordered pattern of the crystalline solid to the disordered arrangement of the liquid state. This transition takes place over a very narrow temperature range, typically of the order of 0.01 K or less.

However, in some materials the transition is not so clearly defined. Instead there is a gradual transition to the liquid state over a temperature range of several tens of degrees. It is also noticeable that even in the solid state, the atoms in these materials form a random, disordered structure, quite unlike the crystalline solids that we discussed in Chapter 2. These materials are known as amorphous solids.

One of the best known amorphous solids is glass. This is a material which almost always solidifies in the amorphous state. In fact, the term 'glass' is sometimes used to describe any material which occurs naturally as an amorphous solid. However, in recent years there has been a growing interest in producing amorphous solids from materials which are normally crystalline in structure. Some of these new amorphous solids have properties which are interesting from a technological point of view, and so we will spend some time looking at this class of solid.

11.2 The melting point

Although the process of melting is something with which we are all familiar, it is an extraordinarily difficult subject to understand from an atomic point of view. Indeed, even the most sophisticated calculations are unable to predict the melting point of a given material with any certainty. Nevertheless, we can obtain a reasonable estimate of the melting point of a solid simply by considering the amplitude of the thermal vibrations of the atoms. By assuming that the vibrating atoms behave as simple harmonic oscillators, as in Section 7.2, we can see from Example 11.1 that in copper the amplitude of the atomic vibrations at the melting point is about 5% of the interatomic separation. By examining the data in Table 11.1 it is found that this value is roughly the same for a wide range of other materials. So by assuming that a solid melts when the amplitude of the thermal vibrations of the atoms is between about 5% and 7.5% of the equilibrium atomic spacing, we can obtain a reasonable estimate of the melting point of the material. Since the force constant, γ, is related to the strength of the bond between the atoms (see Section 7.2), we can also use this result to explain the correlation between the cohesive energy of the crystal and the melting point, as shown in Fig. 1.25.

Although this model of thermal vibration of the atoms appears to give a reasonably good estimate of the melting point, there are several other factors which contribute to the process of melting. For instance, as the temperature of a solid is increased we find that there is an increase in the number of vacancies in the crystal structure—in other words, places in the crystal lattice where an atom is missing. In general, a vacancy is usually formed when an atom moves from its lattice site to an interstitial position, as shown in Fig. 11.1. This is known as a Frenkel defect.

We can understand why Frenkel defects form by considering a statistical argument. The energy required to form a Frenkel defect is equal to the combined potential energy of the vacancy and the interstitial minus the potential energy of the system if the atom occupied the lattice site. If we denote this quantity by ΔE, then from the Boltzmann distribution we expect that the proportion of vacancies at temperature T is given by $e^{-\Delta E/k_BT}$. Consequently, as the temperature is increased, the proportion of vacancies increases exponentially. As the temperature approaches the melting point it is found that typically somewhere in the region of 2% of the atomic sites in the crystal are vacant. So melting does not occur simply because the crystal is shaken apart by

Table 11.1 Values of the amplitude of atomic vibration, x_{max}, at the melting point given as a percentage of the equilibrium lattice spacing. The values of x_{max} and the force constant, γ, are calculated from the Debye temperature, θ_D, the atomic mass, m, and the melting point temperature, T_{melt}, as in Example 11.1. Since the method tends to underestimate the value of the force constant (see Section 7.2), the amplitude of vibration at the melting point is actually likely to be slightly smaller than our calculated value. (Note that for NaCl the value of m represents the average mass of the ions.)

	θ_D (K)	m (u)	γ (N m^{-1})	T_{melt} (K)	x_{melt} ($\times 10^{-11}$ m)	a_0 ($\times 10^{-10}$ m)	$x_{melt}/a_0 \times 100\%$
Al	428	27	143	934	1.34	2.86	4.7
Ag	225	108	158	1235	1.47	2.90	5.1
Na	160	23	17.0	371	2.45	3.82	6.4
Au	165	197	130	1338	1.69	2.88	5.9
Cu	345	64	219	1358	1.31	2.56	5.1
Pb	95	207	24.7	601	2.59	3.50	7.4
Si	650	28	341	1687	1.17	2.34	5.0
NaCl	281	29	66.1	1070	2.11	2.81	7.5

Example 11.1 For copper, estimate the amplitude of the thermal vibrations of the atoms at the melting point as a percentage of the interatomic separation given the following information:

Debye temperature, $\theta_D = 345$ K,
atomic mass $m = 63.8$ u,
melting point $T_{melt} = 1358$ K, and
nearest neighbour separation $a_0 = 2.56 \times 10^{-10}$ m.

Assume that the atoms behave as simple harmonic oscillators, as described in Section 7.2.

Solution
From eqn (7.4) we know that the amplitude of the thermal vibrations of the atoms, x_{max}, at temperature T is given by

$$x_{max} = \left(\frac{2k_B T}{\gamma}\right)^{1/2} \tag{1}$$

where γ is the force constant, which is defined by eqn (7.2) as

$$\gamma = m\omega^2 \tag{2}$$

where ω is the angular frequency of the vibrations and m is the atomic mass.
If we assume that the Einstein temperature, θ_E, is approximately equal to the Debye temperature, then from eqn (7.15) we have

$$\omega = \frac{k_B \theta_E}{\hbar} \approx \frac{k_B \theta_D}{\hbar}$$

Substituting in eqn (2) gives

$$\gamma \approx m\left(\frac{k_B \theta_D}{\hbar}\right)^2 = (63.8)(1.66 \times 10^{-27}\ \text{kg})\left(\frac{(1.38 \times 10^{-23}\ \text{J K}^{-1})(345\ \text{K})}{1.05 \times 10^{-34}\ \text{J s}}\right)^2$$

$$= 219.1\ \text{N m}^{-1}$$

Therefore, from eqn (1), at the melting point the amplitude of the vibration is

$$x_{melt} = \left(\frac{2k_B T_{melt}}{\gamma}\right)^{1/2} = \left(\frac{(2)(1.38 \times 10^{-23}\ \text{J K}^{-1})(1358\ \text{K})}{219.1\ \text{N m}^{-1}}\right)^{1/2}$$

$$= 1.31 \times 10^{-11}\ \text{m}$$

As a percentage of the interatomic spacing this corresponds to

$$\frac{x_{melt}}{a_0} \times 100\ \% = \frac{1.31 \times 10^{-11}\ \text{m}}{2.56 \times 10^{-10}\ \text{m}} \times 100 = 5.1\%$$

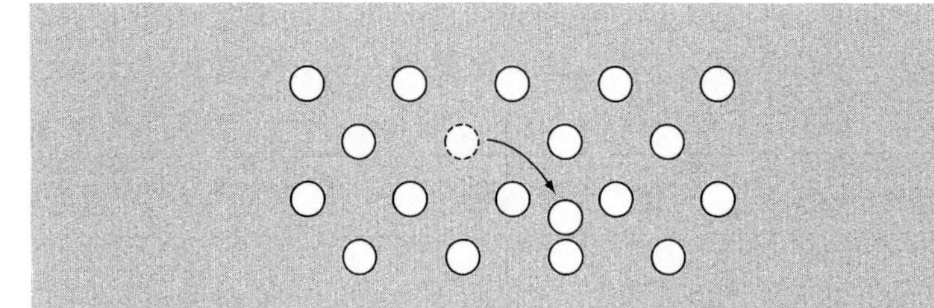

Figure 11.1 A Frenkel defect is formed when an atom moves from a lattice site to an interstitial position, leaving behind a vacancy (as shown by the dashed circle).

the thermal vibration of the atoms, but also because the fabric of the crystal has begun to crumble.

This effect is even more significant near the surface of the crystal. In this case a vacancy can be formed without creating an associated interstitial defect, because the atom can simply move to the surface, as shown in Fig. 11.2. Consequently, the formation energy of a vacancy near the surface is considerably smaller than the amount of energy required to create a vacancy within the crystal, and so the atoms at the surface are far more mobile than those in the bulk of the crystal. Therefore even when the temperature is considerably lower than the bulk melting point, the top few layers of atoms near the surface behave more like a liquid than a solid. This phenomenon is known as surface melting. It can be demonstrated quite simply by placing two ice cubes next to one another in a domestic freezer. Even when the temperature in the freezer remains well below the melting point of ice, it is found after a few hours that the ice cubes have fused together. This is because as the liquid layers at the adjoining surfaces of the ice cubes coalesce, they no longer behave as surface layers, and so this region freezes (see Fig. 11.3).

The melting point is also affected by the external pressure. In most cases the melting point increases with pressure, but in a few materials, such as ice, the melting point is reduced when a pressure is applied. Again we can demonstrate this at home with a large block of ice and a fine wire with weights attached to the ends. If the wire is placed over the block of ice, as in Fig. 11.4, the pressure from the wire causes the ice

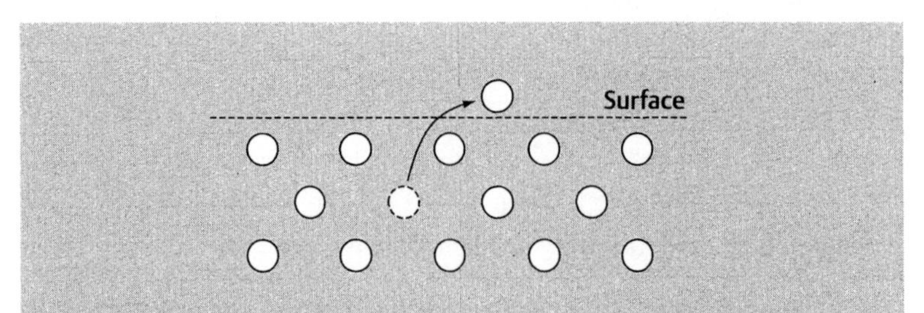

Figure 11.2 A vacancy can be created without producing a corresponding interstitial if the atom migrates to the surface of the crystal.

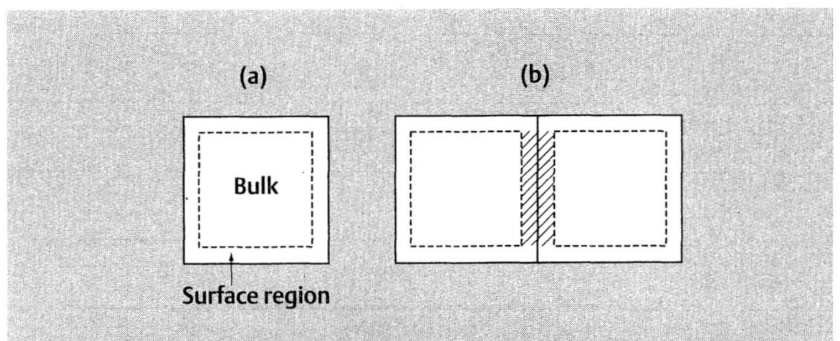

Figure 11.3 (a) A solid at a temperature just below the bulk melting point with the region affected by surface melting shown by the dashed line. (This region is grossly exaggerated.) (b) When two solids are placed in contact at a temperature just below the melting point, the surface layers in the shaded region coalesce and then freeze (because they are no longer surface layers).

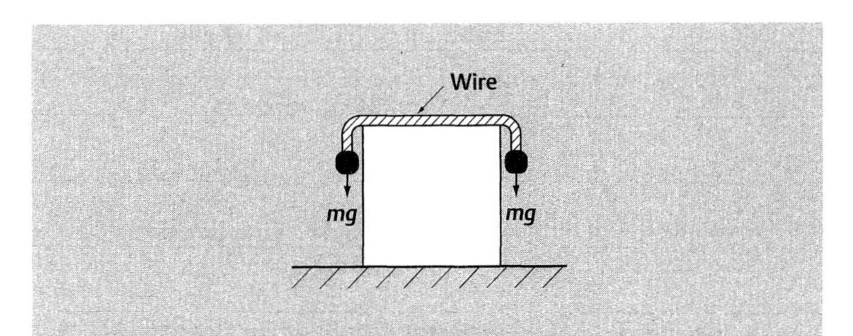

Figure 11.4 Demonstration to show that the melting point of ice is reduced when a pressure is applied. The pressure exerted by the wire leads to localized melting immediately beneath the wire, and so the wire cuts slowly through the block.

immediately beneath the wire to melt, and so the wire moves slowly down through the block of ice. (Note that as the wire cuts through the block, the water above the wire freezes again, so that even when the wire has passed right through the block, the block remains intact!)

The transition from solid to liquid is characterized by a discontinuous change in the density, and therefore in the volume, of the substance, as shown in Fig. 11.5. This is directly due to the change in atomic structure—from the regular structure of a crystal to the disordered structure of a liquid. We have seen in Chapter 2 that if the atoms are treated as spheres then they can occupy 74% of the total available volume if these spheres are arranged in a regular close-packed structure, such as the face-centred cubic or hexagonal close-packed structures. In comparison, it is found that when the spheres are arranged at random they occupy no more than 64% of the total volume. Therefore, when a close-packed crystal melts, the density of the substance should decrease by about 14% (see Question 11.4). In practice this is very close to the figure

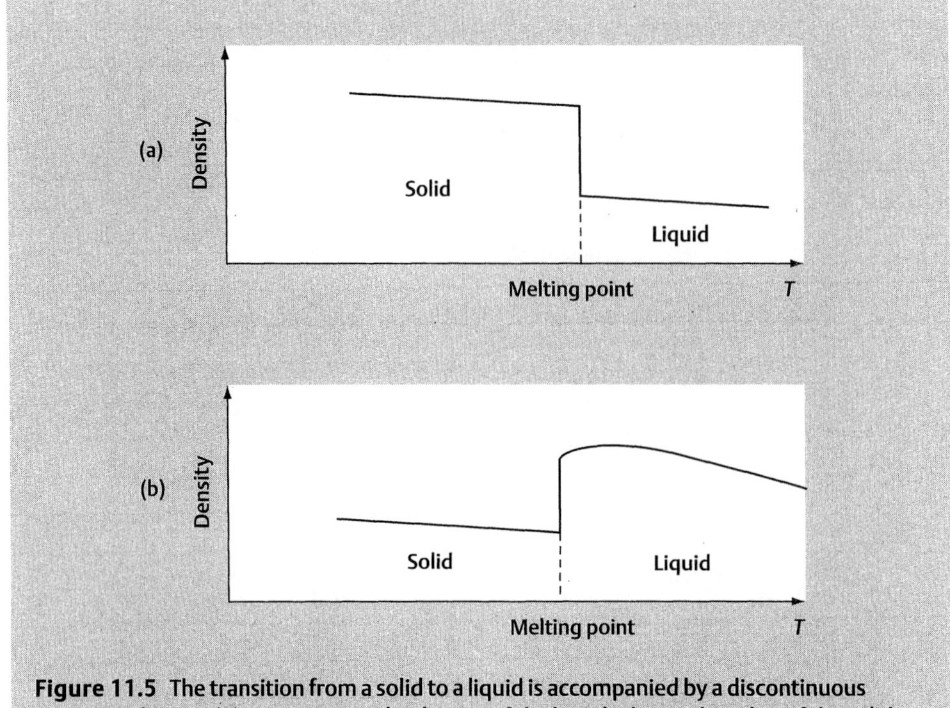

Figure 11.5 The transition from a solid to a liquid is accompanied by a discontinuous change in density. (a) In most cases the density of the liquid is lower than that of the solid, but (b) in a few materials, such as ice, the density increases as the solid melts.

observed with inert elements, but for metals the actual value is closer to 5%. The poor agreement for metals is assumed to be due to the free electrons which cause the ions in the crystal to be positioned slightly further apart than we might otherwise expect. In other words, the hard ionic spheres are not quite in contact with one another in the metallic crystal.

Of course, there are some materials, such as ice, in which the density increases when the solid melts (see Fig. 11.5(b)). In these materials the crystals are found to have a very open structure, and so the molecules actually occupy less volume when they are randomly arranged.

The transition from a solid to a liquid also involves the input of latent heat. This is a quantity of energy which is absorbed but does not change the temperature of the substance. Instead it can be thought of as the energy required to break the bonds between the atoms in the crystal. There is also a discontinuous change in the heat capacity, as we can see from Fig. 11.6. This is to be expected because the heat capacity of a solid arises from the vibrations of the atoms, whereas in a liquid it is due to the translational and rotational motion of the atoms or molecules.

One other point which we should note is that the transition is not always from a solid to a liquid. Under certain circumstances a solid may change directly into a gas, a process known as sublimation. This occurs when the pressure and temperature are below the triple point, as shown in Fig. 11.7. (The triple point is the unique combina-

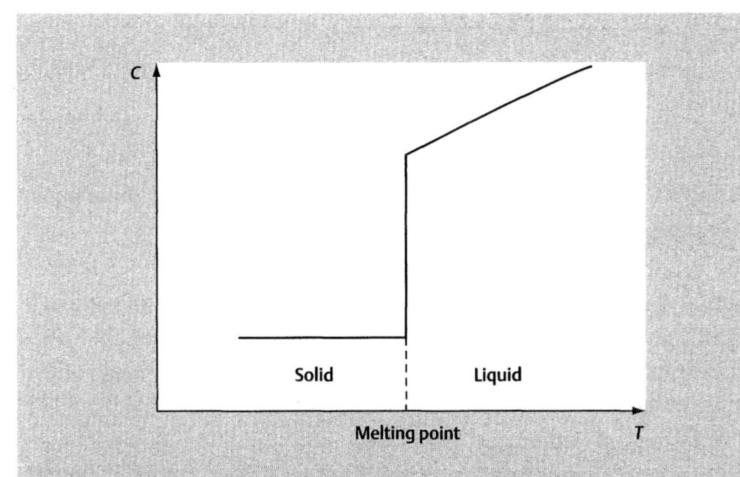

Figure 11.6 An abrupt change in the heat capacity, *C*, is also observed at the transition of a solid to a liquid.

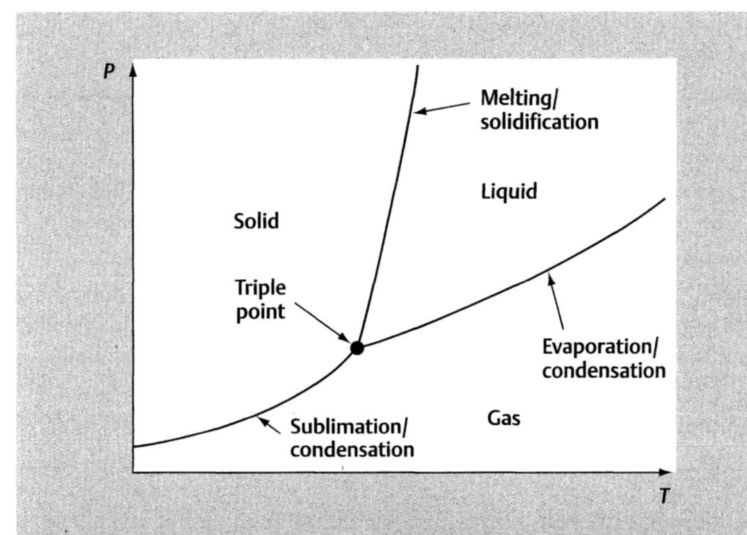

Figure 11.7 A typical phase diagram. The solid lines indicate where phase transitions occur. Along these lines the two different phases are in equilibrium. At the triple point, where these lines meet, all three phases are in equilibrium.

tion of temperature and pressure at which the solid, liquid and gas phases of a particular substance are all in equilibrium.) In most materials the triple point corresponds either to a temperature which is far below room temperature or a pressure which is much less than atmospheric pressure, and so the process of sublimation is less familiar than that of melting. However, in carbon dioxide the triple point is at 217 K and a pressure of about 5 atmospheres. Consequently, when solid carbon dioxide (which is also known as dry ice) is placed at room temperature and normal atmospheric pressure, it changes directly into a gas. The cold carbon dioxide gas that results causes condensation of water molecules in the air, and so produces a fog which is often used for theatrical effect.

11.3 Crystallization

Crystallization is simply the reverse process of melting. However, whilst it is intuitively reasonable to assume that the increased thermal vibration of the atoms should lead to a breakdown of the crystal structure, it is less obvious why a reduction in temperature should cause the atoms spontaneously to rearrange themselves in a regular pattern.

In an attempt to understand this process, let us consider the arrangement of the atoms in a monatomic liquid at a temperature close to the melting point. At any instant in time we are likely to find that there are small groups of atoms, called 'seeds', in which the atoms are arranged in a manner similar to those in a close-packed crystal—as in the region marked by the dashed line in Fig. 11.8. We can think of the formation of these seeds in terms of two competing factors. The atoms tend to form such arrangements because the energy is lower than that of a disorganized structure, but the constant thermal motion of the atoms tends to destroy this ordered arrangement. At temperatures just above the melting point, the thermal motion is the dominant factor and the seeds are destroyed before more atoms have a chance to join the structure.

If we lower the temperature to just below the melting point, then the balance is reversed. Atoms join the cluster at a faster rate than they are removed by thermal motion, and so the crystal grows larger with time. Under carefully controlled conditions it is possible to produce a single crystal, but usually crystallization involves many crystals forming from different seeds so that the result is a polycrystalline solid. In some

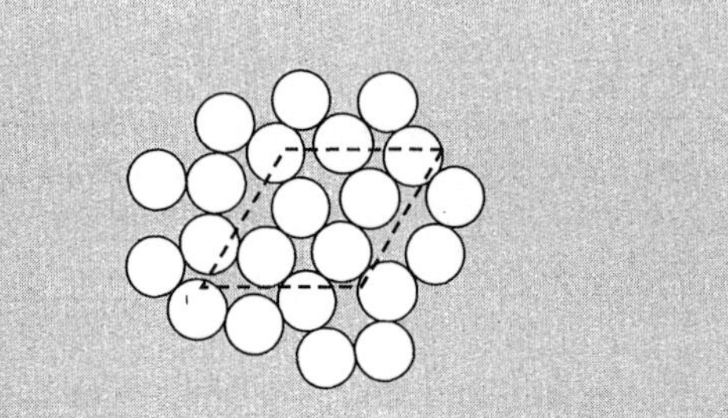

Figure 11.8 The arrangement of the atoms in a monatomic liquid at a temperature just above the melting point. At any instant in time there will be many small 'seeds'—regions in which a few atoms are arranged in a similar structure to a crystal—as in the region enclosed by the dashed lines.

instances a seed does not form spontaneously. In this case we end up with a supercooled liquid. For example, under certain conditions it is possible for pure water at normal atmospheric pressure to remain liquid at temperatures below 0°C. Such supercooled liquids are usually unstable and crystallize if given the slightest opportunity, but in some cases the material does not crystallize no matter how far the temperature is reduced. These materials form amorphous solids and we will discuss the properties of these materials in the remainder of this chapter.

11.4 Amorphous solids

An amorphous solid is a material which behaves like a solid—for example, it has strength and a fixed shape—but it has the atomic structure of a liquid. In other words, the atoms are arranged in a random manner, unlike the neat, ordered structure of a crystalline solid that we discussed in Chapter 2.

In the rest of this chapter we will look at some of the properties of amorphous solids, in particular concentrating on the ways in which the behaviour of an amorphous solid differs from that of a crystalline solid. Let us begin by considering why these materials do not crystallize.

Most amorphous solids are formed from liquids which have very high viscosity. An example is silica, SiO_2, which forms ordinary glass. The material becomes amorphous because the molecules in a viscous liquid cannot move easily past one another, and so as the temperature is lowered below the normal freezing point, T_f, the molecules are unable to move to appropriate positions to form a crystal. At T_f the coefficient of viscosity, η, is typically about 10^5 N s m^{-2} (cf. for water at room temperature $\eta \approx 0.001$ N s m^{-2}), and as the temperature is reduced below T_f the viscosity increases further, as shown in Fig. 11.9, making it even less likely that the molecules will be able to arrange themselves in a crystal structure.

Since there is no change in the atomic structure as an amorphous solid passes through the normal freezing point, T_f, there is no release of latent heat and no discontinuous change in the density, volume or heat capacity of the substance. However, as the temperature is reduced further, a sharp change in the heat capacity is often observed, as shown in Fig. 11.10. This occurs at what is known as the glass transition temperature, T_g, and may typically be about 100 degrees below the normal freezing point, T_f. We can understand this behaviour if we assume that at T_g the material changes from a very viscous liquid to a solid. We can then argue that the heat capacity below T_g is due to vibrations of the atoms, whereas above T_g it is due to translational movement and rotation of the molecules. However, it is important to note that there is no change in the atomic structure at this point—the material remains disordered—and there is no latent heat involved. What has happened is that the viscosity has become so large that the molecules effectively become immobile. In fact, the viscosity at the glass transition temperature is typically about 10^{14} N s m^{-2} (see Fig. 11.9). To all intents and purposes, this value is effectively infinite, and therefore the material behaves as a solid at temperatures below T_g.

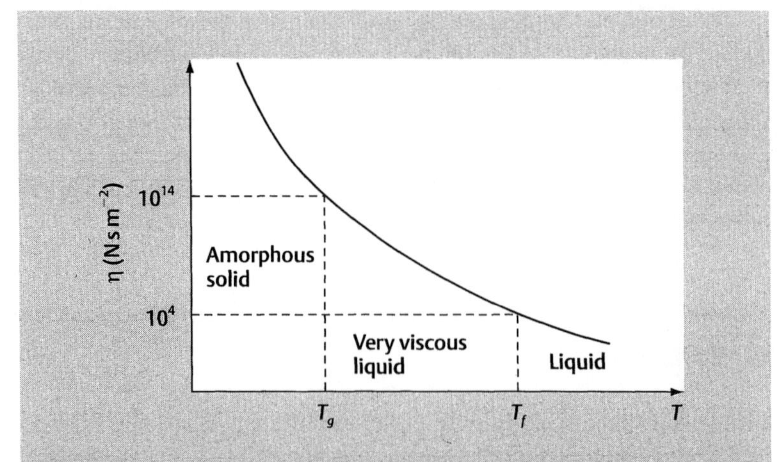

Figure 11.9 The change in the coefficient of viscosity, η, with temperature as a viscous liquid becomes an amorphous solid. T_f denotes the normal freezing temperature for the crystalline material and T_g is the glass transition temperature.

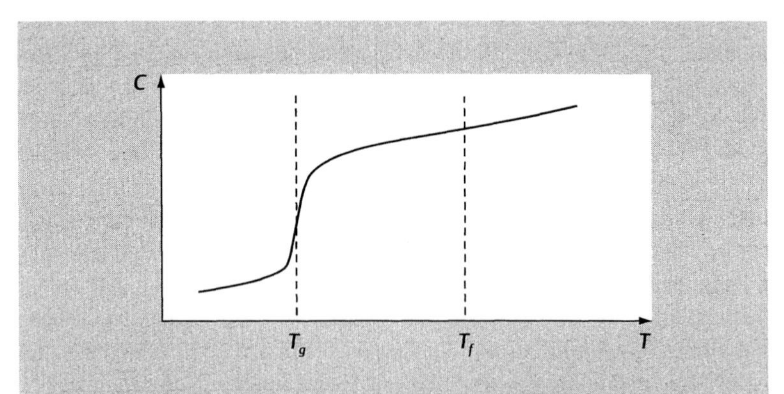

Figure 11.10 The change in heat capacity, C, with temperature for an amorphous solid. T_f is the normal freezing temperature for the crystalline material and T_g is the glass transition temperature.

So far we have discussed materials which naturally solidify in the amorphous state, but amorphous solids can also be produced from many materials which are normally crystalline. To achieve this 'artificial' state it is necessary to cool the material very rapidly so that the atoms do not have chance to move to the appropriate lattice sites. Cooling rates of up to 10^9 K s^{-1} can be achieved by various methods (see, e.g. Chapter 1 of Elliott), which is sufficient to allow many semiconductors and some metals to become amorphous. The incentive for producing these synthetic amorphous solids is that some of these materials have useful properties which they do not possess in the crystalline state. In many cases these materials are also much cheaper to manufacture in amorphous rather than crystalline form. We will look at some applications of these materials in Sections 11.6 and 11.7.

Having obtained an amorphous solid—either naturally or by rapid cooling—we need to consider the stability of the material. We know that the lowest energy state of a material occurs when the atoms are arranged in a regular manner in a crystal. For example, quartz, which is the crystalline form of silica, has a lower total energy than silica glass at room temperature. So why is it that a sheet of glass does not spontaneously become a quartz crystal? The answer is that although the energy of the crystal state is lower, there is an energy barrier ΔE which separates the amorphous state from the crystalline state, as shown schematically in Fig. 11.11. This represents the amount of energy required to move the atoms from the disordered positions in the amorphous solid to the lattice sites of the crystal.

In silica glass, this energy barrier is much larger than the thermal energy, k_BT, at room temperature, and therefore silica is in what is known as a metastable state. As there are specimens of glass which are many centuries old and show no signs of crystallization, we can assume that the lifetime of this metastable state is extremely long. In contrast, in many synthetic amorphous solids the energy barrier, ΔE, is smaller than the average thermal energy at room temperature, and so these materials are only stable at very low temperatures.

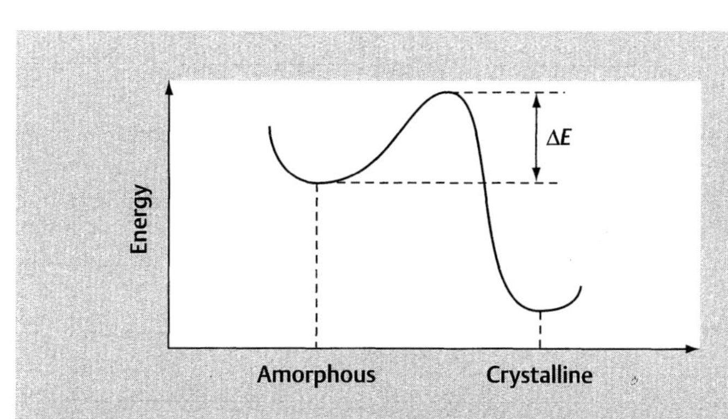

Figure 11.11 In order to convert an amorphous solid into the lower energy crystalline state it is necessary to overcome an energy barrier ΔE, which represents the energy required to move the atoms into the positions of the crystal lattice.

11.5 Optical properties of amorphous solids

The most useful optical property of glass is that it is transparent to visible light. This is due to two factors.

Firstly, as we have seen in Section 10.6, the absorption coefficient at visible wavelengths must be low for a material to be transparent, which implies that the band gap must be greater than about 3 eV (see Question 10.9).

The second factor is the diffraction of light within the material. To see how this affects the transparency, let us consider the behaviour of two parallel rays of light incident on a polycrystalline solid. As the light rays pass through the solid they are

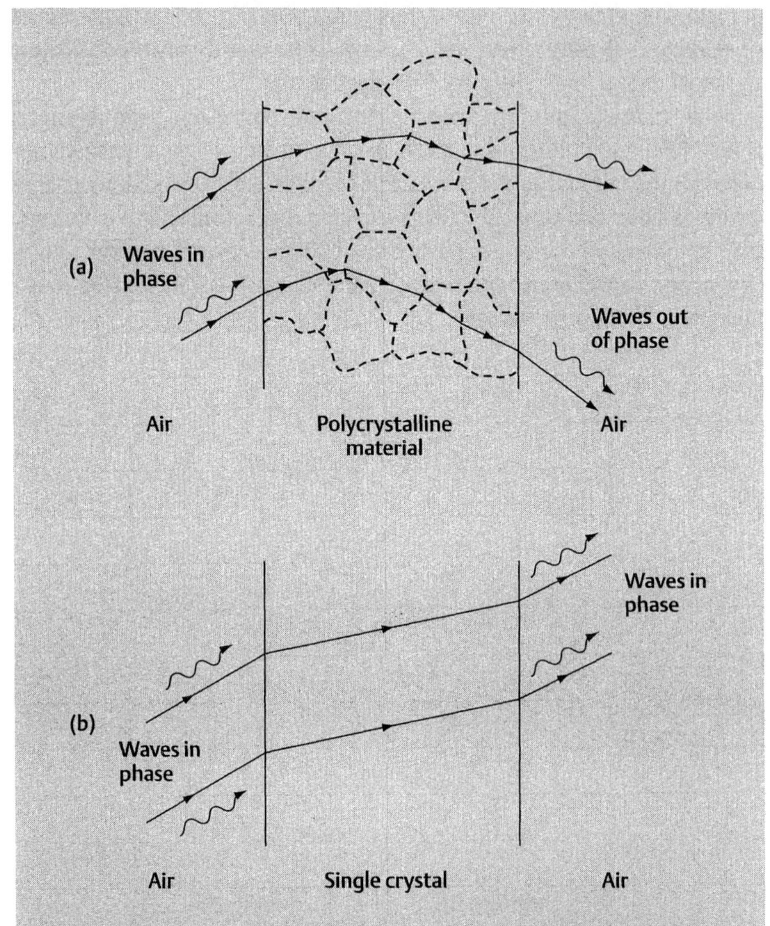

Figure 11.12 (a) As a light ray propagates through a polycrystalline solid it is diffracted at each grain boundary, so the transmitted waves are out of phase with one another. (b) In a single crystal, or in an amorphous solid, light propagates without being diffracted, and so the transmitted waves are in phase.

diffracted at each grain boundary, as shown in Fig. 11.12(a). This means that two light rays which enter the solid close together may follow quite different paths through the material. In particular, these light rays travel different distances, so when they emerge on the far side of the solid they are no longer in phase with one another. As a result, destructive interference occurs and so virtually no light is transmitted, i.e. the material is opaque.

In contrast, if we consider the same scenario with a perfect crystal, then the light travels through the medium without being diffracted (except at the outer surfaces) and so the transmitted light rays interfere constructively (see Fig. 11.12(b)).

What happens if we consider the propagation of light through an amorphous material? Since an amorphous solid does not have any ordered atomic structure, there are no grain boundaries, and therefore no reason why the light should be diffracted. Consequently, we obtain the same result as for a perfect crystal: the transmitted light rays interfere constructively, and so the material is transparent.

It may seem rather strange that a perfect crystal and an amorphous solid—which represent the two extreme examples of atomic order and disorder—exhibit such similar behaviour, whereas a polycrystalline solid—which in a sense is an intermediate case—displays completely opposite behaviour. However, this result is confirmed by observation. For example, if we consider the forms of silica, then we know that it is transparent as a single crystal of quartz and as an amorphous glass, but it is completely opaque in the polycrystalline form as sand.

11.6 Amorphous semiconductors

Many semiconductor materials can be produced synthetically in an amorphous state. In this section we will consider how the properties of these materials differ from those of the crystalline state, and we will take a brief look at a few applications of these materials.

Let us begin by considering the structure of amorphous silicon. We know that silicon is similar to carbon in many respects. Each atom has four valence electrons, and so silicon normally crystallizes in the diamond structure (as shown in Fig. 1.20) with each atom forming covalent bonds with the four nearest neighbours. Similarly, we find that most atoms in amorphous silicon also form covalent bonds with four other atoms, but in this case the distances between nearest neighbours and the angles between the bonds are not uniform. This is most easily demonstrated by considering an imaginary two-dimensional structure, as shown in Fig. 11.13. Of course, as the atoms are arranged in a random manner, it is not possible for every silicon atom to bond with precisely four other atoms, and so in some places we end up with an unpaired electron, which is known as a dangling bond (see Fig. 11.13(b)).

If we look at the energy band diagram then we find that there are so-called band tails, as indicated by the shaded regions in Fig. 11.14. These represent allowed electron and hole states at energies corresponding to what is normally the band gap of the crystalline material. Of course, we should not be too surprised to find allowed states

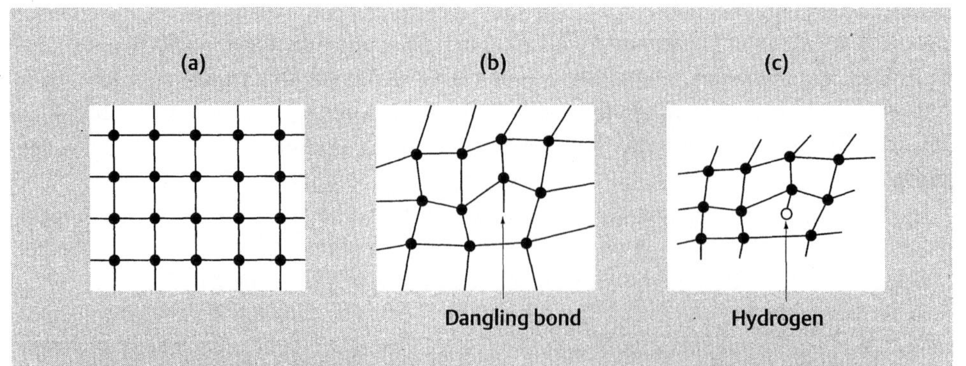

Figure 11.13 A two-dimensional representation of a substance such as silicon in which each atom ideally forms directed (i.e. covalent) bonds with four neighbours. (a) A perfect crystal. (b) The amorphous form. A dangling bond occurs when an atom is bonded with only three nearest neighbours. (c) A hydrogenated amorphous solid. Hydrogen atoms form bonds with unbonded electrons, removing the presence of the dangling bonds.

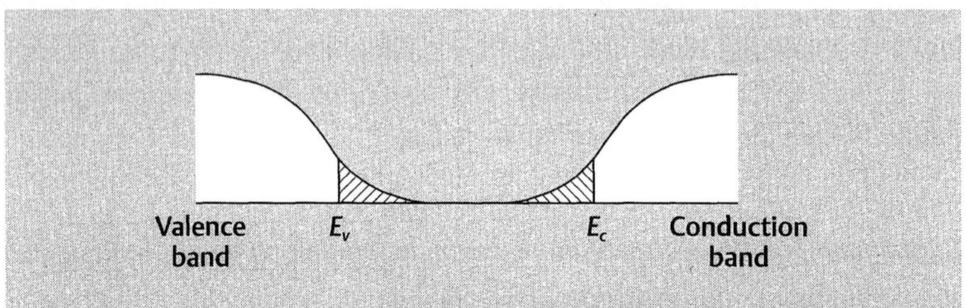

Figure 11.14 The energy diagram of an amorphous semiconductor. The shaded region indicates localized states which do not normally take part in electrical conduction.

at these energies because the band gap applies only if we have a perfect crystal (see Section 5.7), and the amorphous solid is nothing like a perfect crystal. In general, we therefore find that an amorphous semiconductor has a smaller band gap than the corresponding crystalline solid, which means that absorption of light takes place at lower frequencies than in the crystalline case. However, this does not necessarily mean that the conductivity is increased in this case. Since the energy states which lie in the band gap are associated with the dangling bonds, they are localized states, and so electrons and holes in these states are not normally able to take part in electrical conduction. This means that in order to change the electrical conductivity it is necessary to excite an electron into a state with energy greater than E_c, or a hole into a state with energy less than E_v (see Fig. 11.14). Consequently, the frequency of light required to produce photoconductivity is approximately the same as in the crystalline material.

What about doping of amorphous semiconductors? The main problem in this case is that any extra electrons or holes that are introduced by the impurities become trapped by the dangling bonds and so enter the localized states where they do not take part in electrical conduction. Therefore, doping an amorphous semiconductor does not necessarily affect the conductivity of the material. However, we can solve this problem if the sample of amorphous material is formed in the presence of a gas containing hydrogen. In this instance, the hydrogen atoms tend to attach themselves to any unbonded valence electrons, as shown in Fig. 11.13(c). This means that the dangling bonds are no longer able to trap the extra electrons and holes, and we can create n- or p-type amorphous materials by doping with appropriate impurities as in the crystalline case.

One application of amorphous silicon is in solar cells. Panels constructed from large single crystals of silicon are capable of converting solar energy into electrical energy, and have been used in several satellites, but since a large surface area is required to produce a useful power output (see Question 11.5), the cost of producing these panels is prohibitively expensive for any terrestrial application. On the other hand, polycrystalline silicon, which is much cheaper, has too many defects at the grain boundaries which trap the electrons and holes and therefore render the material useless. Hydrogenated amorphous silicon therefore appears to provide the perfect solution. It can be produced cheaply in large sheets and by introducing hydrogen the problems with the troublesome dangling bonds can be avoided.

Another class of important amorphous semiconductors are the chalcogenide glasses, such as amorphous selenium. This material is important in the Xerox process used in photocopier machines.

11.7 Amorphous magnets

Most metals are very difficult, if not impossible, to obtain in an amorphous state, and many of the amorphous metals which can be formed are only stable at very low temperatures. This is because the positions of the atoms in the irregular close-packed structure (which is the arrangement in a liquid or amorphous metal) is not very different to the positions in the regular close-packed structure of a crystalline metal. Consequently, the energy ΔE (see Fig. 11.11) required to change an amorphous metal into a crystalline material is generally very small. Nevertheless, many amorphous metal alloys are stable at room temperature, and some of these are of particular interest for ferromagnetic applications. These ferromagnetic amorphous alloys typically consist of iron, nickel or cobalt alloyed with approximately 20% of carbon, boron, phosphorus or silicon.

From a magnetic point of view, amorphous materials are of interest because they have no grain boundaries. As we mentioned in Section 8.11, the grain boundaries are a very significant factor in hindering the movement of the domain walls and therefore in determining the remanent magnetization and coercive field of the magnet. Consequently, amorphous ferromagnets tend to have a very narrow hysteresis curve and a low energy product and so make very good soft magnets.

As we discussed in Chapter 8, one of the main applications of soft magnets is in electrical motors and transformers. By using soft amorphous materials the motor or transformer operates more efficiently because the time lag between the changing magnetic field and the magnetization of the material is reduced. It has been estimated (see Elliott) that by replacing the cores of power distribution transformers with amorphous magnets, the energy losses would be reduced by 75%, which in the United States alone would amount to a saving of $250 million per annum.

An additional criterion for electrical transformers is that the resistivity of the core should be as high as possible in order to reduce losses due to eddy currents. Again, amorphous metals are ideal in this respect because they have very high values of resistivity. To understand why this is so, we need to think back to our quantum picture of electron transport in a metal (Section 4.4). We know that an electron travels through a perfect crystal lattice without being scattered by the ions. However, in an amorphous material the disordered arrangement of the ions means that the mean free path is very short—hence the high value of resistivity. In fact, typically the mean free path in an amorphous metal corresponds to a distance which is approximately equal to the interatomic spacing. Since it does not make any sense to have a mean free path which is less than the atomic spacing, this represents the maximum possible resistivity for a material which exhibits metallic conduction.

11.8 Summary

In this chapter we have looked at the changes which occur on an atomic scale when a substance melts or freezes. We have seen that the transition between the disordered liquid and the crystalline solid leads to abrupt changes in the density and heat capacity of the material. However, in amorphous materials, crystallization does not occur and the transition from liquid to solid is more gradual.

By studying the behaviour of these amorphous solids we have found, rather surprisingly, that the absence of grain boundaries often leads to properties which are similar to those of a single crystal, with the advantage that amorphous solids are usually much cheaper to produce than single crystals.

The ability to create synthetic amorphous materials by very rapid cooling makes it possible to create new materials with a wide range of different properties. The challenge is to discover new amorphous solids with properties which are superior to those of crystalline materials. Amorphous magnets are a good example, but even in this case scientists have by no means exhausted all of the possibilities for creating even better amorphous magnetic materials.

Questions

11.1 Melting point. For potassium chloride, estimate the amplitude of the thermal vibrations of the atoms at the melting point, x_{melt}, as a percentage of the interatomic spacing, a_0, using the following data:

Debye temperature $\theta_D = 230$ K,
average atomic mass $m = 37$ u,
melting point $T_{melt} = 1063$ K, and
nearest neighbour separation $a_0 = 3.14 \times 10^{-10}$ m.

11.2 **Melting point.** Estimate the melting point of diamond by assuming that the crystal melts when the amplitude of the atomic vibration is 7% of the interatomic separation. (The interatomic spacing in diamond is 1.54×10^{-10} m, the Debye temperature is 1850 K and the mass of a carbon atom is 12.0 u.)

11.3 **Melting point.** A monatomic liquid with a volume of 1.0×10^{-6} m^3 crystallizes to form a face-centred cubic structure. Estimate the volume of the crystal given that the atoms in the liquid occupy 64% of the volume, and those in the crystal occupy 74% of the volume.

11.4 **Melting point.** Determine the percentage change in density that occurs when a monatomic close-packed crystal (in which the atoms occupy 74% of the volume) becomes an irregular close-packed liquid (in which the atoms occupy 64% of the volume).

11.5 **Amorphous semiconductors.** A solar panel is made from amorphous silicon. If the flux of photons (with appropriate energy) incident on the panel is 2×10^{21} m^{-2} s^{-1}, calculate the electrical power produced by a panel of area 10 m^2. Assume that the efficiency of the panel is 25% and that the optical band gap of amorphous silicon is 1.0 eV.

Chapter 12

Polymers

12.1 Introduction

Polymer is a term used to describe a long chain molecule which usually has a backbone of carbon atoms. For example, polyethylene consists of a long string of carbon atoms, each of which is bonded to two other carbon atoms and two hydrogen atoms, as shown schematically in Fig. 12.1(a). The basic building block—which in this case consists of one carbon atom and two hydrogen atoms—is called a monomer, and the polymer is formed by simply repeating this structure over and over again. A single molecule of polyethylene typically consists of many thousands (or even tens of thousands) of these monomers and can measure several micrometres in length (see Question 12.1). The chain is terminated only when a carbon atom bonds with three hydrogen atoms.

Polyethylene is one of the simplest examples of a polymer, and there are many other ways in which we can construct similar giant molecules. For example, we can replace the hydrogen atoms with any of the halogen elements, i.e. those atoms which

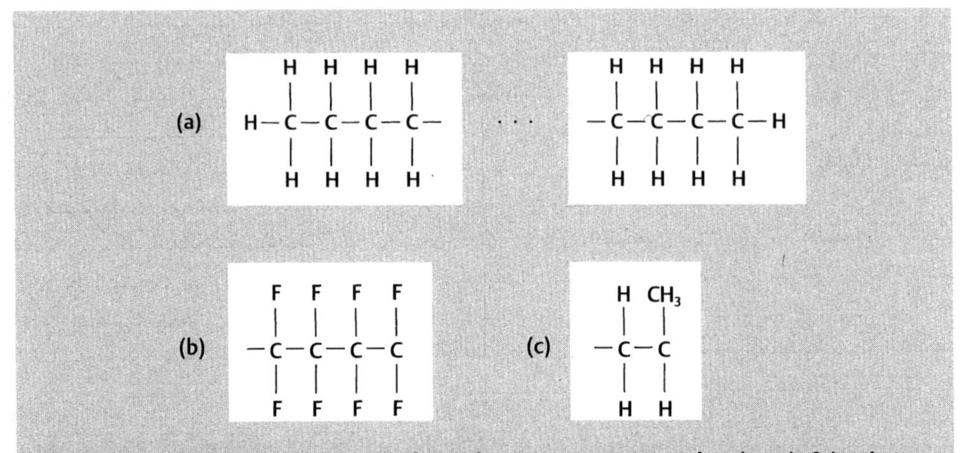

Figure 12.1 (a) A molecule of polyethylene. The monomer is CH_2 and each end of the chain is terminated with a CH_3. (b) A short section of polytetrafluoroethylene (PTFE). (c) A single monomer of polypropylene.

require one electron to achieve a filled subshell. Using fluorine atoms in place of the hydrogen atoms in polyethylene produces polytetrafluoroethylene, which is usually abbreviated to PTFE. (This material is also known by the tradename Teflon.) A short section of PTFE is shown in Fig. 12.1(b).

More complex polymers can be achieved if some of the hydrogen atoms in polyethylene are replaced by molecules. These are known as branched polymers. An example is polypropylene, which has a CH_3 molecule attached to each alternate carbon atom, as shown in Fig. 12.1(c). Yet another possibility is that the carbon chain can be replaced by a chain of silicon atoms.

As there is such a huge variety of polymers, it is perhaps not surprising that these materials also display a wide range of physical properties. However, most polymers share the common properties of being lightweight, flexible, resistant to corrosion and easy to cut or mould into any desired shape. For these reasons polymers have become common in many everyday items, often replacing traditional materials such as wood and metal. Indeed, if anything plastics have been too successful. The low cost of these materials means that in many cases they are treated as poor alternatives to the 'real thing'. Nevertheless, there are many instances in which polymers are indisputably superior to conventional materials. One such material which we will consider in this chapter is Kevlar, which is not only considerably stronger than steel, but also about six times less dense, and so has a strength to weight ratio about 10 times larger than steel.

12.2 Elastic properties of rubber

It is well known that certain polymers, such as rubber, display extraordinary elastic properties. These materials are known as elastomers. Whereas most materials reach the elastic limit at strains of less than 1%, elastomers can exhibit strains of more than 400%.

Another notable property of these elastomers is that they can be deformed very easily. For instance, a rubber band can be doubled in length with very little effort. In fact, the elastic moduli of elastomers may be as low as 10^5 N m^{-2}. Comparing with the values in Table 3.1, we can see that this is at least a hundred thousand times smaller than that of most other materials.

We should also point out that the elastic moduli of elastomers display an anomalous temperature dependence. In most materials a rise in temperature causes the elastic moduli to decrease, but in an elastomer the elastic moduli are found to increase with temperature. This can be demonstrated quite simply by suspending a mass from one end of an elastic band. If the elastic band is heated (e.g. with a hairdryer), the mass moves upwards. This is because the heat increases Young's modulus of the material, and so the strain is reduced.

Therefore, the elastic properties of elastomers can be summarized as follows:

(1) the strain at the elastic limit is very large, typically of the order of 100%;

(2) the elastic moduli are very small; and
(3) the elastic moduli increase with temperature.

How do we explain these extraordinary characteristics?

In most materials an elastic strain is produced by stretching the atomic bonds, but polymers consist of a chain of C—C bonds, which we know are very strong. Such a low value of the elastic moduli could not possibly be achieved by stretching these bonds. Instead, we need to look at the shape of the molecules.

Firstly, we should point out that the monomers in a polymer do not form a straight line. The only criterion governing the shape of a polymer is that the angle between two adjoining C—C bonds is 109.5°. This means that the positions of the carbon atoms are not uniquely determined. For example, if we consider Fig. 12.2(a), we can see that even if the positions of atoms C_2 and C_3 are fixed, the atom C_4 can occupy any position on the dashed circle. (Note that C_4 does not necessarily lie in the same plane as C_2 and C_3.) In the absence of any external factors (such as an applied stress), the position of C_4 is quite arbitrary. Consequently, if we consider a polymer containing several thousand monomers, the overall shape of the polymer is quite random (see Fig. 12.2(b)).

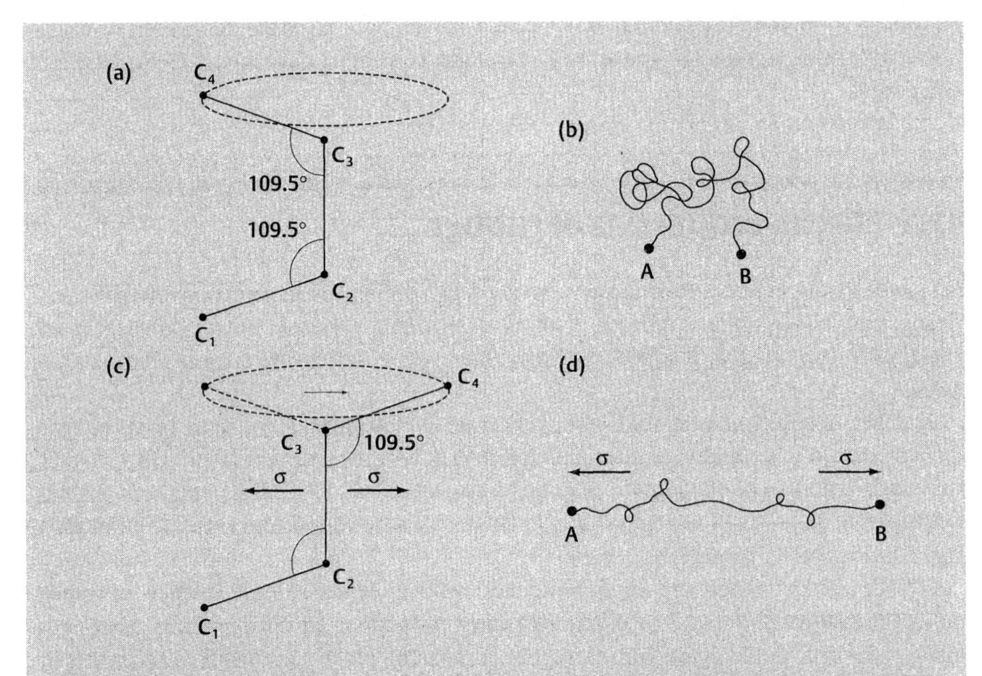

Figure 12.2 (a) A short section of a polymer showing just the carbon atoms for clarity. Since the angle between adjacent carbon atoms is 109.5°, atom C_4 can occupy any of the positions on the dashed circle. (Note that C_4 is not necessarily coplanar with C_2 and C_3.) (b) Typical shape of a polymer containing several thousand monomers. (c) When a stress σ is applied to the polymer, the C_3—C_4 bond tends to rotate so that the bond becomes as closely aligned as possible with the applied stress. (d) As a result, the polymer becomes stretched out along the direction of the applied stress.

When a stress is applied to the polymer, individual C—C bonds have a tendency to rotate so that the bonds are aligned as closely as possible along the direction of the applied stress, as shown in Fig. 12.2(c). Overall, this means that the polymer becomes stretched out, as shown in Fig. 12.2(d). Therefore, the distance A–B between the two ends of the polymer increases dramatically, and so we can see that a large strain can be produced simply by changing the shape of the molecule, rather than by stretching the bonds. Since the magnitude of the stress required to cause a C—C bond to rotate is far smaller than that required to stretch a C—C bond, we can also see why the elastic moduli are so small.

What about the temperature dependence of the elastic moduli? To explain this phenomenon we need to use a thermodynamic argument. In particular, we will consider the entropy, which is a measure of the disorder of a system. By comparing Figs 12.2(b) and 12.2(d), we can see that the disorder—and therefore the entropy—is reduced when a stress is applied. What happens when we heat a stressed polymer? Since the entropy of a system increases as the temperature is raised, this suggests that the polymer chains should tend to coil up, as in Fig. 12.2(b), in order to increase the entropy of the system. By reducing the overall length of the molecule, this leads to a decrease in strain, and so an increase in Young's modulus. Therefore, the increase in the elastic moduli with temperature is directly due to the increase in entropy of the polymer chains.

Having explained the characteristic properties of elastomers by considering the change in shape of the molecule, let us now try to use this model to obtain an estimate of the magnitude of the maximum elastic strain. We can assume that the maximum elastic strain is achieved when the shape of the polymer approximates to that shown in Fig. 12.3. From Question 12.3 we find that the distance between the two ends of the polymer in this case is given by

$$d_{max} = 0.816aN \tag{12.1}$$

where a is the length of a C—C bond and N is the number of carbon atoms in the chain. In order to determine the magnitude of the strain we also need to know the distance between the ends of the polymer in the unstressed state, $d_{unstressed}$. This may seem like an impossible task. Even with a small number of carbon atoms, there are an enormous number of different atomic configurations, and therefore different values of the distance $d_{unstressed}$. However, by using statistical methods it can be shown that the

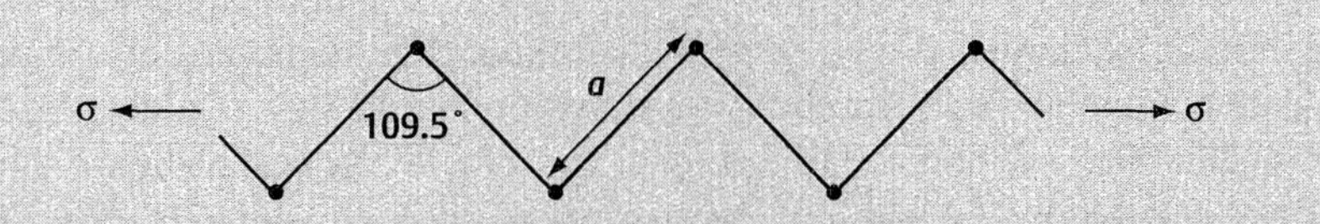

Figure 12.3 Schematic diagram of a polymer chain stretched so that the distance between the two ends of the polymer, d_{max}, is as large as possible. The length of the C—C bonds is denoted by a.

Example 12.1 Estimate the maximum strain that can be achieved by the rotation of C-C bonds in a polymer containing 5000 monomers.

Solution From eqn. (12.2) the natural (unstressed) length of the polymer is

$$d_{unstressed} \approx a\sqrt{N}$$

whereas using eqn. (12.1) the maximum elastic length in the stressed case is

$$d_{max} \approx 0.816aN$$

Since the strain \epsilon is equal to the change in length divided by the original length, we have

$$\epsilon = \frac{(0.816aN - a\sqrt{N})}{a\sqrt{N}} \times 100\ \% = \frac{(0.816N - \sqrt{N})}{\sqrt{N}} \times 100\ \%$$

$$= \frac{(0.816)(5000) - \sqrt{5000}}{\sqrt{5000}} \times 100\ \%$$

$$= 5670\ \%$$

most probable value is approximately equal to the bond length multiplied by the square root of the number of atoms, i.e.

$$d_{unstressed} \approx a\sqrt{N} \qquad (12.2)$$

(A proof of this result is given by Tabor—see Further reading. In fact, if you are familiar with statistical mechanics you may recognize this result as the most probable length of a three-dimensional simple random walk.)

If we apply these results to a polymer containing 5000 monomers, then from Example 12.1 we can see that the maximum elastic strain is of the order of 5500%. However, this is far larger than is observed in practice. What is wrong? The problem is that we have considered an isolated polymer chain. In a solid there are many polymer chains in close proximity, and they interact with one another. The most significant interaction is the existence of strong crosslinks between the chains. These are often due to impurities. For example, sulphur atoms in rubber form a strong bond between two adjoining polymer chains. We therefore have to think of a solid polymer as a network of interconnected chains, as shown in Fig. 12.4(a), rather than as a collection of individual chains. When a stress is applied, the movement of the polymer chains is restricted by these crosslinks, as shown in Fig. 12.4(b), and so the deformation is smaller than we previously assumed. In fact, we can make a reasonable estimate of the maximum elastic strain if we assume that the value of N in eqns (12.1) and (12.2) corresponds to the average number of atoms between each of the crosslinks. If we assume that a typical value of N for rubber is about 250, then by applying the same argument as in Example 12.1 we find that the maximum elastic strain is about 1200% (see Question 12.4). This is still considerably larger than the observed value of about

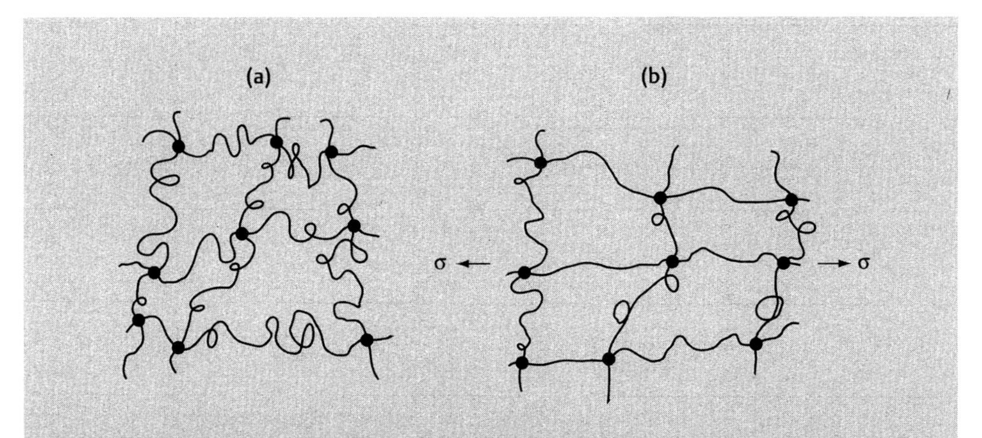

Figure 12.4 (a) The polymer chains in a solid are connected by crosslinks, shown by solid circles. (b) When a stress σ is applied, the movement of the chains is restricted by the presence of the crosslinks.

400% because there are many other factors, such as the van der Waals forces and hydrogen bonds between adjoining polymer chains that we have not taken into account.

12.3 The rubbery and glassy states

In the previous section we have seen that some polymers exhibit extraordinary elastic deformation and very low values of elastic moduli, but these properties change dramatically at low temperatures. For example, if rubber is cooled to liquid nitrogen temperatures it becomes brittle and shatters if we apply a stress. Similarly, some polymers which are brittle at room temperature, such as Perspex, are found to display much greater elastic deformation at higher temperatures.

We can therefore identify two states. At low temperatures the material is in a glassy or vitreous state and at higher temperatures it enters a rubbery or viscoelastic state. The rubbery state is characterized by low elastic moduli and the ability to undergo large elastic deformation. As we have seen in the previous section, this behaviour is due to the rotation of the C—C bonds. In the glassy state the C—C bonds are no longer able to rotate and the material is capable of only a small amount of elastic deformation before brittle fracture occurs. The elastic moduli are also much larger than in the rubbery state, though generally slightly smaller than those of other materials. A typical value of Young's modulus for a polymer in the glassy state is about 10^9 N m^{-2}. The variation of Young's modulus with temperature for a typical elastomer is shown in Fig. 12.5.

The temperature at which the transition from glassy to rubbery behaviour occurs is called the glass transition temperature, T_g. I should point out that the glass transition temperature for a polymer means something very different to the glass transition

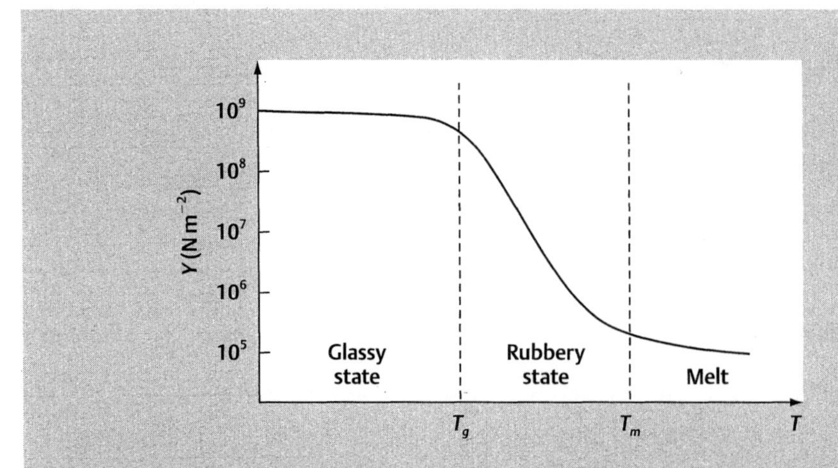

Figure 12.5 The variation of Young's modulus, Y, with temperature, T, for a typical elastomer. T_g denotes the glass transition temperature and T_m is the melting point.

temperature in an amorphous solid that we discussed in the previous chapter. A polymer is most definitely in a solid state at a temperature above T_g. The melting point, at which a polymer becomes a very viscous liquid, typically occurs at a temperature of about 100 K above T_g. Values of T_g for some polymers are given in Table 12.1.

The temperature is not the only factor which determines whether a polymer is in the glassy or rubbery state. The way in which a polymer deforms also depends on the rate at which the stress is applied. A familiar example is the silicon-based polymer dimethyl silicone, commonly known as silicone putty. The material behaves as a very viscous liquid at room temperature under low rates of shear, but at high rates of shear it behaves like a solid polymer in the rubbery state. Consequently, if thrown hard at a solid floor, it bounces back. At still higher rates of shear, it breaks in a brittle manner.

Table 12.1 Values of the glass transition temperature, T_g, for a selection of polymers.

	T_g (K)
Polyethylene	188
Rubber	193
Polypropylene	278
Nylon	320
Polystyrene	380
Perspex	410

12.4 Amorphous and crystalline polymers

The fact that many polymers are amorphous in the glassy state is not surprising. Since we are dealing with large molecules, which are often of quite different lengths even in the same sample, it seems highly unlikely that a crystal structure would be able to form. Nevertheless, under appropriate circumstances some polymers can become crystalline in nature. However, unlike other crystalline solids, a crystalline polymer consists of a mixture of amorphous and crystalline regions. The index of crystallinity of a sample provides a measure of the proportion of a sample that is in the crystalline state.

The structure of the crystalline regions is very different to that of the crystals described in Chapter 2. The regions consist of thin lamellae, typically about 10 nm thick. This length scale at first seems rather surprising since it is much longer than the length of a monomer, but far shorter than the total length of a polymer chain. So how do the polymer chains fit into the lamellae? The answer is that the polymers are folded back on themselves, in a concertina-like fashion, as shown in Fig. 12.6. It is interesting to note that a polymer chain may pass through both crystalline and amorphous regions, so a single polymer chain may be partially crystalline and partially amorphous.

At temperatures above the glass transition temperature the amorphous regions of a crystalline polymer behave like a polymer in the rubbery state, as described in the previous section, but the crystalline regions remain rigid. Therefore the flexibility of the sample in this state depends on the index of crystallinity. For a material with a low index of crystallinity, Young's modulus of the sample drops sharply above T_g, although it does not become as low at that of an elastomer. In contrast, materials which have a high index of crystallinity exhibit virtually no change in the elastic modulus above T_g.

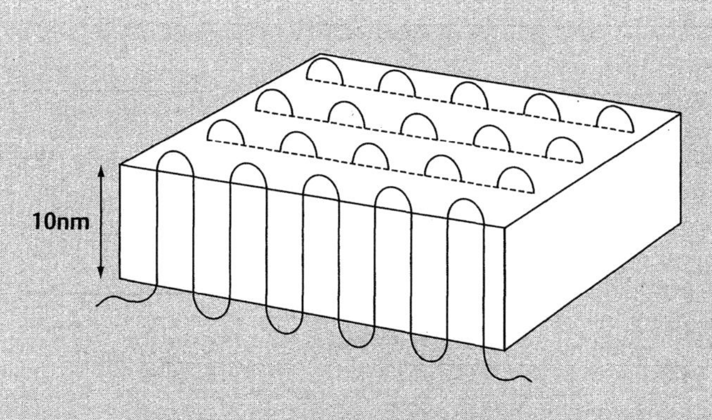

Figure 12.6 The crystalline regions of a polymer consist of lamellae in which the polymer chains are folded. Individual polymer chains may extend through both amorphous and crystalline regions.

The optical properties of polymers are also strongly affected by the amorphous or crystalline nature of the material. As we have seen in the previous chapter, amorphous materials tend to be transparent, whereas polycrystalline materials are opaque. Similarly, it is found that amorphous polymers are transparent so long as the material does not absorb light in the visible region, whereas the presence of the crystalline lamellae causes scattering of the light and so crystalline polymers are opaque (see Section 11.5).

12.5 Oriented crystalline polymers

A special class of crystalline polymers are those in which a significant proportion of the polymer chains are aligned in a common direction. Some natural polymers, such as cotton and silk, fall into this category, but many others are artificial.

The interest in these materials is due to the high tensile strength along the axis of the fibre, coupled with the low density of these materials. For example, the breaking strength of these fibres can be comparable with that of steel cables (see Table 12.2), whilst the density is approximately six times smaller than that of steel. The high strength of these fibres is due to a number of factors which we can demonstrate by considering the properties of a few different oriented crystalline polymers.

First of all let us take another look at polyethylene, which we introduced in Section 12.1. The polymer chains in polyethylene can be aligned by drawing a fine fibre from the molten material. This process causes the polymer chains to stretch, so that when the material solidifies the chains are already in the elongated shape shown in Fig. 12.2(d). Consequently, when a stress is subsequently applied along the axis of the fibre, the only way in which the material can deform is by stretching of the strong C—C bonds. As a result, the value of Young's modulus of oriented polyethylene is comparable with that of a metal—over 30 times larger than that of amorphous polyethylene. However, the breaking strength of oriented polyethylene is still relatively low. This is because the individual chains are only held together by very weak

Table 12.2 The maximum breaking stress and density of some materials. (Note that the breaking stress depends critically on the quality of the material. These values represent typical maximum values for macroscopic samples.)

	Breaking stress (10^6 N m^{-2})	Density$_t$ (10^3 kg m^{-3})
Steel wire	2000	7.8
Nylon	1000	1.15
Kevlar	3000	1.3

van der Waals forces. Therefore, when a large tensile stress is applied, the polyethylene chains simply glide past one another.

Another very familiar oriented polymer is nylon. As we can see from Fig. 12.7, the polymer chains in nylon are very similar to those of polyethylene except for the periodic appearance of the CO—NH structure, which is known as an amide group. The importance of these amides is that they provide a relatively strong link between adjacent polymer chains, as shown in Fig. 12.8. The link is a hydrogen bond which forms between the oxygen and hydrogen, as discussed in Section 1.9. As a result, the tensile breaking strength of nylon along the axis of the fibre can be as much as half that of a steel cable.

Finally, we should mention Kevlar, which has quite remarkable properties. Its high breaking point and low density have been exploited in many applications, including bullet-proof vests which weigh no more than conventional clothing. From Fig. 12.9 we can see that, like nylon, Kevlar also contains amide groups which bond the individual polymer chains together. The other significant feature of the Kevlar chain is the appearance of rings of carbon atoms, called aromatics. The presence of these aromatic groups increases the index of crystallinity of the material, which in turn further increases the tensile strength of the fibres.

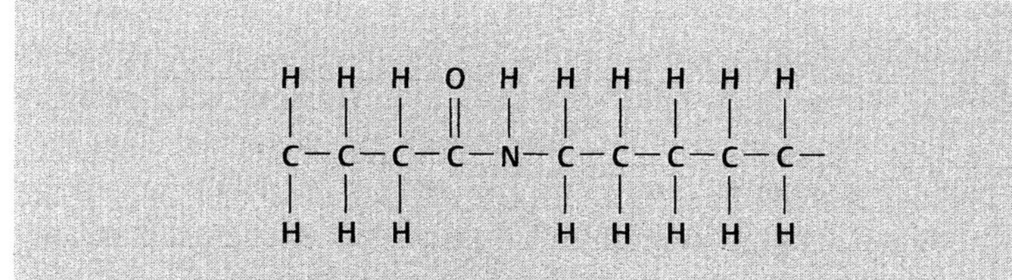

Figure 12.7 A short section of a nylon polymer.

```
 H   H           H   H
 |   |           |   |
—C — C — N — C — C — C—
 |   |   |   ||  |   |
 H   H   H   O   H   H
         :   :
 H   H   O   H   H   H
 |   |   ||  |   |   |
—C — C — C — N — C — C—
 |   |           |   |
 H   H           H   H
```

Figure 12.8 The formation of hydrogen bonds (indicated by the dashed lines) between adjacent nylon chains.

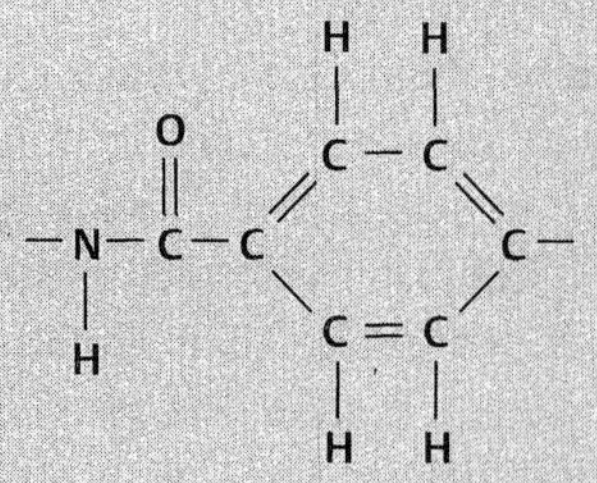

Figure 12.9 A monomer of Kevlar.

12.6 Conducting polymers

Many plastics and other polymers are electrical insulators, a fact which is not surprising given that all of the valence electrons are involved in covalent bonds. However, some polymers have a band gap of only 1 or 2 eV, and so behave like intrinsic semiconductors. As in conventional semiconductors, the conductivity of these materials can be altered by introducing impurities. For example, a reducing agent acts like a donor impurity—the impurity becomes ionized and produces an extra electron on a nearby polymer chain. When a voltage is applied to the polymer, the extra electron travels along the polymer chain towards the positive terminal. However, since an individual polymer chain does not reach from one side of the sample to the other, it is necessary for the electron to jump from one chain to another if it is going to reach the positive terminal. In an amorphous material the polymer chains are only in close proximity with one another over very short distances (see Fig. 12.10); therefore the probability of an electron jumping from one chain to another is small, and so the conductivity is poor. However, in an oriented crystalline polymer, the chains run parallel over relatively large distances, giving a much greater chance for the electrons to move from one chain to another, and therefore producing a much higher conductivity.

The above describes the behaviour of an n-type polymer, but we can also create p-type polymers by introducing oxidizing agents which remove electrons from the chemical bonds and so give rise to holes which move along the polymer chains. With these two types of material we can produce polymer versions of most of the electronic and optoelectronic devices described in Chapter 6. However, as existing semiconductors are already well established in this field, polymer devices must present substantial benefits if they are to be accepted. So although it is possible to fabricate a polymer-based transistor, they are unlikely to become commonplace in integrated circuits because they are slower and more expensive to manufacture than silicon devices. One of the most promising applications appears to be large-area LED displays. To produce such a structure with conventional semiconductor devices requires many small semiconductor crystals, which means a high overall cost, whereas polymers can be easily and cheaply produced in large sheets.

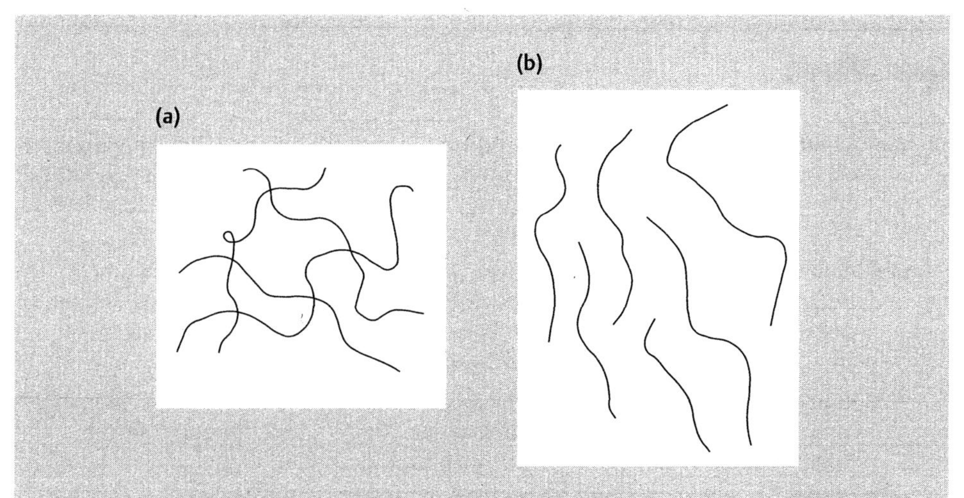

Figure 12.10 (a) The structure of an amorphous polymer. (b) The structure of an oriented polymer. In the latter case adjacent polymer chains are aligned parallel with one another, making it easy for an electron to jump from one chain to another.

In some instances, the conductivity of a doped polymer can be comparable with that of a metal, and so another possibility is that conducting polymers could be used to replace metallic conductors in situations in which the low density or flexibility of the polymers is important.

12.7 Summary

In this short chapter we have only had chance to consider just a small number of the huge range of polymers that are currently available. We have also seen that the physical properties of these materials owes as much to the way in which these polymers are arranged in the solid (i.e. whether it is amorphous, partly crystalline, or with the polymers oriented in a particular direction) as it does to the chemical structure of the polymer. These properties can also be altered dramatically by changing the temperature, the concentration of impurities, and even the rate at which a stress is applied to the material.

Although it is estimated that about 30 000 different polymers have been created and tested in research laboratories around the world, this represents only a tiny proportion of the range of polymers that could feasibly exist. In the ocean of polymers, we have only just begun to explore the coastal waters. Far out in the depths of this ocean there are likely to be many polymers with properties far superior to those of any currently known material.

Questions

12.1 General. Calculate the total length and mass of a polyethylene molecule which consists of 20 000 carbon atoms. (The length of a C—C bond is 0.154 nm, and the carbon and hydrogen atoms have masses of 12 u and 1 u, respectively.) For comparison, estimate the mass of a soot particle (which consists of carbon atoms) of diameter 2 μm.

12.2 Rubber. Calculate Young's modulus of a rubber strip of cross-sectional area 0.1 cm^2 given that a force of 3 N causes the strip to double in length. Determine the strain that would be produced by this force in an aluminium strip with the same cross-sectional area. (Young's modulus of aluminium is 7.0×10^{10} N m^{-2}.)

12.3 Rubber. Determine the maximum distance between the ends of a polymer chain, d_{max}, as shown in Fig. 12.3. Assume that the chain contains N carbon atoms and that the length of a C—C bond is a.

12.4 Rubber. Calculate the maximum elastic strain for rubber assuming that on average there are about 250 monomers between each crosslink.

12.5 Rubbery and glassy states. For each of the materials listed in Table 12.1, state whether you would expect it to be in the rubbery or glassy state at room temperature (i.e. 20°C).

12.6 Crystalline polymers. If a polyethylene chain consists of 25 000 carbon atoms, estimate the number of times that the chain is folded in order to fit into a lamella of thickness 10 nm. Assume that the entire polymer chain is contained within a single lamella. (The length of the C—C bond is 0.154 nm.)

12.7 Oriented crystalline polymers. Using the data in Table 12.2, calculate the maximum length of a nylon cable that can be suspended vertically without breaking under its own weight. Repeat the calculation for a steel cable.

12.8 Oriented crystalline polymers. Using the data in Table 12.2, calculate the minimum cross-sectional area of a Kevlar cable required to suspend a car of mass 800 kg. (You may neglect the mass of the cable.)

Further reading

I hope that after reading all or part of this book you will have the urge to pursue this subject further. In this section I have compiled a list of books and articles which you may find helpful in this task.

Introductory general physics texts

It is assumed that you have a background in physics roughly comparable with that of the college physics or university physics texts. Suitable books in this category include

- Alonso, M. and Finn, E. J., *Physics*, Addison-Wesley, Reading, MA (1992)
- Serway, R. A., *Physics for Scientists and Engineers* (4th edition), Saunders, Philadelphia (1996)

I should also mention my own book, *The Quantum Dot: A Journey into the Future of Microelectronics*, W. H. Freeman/Spektrum, Oxford, England (1995), which deals with many of the topics covered by this book (particularly the content of Chapters 1, 4, 5, 6 and 9) in a non-mathematical manner.

More advanced solid state texts

There is such a vast range of solid state textbooks available that it can be difficult to know where to begin, especially when many of the books aimed at a far higher level than this one describe themselves as 'elementary' texts. I have therefore selected a small number of texts which deal with largely the same subject matter as this book, but at a slightly higher level. Even so, there are many books which fall into this category. The following is just a list of my personal favourites.

- Rosenberg, H. M., *The Solid State* (3rd edition), Oxford University Press, Oxford, England (1988)
  A well-written and compact book (by solid state textbook standards). Most of it should be within the range of a second-year undergraduate.
- Solymar, L. and Walsh, D., *Lectures on the Electrical Properties of Materials* (6th edition), Oxford University Press, Oxford, England (1998)
  A very readable and entertaining book, one of my all-time favourite textbooks. Includes many applications of solid state physics. A familiarity with Schrödinger's equation is assumed in a few sections.
- Kittel, C., *Introduction to Solid State Physics* (7th edition), Wiley, New York (1996)
  Although first published in 1953, this standard text does not show its age. Although at a considerably higher level than this book, I have included it as a useful reference work.

Chapter 1: Bonds between atoms

- Beiser, A., *Concepts of Modern Physics* (5th edition), McGraw-Hill, New York (1973)
  Has a more detailed treatment of the electronic energy levels in an atom.

Chapter 2: Crystals and crystalline solids

- Young, H. D., *University Physics*, Addison-Wesley, Reading MA (1992)
  Gives a good description of the operation of an electron microscope.
- Binnig, G. and Rohrer, H., The scanning tunelling microscope, *Scientific American*, **253**, p. 40 (Aug. 1985)
  The inventors of the scanning tunnelling microscope describe how it works and some applications.
- Wickramasinghe, H. K., Scanned-probe microscopes, *Scientific American*, **261**, p. 74 (Oct. 1989)
  Describes several variations on the scanning tunnelling microscope including the atomic force microscope.

Chapter 3: Mechanical properties of solids

- Guinier, A. and Jullien, R., *The Solid State: From Superconductors to Superalloys*, Oxford University Press, Oxford, England (1989)
  Contains a lot of information about plastic deformation, fracture mechanisms and creep, with a mainly non-mathematical approach.
- Tabor, D., *Gases, Liquids and Solids and other states of matter* (3rd edition), Cambridge University Press, Cambridge, England (1991)
  Contains a fairly simple mathematical treatment of elastic deformation, yield stress and brittle fracture.
- Gordon, J. E., *The New Science of Strong Materials* (2nd edition), Penguin, Harmondsworth (1968)
  A very entertaining look at the mechanical properties of solids including a great deal of historical detail and anecdotes.
- Bradshaw, J., Windscreens: never seen until damaged, *Physics World*, **10**, p. 39 (Aug. 1997)
  An interesting insight into the properties of various types of glass.

Chapter 4: Electrical properties of solids

- Beiser, A., *Concepts of Modern Physics* (5th edition), McGraw-Hill, New York (1995)
  Includes a derivation of the Fermi-Dirac distribution.

Chapter 5: Semiconductors

The quantum Hall effect is discussed in more detail in the following articles:

- Halperin, B. I., The quantized Hall effect, *Scientific American*, p. 40 (Apr. 1986)
- Störmer, H. L. and Tsui, D. C., The quantized Hall effect, *Science*, **220**, p. 1241 (June 1983)

See also Turton *The Quantum Dot* and Kittel *Introduction to Solid State Physics* (see above).

Chapter 6: Semiconductor devices

Poisson's equation is discussed in I. S. Grant and W. R. Phillips, *Electromagnetism* (2nd edition), Wiley, Chichester, England (1990)

For a more detailed, but non-mathematical, treatment of the uses of heterojunctions in electronic and optoelectronic devices, I suggest my own book, *The Quantum Dot* (see above). There are also many shorter articles at a similar level, such as

- Drummond, T. J., Gourley, P. L. and Zipperian, T. E., Quantum-tailored solid-state devices, *IEEE Spectrum*, p. 33 (June 1988)
- Bate, R. T., The quantum-effect device: tomorrow's transistor, *Scientific American*, p. 78 (Mar. 1988)
- Kouwenhoven, L. and Marcus, C., Quantum dots, *Physics World*, **11**, p. 35 (June 1998)

Chapter 7: Thermal properties

A derivation of the coefficient of linear expansion from the form of the potential curve is given in *Properties of Matter* by B. H. Flowers and E. Mendoza, pp. 239–243, Wiley, Chichester, England (1970).

A derivation of eqn (7.23) for the thermal conductivity is given in the same text, pp. 162–163.

Chapter 8: Magnetic properties

A proof that the Landé splitting factor is given by eqn (8.8) is given by Enge, H. A., Wehr, M. R. and Richards, J. A., *Introduction to Atomic Physics* p.266, Addison-Wesley, Reading MA (1972).

New approaches to the design of better hard magnets is described by Hywel Davies in Ultrafine alloys make their mark, *Physics World*, **7**, p.40 (Nov. 1994).

Articles on giant magnetoresistance (GMR) and other properties and applications of novel magnetic materials can be found in

- Smits, J. W., Magnetic multilayers, *Physics World*, **5**, p.48 (Nov. 1992)
- Rodgers, P. Giants in their field, *New Scientist*, p.34 (10 Feb. 1996)
- *Physics Today*, Special edition on magnetoelectronics (Apr. 1995)
- Barthélémy, A., Fert, A., Morel, R. and Steren, L. Giant steps with tiny magnets, *Physics World*, **7**, p.34 (Nov. 1994)

Chapter 9: Superconductivity

There are numerous popular articles dealing with various aspects of superconductivity. The following list is by no means exhaustive.

- Simon, R. and Smith, A., *Superconductors: conquering technology's new frontier*, Plenum Press, New York (1988)
  A non-mathematical book which makes a pleasant read. Includes the discovery of high-temperature superconductors and has detailed descriptions of many potential applications.
- Little, W. A., Superconductivity at room temperature, *Scientific American*, **212**, p.21 (Feb. 1965)

It is doubtful if the materials described here would actually superconduct at room temperature, but the article is well worth reading for the beautiful analogies used to explain BCS theory.

- Essman, U. and Träuble, H., The magnetic structure of superconductors, *Scientific American*, **224**, p.74 (Mar. 1971)
  Description of properties of type II superconductors.
- Matisso, J., The superconducting supercomputer, *Scientific American*, **242**, p.38 (May 1980)
  Detailed description of Josephson junctions and how they can be used to perform transistor-like functions.
- Fitzgerald, K., Superconductivity: fact vs. fancy, *IEEE Spectrum*, **258**, p.30 (May 1988)
  Concentrates on the properties of high-temperature superconductors.
- Wolsky, A. M., Giese, R. F. and Daniels, E. J., The new superconductors: prospects for applications, *Scientific American*, **260**, p.45 (Feb. 1989)
  The title says it all—deals with potential applications of high-temperature superconductors.
- Cava, R. J., Superconductors beyond 1-2-3, *Scientific American*, **263**, p.42 (Aug. 1990)
  A detailed discussion of the crystal structure of high-temperature superconductors.

For a more mathematical treatment consult any of the books listed under **More Advanced Solid State Texts**. I also recommend an article by M. Tinkham and C. J. Lobb, p.91 in volume 42 of the series *Solid State Physics* (Academic Press, New York, eds H. Ehrenreich and D. Turnbull, 1993) which deals with high-temperature superconductors.

If you are unfamiliar with the electromagnetism used in this chapter (e.g. electromagnetic induction, the magnetic field due to a current in a wire, etc.), this is covered by most of the college physics texts, e.g. H. D. Young, *University Physics*, Addison-Wesley, (1992), or standard undergraduate electromagnetism texts, e.g. I. S. Grant and W. R. Phillips, *Electromagnetism*, Wiley, (1990).

Chapter 10: Dielectrics

For a discussion on the macroscopic properties of dielectrics, see R. P. Feynman, R. B. Leighton and M. Sands, *The Feynman Lectures on Physics, Volume II* Addison-Wesley, Redwood City, CA (1989).

A comprehensive treatment of forced damped harmonic oscillators (used in Section 10.5) is given by French, *Vibrations and waves* Norton, New York (1971). A brief description is given in Alonso and Finn (see above).

Other interesting articles dealing with topics discussed in Chapter 10 include

- Geis, M. W. and Angus, J. C., Diamond film semiconductors, *Scientific American*, p.64 (Oct. 92)
  Discusses the possibility and advantages of doping diamond films to produce transistor-like devices. (The article also discusses methods of artificially fabricating diamonds.)

- Scott, J.F., Ferroelectric memories, *Physics World*, **8**, p.46 (Feb. 95)
  Examines the advantages and disadvantages of using ferroelectric materials in computer memories and other related applications.

Chapter 11: Crystallization and amorphous solids

A far more detailed description description of amorphous solids is given by S. R. Elliott in *Physics of Amorphous Solids* (2nd edition), Longman, Harlow (1990). There are some very technical sections, but most of the book is accessible to an undergraduate reader.

A somewhat older, but more readable, account of the applications of amorphous solids is given by David Adler in Amorphous semiconductor devices, *Scientific American*, p.36 (May 1977).

Chapter 12: Polymers

A detailed discussion of polymers is given by Ulf Gedde in *Polymer Physics* (Chapman and Hall, London, 1995). Some of the sections are rather technical.

Articles dealing with the applications of conducting polymers include

- Yam, P., Plastics get wired, *Scientific American*, p.74 (July 1995).
- Kaner, R. B. and McDiarmid, A. G., Plastics that conduct electricity, *Scientific American*, p. 60 (Feb. 1988).

Appendix A: Introduction to quantum concepts

One of the key concepts in quantum theory is that of wave–particle duality. In this appendix we will explain briefly what this means by considering the behaviour of light and electrons.

Let us first consider light. Does light behave as a wave or as a stream of particles? Historically, both views have had strong advocates. For example, in the seventeenth century Sir Isaac Newton argued that light must consist of small particles, or corpuscles, whereas Christian Huygens proposed that light can be best understood if treated as a wave. The debate continued for over a hundred years without either side being able to offer any conclusive evidence. The argument finally appeared to be settled in 1801 when Thomas Young showed that interference fringes are produced when light is passed through two narrow slits. This result could only be explained by treating light as a wave, and so it seemed that the problem had been solved.

However, a hundred years later, Max Planck and Albert Einstein showed that in some cases light most definitely behaves as a particle, or rather as a small packet of energy, which they called a photon. So it appeared that the wave–particle argument had reared its head again.

In fact, we now know that light is not really a wave or a particle, it is just that we can understand some phenomena by using a wavelike description, whereas in other cases we need to consider the particle-like aspects of light. So the two descriptions complement one another. This is demonstrated by the fact that the energy, E, of a photon is defined in terms of the frequency, f, of the corresponding wave, i.e

$$E = hf \tag{A.1}$$

where h is called Planck's constant. Since we know that the frequency f, wavelength λ, and speed of a wave c are related by the equation

$$c = f\lambda \tag{A.2}$$

we can also write

$$E = \frac{hc}{\lambda} \tag{A.3}$$

In 1923, Louis de Broglie postulated that matter also has corresponding wavelike properties and suggested that the wavelength of an object is related to the momentum of the object, p, by the relationship

$$\lambda = \frac{h}{p} \tag{A.4}$$

In many cases it is more useful to relate the wavelength to the energy of the particle. We can do this by noting that the momentum of a particle is mv where m is the mass and v is the velocity of the particle. If we express the kinetic energy by $E = \frac{1}{2}mv^2$, this can be rearranged to give

$$v = \sqrt{\frac{2E}{m}}$$

and therefore

$$p = mv = \sqrt{2mE}$$

Substituting into eqn (A.4) gives

$$\lambda = \frac{h}{\sqrt{2mE}} \tag{A.5}$$

For a macroscopic object the de Broglie wavelength is far too small to be observed – for example, a golf ball travelling at 5 m s^{-1} has a de Broglie wavelength of about 10^{-33}, i.e. approximately 10^{18} times smaller than the diameter of an atomic nucleus – but for a subatomic particle, such as an electron, the de Broglie wavelength is of the order of atomic dimensions. The existence of matter waves was first demonstrated by Clinton Davisson and Lester Germer in 1927 by scattering low-energy electrons off a crystal surface. The results were similar to those observed in Bragg reflection of X-rays from a crystal (see Chapter 2), and so showed that electrons can produce interference effects which we associate with wavelike behaviour.

Appendix B: Relationship between interatomic force and potential energy

There are many physical systems in which we are interested in the relationship between force and potential energy. For instance, common examples include a particle in a gravitational or an electrostatic field, or a particle executing simple harmonic motion. These examples are often treated in introductory physics texts (see Further reading). In Chapter 1 of this book we are particularly concerned with the forces between atoms and the total potential energy of the system. In this case the relationship is exactly the same as in the other instances: we obtain the result that the potential energy E is related to the force F by the equation

$$E(r) = -\int F(r)\,dr \tag{B.1}$$

The purpose of this appendix is to provide a brief proof of this result.

Let us consider a force F acting on a particle. If the work done by the force on the particle is dependent only on the difference between the initial and final positions of the particle then the force is a conservative force and the work done on the particle is given by

$$W = \int_{r_i}^{r_f} F(r)\,dr = -E_f + E_i \tag{B.2}$$

where r_i and r_f are the initial and final positions of the particle and E_i and E_f are the initial and final energies of the system. Since the change in energy of the system is $\Delta E = E_f - E_i$, we can write

$$\Delta E = -W = -\int_{r_i}^{r_f} F(r)\,dr \tag{B.3}$$

It is usual to define the potential energy of the system to be zero at $r = \infty$, i.e. for a system of two particles the potential energy is zero when the separation of the particles is infinite. So if we move the particles from an infinite distance apart to a separation r, then the difference in energy ΔE is simply equal to the potential energy of the system, E, i.e.

$$E(r) = -\int_{\infty}^{r} F(r)\,dr \tag{B.4}$$

By rearranging the above equation we can also obtain an expression for the force in terms of the potential energy, i.e.

$$F(r) = -\frac{dE(r)}{dr} \tag{B.5}$$

Solutions to questions

Chapter 1

1.1 The radii and energy levels of the hydrogen atom are given by eqns (1.1) and (1.2), respectively.

For $n = 2$

$$r_2 = \frac{n^2 h^2 \varepsilon_0}{\pi m_e e^2} = \frac{(2)^2(6.63\times 10^{-34}\ \text{J s})^2(8.85\times 10^{-12}\ \text{F m}^{-1})}{(3.14)(9.11\times 10^{-31}\ \text{kg})(1.60\times 10^{-19}\ \text{C})}$$

$$= 2.12\times 10^{-10}\ \text{m}$$

$$E_2 = -\frac{m_e e^4}{8\varepsilon_0^2 h^2 n^2} = -\frac{(9.11\times 10^{-31}\ \text{kg})(1.60\times 10^{-19}\ \text{C})^4}{(8)(8.85\times 10^{-12}\ \text{F m}^{-1}(6.63\times 10^{-34}\ \text{J s})^2(2)^2}$$

$$= -5.42\times 10^{19}\ \text{J or } -3.39\ \text{eV}$$

Similarly, for $n = 3$

$$r_3 = 4.76\times 10^{-10}\ \text{m}$$
$$E_3 = -2.41\times 10^{-19}\ \text{J or } -1.51\ \text{eV}$$

Therefore higher values of n correspond to larger orbits and smaller binding energies.

1.2 The binding energy is given by eqn (1.3).

For helium with n=1 and Z=2,

$$E_1 = -\frac{Z^2}{n^2} 13.6\ \text{eV} = -54.4\ \text{eV}$$

Note that the binding energy is a positive quantity equal to $-E_1$, i.e. the binding energy of these electrons is +54.5 eV.

For lithium with n=1 and Z=3.

$$-E_1 = 122.4\ \text{eV}$$

Therefore, if all of the electrons are in the n=1 state, this equation predicts that the ionization energy of Li is greater than that of He.

Why is this result incorrect? The exclusion principle tells us that only two electrons in Li can occupy the n=1 state; therefore one of the electrons must be in the n=2 state.

Also the electron in the n=2 state experiences a reduced nuclear charge due to the screening of the n=1 electrons. Since the net charge of the nucleus plus the n=1 electrons is $+1e$, we can estimate the binding energy of the n=2 electron by replacing the nuclear charge Ze in eqn (1.3) by a screened nuclear charge $Z'e$ where Z' is equal to +1.

The ionization energy of Li (i.e. the binding energy of the n=2 electron) is therefore

$$-E_2 = \frac{(Z')^2}{n^2} 13.6 \text{ eV} = 3.40 \text{ eV}$$

(For comparison, the experimental ionization energy is 5.39 eV. This suggests that the screened nuclear charge $Z'e$ is actually about $+1.3e$.)

1.3 The atomic energy levels up to and including the 4p level are as follows:

1s
2s, 2p
3s, 3p, 3d
4s, 4p

Note that the 4s state is usually at a lower energy than the 3d state. However, we are told in the question that the 4p state is at a higher energy than the 3d state and at a lower energy than the 5s state. So when there are just sufficient electrons to fill the 4p state, this means that all of the above listed states should be filled and all of the others should be empty.
Applying the exclusion principle we find that the maximum occupancy of these states is as follows:

1s (2), 2s (2), 2p (6), 3s (2), 3p (6), 3d (10), 4s (2), 4p (6)

giving a total of 36 electrons. This corresponds to the electronic configuration of krypton, which is an inert element.

1.4 For an explanation of why the electrons in the filled electron shells do not usually take part in bonding refer to Section 1.4.

1.5 From Appendix B, eqn (B.4), the potential energy $E(r)$ is related to the force $F(r)$ by the equation

$$E(r) = -\int_{\infty}^{r} F(r)\, dr$$

Using the form of $F(r)$ given in eqn (1.4) the potential energy is therefore

$$E(r) = -\int_{\infty}^{r} -\frac{e^2}{4\pi\varepsilon_0 r^2} dr = -\left[\frac{e^2}{4\pi\varepsilon_0 r}\right]_{\infty}^{r}$$

$$= -\frac{e^2}{4\pi\varepsilon_0 r} + 0 = -\frac{e^2}{4\pi\varepsilon_0 r}$$

Q.E.D.

1.6 The potential energy of an ion in a sodium chloride crystal is given by eqn (1.9) with $a_0 = 0.281\times10^{-9}$ m, i.e.

$$E_{ion} = -1.748\frac{e^2}{4\pi\varepsilon_0 a_0}$$

$$= -1.748\frac{(1.60\times 10^{-19}\text{C})^2}{(4)(3.14)(8.85\times 10^{-12}\text{ F m}^{-1})(0.281\times 10^{-9}\text{ m})}$$

$$= 1.43\times 10^{-18}\text{ J} = 8.95 \text{ eV}$$

To find the lattice energy of 1 mole of sodium chloride we need to multiply this value by the number of ions in 1 mole and divide by 2 to avoid using each ion twice (see Section 1.5).
Since 1 mole of sodium chloride contains N_A sodium ions and N_A chlorine ions, the total number of ions is $2N_A$. The total lattice energy is therefore

$$E_{lattice} = 2N_A.\frac{1}{2}.E_{ion} = (2)(6.02\times10^{23}\ \text{mol}^{-1}).\frac{1}{2}.(1.43\times10^{-18}\ \text{J})$$

$$= 8.61\times10^{5}\ \text{J mol}^{-1}$$

1.7 From eqn (1.10) the repulsive force is given by

$$F_{repel} = \frac{Ba_0^{10}}{r^{10}}$$

where B is a constant.

At r = a_0, $F_{repel} = B$.

At $r = 2a_0$, $F_{repel} = \frac{B}{2^{10}} = \frac{B}{1024}$.

Therefore the repulsive force decreases in magnitude by a factor of about 1000 when the separation of the ions doubles.
The Coulomb force is given by

$$F_{Coulomb} = -\frac{Aa_0^2}{r^2}$$

At $r = 2a_0$, $F_{Coulomb} = -A$.

At $r = 2a_0$, $F_{Coulomb} = \frac{A}{a^2} = -\frac{A}{4}$.

The Coulomb force decreases only by a factor of 4 for the same increase in separation.
The repulsive force is therefore a short-range force because a small increase in separation leads to a dramatic decrease in the magnitude of this force, whereas the Coulomb force is a long-range force because of the comparatively small change in the force as the separation increases.

1.8 (i) From Example 1.4 we know that the constants A and B in eqn (1.10) are equal to one another and that the potential energy in this case is therefore

$$E(r) = -\frac{Aa_0^2}{r} + \frac{Aa_0^{10}}{9r^9}$$

The minimum potential energy occurs when $dE/dr = 0$, i.e.

$$\frac{Aa_0^2}{r^2} - \frac{Aa_0^{10}}{r^{10}} = 0 \Rightarrow r^8 = a_0^8 \Rightarrow r = a_0$$

Q.E.D

(ii) At the equilibrium separation $r = a_0$ the potential energy is therefore

$$E(a_0) = -\frac{Aa_0^2}{a_0} + \frac{Aa_0^{10}}{9a_0^9} = -Aa_0 + \frac{Aa_0}{9}$$

The first term corresponds to the attractive Coulomb potential and the second term is due to the repulsive force. We can see that the magnitude of the potential energy due to the repulsive term is one-ninth, i.e. about 11%, of that due to the attractive term; therefore the repulsive term decreases the lattice energy by 11%.

1.9 To find the value of the constant A:
The binding energy E_{be} is equal to the negative of the potential energy at $r = a_0$, i.e. $E_{be} = -E(a_0)$ (see Fig. 1.9b). So using the expression for $E(r)$ from Example 1.4 we have

$$E_{be} = -E(a_0) = +\frac{Aa_0^2}{a_0} - \frac{Aa_0^{10}}{9a_0^9} = +Aa_0 - \frac{Aa_0}{9} = \frac{8Aa_0}{9}$$

Rearranging, we have

$$A = \frac{9E_{be}}{8a_0} = \frac{(9)(1.29\times10^{-18}\ \text{J})}{(8)(0.281\times10^{-9}\ \text{m})} = 5.16\times10^{-9}\ \text{N}$$

1.10 In sodium chloride, $r_{Na} + r_{Cl} = 0.281$ nm. Since $r_{Cl} = 0.181$ nm, $r_{Na} = 0.100$ nm. In metallic sodium, $2r_{Na} = 0.372$ nm, so $r_{Na} = 0.186$ nm.

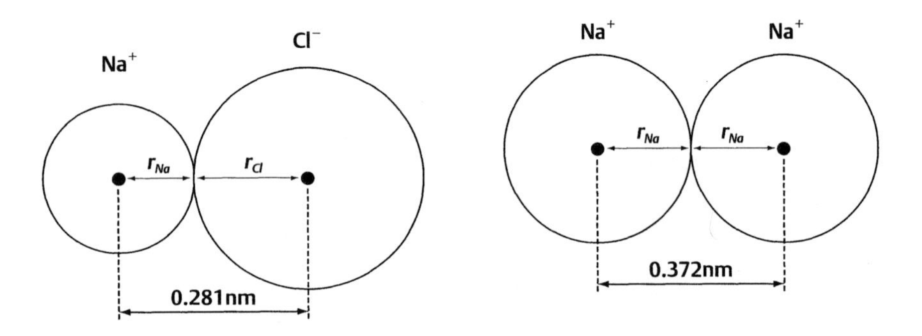

The sodium ions in metallic sodium have a larger radius than in sodium chloride because the metallic bond is much weaker than the ionic bond.

1.11 The length of a polymer chain is determined by the length of the C—C bond between two monomers.
The length of a polyethylene chain containing 25 000 CH_2 monomers is therefore $(25\ 000)(0.154\times10^{-9}\ \text{m}) = 3.85\times10^{-6}$ m.

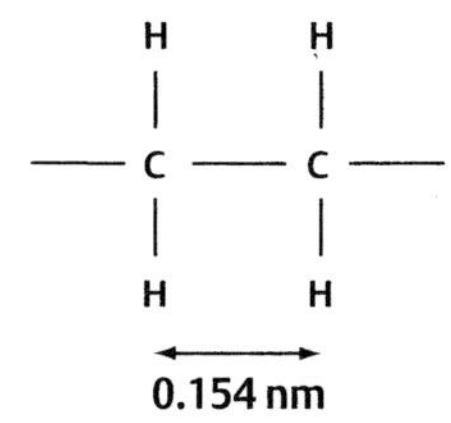

1.12 The number of C atoms in 12.0 g = Avogadro's number, N_A.
Therefore, the number of C atoms in 0.02 g is

$$\frac{0.02\ \text{g}}{12.0\ \text{g mol}^{-1}}.N_A = (1.67\times10^{-3}\ \text{mol})(6.02\times10^{23}\ \text{mol}^{-1}) = 1.01\times10^{21}$$

(Note that a mass of 0.2 g corresponds to a 1 carat diamond.)

1.13 Consider Fig. 1.18. If we denote the length of a side of the cube as a, then the distance from H1 to H2 is $a\sqrt{2}$. and the distance from each of the hydrogen atoms to the central carbon atom is equal to half the body diagonal of the cube, i.e. $a\sqrt{3}/2$.

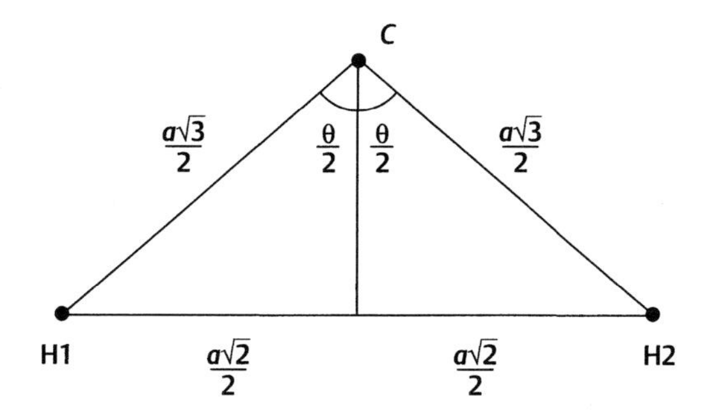

If the angle between the bonds is denoted by θ, then we can see from the diagram above that

$$\sin\frac{\theta}{2} = \frac{a\sqrt{2}/2}{a\sqrt{3}/2} = \sqrt{\frac{2}{3}}$$

Therefore $\theta/2 = 54.74°$, and so $\theta = 109.5°$.

1.14 (i) Using Appendix B, eqn (B.5), the force is given by

$$F(r) = -\frac{dE(r)}{dr} = -\frac{6Aa_0^6}{r^7} + \frac{12Ba_0^{12}}{r^{13}}$$

Since the net force is zero at $r = a_0$, we have

$$F(a_0) = 0 = -\frac{6Aa_0^6}{a_0^7} + \frac{12Ba_0^{12}}{a_0^{13}} = -\frac{6A}{a_0} + \frac{12B}{a_0}$$

Therefore, $A = 2B$.

(ii) To obtain σ in terms of a_0 we can use the fact that dE/dr is zero when $r = a_0$.

$$\frac{dE(r)}{dr} = 4\epsilon\left(\frac{12\sigma^{12}}{r^{13}} - \frac{6\sigma^6}{r^7}\right)$$

Substituting $r = a_0$ we have

$$\frac{dE(a_0)}{dr} = 0 = 4\epsilon\left(\frac{12\,\sigma^{12}}{a_0^{13}} - \frac{6\,\sigma^6}{a_0^{7}}\right)$$

$$\Rightarrow 12\,\sigma^6 = 6a_0^6$$

$$\Rightarrow \sigma = a_0\,2^{-1/6}$$

Next we can substitute this expression into the equation for $E(r)$ to find the potential energy at $r = a_0$, i.e.

$$E(a_0) = 4\epsilon\left[\left(\frac{a_0\,2^{-1/6}}{a_0}\right)^{12} - \left(\frac{a_0\,2^{-1/6}}{a_0}\right)^{6}\right] = 4\epsilon\left[\frac{1}{4} - \frac{1}{2}\right]$$

$$= -\epsilon$$

Therefore ϵ is the binding energy.

Chapter 2

2.1 The solution follows similar lines to that in Example 2.2.

First we determine the total volume of the parts of the spheres contained within one unit cell (see Fig. 2.5). The sphere at the centre of the cube lies entirely within the cell, plus there is one-eighth of each of the eight spheres at the vertices (see Example 2.2), making a total of two spheres. If the spheres are of radius r then the total volume occupied by the spheres is $V_s = 2.\frac{4}{3}\pi r^3 = \frac{8}{3}\pi r^3$.

Next we determine the relationship between r and the length of one edge of the cube, d.

Since the spheres are in contact along the body diagonal, PQ, the length of PQ is $4r$ (see the second diagram).

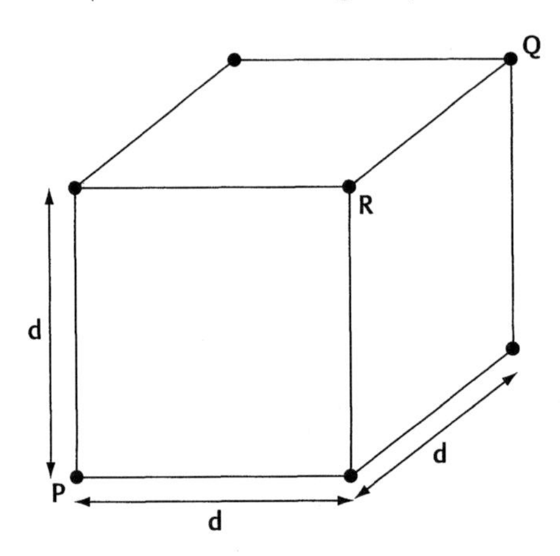

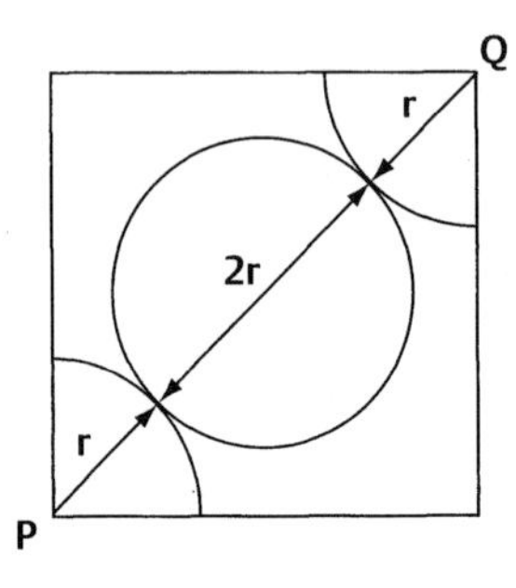

Also, from the first diagram, length of PQ = $\sqrt{3}d$.

Therefore $4r = \sqrt{3}d$, i.e. $d = \dfrac{4r}{\sqrt{3}}$

The volume of the cube is therefore $V_c = d^3 = \dfrac{64r^3}{3\sqrt{3}}$

The percentage volume occupied by the spheres is therefore

$$\frac{V_s}{V_c} \times 100\ \% = \frac{\frac{8}{3}\pi r^3}{\frac{64r^3}{3\sqrt{3}}} \times 100\ \% = 68.0\ \%$$

2.2 The solution is similar to that given for Question 2.1, with the following differences: in addition to the spheres at the vertices, there are six spheres (one at the centre of each face) which are shared between two unit cells; therefore the total number of spheres in a unit cell is

$$\left(\frac{1}{8} \times 8\right) + \left(\frac{1}{2} \times 6\right) = 4$$

(Note that there is no sphere at the centre of the cube.)
The spheres are in contact along the face diagonal, PR (see first diagram in Question 2.1), which is of length $\sqrt{2}d$; therefore $4r = \sqrt{2}d$, i.e. $d = \dfrac{4r}{\sqrt{2}}$.
Therefore the percentage volume occupied by the spheres is

$$\frac{V_s}{V_c} \times 100\ \% = \frac{4.\frac{4}{3}\pi r^3}{\left(\frac{4r}{\sqrt{2}}\right)^3} \times 100\ \% = 74.0\ \%$$

2.3 From the solution to Question 2.1 we know that one side of the unit cell is

$$d = \frac{4r}{\sqrt{3}} = 2.309r$$

From the diagram below the maximum diameter of the interstitial B is

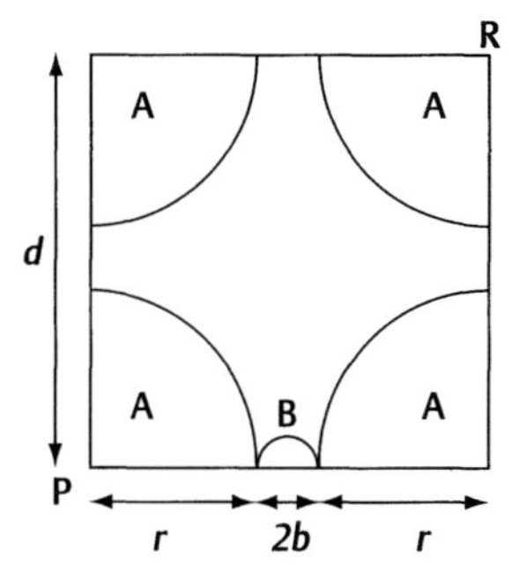

$2b = d - 2r = 0.309r$.

Therefore the maximum radius is $b = 0.155r$.

2.4 From Fig. 2.24,

$$\mathbf{a} = \frac{d}{2}(\hat{\mathbf{x}} + \hat{\mathbf{y}}) \qquad \mathbf{b} = \frac{d}{2}(\hat{\mathbf{x}} + \hat{\mathbf{z}}) \qquad \mathbf{c} = \frac{d}{2}(\hat{\mathbf{y}} + \hat{\mathbf{z}})$$

The position vector for A is $d(\hat{\mathbf{x}} + \hat{\mathbf{y}}) = 2\mathbf{a}$.

B is $\frac{d}{2}(2\hat{\mathbf{x}} + \hat{\mathbf{y}} + \hat{\mathbf{z}}) = \mathbf{a} + \mathbf{b}$.

C is $d(\hat{\mathbf{x}} + \hat{\mathbf{y}} + \hat{\mathbf{z}}) = \mathbf{a} + \mathbf{b} + \mathbf{c}$.

D is $2\hat{\mathbf{x}} = \mathbf{a} + \mathbf{b} - \mathbf{c}$.

2.5 Following Example 2.4, the intercepts are at 2, −1 and 1.
Taking reciprocals we obtain $\frac{1}{2}$, −1 and 1.
The smallest set of integers with the same common ratio is 1, −2, 2; therefore the plane is $(1\bar{2}2)$.
By definition, the direction normal to this plane is $[1\bar{2}2]$.

2.6 The solution to this problem follows the reverse procedure of Example 2.4.
The (012) plane has intercepts at ∞, 1, and $\frac{1}{2}$, which is equivalent to ∞, 2, and 1.

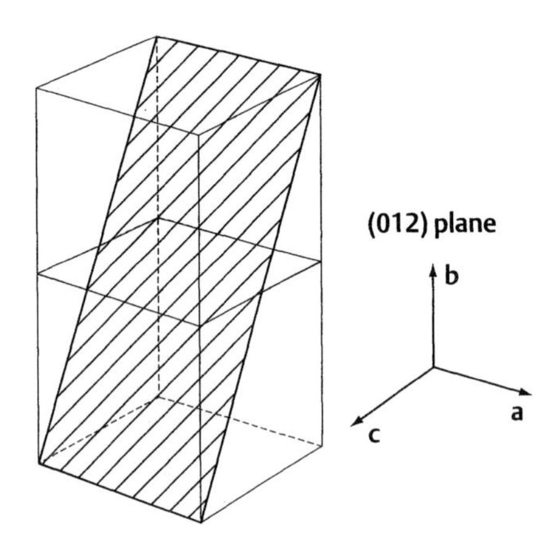

Using the axis convention from Fig. 2.14 the plane looks like this:

2.7 In Fig. 2.19 the distance XV = d and the angle VXU = θ.
Therefore UV = $d \sin \theta$.
Similarly, VW = UV.
Path difference = UV + VW = $2d \sin \theta$.

2.8 Using eqn (2.2), $n\lambda = 2d \sin \theta$, the maximum possible value of λ is when the first-order ($n = 1$) reflection is at $\theta = 90°$.
That is, $\lambda = 2d = 0.4$ nm.

2.9 From eqn (2.2),

$$\theta = \sin^{-1}\left(\frac{n\lambda}{2d}\right)$$

Therefore in each pair of peaks, the longer wavelength radiation (K_α) corresponds to the larger angle, θ.
From Fig 2.20, the $n = 1$ K_α peak is at $\theta \approx 15.8°$; therefore $d = 0.2827$ nm.
(Note: because the NaCl lattice is face-centred cubic, the separation between the lattice planes in the (001) direction is equal to half the length of the unit cell, i.e. the value of d corresponds to the distance between the centres of

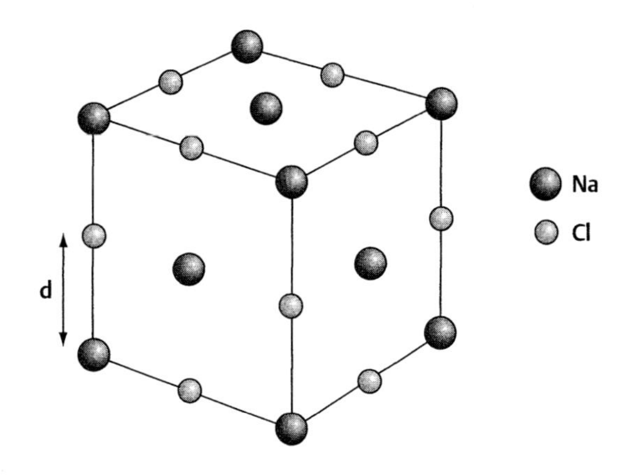

neighbouring Na and Cl ions—see the diagram.)
An uncertainty in θ of $\pm 0.05°$ gives the following limits on d:

$\theta = 15.75°$ $\quad d = 0.2835$ nm
$\theta = 15.85°$ $\quad d = 0.2818$ nm

Therefore the uncertainty in d is ≈ 0.0009 nm.

2.10 The smallest angle corresponds to the largest separation, i.e. the {100} planes. Using eqn (2.2) with $n = 1$, $\lambda = 0.09$ nm and $\theta = 8.9°$ gives $d_{100} = 0.291$ nm.

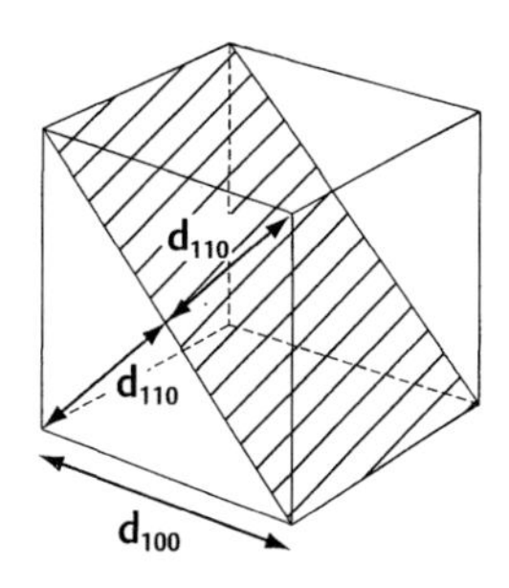

From the above diagram, the separation between adjacent (110) planes

$d_{110} = \dfrac{d_{100}}{\sqrt{2}} = 0.206$ nm.

Therefore, using eqn (2.2), with $n = 1$, $\lambda = 0.09$ nm and d $= 0.206$ nm gives $\theta_{110} = 12.6°$.

2.11 An energy of 100 keV equals 1.6×10^{-14} J. Using eqn (2.3), electron wavelength is

$$\lambda = \frac{h}{\sqrt{2mE}} = \frac{(6.63 \times 10^{-34}\ \text{J s})}{[(2)(9.11 \times 10^{-31}\ \text{kg})(1.6 \times 10^{-14}\ \text{J})]^{1/2}} = 3.88 \times 10^{-12}\ \text{m}$$

2.12 The main advantage of electron microscopes is that they produce a direct image of the crystal, whereas X-ray methods produce a diffraction pattern from which the crystal structure must be calculated. Having a direct image means that it is easy to identify faults in the crystal structure, as shown in Fig. 2.21. The main disadvantage of an electron microscope is that only very thin samples can be used, whereas X-ray methods can be used with bulk samples. In addition, X-ray methods allow quantitative measurements of the atomic spacings to be made with great accuracy. Such information cannot be obtained from an electron microscope.

2.13 For a body-centred cubic structure the spheres occupy 68.0% of the volume (see Question 2.1).
For a face-centered cubic structure the spheres occupy 74.0% of the volume (see Question 2.2).
Consider a sample of iron of volume 10.0 cm^3.
Below 912°C the actual volume of the atoms is 6.80 cm^3.
Assuming that this value is unchanged, above 912°C the volume of the sample is

$$(6.80\ \text{cm}^3)\left(\frac{100}{74}\right) = 9.19\ \text{cm}^3$$

The percentage change is therefore

$$\frac{(9.19\ \text{cm}^3) - (10.0\ \text{cm}^3)}{10.0\ \text{cm}^3} \times 100\,\% = -8.1\%$$

where the minus sign indicates that the volume decreases.
(Note: in practice, the magnitude of the volume change is only about 2%, indicating that the atomic spheres undergo an increase in size.)

Chapter 3

3.1 From eqn (3.1), stress is

$$\sigma = \frac{F}{A} = \frac{mg}{\pi r^2} = \frac{(1000\ \text{kg})(9.81\ \text{ms}^{-2})}{(3.14)(5 \times 10^{-3}\ \text{m})^2}$$

$$= 1.25 \times 10^{8}\ \text{Nm}^{-2}$$

Rearranging eqn (3.3) and using the value of Young's modulus for steel ($= 210 \times 10^9\ \text{Nm}^{-2}$) we have

$$\epsilon = \frac{\sigma}{Y} = \frac{1.25\times 10^{8}\ \text{Nm}^{-2}}{210\times 10^{9}\ \text{Nm}^{-2}} = 5.95\times 10^{-4}$$

3.2 Volume of the cube in Fig. 3.4 is

$$(l_1 + \Delta l_1)(l_2 + \Delta l_2)(l_3 + \Delta l_3) \approx l_1 l_2 l_3 + \Delta l_1 l_2 l_3 + l_1 \Delta l_2 l_3 + l_1 l_2 \Delta l_3$$

The approximation assumes that the changes in length Δl_1, Δl_2 and Δl_3 are small in comparison with l_1, l_2 and l_3; therefore we can neglect terms such as Δl_1, Δl_2 l_3 etc. The change in volume due to the stress is therefore

$$\begin{aligned}\Delta l_1 l_2 l_3 + l_1 \Delta l_2 l_3 + l_1 l_2 \Delta l_3 &= \Delta l_1 l_2 l_3 + l_1\left(-\nu \frac{\Delta l_1}{l_1} l_2\right) l_3 + l_1 l_2\left(-\nu \frac{\Delta l_1}{l_1} l_3\right)\\ &= (1-2\nu)\Delta l_1 l_2 l_3\end{aligned}$$

where we have made use of eqn (3.4).
Since $\Delta l_1 > 0$,we can see that $(1-2\nu)\ \Delta l_1\ l_2\ l_3$ is positive if $\nu < 0.5$.

3.3 Using the equations

$$G = \frac{Y}{2(1+\nu)} \quad \text{and} \quad K = \frac{Y}{3(1-2\nu)}$$

and the data for Y and ν from Table 3.1, we obtain the following results:

	G ($10^9\,\text{N m}^{-2}$)	K ($10^9\,\text{N m}^{-2}$)
Al	26.3	68.6
Fe	51.3	60.6

The values of G and K for Al are within a few per cent of the experimental results in Table 3.1. The agreement for Fe is much worse.
A possible explanation for the discrepancies is that the materials are not isotropic (as assumed in the derivation of eqns (3.7) and (3.8)).

3.4 Rearanging the equation we have

$$C = \frac{\sigma}{\epsilon - 2\epsilon^2}$$

whereas Young's modulus is given by eqn (3.3), i.e.

$$Y = \frac{\sigma}{\epsilon}$$

For $\epsilon = 0.001$

$$C = \frac{\sigma}{9.98\times 10^{-4}} = 1002\,\sigma$$

$$Y = \frac{\sigma}{\epsilon} = 1000\,\sigma$$

Therefore the fractional difference is

$$\frac{C-Y}{Y}=\frac{2}{1000}=0.002$$

For $\epsilon = 0.04$

$$C=\frac{\sigma}{0.0368}=27.2\sigma$$

$$Y=\frac{\sigma}{\epsilon}=25.0\sigma$$

Therefore the fractional difference is

$$\frac{C-Y}{Y}=\frac{2.2}{25.0}=0.087$$

These results show that for most samples (which have an elastic limit of about 0.1% strain) Young's modulus is virtually constant and therefore it is a good approximation to assume that stress is directly proportional to strain.

3.5 Maximum operating stress $\sigma_{max} = (0.2)\,(2 \times 10^8\ \mathrm{N\ m^{-2}}) = 0.4 \times 10^8\ \mathrm{N\ m^{-2}}$
Maximum force on cable is

$$F_{max}=\sigma_{max}A=(0.4\times 10^8\ \mathrm{N\ m^{-2}})(3.14)(0.01\ \mathrm{m})^2=1.26\times 10^4\ \mathrm{N}$$

Maximum mass (lift and people) is

$$m_{max}=\frac{F_{max}}{g}=\frac{1.26\times 10^4\ \mathrm{N}}{9.81\ \mathrm{m\ s^{-2}}}=1.28\times 10^3\ \mathrm{kg}$$

Subtracting the mass of the lift, maximum carrying capacity is 1.13×10^3 kg; therefore maximum number of people is

$$\frac{1.13\times 10^3\ \mathrm{kg}}{70.0\ \mathrm{kg}}=16$$

3.6 Yield shear stress is

$$\tau_Y=\frac{F}{A}=\frac{mg}{\pi r^2}=\frac{(2000\ \mathrm{kg})(9.81\ \mathrm{m\ s^{-2}})}{(3.14)(0.015\ \mathrm{m})^2}=2.78\times 10^7\ \mathrm{N\ m^{-2}}$$

Theoretical estimate is $G/30 = 2.8 \times 10^9\ \mathrm{N\ m^{-2}}$ (using data from Table 3.1). The difference (a factor of 100) is due to the presence of dislocations in the real sample.
The angle of shear at the elastic limit is (from eqn (3.5))

$$\alpha_Y=\frac{\tau_Y}{G}=\frac{2.78\times 10^7\ \mathrm{N\ m^{-2}}}{84\times 10^9\ \mathrm{N\ m^{-2}}}=3.3\times 10^{-4}\ \mathrm{rad}$$

3.7 From Table 3.1, Young's modulus for Al is $70 \times 10^9\ \mathrm{N\ m^{-2}}$; therefore the theoretical breaking stress in the absence of cracks is (see Section 3.7)

$$\sigma_B \approx 0.1Y \approx 7.0 \times 10^9 \text{ N m}^{-2}$$

The stress concentration factor at the tip of the crack is

$$2\left(\frac{l}{r}\right)^{1/2} = 2\left(\frac{1.0 \times 10^{-6} \text{ m}}{1.0 \times 10^{-9} \text{ m}}\right)^{1/2} = 31.6$$

Therefore the actual breaking stress is

$$\frac{\sigma_B}{31.6} \approx 2 \times 10^8 \text{ N m}^{-2}$$

Since the yield stress for plastic deformation is about 2.6×10^7 N m^{-2} (see Table 5.2), the material deforms plastically rather than by brittle fracture.

3.8 The problem with using glass cables is that they undergo brittle fracture; therefore the result of exceeding the yield point in a load-bearing application would be catastrophic. In contrast, steel wires do not break until well beyond the yield point.

3.9 For Al, breaking strain $\gg$ yield strain; therefore the material undergoes plastic deformation. But breaking stress $\approx$ yield stress; therefore no significant work hardening occurs. A typical stress–strain curve for such a material is shown in Fig. 3.1.
For glass, breaking strain = yield strain; therefore the material is brittle, i.e. it undergoes elastic deformation only. A typical stress–strain curve is shown in Fig. 3.18.
For steel, breaking strain $\gg$ yield strain and breaking stress $\gg$ yield stress; therefore the material undergoes plastic deformation and work hardening. A typical stress–strain curve is shown in Fig. 3.14.

Chapter 4

4.1 Rearranging eqn (4.1) the average thermal velocity of an electron is

$$v_t = \sqrt{\frac{3k_BT}{m_e}} = \left(\frac{(3.0)(1.38 \times 10^{-23} \text{ J K}^{-1})(295 \text{ K})}{9.11 \times 10^{-31} \text{ kg}}\right)^{1/2} = 1.16 \times 10^5 \text{ m s}^{-1}$$

If the mean free path, λ, is 1 nm, then from eqn (4.2) the average time between collisions τ is

$$\tau = \frac{\lambda}{v_t} = \frac{1.0 \times 10^{-9} \text{ m}}{1.16 \times 10^5 \text{ m s}^{-1}} = 8.62 \times 10^{-15} \text{ s}$$

4.2 From eqn (4.7) the mobility is

$$\mu = \frac{e\tau}{m_e} = \frac{(1.60 \times 10^{-19} \text{ C})(8.62 \times 10^{-15} \text{ s})}{9.11 \times 10^{-31} \text{ kg}} = 1.51 \times 10^{-3} \text{ m}^2 \text{ V}^{-1} \text{ s}^{-1}$$

From eqn (4.5) the drift velocity is

$$\overline{\Delta v} = \frac{e\mathcal{E}\tau}{m_e} = \frac{(1.60\times 10^{-19}\ \text{C})(10.0\ \text{V m}^{-1})(8.62\times 10^{-15}\ \text{s})}{9.11\times 10^{-31}\ \text{kg}} = 0.015\ \text{m s}^{-1}$$

The drift velocity represents the net velocity of the electrons in the direction of the electric field. Since the electrons undergo many collisions with the ions, the direction of motion of the electrons is more or less random even when an electric field is applied, i.e. the probability of finding the electron travelling in the direction of the field is only slightly higher than that of finding it travelling in the opposite direction. Consequently, the drift velocity is very small in comparison with the thermal velocity.

4.3 The current density J, electric field strength $\mathcal{E}$ and resistance R can be written as

$$J = \frac{I}{A} \qquad \mathcal{E} = \frac{V}{L} \qquad R = \frac{\rho L}{A}$$

where A and L are the cross-sectional area and the length of the sample, respectively. Substituting into eqn (4.11) we have

$$J = \frac{\mathcal{E}}{\rho}$$

$$\Rightarrow \frac{I}{A} = \frac{V}{L}\cdot\frac{L}{RA}$$

$$\Rightarrow I = \frac{V}{R}$$

4.4 Number of lithium atoms in 1 kg is

$$\frac{\text{Avogadro's number}}{\text{molar mass}} = \frac{6.02\times 10^{23}\ \text{mol}^{-1}}{7\times 10^{-3}\ \text{kg mol}^{-1}} = 8.6\times 10^{25}\ \text{kg}^{-1}$$

Number of lithium atoms in 1 m^3 is

$$(\text{density}).(\text{number of atoms in 1 kg}) = (530\ \text{kg m}^{-3}).(8.6\times 10^{25}\ \text{kg}^{-1})$$
$$= 4.56\times 10^{28}\ \text{m}^{-3}$$

Since lithium is monovalent, the number density of electrons n is equal to the number density of lithium atoms, i.e.

$$n = 4.56\times 10^{28}\ \text{m}^{-3}$$

If the mean free path is 1 nm, then from Question 4.1 the relaxation time at 295 K is 8.62×10^{-15} s. Using eqn (4.7) the mobility is therefore

$$\mu = \frac{e\tau}{m_e} = \frac{(1.60\times 10^{-19}\ \text{C})(8.62\times 10^{-15}\ \text{s})}{9.11\times 10^{-31}\ \text{kg}} = 1.51\times 10^{-3}\ \text{m}^2\ \text{V}^{-1}\ \text{s}^{-1}$$

From eqn (4.9) the conductivity is

$$\sigma = n\mu e = (4.56\times 10^{28}\ \text{m}^{-3})(1.51\times 10^{-3}\ \text{m}^2\ \text{V}^{-1}\ \text{s}^{-1})(1.60\times 10^{-19}\ \text{C})$$
$$= 1.10\times 10^{7}\ \Omega^{-1}\ \text{m}^{-1}$$

4.5 From eqns (4.9) and (4.7) we have

$$\sigma = n\mu e = \frac{ne^2\tau}{m_e}$$

Rearranging and using data from Table 4.1, we can determine the relaxation time:

$$\tau = \frac{m_e\sigma}{ne^2} = \frac{(9.11\times10^{-31}\ \text{kg})(4.00\times10^{7}\ \Omega^{-1}\ \text{m}^{-1})}{(18.1\times10^{28}\ \text{m}^{-3})(1.60\times10^{-19}\ \text{C})^2} = 7.86\times10^{-15}\ \text{s}$$

At 273 K the thermal velocity of the electrons is 1.11×10^5 m s^{-1} (following the solution to Question 4.1); therefore the mean free path is (from eqn (4.2))

$$\lambda = v_t\tau = (1.11\times10^5\ \text{m s}^{-1})(7.86\times10^{-15}\ \text{s}) = 8.72\times10^{-10}\ \text{m}$$

4.6 Using eqn (4.9) and the data in Table 4.1, the mobility is given by

$$\mu = \frac{\sigma}{ne} = \frac{(6.45\times10^{7}\ \Omega^{-1}\ \text{m}^{-1})}{(8.50\times10^{28}\ \text{m}^{-3})(1.60\times10^{-19}\ \text{C})} = 4.74\times10^{-3}\ \text{m}^2\ \text{V}^{-1}\ \text{s}^{-1}$$

Since the electric field strength is

$$\mathcal{E} = \frac{\text{potential difference}}{\text{length}} = \frac{0.3\ \text{V}}{5.0\ \text{m}} = 0.06\ \text{V m}^{-1}$$

then from eqn (4.6) the drift velocity is

$$\overline{\Delta v} = \mu\mathcal{E} = (4.74\times10^{-3}\ \text{m}^2\ \text{V}^{-1}\ \text{s}^{-1})(0.06\ \text{V m}^{-1}) = 2.84\times10^{-4}\ \text{m s}^{-1}$$

From Fig. 4.3, the net number of electrons passing through area A in 1 second is

$$n\overline{\Delta v}A = (8.50\times10^{28}\ \text{m}^{-3})(2.84\times10^{-4}\ \text{m s}^{-1})(2.5\times10^{-6}\ \text{m}^2) = 6.04\times10^{19}\ \text{s}^{-1}$$

4.7 Using eqns (4.9), (4.7) and (4.2) we have

$$\sigma = n\mu e = \frac{ne^2\tau}{m_e} = \frac{ne^2\lambda}{m_e v_t}$$

From Question 4.5, v_t at 273 K is 1.11×10^5 m s^{-1}.

For silver

$$\sigma = \frac{(5.86\times10^{28}\ \text{m}^{-3})(1.60\times10^{-19}\ \text{C})^2(1.0\times10^{-9}\ \text{m})}{(9.11\times10^{-31}\ \text{kg})(1.11\times10^{5}\ \text{m s}^{-1})}$$

$$= 1.48\times10^{7}\ \Omega^{-1}\ \text{m}^{-1}$$

For lead

$$\sigma = \frac{(13.2\times10^{28}\ \text{m}^{-3})(1.60\times10^{-19}\ \text{C})^2(1.0\times10^{-9}\ \text{m})}{(9.11\times10^{-31}\ \text{kg})(1.11\times10^{5}\ \text{m s}^{-1})}$$

$$= 3.34\times10^{7}\ \Omega^{-1}\ \text{m}^{-1}$$

In comparison, the experimental values of conductivity given in Table 4.1 for silver and lead are $6.80 \times 10^7\ \Omega^{-1}\ \text{m}^{-1}$ and $0.52 \times 10^7\ \Omega^{-1}\ \text{m}^{-1}$, respectively. So although the magnitudes of the calculated values are approximately correct in each case, the results are the wrong way round. Lead is predicted to exhibit better conductivity than silver on the basis that it has a higher concentration of delocalized electrons, but in fact the conductivity of silver is over 10 times better than that of lead. This demonstrates that Drude's model is rather over-simplified and that there are other factors besides the electron concentration which also affect the conductivity.

4.8 According to eqn (4.1) the thermal velocity v_t is proportional to $T^{1/2}$, i.e.

$$v_t = \left(\frac{3k_BT}{m}\right)^{1/2} \propto T^{1/2}$$

If the mean free path λ is constant with temperature, then eqn (4.2) shows that the relaxation time τ is proportional to $T^{-1/2}$, i.e.

$$\tau = \frac{\lambda}{v_t} \propto T^{-1/2}$$

Combining eqns (4.7) and (4.10) we have

$$\rho = \frac{m_e}{ne^2\tau}$$

Since the electron concentration, n, is independent of temperature in Drude's model, this predicts $\rho \propto T^{1/2}$.

4.9 Following the discussion in Example 4.2, the vibrational energy per ion is $3kT$ and the translational energy per electron is $\frac{3}{2}k_BT$. Since there are two valence electrons per ion, the total energy of 1 mole of a divalent metal is

$$3N_A k_BT + 2N_A\left(\frac{3}{2}k_BT\right) = 6N_A k_BT = 6RT$$

Therefore, Drude theory predicts that the energy required to raise the temperature of 1 mole of a divalent metal by 1 K is $6R$.

4.10 In the quantum model of conductivity the resistivity of a metal is due to the thermal vibration of the ions and the presence of impurities and imperfections. As the temperature is reduced towards absolute zero, the thermal vibrations of the atoms decrease, and so the resistivity decreases. However, the resistivity does not go to zero because the presence of impurities and imperfections in the sample always gives rise to some scattering of the electrons.

(Actually, quantum theory shows that even at absolute zero the thermal vibration of the ions also is non-zero.) Consequently, there is always a residual resistivity even at $T = 0$ K.

(The behaviour of superconductors, which do have zero resistivity, requires a rather different explantion. This is dealt with in Chapter 9.)

4.11 If we have N electrons, then at $T = 0$ K these electrons fill the lowest $N/2$ states. We also know that the energy of the highest filled state at 0 K is the Fermi energy, E_F. So the average spacing between the energy levels is

$$\text{average spacing} = \frac{\text{energy range}}{\text{number of quantum states}} = \frac{E_F}{N/2}$$

For 1 cm^3 of sodium, $N = nV = (2.65 \times 10^{28}\ \text{m}^{-3})\,(1.0 \times 10^{-6}\ \text{m}^3) = 2.65 \times 10^{22}$. Since we can have two electrons in each quantum state, the number of quantum states required to accommodate these electrons is $2.65 \times 10^{22}/2$. Therefore

$$\text{average spacing} = \frac{3.22\ \text{eV}}{(2.65 \times 10^{22})/2} = 2.43 \times 10^{-22}\ \text{eV}$$

Therefore we are justified in treating these states as a continuous band of energies.

4.12 If $E < E_F$, then as $T \to 0$

$$(E - E_F)/k_B T \to -\infty$$

So eqn (4.12) becomes

$$f = \frac{1}{e^{-\infty} + 1} = \frac{1}{0 + 1} = 1.0$$

If $E > E_F$, then as $T \to 0$

$$(E - E_F)/k_B T \to +\infty$$

So eqn (4.12) becomes

$$f = \frac{1}{e^{+\infty} + 1} = \frac{1}{\infty + 1} = 0.0$$

This agrees with Fig. 4.16.

4.13 Rearranging eqn (4.12) we have

$$E - E_F = k_B T\ \ln\left(\frac{1}{f(E)} - 1\right)$$

When $f(E) = 0.95$

$$E_F - E = K_B T \ln\left(\frac{1}{0.95} - 1\right) = (1.38\times10^{-23}\ \mathrm{J\ K^{-1}})(300\ \mathrm{K})(2.94)$$

$$= 1.22\times10^{-20}\ \mathrm{J} = 0.076\ \mathrm{eV}$$

When $f(E) = 0.05$,

$$E_F - E = -0.076\ \mathrm{eV}$$

4.14 For an infinitely deep 3-D well Schrödinger's equation is

$$-\frac{\hbar^2}{2m}\left(\frac{\partial^2\psi}{dx^2} + \frac{\partial^2\psi}{dy^2} + \frac{\partial^2\psi}{dz^2}\right) = E\psi$$

Using a trial function

$$\psi = A\sin(k_x x)\sin(k_y y)\sin(k_z z)$$

we have

$$\frac{\partial^2\psi}{dx^2} = -k_x^2\psi \qquad \frac{\partial^2\psi}{dy^2} = -k_y^2\psi \qquad \frac{\partial^2\psi}{dz^2} = -k_z^2\psi$$

Substituting in Schrödinger's equation we get

$$E = \frac{\hbar^2}{2m}(k_x^2 + k_y^2 + k_z^2)$$

Boundary conditions are $\psi = 0$ at $x = 0$, $y = 0$, $z = 0$, $x = L$, $y = L$ and $z = L$. The first three conditions are satisfied by the choice of trial wavefunction. To satisfy the three latter conditions we must have

$$k_x = \frac{n_x\pi}{L} \qquad k_y = \frac{n_y\pi}{L} \qquad k_z = \frac{n_z\pi}{L}$$

Therefore the allowed energy levels are

$$E = \frac{\hbar^2}{2m}\left(\frac{\pi}{L}\right)^2(n_x^2 + n_y^2 + n_z^2)$$

4.15 First note that the derivation of eqn (4.17) in Example 4.5 not only applies for the Fermi energy, but also proves the more general result that the number of electrons N with an energy less than E is given by

$$E = \frac{\hbar^2}{2m_e}\left(\frac{3\pi^2 N}{V}\right)^{2/3}$$

Rearranging we have

$$N = \frac{V}{3\pi^2}\left(\frac{2m_e E}{\hbar^2}\right)^{3/2}$$

The density of states is obtained by differentiating with respect to E, i.e.

$$g(E)=\frac{dN}{dE}=\frac{V}{2\pi^2}\left(\frac{2m_e}{\hbar^2}\right)^{3/2}E^{1/2}$$

4.16 Using eqn (4.17) with $N/V = 8.50 \times 10^{28}$ m^{-3} gives

$$E_F=\frac{\hbar^2}{2m_e}(3\pi^2\frac{N}{V})^{2/3}$$
$$=\frac{(1.05\times10^{-34}\text{ J s})^2}{(2)(9.11\times10^{-31}\text{ kg})}\left[(3.0)(3.14)^2(8.50\times10^{28}\text{ m}^{-3})\right]^{2/3}$$
$$=1.12\times10^{-18}\text{ J}=7.00\text{ eV}$$

The velocity at the Fermi energy, v_F, is given semi-classically by eqn (4.20),

$$\frac{1}{2}m_e v_F^2=E_F$$

or in a quantum description by

$$m_e v_F=\hbar k_F$$

where k_F is the value of k corresponding to the Fermi energy.
In either case we obtain

$$v_F=\sqrt{\frac{2E_F}{m_e}}=\left(\frac{(2.0)(1.12\times10^{-18}\text{ J})}{9.11\times10^{-31}\text{ kg}}\right)^{1/2}=1.57\times10^6\text{ m s}^{-1}$$

Since the density of states function $g(E)$ describes the number of states per unit energy, the average spacing between the states is given by $1/g(E)$, i.e. energy spacing at E_F is

$$\frac{1}{g(E_F)}=\frac{2\pi^2}{V}\left(\frac{\hbar^2}{2m_e}\right)^{3/2}E_F^{-1/2}$$
$$=\frac{(2)(3.14)^2}{1.0\times10^{-6}\text{ m}^3}\left(\frac{(1.05\times10^{-34}\text{ J s})^2}{(2)(9.11\times10^{-31}\text{ kg})}\right)^{3/2}(1.12\times10^{-18}\text{ J})^{-1/2}$$
$$=8.78\times10^{-42}\text{ J}=5.49\times10^{-23}\text{ eV}$$

An alternative (but cruder) estimate of the spacing between the energy levels in a band can be obtained by dividing the energy range of the occupied states at T = 0 K (which is equal to the Fermi energy) by the number of occupied states (which is equal to $N/2$). For 1 cm^3 of copper,

$$\frac{N}{2}=(8.50\times10^{28}\text{ m}^{-3})(1.0\times10^{-6}\text{ m}^{-3})/2=4.25\times10^{22}$$

So the average spacing between the states is

$$\frac{E_F}{N/2}=\frac{1.12\times10^{-18}\text{ J}}{4.25\times10^{22}}=2.64\times10^{-41}\text{ J}=1.65\times10^{-22}\text{ eV}$$

In either case we can see that the spacing between energy levels is so small that we are justified in treating the energy band as a continuous range of allowed energies.

4.17 Using the approximation that the number of electrons participating in conduction, N_c, is found by integrating the function $g(E)$ between $E_F - 2.2\,k_BT$ and E_F, and substituting for $g(E)$ from eqn (4.19), we have

$$N_c = \int_{E_F-2.2\,k_BT}^{E_F} g(E)\,dE = \int_{E_F-2.2\,k_BT}^{E_F} \frac{V}{2\pi^2}\left(\frac{2m_e}{\hbar^2}\right)^{3/2} E^{1/2}\,dE$$

$$= \frac{V}{2\pi^2}\left(\frac{2m_e}{\hbar^2}\right)^{3/2}\left[\frac{2}{3}E^{3/2}\right]_{E_F-2.2\,k_BT}^{E_F}$$

$$= \frac{V}{3\pi^2}\left(\frac{2m_e}{\hbar^2}\right)^{3/2}\left[E_F^{3/2} - (E_F - 2.2\,k_BT)^{3/2}\right]$$

Using the values for sodium from Table 4.2 we have $E_F = 3.22$ eV ($= 5.152 \times 10^{-19}$ J) and $E_F - 2.2\,k_BT = 5.0609 \times 10^{-19}$ J at 300 K. Therefore

$$N_c = \frac{V}{(3.0)(3.14)^2}\left(\frac{(2.0)(9.11\times10^{-31}\ \text{kg})}{(1.05\times10^{-34}\ \text{J s})^2}\right)^{3/2} \times\left[(5.152\times10^{-19}\ \text{J})^{3/2} - (5.0609\times10^{-19}\ \text{J})^{3/2}\right]$$

$$= 7.01\times10^{26}\,V$$

where V is the volume of the crystal.
Since from Table 4.1 the total number of valence electrons in sodium is $N = 2.65 \times 10^{28}\,V$ the proportion of valence electrons that actually take part in conduction is

$$\frac{7.01\times10^{26}}{2.65\times10^{28}} = 0.026$$

i.e. 2.6%.

4.18 From eqn (4.20) the velocity of an electron at the Fermi energy is given by

$$v_F = \sqrt{\frac{2E_F}{m_e}}$$

For sodium the Fermi energy is 3.22 eV = 5.15×10^{-19} J. Therefore

$$v_F = \left(\frac{(2.0)(5.15\times10^{-19}\ \text{J})}{9.11\times10^{-31}\ \text{kg}}\right)^{1/2} = 1.06\times10^{6}\ \text{m s}^{-1}$$

Similarly for copper we have
$E_F = 7.00$ eV = 1.12×10^{-18} J, so $v_F = 1.57 \times 10^6$ m s^{-1},
and for aluminium,
$E_F = 11.58$ eV = 1.85×10^{-18} J, so $v_F = 2.02 \times 10^6$ m s^{-1}.

4.19 From eqns (4.7) and (4.9) we can write the relaxation time as

$$\tau = \frac{m_e\sigma}{n_c e^2}$$

where n_c is the actual number of electrons that participate in electrical conduction. Since this is only 3% of the total number of valence electrons,

$$n_c = (0.03)(8.50\times10^{28}\ \text{m}^{-3}) = 2.55\times10^{27}\ \text{m}^{-3}$$

Therefore

$$\tau = \frac{(9.11\times 10^{-31}\ \text{kg})(6.45\times 10^{7}\ \Omega^{-1}\ \text{m}^{-1})}{(9.00\times 10^{27}\ \text{m}^{-3})(1.60\times 10^{-19}\ \text{C})^2}$$
$$= 9.00\times 10^{-13}\ \text{s}$$

To estimate the mean free path we use eqn (4.2) but replace the thermal velocity v_t with the Fermi velocity v_F. From Question 4.18 we know that v_F for copper is 1.57×10^6 m s^{-1}. Therefore, the mean free path is

$$\lambda = \tau v_F = (8.21\times 10^{-13}\ \text{s})\,(1.57\times 10^{6}\ \text{m s}^{-1}) = 1.29\times 10^{-6}\ \text{m}$$

4.20 If only 3% of valence electrons contribute to the heat capacity at 300 K, then the number of thermally excited valence electrons in 1 mole of a monovalent metal is 0.03 N_A (where N_A is Avogadro's number), and the thermal energy of these electrons is 0.03 $N_A k_BT$ = 0.03 RT.
The contribution of the electrons to the molar specific heat capacity is equal to the increase in energy of the electrons when the temperature is raised by 1 K, i.e. 0.03R.
In comparison, from Example 4.2, the molar specific heat capacity of the ions is 3R. Therefore the electronic contribution to the heat capacity at room temperature is only about 1% of the total heat capacity.

Chapter 5

5.1 The probability of an electron having an energy equal to the lowest energy in the conduction band is given by eqn (5.2), i.e.

$$f(E) = [e^{E_g/2k_BT} + 1]^{-1}$$

The band gap of Si is E_g = 1.11 eV = 1.78×10^{-19} J, so at 300 K the exponential term in the above equation is

$$e^{E_g/2k_BT} = \exp\left(\frac{1.78\times 10^{-19}\ \text{J}}{(2)(1.38\times 10^{-23}\ \text{J K}^{-1})(300\ \text{K})}\right)$$
$$= 2.17\times 10^{9}$$

The factor of 1 in the square brackets in eqn (5.2) is negligible in comparison so we can write

$$f(E) \approx [e^{E_g/2k_BT}]^{-1}$$

5.2 If the intrinsic carrier concentration is given by eqn (5.4), then the constant C is given by

$$C = ne^{E_g/2k_BT}$$

For Ge, $n = 3 \times 10^{19}$ m^{-3}, $E_g = 0.67$ eV $= 1.07 \times 10^{-19}$ J. Therefore

$$C = (3\times 10^{19}\ \mathrm{m}^{-3})\exp\left(\frac{1.07\times 10^{-19}\ \mathrm{J}}{(2)(1.38\times 10^{-23}\ \mathrm{J\ K}^{-1})(300\ \mathrm{K})}\right) = 1.23\times 10^{25}\ \mathrm{m}^{-3}$$

For Si, $n = 1 \times 10^{16}$ m^{-3}, $E_g = 1.11$ eV $= 1.78 \times 10^{-19}$ J. Therefore $C = 2.17 \times 10^{25}$ m^{-3}. For GaAs, $n = 2 \times 10^{13}$ m^{-3}, $E_g = 1.43$ eV $= 2.29 \times 10^{-19}$J. Therefore $C = 2.05 \times 10^{25}$ m^{-3}. So the value of C for these three materials is remarkably similar considering that the value of n varies by several orders of magnitude.

5.3 From Table 5.1, the band gap of diamond is 5.5 eV $=8.8 \times 10^{-19}$ J. Substituting in eqn (5.4) and putting $C = 10^{25}$ m^{-3} gives a carrier concentration

$$n = Ce^{-E_g/2k_BT} = (10^{25}\ \mathrm{m}^{-3})\exp\left(-\frac{8.8\times 10^{-19}\ \mathrm{J}}{(2)(1.38\times 10^{-23}\ \mathrm{J\ K}^{-1})(300\ \mathrm{K})}\right)$$

$$= 6.97\times 10^{-22}\ \mathrm{m}^{-3}$$

Therefore in a volume V of 10^{-6} m^3, the number of carriers is $nV = (6.97 \times 10^{-22}$ m$^{-3})(10^{-6}$ m$^3) \approx 7 \times 10^{-28}$, i.e. the probability of finding a single carrier in 1 cm^3 of diamond at 300 K is virtually zero.

5.4 From eqn (4.9) the conductivity is given by

$$\sigma = n\mu e$$

where μ is the mobility and n is given by eqn (5.4). The band gap of Si is $E_g = 1.11$ eV $= 1.78 \times 10^{-19}$ J, so at 30 K, the carrier concentration is

$$n = Ce^{-E_g/2k_BT} = (2\times 10^{25}\ \mathrm{m}^{-3})\exp\left(-\frac{1.78\times 10^{-19}\ \mathrm{J}}{(2)(1.38\times 10^{-23}\ \mathrm{J\ K}^{-1})(300\ \mathrm{K})}\right)$$

$$= 8.67\times 10^{-69}\ \mathrm{m}^{-3}$$

Therefore $\sigma = (8.67 \times 10^{-69}$ m$^{-3})(0.135$ m^2 V^{-1} s$^{-1})(1.60 \times 10^{-19}$ C) $= 1.87\times 10^{-88}$ Ω^{-1} m^{-1}.
At 300 K, $n = 9.22 \times 10^{15}$ m^{-3}, and $\sigma = 1.99 \times 10^{-4}$ Ω^{-1} m^{-1}.
At 1000 K, $n = 3.16 \times 10^{22}$ m^{-3}, and $\sigma = 683$ Ω^{-1} m^{-1}.
Comparing these values with the data in Fig. 5.1, at 1000 K the conductivity of Si is not much lower than that of a poor metallic conductor, but at 30 K the conductivity of Si is far worse than that of even the best insulator at 300 K. (Note that these calculations are for a perfectly pure sample of silicon. The presence of impurities means that the conductivity of a real sample of silicon at 30 K will actually be much higher than our predicted value.)

5.5 The conductivity is given by $\sigma = n\mu e$.
For a metal we expect that μ decreases with increasing temperature because of the increased thermal vibration of the ions (see Section 4.4) and the num-

ber of valence electrons n is constant. Therefore we expect the conductivity to decrease with increasing temperature.

For a semiconductor, μ is also expected to decrease with increasing temperature for the same reasons as in a metal, but the number of conduction electrons increases exponentially—see eqn (5.4). Therefore the conductivity of a semiconductor increases exponentially with temperature, as shown in Fig. 5.6.

5.6 Using the mobility data given in the question and recognizing that for an intrinsic material we have $n = p$, the conductivity due to the combined effect of the electrons and holes is given by eqn (5.5), i.e.

$$\sigma = |e|(n\mu_e + p\mu_h) = |e|n(\mu_e + \mu_h)$$
$$= (1.60\times 10^{-19}\ \text{C})(1\times 10^{16}\ \text{m}^3)[(0.135\ \text{m}^2\ \text{V}^{-1}\ \text{s}^{-1}) + (0.048\ \text{m}^2\ \text{V}^{-1}\ \text{s}^{-1})]$$
$$= 2.93\times 10^{-4}\ \Omega^{-1}\ \text{m}^{-1}$$

5.7 From eqn (4.9) the conductivity is given by

$$\sigma = n\mu e$$

If μ is constant with temperature, then the conductivity doubles if n doubles. (Note that we get the same result if we consider the holes, since the concentration of holes is equal to the concentration of conduction electrons.)

From Table 5.1 the band gap of Ge is 0.67 eV = 1.07×10^{-19} J; therefore at 300 K, the electron concentration n_1 is given by eqn (5.4),

$$n_1 = Ce^{-E_g/2k_BT} = C\exp\left(-\frac{1.07\times 10^{-19}\ \text{J}}{(2)(1.38\times 10^{-23}\ \text{J K}^{-1})(300\ \text{K})}\right) = 2.44\times 10^{-6}C$$

If the electron concentration is equal to $2n_1$ at a temperature T_2, then from eqn (5.4)

$$2n_1 = Ce^{-E_g/2k_BT_2}$$

Rearranging we have

$$T_2 = \frac{E_g}{2k_B\ln(C/2n_1)}$$

The log term is

$$\ln\left(\frac{C}{2n_1}\right) = \ln\left(\frac{C}{(2)(2.44\times 10^{-6}C)}\right) = 12.2$$

Therefore

$$T_2 = \frac{1.07\times 10^{-19}\ \text{J}}{(2)(1.38\times 10^{-23}\ \text{J K}^{-1})(12.2)} = 318\ \text{K}$$

5.8 The band gap of a semiconductor is related to the wavelength of the emitted light by eqn (5.7), so to emit light with a wavelength of 1 μm requires a band gap of

$$E_g = \frac{hc}{\lambda} = \frac{(6.63\times 10^{-34}\ \text{J s})(3.00\times 10^{8}\ \text{m s}^{-1})}{1.00\times 10^{-6}\ \text{m}^{-1}} = 1.99\times 10^{-19}\ \text{J} = 1.24\ \text{eV}$$

The indium content, x, for an alloy of GaInAs with a band gap of 1.24 eV is found by linearly interpolating between the values for GaAs and InAs:

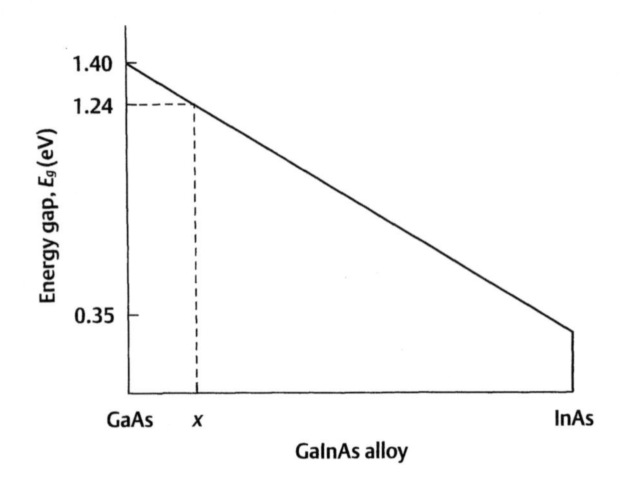

$$x = \frac{E_g(GaAs) - (1.24\ \text{eV})}{E_g(GaAs) - E_g(InAs)} = \frac{(1.40\ \text{eV}) - (1.24\ \text{eV})}{(1.40\ \text{eV}) - (0.35\ \text{eV})} = 0.15$$

Therefore the required alloy is 15% In and 85% Ga, i.e. $Ga_{0.85}In_{0.15}As$.

5.9 The minimum frequency of light absorbed by a semiconductor or insulator is given by eqn (5.7).
For silicon, $E_g = 1.11$ eV $= 1.78\times 10^{-19}$ J, so

$$\nu_{min} = \frac{E_g}{h} = \frac{1.78\times 10^{-19}\ \text{J}}{6.63\times 10^{-34}\ \text{J s}} = 2.68\times 10^{14}\ \text{Hz}$$

For diamond, $E_g = 5.5$ eV $= 8.8\times 10^{-19}$ J, so

$$\nu_{min} = \frac{8.80\times 10^{-19}\ \text{J}}{6.63\times 10^{-34}\ \text{J s}} = 1.33\times 10^{15}\ \text{Hz}$$

In comparison, visible light has frequencies from about 4×10^{14} Hz to 8×10^{14} Hz. Therefore silicon absorbs all frequencies of visible light, whereas diamond does not absorb any visible light.

5.10 The difference in wavevector between the electron and hole in silicon is approximately

$$k_{diff} \approx \frac{\pi}{a} = \frac{3.14}{0.543\times 10^{-9}\ \text{m}} \approx 6\times 10^{9}\ \text{m}^{-1}$$

For a photon the wavevector is (see Appendix A)

$$k_{photon} = \frac{2\pi}{\lambda} = 2\pi \frac{E_g}{hc}$$

In silicon the band gap is 1.11 eV = 1.78 × 10^{-19} J. Therefore

$$k_{photon} = (2)(3.14)\frac{1.78\times 10^{-19}\ \text{eV}}{(6.63\times 10^{-34}\ \text{J s})(3.00\times 10^{8}\ \text{m})} = 5.62\times 10^{6}\ \text{m}^{-1}$$

Therefore the wavevector of the photon is about 1000 times smaller than that required to allow the conduction electron and the hole to recombine.

5.11 Using eqn (5.8), the cyclotron frequency for a free electron is

$$\nu_c = \frac{Be}{2\pi m_e} = \frac{(0.05\ \text{T})(1.60\times 10^{-19}\ \text{C})}{(2)(3.14)(9.11\times 10^{-31}\ \text{kg})} = 1.40\times 10^{9}\ \text{Hz}$$

For an electron in GaAs we use eqn (5.9) with $m_e^* = 0.065\ m_e$. Therefore

$$\nu_c = \frac{Be}{2\pi m_e^*} = \frac{(0.05\ \text{T})(1.60\times 10^{-19}\ \text{C})}{(2)(3.14)(0.065)(9.11\times 10^{-31}\ \text{kg})} = 2.15\times 10^{10}\ \text{Hz}$$

5.12 (i) From eqn (1.1), the radius of the electron orbit in a hydrogen atom (with $n = 1$) is

$$r_1 = \frac{h^2\varepsilon_0}{\pi m_e e^2} = 0.529\times 10^{-9}\ \text{m}$$

For the donor electron we replace m_e by m_e^* and ε_0 by $\varepsilon\varepsilon_0$, so

$$r_d = \frac{h^2\varepsilon\varepsilon_0}{\pi m_e^* e^2} = \frac{11.7}{0.43}\frac{h^2\varepsilon_0}{\pi m_e e^2} = (27.2)(0.529\times 10^{-9}\ \text{m}) = 1.44\times 10^{-8}\ \text{m}$$

(ii) Assume the volume occupied by each silicon atom is approximately a cube with sides of 0.234 nm; then

$$V_{atom} = (0.234\times 10^{-9})^3 = 1.28\times 10^{-27}\,\text{m}^3$$

The volume enclosed by the donor electron orbit is

$$V_{orbit} = \frac{4\pi r_d^3}{3} = \frac{(4)(3.14)(1.44\times 10^{-8}\ \text{m})^3}{3} = 1.25\times 10^{-23}\ \text{m}^3$$

The number of silicon atoms in this volume is

$$\frac{V_{orbit}}{V_{atom}} = \frac{1.25\times 10^{-23}\ \text{m}^3}{1.28\times 10^{-27}\ \text{m}^3} \approx 10^4$$

5.13 Following Example 5.3 and using data from Table 5.2,

$$E_d = -\frac{m_e^* e^4}{8\varepsilon^2\varepsilon_0^2 h^2} = -\frac{0.60}{15.8^2}\frac{m_e e^4}{8\varepsilon_0^2 h^2} = -(2.40\times 10^{-3}).(13.6\ \text{eV})$$
$$= 0.033\ \text{eV}$$

i.e. the donor state is 33 meV below the conduction band edge.

5.14 The volume of 1 mole of silicon is

$$\frac{\text{molar mass}}{\text{density}} = \frac{0.028 \text{ kg mol}^{-1}}{2300 \text{ kg m}^{-3}} = 1.22 \times 10^{-5} \text{ m}^3$$

The number of silicon atoms in 1 m^3 is therefore

$$\frac{\text{Avogadro's number}}{\text{Volume of 1 mole}} = \frac{6.02 \times 10^{23} \text{ mol}^{-1}}{1.22 \times 10^{-5} \text{ m}^3} = 4.93 \times 10^{28} \text{ m}^{-3}$$

If one in 10^6 atoms is replaced by an impurity, then the concentration of impurities is

$$\frac{(4.93 \times 10^{28} \text{ m}^{-3})}{(1 \times 10^6)} = 4.93 \times 10^{22} \text{ m}^{-3}$$

5.15 If all of the donors are ionized, then the concentration of conduction electrons is $n = 5 \times 10^{22}$ m^{-3}. Using eqn (4.9) the conductivity is

$$\sigma = n\mu e = (5 \times 10^{22} \text{m}^{-3})(0.135 \text{m}^2 \text{V}^{-1} \text{s}^{-1})(1.60 \times 10^{-19} \text{C}) = 1080\ \Omega^{-1} \text{m}^{-1}$$

Comparing with Question 5.4, the conductivity is about 5×10^6 times larger than the conductivity of intrinsic silicon at 300 K.

5.16 We have defined the zero of energy to coincide with the top of the valence band, so the number of electrons at the valence band edge is proportional to the Fermi–Dirac distribution in eqn (5.1) with $E = 0$, i.e.

$$f(0) = \frac{1}{e^{-E_F/k_B T} + 1}$$

Since a hole is a state which is not occupied by an electron, the probability of finding a hole at the valence band edge is

$$1 - f(0) = 1 - \frac{1}{e^{-E_F/k_B T} + 1} = \frac{e^{-E_F/k_B T}}{e^{-E_F/k_B T} + 1}$$

Since the Fermi energy is typically much larger than the thermal energy $k_B T$ (except at very high levels of p-type doping), the exponent is large and negative, i.e. $e^{-E_F/k_B T} \ll 1$. Therefore

$$1 - f(0) \approx \frac{e^{-E_F/k_B T}}{1} = e^{-E_F/k_B T}$$

Therefore the concentration of holes is

$$p = Ce^{-E_F/k_B T}$$

where C is a constant.

5.17 Since the concentration of donors, N_d, is greater than the concentration of acceptors, N_a, the concentration of conduction electrons is equal to the net concentration of donors, i.e.

$$n = N_d - N_a = (6 \times 10^{22} \text{ m}^{-3}) - (1 \times 10^{22} \text{ m}^{-3}) = 5 \times 10^{22} \text{ m}^{-3}$$

The concentration of holes is found by using the law of mass action, eqn (5.16):

$$p=\frac{n_i p_i}{n}=\frac{(5\times 10^{15}\ \mathrm{m}^{-3})^2}{5\times 10^{22}\ \mathrm{m}^{-3}}=5\times 10^{8}\ \mathrm{m}^{-3}$$

Since $n \gg p$ the conduction electrons are the majority carriers and the holes are the minority carriers.

5.18 From eqns (5.17) and (5.19),

$$\mathcal{E}_H=R_H J_x B_z=\frac{J_x B_z}{Ne}$$

The concentration of conduction electrons is therefore

$$N=\frac{J_x B_z}{\mathcal{E}_H e}$$

To find J_x, the cross-sectional area of the sample is

$$A=(\text{width})(\text{thickness})=(0.01\ \mathrm{m})(0.002\ \mathrm{m})=2\times 10^{-5}\ \mathrm{m}^2$$

The current density is therefore

$$J_x=\frac{\text{current}}{A}=\frac{0.001\ \mathrm{A}}{2\times 10^{-5}\ \mathrm{m}^2}=50\ \mathrm{A\ m}^{-2}$$

So

$$N=\frac{(50\ \mathrm{A\ m}^{-2})(0.2\ \mathrm{T})}{(3\times 10^{-4}\ \mathrm{V})(1.60\times 10^{-19}\ \mathrm{C})}=2.08\times 10^{23}\ \mathrm{m}^{-3}$$

5.19 In the free electron model the energy is given by eqn (5.22):

$$E=\frac{\hbar^2 k^2}{2m_e}$$

Therefore

$$\frac{d^2E}{dk^2}=\frac{\hbar^2}{m_e}$$

Substituting in eqn (5.21), the effective mass is

$$m_e^*=\hbar^2\left(\frac{d^2E}{dk^2}\right)^{-1}=\hbar^2\frac{m_e}{\hbar^2}=m_e$$

Therefore the effective mass, m_e^*, is identical to the free electron mass, m_e.

5.20 From eqn (5.27) the conduction electron concentration is

$$n=2\left(\frac{m_e^* k_B T}{2\pi\hbar^2}\right)^{3/2} e^{(E_F-E_g)/k_B T}$$

Using $E_g = 1.11$ eV $= 1.78\times 10^{-19}$ J, $m_e^* = 0.43\, m_e$ and assuming $E_F = E_g/2$,

$$n = 2\left(\frac{m_e^* k_B T}{2\pi\hbar^2}\right)^{3/2} e^{-E_g/2k_B T}$$

$$= 2\left(\frac{(0.43)(9.11\times10^{-31}\ \text{kg})(1.38\times10^{-23}\ \text{J K}^{-1})(300\ \text{K})}{(2)(3.14)(1.05\times10^{-34}\ \text{J s})^2}\right)^{3/2}$$

$$\times \exp\left(-\frac{1.78\times10^{-19}\ \text{J}}{(2)(1.38\times10^{-23}\ \text{J K}^{-1})(300\ \text{K})}\right)$$

$$= 3.31\times10^{15}\ \text{m}^{-3}$$

5.21 Rearranging eqn (5.27), the Fermi energy is given by

$$E_F = E_g + k_B T\ \ln\left[\frac{n}{2}\left(\frac{2\pi\hbar^2}{m_e^* k_B T}\right)^{3/2}\right]$$

If all of the donors are ionized then $n = 10^{22}\ \text{m}^{-3}$. For silicon $m_e^* = 0.43\ m_e$. Therefore

$$E_F = E_g + (1.38\times10^{-23}\ \text{J K}^{-1})(300\ \text{K})$$

$$\times \ln\left[\frac{1\times10^{22}\ \text{m}^{-3}}{2}\left(\frac{(2)(3.14)(1.05\times10^{-34}\ \text{J s})^2}{(0.43)(9.11\times10^{-31}\text{kg})(1.38\times10^{-23}\ \text{J K}^{-1})(300\ \text{K})}\right)^{3/2}\right]$$

$$= E_g - (2.72\times10^{-20}\ \text{J}) = E_g - (0.170\ \text{eV})$$

i.e. the Fermi energy is 0.17 eV below the conduction band edge.

5.22 Since the Fermi energy is typically much larger than the thermal energy $k_B T$ (except at very high levels of p-type doping), we can write

$$1 - f(E) \approx e^{(E-E_F)/k_B T}$$

(See solution to Question 5.16.) Using this approximation and the density of states function in eqn (5.29), the expression in eqn (5.28) becomes

$$p = \frac{1}{V}\int_{-\infty}^{0} g(E)[1-f(E)]dE = \frac{1}{V}\int_{-\infty}^{0}\frac{V}{2\pi^2}\left(\frac{2m_h^*}{\hbar^2}\right)^{3/2}(-E)^{1/2}e^{(E-E_F)/k_B T}dE$$

$$= C_h\int_{-\infty}^{0}(-E)^{1/2}e^{(E-E_F)/k_B T}dE$$

where

$$C_h = \frac{1}{2\pi^2}\left(\frac{2m_h^*}{\hbar^2}\right)^{3/2}$$

By making the substitution $x = -E/k_B T$ and using the standard integral

$$\int_{-\infty}^{0} -x^{1/2}e^{-x}dx = \frac{1}{2}\sqrt{\pi}$$

$$p = C_h (k_B T)^{3/2} e^{-E_F/k_B T} \int_{-\infty}^{0} -x^{1/2} e^{-x} dx$$

$$= \frac{1}{2\pi^2} \left(\frac{2m_h^*}{\hbar^2} \right)^{3/2} (k_B T)^{3/2} e^{-E_F/k_B T} \frac{1}{2}\sqrt{\pi}$$

$$= 2 \left(\frac{m_h^* k_B T}{2\pi\hbar^2} \right)^{3/2} e^{-E_F/k_B T}$$

5.23 From eqn (5.30) the concentration of holes is given by

$$p = 2 \left(\frac{m_h^* k_B T}{2\pi\hbar^2} \right)^{3/2} e^{-E_F/k_B T}$$

Using $E_g = 1.11$ eV $= 1.78 \times 10^{-19}$ J, $m_h^* = 0.54\, m_e$ and assuming $E_F = E_g/2$,

$$p = 2 \left(\frac{m_h^* k_B T}{2\pi\hbar^2} \right)^{3/2} e^{-E_F/2k_B T}$$

$$= 2 \left(\frac{(0.54)(9.11 \times 10^{-31}\ \text{kg})(1.38 \times 10^{-23}\ \text{J K}^{-1})(300\ \text{K})}{(2)(3.14)(1.05 \times 10^{-34}\ \text{J s})^2} \right)^{3/2}$$

$$\times \exp\left(-\frac{1.78 \times 10^{-19}\ \text{J}}{(2)(1.38 \times 10^{-23}\ \text{J K}^{-1})(300\ \text{K})} \right)$$

$$= 4.65 \times 10^{15}\ \text{m}^{-3}$$

According to eqn (5.6), this value should be exactly the same as the concentration of conduction electrons calculated in Question 5.20. The slight difference in the values is due to the fact that the Fermi energy is not quite at the centre of the band gap as we assumed in these calculations.

5.24 For an intrinsic material the values of n and p in eqns (5.27) and (5.30) should be identical. Therefore

$$2 \left(\frac{m_e^* k_B T}{2\pi\hbar^2} \right)^{3/2} e^{(E_F - E_g)/k_B T} = 2 \left(\frac{m_e^* k_B T}{2\pi\hbar^2} \right)^{3/2} e^{-E_F/k_B T}$$

By cancelling and rearranging we get

$$e^{(2E_F - E_g)/k_B T} = \left(\frac{m_h^*}{m_e^*} \right)^{3/2}$$

Therefore

$$E_F = \frac{E_g}{2} + \frac{3}{4} k_B T \ln\left(\frac{m_h^*}{m_e^*} \right)$$

For indium antimonide $m_e^* = 0.014 m_e$ and $m_h^* = 0.4 m_e$, so at 300 K the shift in the Fermi energy from the centre of the band gap is

$$\Delta E = \frac{3}{4} k_B T \ln\left(\frac{m_h^*}{m_e^*}\right) = \frac{3}{4}(1.38\times 10^{-28}\ \text{J K}^{-1})(300\ \text{K})\ln\left(\frac{0.4\ m_e}{0.014\ m_e}\right)$$

$$= (3.105\times 10^{-21}\ \text{J})\ln(28.6)$$

$$= 1.04\times 10^{-20}\ \text{J} = 0.065\ \text{eV}$$

Since the band gap of indium antimonide is 0.18 eV (see Table 5.1), the Fermi energy is at

$$E_F = \frac{E_g}{2} + \Delta E = 0.156\ \text{eV}$$

i.e. the Fermi energy is much closer to the conduction band than the valence band. For silicon $m_e^* = 0.43m_e$ and $m_h^* = 0.54m_e$, so at 300 K the shift in the Fermi energy from the centre of the band gap is

$$\Delta E = 7.07 \times 10^{-22}\ \text{J} = 4.42 \times 10^{-3}\ \text{eV}$$

So in silicon the difference in energy between the Fermi energy and the centre of the band gap is negligible in comparison with the magnitude of the band gap (i.e. 1.11 eV), and so it is a good assumption to assume that the Fermi energy lies at the centre gap. However, in indium antimonide the shift in energy is much larger than in silicon and the band gap is much smaller, so this assumption is not valid.

Chapter 6

6.1 Consider a single hole diffusing from the p-type side to the n-type side. It recombines with a conduction electron shortly after entering the n-type material. As more holes diffuse across the junction and recombine, there is a build-up of positive charge on the n-type side of the junction, whilst on the p-type side there are many uncompensated acceptor ions producing a net negative charge. We therefore end up with the same charge distribution as shown in Fig. 6.2(b).

6.2 The energy band profile is as follows:

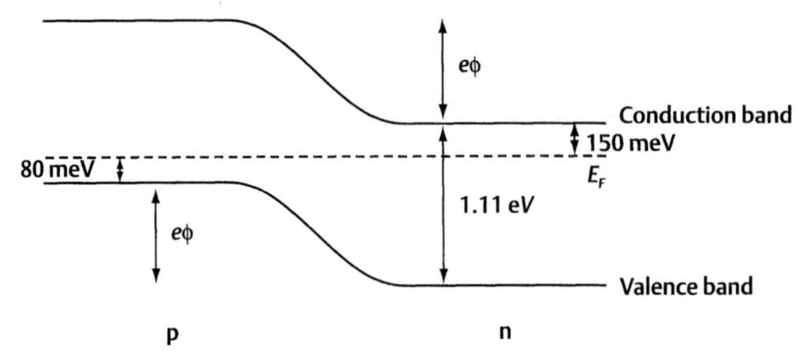

From the diagram, the value of $e\phi$ is

$$e\phi = (1.11\ \text{eV}) - (0.15\ \text{eV}) - (0.08\ \text{eV}) = 0.88\ \text{eV}$$

Therefore the contact potential is $\phi = 0.88$ V.

6.3 Let us denote the potential as U_0, U_1 and U_2 at $x = 0$, l_n and $-l_p$, respectively, as shown in Fig. 6.3(b).
Integrating eqn (1) from Example 6.1 we have

$$\int_{U_0}^{U_1} dU = \frac{\rho_n}{\varepsilon\varepsilon_0}\int_0^{l_n}(l_n - x)\,dx$$

$$[U]_{U_0}^{U_1} = \frac{\rho_n}{\varepsilon\varepsilon_0}\left[l_n x - \frac{x^2}{2}\right]_0^{l_n}$$

$$U_1 - U_0 = \frac{\rho_n}{\varepsilon\varepsilon_0}\frac{l_n^2}{2}$$

Substituting for ρ_n from eqn (6.3) gives

$$U_0 - U_1 = \frac{eN_d l_n^2}{2\,\varepsilon\varepsilon_0}$$

Similarly, using eqn (2) from Example 6.1 we obtain

$$U_2 - U_0 = \frac{eN_a l_p^2}{2\,\varepsilon\varepsilon_0}$$

From Fig. 6.3(b) we can see that the contact potential is therefore

$$\phi = U_2 - U_1 = (U_2 - U_0) + (U_0 - U_1)\frac{e}{2\,\varepsilon\varepsilon_0}\left(N_d l_n^2 + N_a l_p^2\right)$$

6.4 From eqn (6.5) we have

$$l_n = \frac{N_a l_p}{N_d}$$

Substituting in eqn (6.6) gives

$$\phi = \frac{e}{2\,\varepsilon\varepsilon_0}\left(\frac{N_a^2 l_p^2}{N_d} + N_a l_p^2\right)$$
$$= \frac{eN_a}{2\,\varepsilon\varepsilon_0}\frac{N_a + N_d}{N_d}l_p^2$$

Rearranging, we get an expression for the depletion layer width on the p-type side:

$$l_p = \left(\frac{\phi 2\,\varepsilon\varepsilon_0}{eN_a}\frac{N_d}{N_a + N_d}\right)^{1/2}$$

Performing a similar analysis for the n-type side we can show that

$$l_n = \left(\frac{\phi 2\,\varepsilon\varepsilon_0}{eN_d}\frac{N_a}{N_a + N_d}\right)^{1/2}$$

6.5 We can obtain expressions for the carrier concentrations, n_n, p_n and p_p, by integrating the relevant occupied density of states function, as in Examples 5.9 and 6.2. However, in general we can see that the concentration of carriers with an energy greater than E is proportional to $e^{-\Delta E/k_BT}$ where ΔE is the difference in energy between E and the Fermi energy E_F.

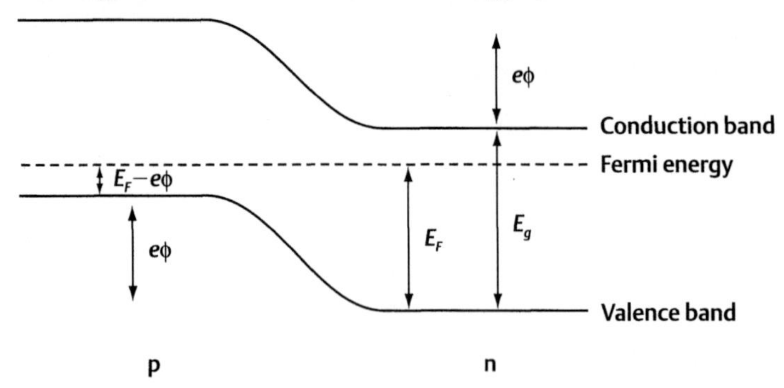

Using the above diagram, when E is the conduction band edge in the n-type region, $\Delta E = E_g - E_F$. Therefore

$$n_n = N_c e^{(E_F - E_g)/k_B T}$$

where N_c is a constant.
Similarly, for the valence band edge on the n-type side, $\Delta E = E_F$. Therefore

$$p_n = N_v e^{-E_F/k_B T}$$

For the valence band edge on the p-type side, $\Delta E = E_F - e\phi$. Therefore

$$p_p = N_v e^{(e\phi - E_F)/k_B T}$$

We can also see that for the conduction band edge on the p-type side, $\Delta E = E_g + e\phi - E_F$. Therefore

$$n_p = N_c e^{(E_F - E_g - e\phi)/k_B T}$$

as shown in Example 6.2.

6.6 Assuming that all of the impurities are ionized, eqn (6.12) becomes

$$p_p = N_a = N_v e^{(e\phi - E_F)/k_B T}$$

Rearranging gives

$$e\phi - E_F = k_B T \ln\left(\frac{N_a}{N_v}\right)$$

Similarly, using eqn (6.10) and setting $n_n = N_d$ we have

$$E_F - E_g = k_B T \ln\left(\frac{N_d}{N_c}\right)$$

Adding these two expressions gives

$$e\phi - E_g = k_B T \ln\left(\frac{N_d N_a}{N_c N_v}\right) \tag{1}$$

Now we know that the product of the electron and hole concentrations in a material is a constant (see Example 5.6), so using eqns (6.10) and (6.11) we have

$$n_n p_n = N_c N_v e^{-E_g/k_B T} = n_i^2$$

Rearranging gives

$$N_c N_v = n_i^2 e^{E_g/k_B T}$$

Substituting in eqn (1)

$$e\phi - E_g = k_B T\left[\ln\left(\frac{N_d N_a}{n_i^2}\right) - \frac{E_g}{k_B T}\right]$$

Therefore

$$e\phi = k_B T \ln\left(\frac{N_d N_a}{n_i^2}\right)$$

6.7 From eqn (6.13) the contact potential is

$$\phi = \frac{k_B T}{e}\ln\left(\frac{N_a N_d}{n_i^2}\right)$$

$$= \frac{(1.38\times10^{-23}\ \mathrm{J\ K^{-1}})(300\ \mathrm{K})}{1.60\times10^{-19}\ \mathrm{C}}\ln\left(\frac{(4\times10^{21}\ \mathrm{m^{-3}})(2\times10^{22}\ \mathrm{m^{-3}})}{(1.0\times10^{16}\ \mathrm{m^{-3}})^2}\right)$$

$$= 0.709\ \mathrm{V}$$

Using eqns (6.7) and (6.8) the depletion layer widths are

$$l_p = \left(\frac{\phi 2\varepsilon\varepsilon_0}{eN_a}\frac{N_d}{N_a+N_d}\right)^{1/2}$$

$$= \left(\frac{(0.709\ \mathrm{V})(2.0)(11.7)(8.85\times10^{-12}\ \mathrm{F\ m^{-1}})}{(1.60\times10^{-19}\ \mathrm{C})(4\times10^{21}\ \mathrm{m^{-3}})}\times\frac{2\times10^{22}\ \mathrm{m^{-3}}}{(4\times10^{21}\ \mathrm{m^{-3}})+(2\times10^{22}\ \mathrm{m^{-3}})}\right)^{1/2}$$

$$= 4.37\times10^{-7}\ \mathrm{m}$$

and

$$l_p = \left(\frac{\phi 2\varepsilon\varepsilon_0}{eN_d}\frac{N_a}{N_a+N_d}\right)^{1/2}$$

$$= \left(\frac{(0.709\ \mathrm{V})(2.0)(11.7)(8.85\times10^{-12}\ \mathrm{F\ m^{-1}})}{(1.60\times10^{-19}\ \mathrm{C})(2\times10^{22}\ \mathrm{m^{-3}})}\times\frac{4\times10^{21}\ \mathrm{m^{-3}}}{(4\times10^{21}\ \mathrm{m^{-3}})+(2\times10^{22}\ \mathrm{m^{-3}})}\right)^{1/2}$$

$$= 8.74\times10^{-8}\ \mathrm{m}$$

6.8 As in Example 6.3, the drift current is the same as in the zero-voltage case, i.e.

$$I_{drift} = Ce^{(E_F - E_g - e\phi)/k_B T} = I_0$$

The diffusion current is given by eqn (6.18):

$$I_{diffusion} = Ce^{(E_F - E_g - e\phi - eV)/k_B T} = I_0 e^{-eV/k_B T}$$

Therefore, the net current is

$$I = I_{diffusion} - I_{drift} = I_0 e^{-eV/k_B T} - I_0 = I_0 (e^{-eV/k_B T} - 1)$$

6.9 For a forward bias use eqn (6.17):

$$I = I_0 (e^{eV/k_B T} - 1)$$

and for a reverse bias use eqn (6.19):

$$I = I_0 (e^{-eV/k_B T} - 1)$$

(Note – in each case V represents the magnitude of the applied voltage.)
This should give the graph shown in Fig. 6.6.

6.10 When a reverse bias is applied the current is given by eqn (6.19):

$$I = I_0 (e^{-eV/k_B T} - 1)$$

When V is 0.3 V, the term e^{-eV/k_BT} is equal to 9.2×10^{-6}, so $I \approx -I_0$. Therefore, from the information in the question, $I_0 = 50 \times 10^{-9}$ A.
When a forward bias of 0.3 V is applied, the current is given by eqn (6.17), i.e.

$$I = I_0 (e^{eV/k_B T} - 1)$$

$$= (50 \times 10^{-9}\ \text{A}) \left[\exp\left(\frac{(1.60 \times 10^{-19}\ \text{C})(0.3\ \text{V})}{(1.38 \times 10^{-23}\ \text{J K}^{-1})(300\ \text{K})} \right) - 1 \right]$$

$$= 5.42 \times 10^{-3}\ \text{A}$$

6.11 From eqn (6.15) we see that the current I_0 is proportional to $e^{(E_F - E_g - e\phi)/k_BT}$.
If we assume that $E_g + e\phi - E_F = 1$ eV (i.e. 1.60×10^{-19} J), then at 300 K we have

$$\exp\left(\frac{E_F - E_g - e\phi}{k_B T} \right) = \exp\left(-\frac{1.60 \times 10^{-19}\ \text{J}}{(1.38 \times 10^{-23}\ \text{J K}^{-1})(300\text{K})} \right) = 1.64 \times 10^{-17}$$

and at 310 K the term is equal to 5.72×10^{-17}.
Therefore we expect I_0 to increase by a factor of

$$\frac{5.72 \times 10^{-17}}{1.64 \times 10^{-17}} \approx 3.5$$

i.e. $I_0 \approx (3.5)(50 \times 10^{-9}\ \text{A}) = 1.75 \times 10^{-7}\ \text{A}$.
(Note that although the value of the exponential is highly sensitive to the value we assume for $E_g + e\phi - E_F$, the ratio of these quantities does not vary significantly and so the value of I_0 should be reasonably accurate.)
As shown in the solution to Question 6.10, the reverse bias current is approximately equal to $-I_0$.
For the forward bias current we use eqn (6.17). Therefore

$$I = I_0(e^{eV/k_BT} - 1)$$

$$= (1.75\times 10^{-7}\ \text{A})\left[\exp\left(\frac{(1.60\times 10^{-19}\ \text{C})(0.3\ \text{V})}{(1.38\times 10^{-23}\ \text{J K}^{-1})(310\ \text{K})}\right) - 1\right]$$

$$= 1.31\times 10^{-2}\ \text{A}$$

i.e. the forward bias current increases by a factor of

$$\frac{1.31\times 10^{-2}\ \text{A}}{5.42\times 10^{-3}\ \text{A}} = 2.4$$

6.12 From eqn (6.20), the current gain is given by

$$\beta = \frac{\Delta i_C}{\Delta i_B}$$

If 0.3% of the electrons from the emitter recombine in the base, then $i_B = 0.003 i_E$ and $i_C = 0.997 i_E$. Assuming that both the base and collector currents are zero when i_E is zero, we can write

$$\beta = \frac{\Delta i_C}{\Delta i_B} = \frac{i_C}{i_B} = \frac{0.997 i_E}{0.003 i_E} = 332$$

6.13 We want to find out how long it takes for the minimum feature size to decrease to 0.1 μm, i.e. to reduce by a factor of

$$\frac{0.1\ \mu\text{m}}{0.35\ \mu\text{m}} = 0.286$$

A decrease of 11% means that the minimum feature size changes by a factor of 0.89 per annum. Therefore, if it takes N years for the minimum feature size to decrease by a factor of 0.286, then

$$(0.89)^N = 0.286$$

Therefore, $N = 10.7 \approx 11$ years, i.e. the minimum feature size will reach 0.1 μm by the year 2006.
During this time the area of the circuit is expected to increase by a factor of

$$(1.09)^{11} = 2.58$$

Since the number of devices per unit area increases by a factor of

$$\left(\frac{0.35\,\mu\text{m}}{0.1\,\mu\text{m}}\right)^2 = 12.3$$

the number of devices per circuit in 2006 is expected to be

$$(64\times 10^6)(2.58)(12.3) = 2.0\times 10^9$$

6.14 Following Example 6.4, we know that the switching time, t, is given by

$$t \propto \frac{L}{\overline{\Delta v}}$$

where L is the gate length and $\overline{\Delta v}$ is the drift velocity of the electrons. Scaling linear dimensions by $1/S$ means that

$$L \Rightarrow \frac{L}{S}$$

As in Example 6.4 we have

$$\overline{\Delta v} = \frac{e\mathcal{E}\tau}{m}$$

We assume that the relaxation time τ is constant, but in this instance the electric field $\mathcal{E}$ changes with scaling. Since the voltage remains constant as the dimensions change by $1/S$, we have $\mathcal{E} \Rightarrow S\mathcal{E}$; therefore $\overline{\Delta v} \Rightarrow S\,\overline{\Delta v}$. We therefore have

$$t \Rightarrow \frac{1}{S^2}$$

The scaling of the power per unit area also follows the discussion in Example 6.4. The power dissipated per device is

$$P = IV$$

where

$$I = n\overline{\Delta v}eA$$

Since $n \Rightarrow Sn$ and $A \Rightarrow A/S^2$,

$$I = Sn.S\overline{\Delta v}.e.\frac{A}{S^2} \Rightarrow I$$

In this instance the voltage is constant, and therefore the power per device remains constant. However, since the number of devices per unit area increases by S^2, the power dissipated per unit area also inceases by S^2.

6.15 Following the discussion in Example 6.4, if $n \Rightarrow S^3n$ then the current is given by

$$I \Rightarrow S^3 n.\overline{\Delta v}.e.\frac{A}{S^2} \Rightarrow SI$$

where we have assumed that constant field scaling results in the drift velocity $\overline{\Delta v}$ remaining constant with scaling (see Example 6.4).
Since $V \Rightarrow V/S$, the power dissipated per device is

$$P \Rightarrow SI\frac{V}{S} \Rightarrow P$$

i.e. the power dissipated per device is constant, and therefore the power dissipated per unit area increases as S^2.

6.16 Following the solution to Question 6.7, the contact potential is $\phi = 0.840$ V, and the depletion layer widths are $l_p = 2.04 \times 10^{-7}$ m and $l_n = 1.02 \times 10^{-8}$ m. The total width of the depletion layer in the gate region is $2l_p = 4.08 \times 10^{-7}$ m, as shown below.

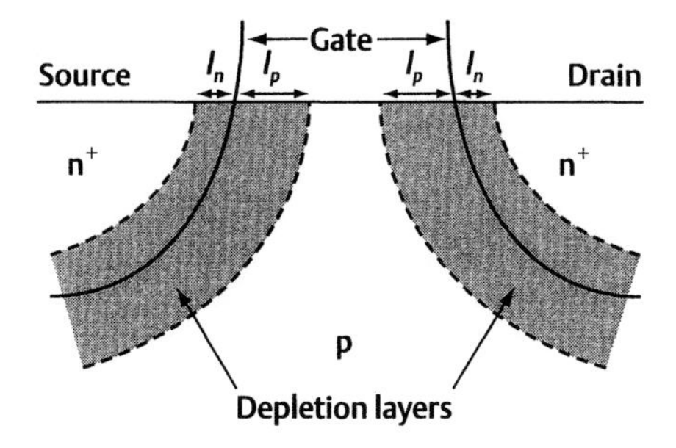

If the doping concentrations are scaled by S, i.e. $N_a \Rightarrow S\,N_a$ and $N_d \Rightarrow S\,N_d$, and if ϕ is assumed to be constant with scaling, then from eqn (6.7) the scaling of the depletion layers in the gate region is given by

$$l_p \Rightarrow \left(\text{const.}\frac{1}{SN_a}\frac{SN_d}{SN_a + SN_d}\right)^{1/2} \Rightarrow \frac{l_p}{S^{1/2}}$$

The minimum gate length occurs when the total depletion layer width, $2l_p$, is equal to the gate width. So we want to find the value of S such that

$$\frac{\text{total depletion layer width}}{\text{depletion layer scaling}} = \frac{\text{gate width}}{\text{length scaling}}$$

i.e.

$$\frac{0.408\,\mu\text{m}}{S^{1/2}} = \frac{1.0\,\mu\text{m}}{S}$$

Therefore

$$S^{1/2} = \frac{1.0\ \mu m}{0.408\ \mu m} \Rightarrow S = 6.01$$

i.e. the minimum gate width is $(1.0\ \mu\text{m})/(6.01) = 0.166\ \mu\text{m}$

6.17 (i) From the first diagram below the conduction band offset, ΔE_c, is given by

$$\Delta E_c = E_g\,(\text{AlGaAs}) - E_g\,(\text{GaAs}) - \Delta E_v$$
$$= (1.80\ \text{eV}) + (1.42\ \text{eV}) - (0.14\ \text{eV}) = 0.24\ \text{eV}$$

The junction is type I.

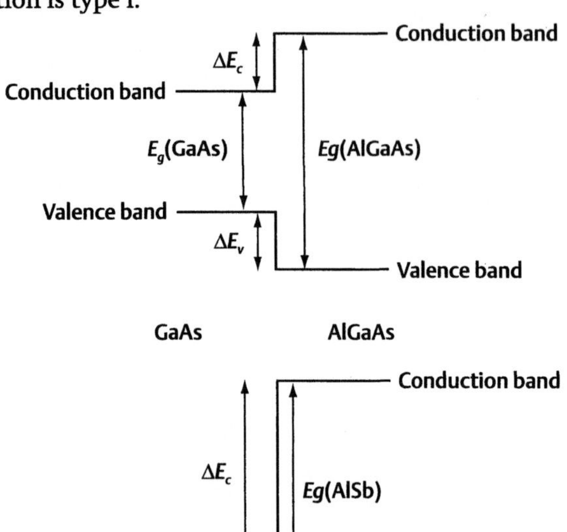

Conduction band
ΔE_c
Eg(AlSb)
Conduction band
Valence band
E_g(InAs)
ΔE_v
Valence band
InAs AlSb

(ii) From the second diagram the valence band offset, ΔE_v, is

$$\Delta E_v = E_g\,(\text{InAs}) + \Delta E_c - E_g\,(\text{AlSb})$$
$$= (0.36\ \text{eV}) + (1.35\ \text{eV}) - (1.58\ \text{eV}) = 0.13\ \text{eV}$$

The junction is type II.

6.18 There are two factors which favour the use of a reverse-biased p–n junction. Firstly, from Question 6.10 we know that the magnitude of the current across a reverse-biased p–n junction in the absence of light is typically about 50×10^{-9} A. When illuminated the current increases to about 0.1×10^{-3} A, an increase by a factor of 20 000. It is therefore very easy to detect the change in current due to the presence of light.

In contrast, if the photodiode is forward biased then the typical current flowing in the absence of light is a few milliamps (see solution to Question 6.10). The change in current when the diode is illuminated is therefore less than 10%, and so it is much harder to determine whether or not the diode is illuminated.

Secondly, we can see from the solution to Question 6.11 that a small change in temperature produces a change in the forward bias current of a few milliamps, i.e. this change is much larger than that produced by absorption of

light. Therefore, we would have to keep the temperature of the device almost constant in order to use it to detect light.
In comparison, although the percentage change in the reverse bias current is also large, the magnitude of the change is much smaller than the photocurrent.

6.19 (i) From eqn (5.7), the band gap required to emit light of wavelength 680 nm is

$$E_g = \frac{hc}{\lambda} = \frac{(6.63\times 10^{-34}\,\text{J s})(3.00\times 10^{8}\,\text{m s}^{-1})}{680\times 10^{-9}\,\text{m}} = 2.93\times 10^{-19}\,\text{J} = 1.83\ \text{eV}$$

If the band gap is given by $E_g = (1.42 + 1.3x)$ eV, then

$$x = \frac{E_g - (1.42\ \text{eV})}{1.3} = 0.32$$

(ii) The maximum band gap of a $GaAs_{1-x}P_x$ alloy is

$$E_g = (1.42 + (1.3 \times 0.44))\ \text{eV} = 1.99\ \text{eV} = 3.18 \times 10^{-19}\ \text{J}$$

The corresponding maximum wavelength of light which can be emitted from this material is therefore

$$\lambda_{max} = \frac{hc}{E_g} = \frac{(6.63\times 10^{-34}\ \text{J s})(3.00\times 10^{8}\ \text{m s}^{-1})}{(3.18\times 10^{-19}\ \text{J})} = 6.25\times 10^{-7}\ \text{m}$$

Chapter 7

7.1 Equation (7.1) has solutions of the form $x = A\cos(\omega t)$. Substituting into eqn (7.1) gives

$$-mA\omega^2\cos(\omega t) = -\gamma A\cos(\omega t)$$

Cancelling and rearranging gives

$$\omega = \sqrt{\gamma / m}$$

7.2 For diamond: from eqn (7.3), γ is given by

$$\gamma = Ya_0 = (9.5\times 10^{11}\ \text{N m}^{-2})(0.154\times 10^{-9}\ \text{m}) = 146\ \text{N m}^{-1}$$

and from eqn (7.2) we can find ω. The mass of a carbon atom is 12 u = 12 × (1.66 × 10^{-27} kg) = 2.00 × 10^{-26} kg. Therefore

$$\omega = \sqrt{\frac{\gamma}{m}} = \sqrt{\frac{146\ \text{N m}^{-1}}{2.00\times 10^{-26}\,\text{kg}}} = 8.54\times 10^{13}\,\text{rad s}^{-1}$$

Similarly, substituting values for copper we have $\gamma = 33.3$ N m^{-1}, $\omega = 1.77 \times 10^{13}$ rad s^{-1}, and for lead $\gamma = 5.25$ N m^{-1}, $\omega = 3.90 \times 10^{12}$ rad s^{-1}.

7.3 For any harmonic oscillator the potential energy at a distance x from the equilibrium position is $\frac{1}{2}\gamma x^2$, where γ is the force constant. At the maximum amplitude, x_{max}, all of the energy of the oscillator is potential energy. Therefore

$$\frac{1}{2}\gamma x_{max}^2 = k_B T$$

Rearranging gives

$$x_{max} = \left(\frac{2k_B T}{\gamma}\right)^{1/2}$$

7.4 Using eqn (7.4) the amplitude of the vibration is

$$x_{max} = \left(\frac{2k_B T}{\gamma}\right)^{1/2} = \left(\frac{(2)(1.38\times 10^{-23}\,\mathrm{J\,K^{-1}})(300\,\mathrm{K})}{100\,\mathrm{N\,m^{-1}}}\right)^{1/2} = 9.10\times 10^{-12}\,\mathrm{m}$$

As a fraction of the atomic spacing this is

$$\frac{9.10\times 10^{-12}\,\mathrm{m}}{0.256\times 10^{-9}\,\mathrm{m}} \times 100\,\% \approx 3.5\%$$

7.5 Using eqn (7.5) the change in length is

$$\Delta L = \alpha L_0 \Delta T = (1.1\times 10^{-5}\,\mathrm{K^{-1}})(20\,\mathrm{m})(30\,\mathrm{K}) = 6.6\times 10^{-3}\,\mathrm{m}$$

Using eqn (3.2) the strain ϵ is

$$\epsilon = \frac{\Delta L}{L_0} = \frac{6.6\times 10^{-3}\,\mathrm{m}}{20\,\mathrm{m}} = 3.3\times 10^{-4}$$

and from eqn (3.3) the stress σ is

$$\sigma = Y\epsilon = (2.1\times 10^{11}\,\mathrm{N\,m^{-2}})(3.3\times 10^{-4}) = 6.93\times 10^{7}\,\mathrm{N\,m^{-2}}$$

Since this value exceeds the yield stress it will produce a permanent deformation.

7.6 From eqn (7.5) and using the data in Table 7.1, the extension of the steel strip is

$$\Delta L_s = \alpha_s L_0 \Delta T = (1.1\times 10^{-5}\,\mathrm{K^{-1}})(0.2\,\mathrm{m})(10\,\mathrm{K}) = 2.2\times 10^{-5}\,\mathrm{m}$$

Similarly, for the aluminium strip, $\Delta L_a = 4.8\times 10^{-5}$ m.
If the radius of curvature of the strip is r, then from Fig. 7.15 the length of each strip is equal to the length of the arc which subtends an angle θ, i.e.

$$r\theta = L_0 + \Delta L_s$$

and

$$(r+\Delta r)\theta = L_0 + \Delta L_a$$

Eliminating θ from these two equations we have

$$r=\left(\frac{L_0+\Delta L_s}{\Delta L_a-\Delta L_s}\right)\Delta r=\left(\frac{(0.2\ \text{m})+(2.2\times10^{-5}\ \text{m})}{(4.8\times10^{-5}\ \text{m})-(2.2\times10^{-5}\ \text{m})}\right)(1\times10^{-3}\ \text{m})$$

$$=7.69\ \text{m}$$

The angle θ is then

$$\theta=\frac{L_0+\Delta L_s}{r}=\frac{(0.2\ \text{m})+(2.2\times10^{-5}\ \text{m})}{7.69\ \text{m}}=0.026\ \text{rad}$$

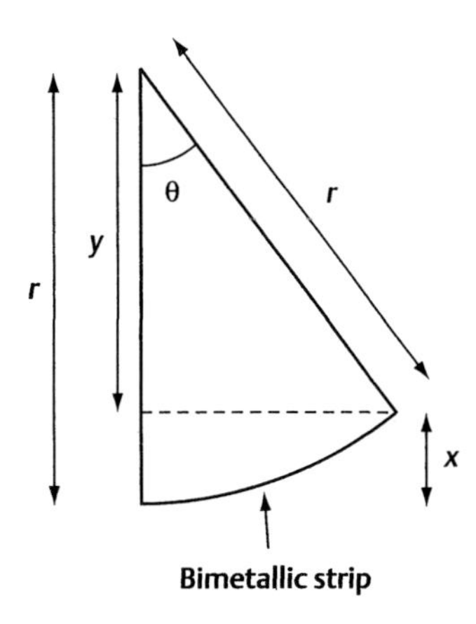

From this diagram we can see that the deflection of the free end, x, is

$$x=r-y=r(1-\cos\theta)=(7.69\ \text{m})[1-\cos(0.026\ \text{rad})]=2.60\times10^{-3}\,\text{m}$$

7.7 Using eqn (2.2), at 300 K the atomic spacing is

$$d_{300}=\frac{n\lambda}{2\sin\theta}=\frac{n\lambda}{2\sin(15.46^\circ)}=1.8757\,n\lambda$$

where λ is the wavelength of the X-rays.
Similarly, at 800 K we have d_{800} = 1.8985$n\lambda$.
The difference is therefore $\Delta d=d_{800}-d_{300}=(1.8985n\lambda)-(1.8757n\lambda)=0.0228n\lambda$
To find the expansion coefficient we assume that the fractional change in length of the sample is equal to the fractional change in the atomic spacing, i.e.

$$\frac{\Delta L}{L_0}=\frac{\Delta d}{d_{300}}$$

Therefore, from eqn (7.5),

$$\alpha=\frac{\Delta L}{L_0}\frac{1}{\Delta T}=\frac{\Delta d}{d_{300}}\frac{1}{\Delta T}=\frac{0.0228n\lambda}{1.8757n\lambda}\frac{1}{500\ \text{K}}=2.43\times10^{-5}\,\text{K}^{-1}$$

7.8 From eqn (7.13) the internal energy is

$$U=3N_A\frac{\hbar\omega}{e^{\hbar\omega/k_BT}-1}$$

To obtain the specific heat capacity we take the partial derivative

$$C=\frac{\partial U}{\partial T}$$

To obtain the derivative let us begin by considering the denominator of this expression. If we write $r = e^{\hbar\omega/k_BT} - 1$ and $s = \hbar\omega/k_BT$, so $r = e^s - 1$, then

$$\frac{\partial r}{\partial s}=e^s \quad \text{and} \quad \frac{\partial s}{\partial T}=-\frac{\hbar\omega}{k_BT^2}$$

Therefore

$$\frac{\partial r}{\partial T}=\frac{\partial r}{\partial s}\frac{\partial s}{\partial T}=-\frac{\hbar\omega}{k_BT^2}e^{\hbar\omega/k_BT}$$

If we now consider the expression for U we have $U = 3N_A\hbar\omega r^{-1}$ and so

$$\frac{\partial U}{\partial r}=-3N_A\hbar\omega r^{-2}$$

Therefore

$$\frac{\partial U}{\partial T}=\frac{\partial U}{\partial r}\frac{\partial r}{\partial T}=3N_A\hbar\omega(e^{\hbar\omega/k_BT}-1)^{-2}\frac{\hbar\omega}{k_BT^2}e^{\hbar\omega/k_BT}$$

$$=3N_Ak_B\left(\frac{\hbar\omega}{k_BT}\right)^2\frac{e^{\hbar\omega/k_BT}}{(e^{\hbar\omega/k_BT}-1)^2}$$

Substituting $R = N_a k_B$ gives the same form as in eqn (7.14).

7.9 When $k_BT \gg \hbar\omega$, then

$$e^{\hbar\omega/k_BT}\approx 1+\frac{\hbar\omega}{k_BT}$$

Therefore

$$\frac{\hbar\omega}{e^{\hbar\omega/k_BT}-1}\approx k_BT$$

If we substitute this expression in eqn (7.13) we obtain

$$U=3N_A\frac{\hbar\omega}{e^{\hbar\omega/k_BT}-1}\approx 3N_Ak_BT=3RT$$

which is the classical value.
The classical value is obtained because under these conditions the distribution of the discrete quantum states becomes approximately the same as the continuous classical distribution, as shown in Fig. 7.8.

7.10 At the Einstein temperature

$$\frac{\hbar w}{k_B T} = 1$$

Substituting in eqn (7.14) we have

$$C = 3R\left(\frac{\hbar\omega}{k_B T}\right)^2 \frac{e^{\hbar\omega/k_B T}}{(e^{\hbar\omega/k_B T} - 1)^2} = 3R(1)^2 \frac{e^1}{(e^1 - 1)^2} = 0.921 \times 3R$$

7.11 Substituting into eqn (7.14) we obtain the following sample values.

Temperature (K)	50	75	100	200	300	500
$C(\mathrm{J\,K^{-1}})$	0.72	4.07	8.41	18.60	21.84	23.76

Plotting these values gives the curve shown in Fig. 7.7.

7.12 From eqn (7.2) the force constant γ is

$$\gamma = \omega^2 m$$

We can obtain a value for ω from eqn (7.15). Assuming that $\theta_E = \theta_D$, then

$$\omega = \frac{\theta_D k_B}{\hbar}$$

Therefore

$$\gamma = \left(\frac{\theta_D k_B}{\hbar}\right)^2 m$$

For diamond, $m = 12\ \mathrm{u} = 12 \times (1.66 \times 10^{-27}\ \mathrm{kg}) = 2.00 \times 10^{-26}\ \mathrm{kg}$; therefore using data from Table 7.3,

$$\gamma = \left(\frac{(1850\ \mathrm{K})(1.38 \times 10^{-23}\ \mathrm{J\,K^{-1}})}{(1.05 \times 10^{-34}\ \mathrm{J\,s})}\right)^2 (2.00 \times 10^{-26}\ \mathrm{kg}) = 1182\ \mathrm{N\,m^{-1}}$$

Similarly, for copper, $m = 64$ u and $\theta_D = 345$ K; therefore $\gamma = 218\ \mathrm{N\,m^{-1}}$.
For lead, $m = 207$ u and $\theta_D = 95$ K; therefore $\gamma = 53.6\ \mathrm{N\,m^{-1}}$.
Note that these values are approximately a factor of 10 times larger than those determined in Question 7.2. Considering the approximations made in Example 7.1, it is expected that these values are more accurate than those obtained in Question 7.2.

7.13 Rearranging eqn (7.14) we have

$$C = 3R\left(\frac{\hbar\omega}{k_B T}\right)^2 e^{\hbar\omega/k_B T}(e^{\hbar\omega/k_B T} - 1)^{-2}$$

$$= 3R\left(\frac{\hbar\omega}{k_B T}\right)^2 e^{-\hbar\omega/k_B T}(1 - e^{-\hbar\omega/k_B T})^{-2}$$

$$= 3R\left(\frac{\hbar\omega}{k_B T}\right)^2 e^{-\hbar\omega/k_B T}(1 + 2e^{-\hbar\omega/k_B T} - e^{-2\hbar\omega/k_B T} + \cdots)$$

When T is small, $e^{-\hbar\omega/k_BT} \ll 1$. Therefore

$$C \approx 3R\left(\frac{\hbar\omega}{k_B T}\right)^2 e^{-\hbar\omega/k_B T}$$

As $T \to 0$, the exponential term goes to zero much faster than the T^{-2} term; therefore

$$C \propto e^{-\hbar\omega/k_B T}$$

7.14 From Fig. 7.9, $\psi = \sin n_x\pi/L$ has

- two nodes (at $x = 0$ and $x = L$) when $n_x = 1$,
- three nodes (at $x = 0$, $x = L/2$ and $x = L$) when $n_x = 2$, and
- four nodes (at $x = 0$, $x = L/3$, $x = 2L/3$ and $x = L$) when $n_x = 3$.

7.15 From eqn (7.19), the maximum vibrational frequency of the ions is

$$\omega_{max} = \frac{\theta_D k_B}{\hbar} = \frac{(345\ \text{K})(1.38\times 10^{-23}\ \text{J K}^{-1})}{(1.05\times 10^{-34}\ \text{J s})} = 4.53\times 10^{13}\ \text{rad s}^{-1}$$

The wavelength λ is related to the wavevector k by $\lambda = 2\pi/k$, and the wavevector is related to the angular frequency by $k = \omega/v$, where v is the velocity of the wave. Therefore, the minimum vibrational wavelength is

$$\lambda_{min} = \frac{2\pi}{k_{max}} = \frac{2\pi v}{\omega_{max}} = \frac{(2)(3.14)(4000\ \text{m s}^{-1})}{4.53\times 10^{13}\ \text{rad s}^{-1}} = 5.54\times 10^{-10}\ \text{m}$$

This is approximately equal to twice the atomic spacing.

7.16 From eqns (7.17) and (7.16) the internal energy is

$$U = 3\int_0^{\omega_{max}} \frac{g(\omega)\hbar\omega}{e^{\hbar\omega/k_B T} - 1}\,d\omega = \frac{3V}{2\pi^2 v^3}\int_0^{\omega_{max}} \frac{\hbar\omega^3}{e^{\hbar\omega/k_B T} - 1}\,d\omega$$

Using eqn (1) from Example 7.4 we can write this as

$$U = \frac{9N}{\omega_{max}^3}\int_0^{\omega_{max}} \frac{\hbar\omega^3}{e^{\hbar\omega/k_B T} - 1}\,d\omega$$

Substituting $x = \hbar\omega/k_BT$ we get

$$U = \frac{9Nk_B^4T^4}{\omega_{max}^3\hbar^3}\int_0^{x_{max}} \frac{x^3}{e^x - 1}\,dx$$

Using the definition for θ_D from eqn (7.19), i.e. $\theta_D = \hbar\omega_{max}/k_B$, the upper limit on the integral is

$$x_{max} = \frac{\hbar\omega_{max}}{k_B T} = \frac{\theta_D}{T}$$

As $T \to 0$, $x_{max} \to \infty$, and therefore the integral approximates to the result for the standard integral:

$$\int_0^{x_{max}} \frac{x^3}{e^x - 1} dx \approx \int_0^{\infty} \frac{x^3}{e^x - 1} dx = \frac{\pi^4}{15}$$

So at low temperatures we have

$$U \approx \frac{9Nk_B^4T^4}{\omega_{max}^3 \hbar^3} \cdot \frac{\pi^4}{15} = \frac{3Nk_B T^4 \pi^4}{5\theta_D^3}$$

where we have again used eqn (7.19) to write ω_{max} in terms of θ_D.
The specific heat capacity is found by taking the partial derivative with respect to T and putting $N = N_A$. Therefore

$$C = \frac{\partial U}{\partial T} = \frac{12N_A k_B T^3 \pi^4}{5\theta_D^3} = \frac{12R\pi^4}{5}\left(\frac{T}{\theta_D}\right)^3$$

where we have used $R = N_A k_B$.

7.17 From eqn (4.19) the density of states at the Fermi energy is

$$g(E_F) = \frac{V}{2\pi^2}\left(\frac{2m_e}{\hbar^2}\right)^{3/2} E^{1/2}$$

Susbtituting for V from eqn (4.17), i.e.

$$V = \left(\frac{\hbar^2}{2m_e E_F}\right)^{3/2} 3\pi^2 N$$

gives

$$g(E_F) = \left(\frac{\hbar^2}{2m_e E_F}\right)^{3/2} \frac{3\pi^2 N}{2\pi^2}\left(\frac{2m_e}{\hbar^2}\right)^{3/2} E_F^{1/2} = \frac{3N}{2E_F}$$

7.18 From eqn (7.20) we have

$$C_e = \frac{9}{2}\frac{k_B^2 TN}{E_F}$$

For 1 mole of copper, $N = N_A = 6.02 \times 10^{23}$ mol^{-1} and $E_F = 7.00$ eV $= 1.12 \times 10^{-18}$ J; therefore

$$C_e = \frac{9}{2}\frac{(1.38 \times 10^{-23}\ \text{J K}^{-1})^2 (295\ \text{K})(6.02 \times 10^{23}\ \text{mol}^{-1})}{1.12 \times 10^{-18}\ \text{J}} = 0.14\ \text{J mol}^{-1}\text{K}^{-1}$$

For 1 mole of zinc, $N = 2N_A$ and $E_F = 9.38$ eV; therefore $C_e = 0.20$ J mol^{-1} K^{-1}.
For 1 mole of aluminium, $N = 3N_A$ and $E_F = 11.58$ eV; therefore $C_e = 0.25$ J mol^{-1} K^{-1}.

7.19 The electronic specific heat capacity for a monovalent metal (i.e. with $N = N_A$) is given by eqn (7.20):

$$C_e = \frac{9}{2}\frac{k_B^2 T N_A}{E_F}$$

At room temperature we assume that the ionic contribution to the heat capacity is $C = 3R = 3N_A k_B$. Therefore the ratio is

$$\frac{C_e}{C} = \frac{9}{2}\frac{k_B^2 T N_A}{E_F}\frac{1}{3N_A k_B} = \frac{3}{2}\frac{k_B T}{E_F}$$

7.20 Assuming that the electronic heat capacity is equal to the ionic heat capacity at a very low temperature, then we can use eqn (7.18) for the heat capacity of the ions. The temperature at which the electronic and ionic contributions are equal is therefore found by equating C in eqn (7.18) with C_e in eqn (7.20), i.e.

$$\frac{12R\pi^4}{5}\left(\frac{T}{\theta_D}\right)^3 = \frac{9}{2}\frac{k_B^2 T N}{E_F}$$

The Fermi energy of copper is 7.00 eV = 1.12 = 10^{-18} J and since copper is monovalent, $N = N_A$. Rearranging we have

$$T^2 = \frac{9}{2}\frac{k_B^2 N_A}{E_F}\frac{5\theta_D^3}{12R\pi^4}$$

$$= \frac{9}{2}\frac{(1.38\times10^{-23}\ \mathrm{J\ K^{-1}})^2(6.02\times10^{23}\ \mathrm{mol^{-1}})}{1.12\times10^{-18}\ \mathrm{J}} \times \frac{(5)(345\ \mathrm{K})^3}{(12)(8.31\ \mathrm{J\ mol^{-1}\ K^{-1}})(3.14)^4}$$

$$= 9.76\ \mathrm{K}^2$$

Therefore the temperature at which these contributions are equal is $T = 3.12$ K.

7.21 Electrical conductivity is due only to movement of electrons; therefore materials which have virtually no delocalized electrons have an electrical conductivity which may be 10^{20} times lower than that of a metal.
Thermal conductivity is due to the movement of electrons and phonons. Since all materials possess phonons, thermal conductivity is always reasonably good.

7.22 The Wiedemann–Franz constant, L, is given by eqn (7.21):

$$L = \frac{\kappa}{\sigma T}$$

For Cu, $\kappa = 400\ \mathrm{W\ m^{-1}\ K^{-1}}$ and $\sigma = 6.45\times10^7\ \Omega^{-1}\ \mathrm{m^{-1}}$ at 273 K, so

$$L = \frac{400\ \mathrm{W\ m^{-1}\ K^{-1}}}{(6.45\times10^7\ \Omega^{-1}\ \mathrm{m^{-1}})(273\mathrm{K})} = 2.27\times10^{-8}\ \mathrm{W\ \Omega\ K^{-2}}$$

For Au, $\kappa = 310$ W m^{-1} K^{-1} and $\sigma = 4.88 \times 10^7\ \Omega^{-1}$ m^{-1} at 273 K, so $L = 2.33 \times 10^{-8}$ W Ω K^{-2}.
For Al, $\kappa = 230$ W m^{-1} K^{-1} and $\sigma = 4.00 \times 10^7\ \Omega^{-1}$ m^{-1} at 273 K, so $L = 2.11 \times 10^{-8}$ W Ω K^{-2}.
For Na, $\kappa = 140$ W m^{-1} K^{-1} and $\sigma = 2.38 \times 10^7\ \Omega^{-1}$ m^{-1} at 273 K, so $L = 2.15 \times 10^{-8}$ W Ω K^{-2}.
So the values of L for these metals at 273 K are constant to within about 10%.

7.23 Using eqns (7.23) and (7.24), the ratio of thermal conductivity of the electrons to that of the phonons is

$$\frac{\kappa_e}{\kappa_p} = \frac{\frac{1}{3} nC_e \lambda_e v_e}{\frac{1}{3} nC\lambda_{ph} v_{ph}} = \frac{C_e v_e}{Cv_{ph}}$$

where we have assumed that $\lambda_e \approx \lambda_{ph}$.
The electronic specific heat capacity is given by eqn (7.20) with $N = N_A$ and $E_F = 5.0$ eV $= 8.0 \times 10^{-19}$ J, i.e.

$$C_e = \frac{9}{2}\frac{k_B^2 TN_A}{E_F} = \frac{9}{2}\frac{(1.38\times10^{-23}\ \text{J K}^{-1})^2(300\ \text{K})(6.02\times10^{23}\ \text{mol}^{-1})}{8.0\times10^{-19}\ \text{J}}$$

$$= 0.19\ \text{J mol}^{-1}\ \text{K}^{-1}$$

The electron velocity is given by eqn (4.20):

$$v_e = \sqrt{\frac{2E_F}{m_e}} = \sqrt{\frac{(2)(8.0\times10^{-19}\ \text{J})}{9.11\times10^{-31}\ \text{kg}}} = 1.33\times10^6\ \text{m s}^{-1}$$

The ionic specific heat capacity is assumed to be equal to the classical value of $3R$, i.e. $C = 24.9$ J mol^{-1} K^{-1}, and the velocity of the phonons is assumed to be equal to the speed of sound, i.e. $v_{ph} = 4000$ m s^{-1}. Therefore the ratio is

$$\frac{\kappa_e}{\kappa_p} = \frac{C_e v_e}{Cv_{ph}} = \frac{(0.19\ \text{J mol}^{-1}\ \text{K}^{-1})(1.33\times10^6\ \text{m s}^{-1})}{(24.9\ \text{J mol}^{-1}\ \text{K}^{-1})(4000\ \text{m s}^{-1})} = 2.5$$

7.24 From eqn (7.23) the phonon mean free path is given by

$$\lambda_{ph} = \frac{3\kappa_p}{nCv_{ph}}$$

where n is the number of moles per unit volume, therefore

$$n = \frac{\text{density}}{\text{molar mass}} = \frac{3.52\times10^3\ \text{kg m}^{-3}}{12\times10^{-3}\ \text{kg mol}^{-1}} = 2.93\times10^5\ \text{mol m}^{-3}$$

Assume that the specific heat capacity is approximately the same at 295 K as at 273 K, i.e. from Table 7.2 $C = 5.0$ J K^{-1}, and assume that the thermal conductivity is entirely due to phonons, i.e. $\kappa_p = \kappa = 2000$ W m^{-1} K^{-1}. Also assume that the phonon velocity is equal to the speed of sound, so $v_{ph} = 5000$ m s^{-1}. Then

$$\lambda_{ph} = \frac{3\kappa_p}{nCv_{ph}} = \frac{(3)(2000\ \text{W m}^{-1}\ \text{K}^{-1})}{(2.93\times10^5\ \text{mol m}^{-3})(5.0\ \text{J K}^{-1})(5000\text{m s}^{-1})} = 8.19\times10^{-7}\ \text{m}$$

Chapter 8

8.1 Equation (8.1) is

$$\chi_m = \frac{\mu_0 \mathbf{M}}{\mathbf{B}_0}$$

$\mathbf{B}_0$ has units of T, **M** has units of J T^{-1} m^{-3}, μ_0 has units of T^2 m^3 J^{-1} and χ_m is dimensionless. Considering units (denoted by square brackets) of terms on the right hand side of the equation we have

$$\frac{[\mu_0][\mathbf{M}]}{[\mathbf{B}_0]} = \frac{[T^2\ m^3\ J^{-1}][J\ T^{-1}\ m^{-3}]}{[T]} = [0]$$

Therefore, both sides of eqn (8.1) are dimensionless, so the equation is dimensionally correct.
Equation (8.2) is

$$\mathbf{B} = \mathbf{B}_0 + \mu_0 \mathbf{M}$$

From the above we see that $\mu_0\mathbf{M}$ has units of T, which is the same as the units of **B** and $\mathbf{B}_0$, so the equation is dimensionally correct.

8.2 In a filled electron subshell there are equal numbers of electrons with $m_s = +\frac{1}{2}$ and $m_s = -\frac{1}{2}$; therefore the sum of the spins, S, is zero.
If the subshell has angular momentum quantum number l, then the orbital magnetic quantum number, m_l, has allowed values from $-l$ to $+l$. If the subshell is filled, then for each pair of electrons ($m_s = \pm\frac{1}{2}$) with a particular value of m_l there is a corresponding pair with a value of $-m_l$, so the sum over m_l is also zero, i.e. $L = 0$.

8.3 When a subshell is half filled all of the electrons have the same value of m_s so that S is maximized, in accordance with Hund's rule (1); therefore all of the electrons have different values of m_l. Since there is one electron for each value of m_l (from $-l$ to $+l$), the sum over m_l is zero, i.e. $L = 0$.

8.4 (a) The 3d orbital has $l = 2$, so there are five allowed values of m_l (i.e. $l = -2, -1, 0, +1$ and $+2$). According to Hund's rules, the eight 3d electrons in Ni^{2+} must be arranged as follows:

	$m_s = +\frac{1}{2}$	$m_s = -\frac{1}{2}$
$m_l = +2$	*	*
+1	*	*
0	*	*
−1	*	
−2	*	

There are five electrons with $m_s = +\frac{1}{2}$ and three with $m_s = -\frac{1}{2}$, so $S = 1$.
Adding the m_l we have $L = 3$.

(b) The 4f subshell corresponds to $l = 3$, so there are seven allowed values of m_l (from −3 to +3). Using Hund's rule (1), the seven 4f electrons in Gd^{3+} occupy states with the same spin; therefore $S = \frac{7}{2}$.
Since this corresponds to one electron in each m_l state, the sum over m_l is zero, i.e. $L = 0$.

8.5 The Landé splitting factor is given by eqn (8.8):

$$g = \frac{3J(J+1) + S(S+1) - L(L+1)}{2J(J+1)}$$

(a) When $S = 0$, then $J = L$ and

$$g = \frac{3J(J+1) + 0 - J(J+1)}{2J(J+1)} = \frac{2J(J+1)}{2J(J+1)} = 1$$

(b) When L=0, then J=S

$$g = \frac{3J(J+1) + J(J+1) - 0}{2J(J+1)} = \frac{4J(J+1)}{2J(J+1)} = 2$$

8.6 For sodium, using Tables 4.1 and 4.2 we have $N = 2.65 \times 10^{28}\ \text{m}^{-3}$ and $E_F = 3.22$ eV $= 5.15 \times 10^{-19}$ J, so using eqn (8.11) the magnetic susceptibility due to Pauli paramagnetism is

$$\chi_m = \frac{3N\mu_B^2\mu_0}{2E_F}$$

$$= \frac{(3)(2.65\times10^{28}\ \text{m}^{-3})(9.27\times10^{-24}\ \text{J T}^{-1})^2(1.26\times10^{-6}\ \text{T}^2\ \text{m}^3\ \text{J}^{-1})}{(2)(5.15\times10^{-19}\ \text{J})}$$

$$= 8.4\times10^{-6}$$

Comparing with Table 8.2, the measured magnetic susceptibility of sodium is 7.2×10^{-6}, so the diamagnetic contribution is

$$(8.4\times10^{-6}) - (7.2\times10^{-6}) = 1.2\times10^{-6}$$

For gold, the values are $N = 5.90 \times 10^{28}\ \text{m}^{-3}$ and $E_F = 5.49$ eV $= 8.78 \times 10^{-19}$ J, so $\chi_m = 1.1 \times 10^{-5}$.
Since the measured magnetic susceptibility of gold is -3.6×10^{-5}, the diamagnetic contribution is

$$(1.1\times10^{-5}) - (-3.6\times10^{-5}) = 4.7\times10^{-5}$$

8.7 The energy change due to a magnetic field $\mathbf{B}_0$ is

$$m_m\mathbf{B}_0 = \mu_B\mathbf{B}_0 = (9.27\times10^{-24}\ \text{J T}^{-1})(1\ \text{T}) = 9.27\times10^{-24}\ \text{J}$$

i.e. 5.79×10^{-5} eV. (Note: in comparison, the mean thermal energy at room temperature is $k_BT \approx 0.025$ eV.)

Since the range of energy levels occupied in the valence band is equal to the Fermi energy, E_F, then for $E_F = 3.0$ eV ($= 4.8 \times 10^{-19}$ J), the proportion of electrons that change their spin, as shown by the shaded region in Fig. 8.4(c), is equal to

$$\frac{m_m \mathbf{B}_0}{E_F} = \frac{9.27\times10^{-24}\ \text{J}}{4.8\times10^{-19}\ \text{J}} = 1.93\times10^{-5}$$

i.e. about 0.002%.

8.8 A 3d electron has values of m_l between −2 and +2. For scandium, with a single 3d electron, we have (from Hund's rules) $S = \frac{1}{2}$, $L = 2$ and so $J = L - S = \frac{3}{2}$. Using eqn (8.8) the Landé splitting factor is therefore

$$g = \frac{3J(J+1)+S(S+1)-L(L+1)}{2J(J+1)} = \frac{3.\frac{3}{2}(\frac{3}{2}+1)+\frac{1}{2}(\frac{1}{2}+1)-2(2+1)}{2\frac{3}{2}(\frac{3}{2}+1)}$$

$$= 0.8$$

The component of the magnetic moment along the direction of $\mathbf{B}_0$ can take values

$$m_J = +\frac{3}{2}g\mu_B, +\frac{1}{2}g\mu_B, -\frac{1}{2}g\mu_B, -\frac{3}{2}g\mu_B$$

and the corresponding energies are

$$-m_J B_0 = -\frac{3}{2}g\mu_B B_0, -\frac{1}{2}g\mu_B B_0, +\frac{1}{2}g\mu_B B_0, +\frac{3}{2}g\mu_B B_0$$

The lowest energy state is therefore

$$-\frac{3}{2}g\mu_B B_0 = -\frac{3}{2}(0.8)(9.27\times10^{-24}\ \text{J T}^{-1})(0.5\ \text{T}) = -5.56\times10^{-24}\ \text{J}$$

From Example 8.3, the probability that an atomic dipole has a dipole moment m_J is given by

$$Ae^{m_J B_0/k_B T}$$

where A is a constant. Therefore, at $T = 1$ K, the probability that the lowest energy state is occupied is

$$A\ \exp\left(\frac{\frac{3}{2}g\mu_B B_0}{k_B T}\right) = A\ \exp\left(\frac{5.56\times10^{-24}\ \text{J}}{(1.38\times10^{-23}\ \text{J K}^{-1})(1\ \text{K})}\right) = 1.50A$$

For the highest energy state (with $-m_J B_0 = +\frac{3}{2}g\mu_B B_0$) we obtain a value of 0.67 A. Therefore the ratio of these probabilities is

$$\frac{1.50A}{0.67A} = 2.24$$

i.e. there are more than twice as many dipoles in the lowest energy state than in the highest energy state.

At 300 K, the probabilities become 1.001A and 0.999A for the lowest and highest energy states, respectively. Since these probabilities are almost identical, there are roughy equal numbers of dipoles in each state, i.e. the susceptibility is approximately zero.

8.9 From Example 8.3 we have an expression for the magnetization

$$M = \frac{NAB_0}{k_B T}\sum_{-J}^{J} m_J^2$$

By substituting $m_J = g\mu_B J$ we get

$$M = \frac{NAB_0 g^2 \mu_B^2}{k_B T}\sum_{-J}^{J} J^2 \qquad (1)$$

To obtain an expression for A we use the fact that the total probability of a dipole having one of the $2J + 1$ orientations is equal to 1.0, i.e.

$$\sum_{-J}^{J} Ae^{m_J B_0 / k_B T} = 1.0$$

As in Example 8.3, we assume $m_J B_0 \ll k_B T$, so we can rewrite the above condition as

$$A\sum_{-J}^{J}\left(1 + \frac{m_J B_0}{k_B T}\right) = 1$$

Since the sum over m_J from $+J$ to $-J$ is zero, this becomes

$$A\sum_{-J}^{J} 1 = 1.0$$

i.e.

$$A = \frac{1.0}{2J+1}$$

Substituting this expression and the identity $\Sigma_{-J}^{J} J^2 = \frac{1}{3}J(J + 1)(2J + 1)$ into eqn (1) gives

$$M = \frac{NB_0 g^2 \mu_B^2}{k_B T}\frac{1.0}{2J+1}\frac{1}{3}J(J+1)(2J+1) = \frac{NB_0 g^2 \mu_B^2 J(J+1)}{3k_B T}$$

Since the susceptibility is given by

$$\chi_m = \frac{\mu_0 M}{B_0} = \frac{C}{T}$$

we have

$$C = \frac{\mu_0 N g^2 \mu_B^2 J(J+1)}{3k_B}$$

8.10 The 4f subshell corresponds to $l = 3$, so there are seven values of m_l ranging from +3 to −3.

Nd^{3+} has three 4f electrons, so from Hund's rules we know that these will all have the same value of spin and will occupy the states $m_l = +3, +2$ and $+1$. So $S = \frac{3}{2}$ and $L = 6$. As the subshell is less than half filled, $J = L - S = \frac{9}{2}$, and using eqn (8.8) gives $g = 0.727$.
The magnetic susceptibility can therefore be calculated by using eqns (8.14) and (8.15), i.e.

$$\chi_m = \frac{\mu_0 N g^2 \mu_B^2 J(J+1)}{3k_B T}$$

At T = 1 K we have

$$\chi_m = \frac{(1.26\times 10^{-6}\ \mathrm{T^2\ m^3\ J^{-1}})(10^{25}\ \mathrm{m^{-3}})(0.727)^2(9.27\times 10^{-24}\ \mathrm{J\ T^{-1}})^2(4.5)(5.5)}{(3)(1.38\times 10^{-23}\ \mathrm{J\ K^{-1}})(1\ \mathrm{K})}$$

$$= 3.42\times 10^{-4}$$

Similarly, at T = 300 K the value is $\chi_m = 1.14\times 10^{-6}$.

8.11 The effective Bohr magneton, p, is given by eqn (8.17):

$$p = g\sqrt{J(J+1)}$$

Using Hund's rules and eqn (8.8) we can construct the following table:

	S	L	J	g	p
Nd^{3+}	$\frac{3}{2}$	6	$\frac{9}{2}$	0.727	3.6
Gd^{3+}	$\frac{7}{2}$	0	$\frac{7}{2}$	2.0	7.9
Cr^{3+}	$\frac{3}{2}$	3	$\frac{3}{2}$	0.4	0.8
Cr^{2+}	2	2	0	0	0
Fe^{2+}	2	2	4	1.5	6.7
Ni^{2+}	1	3	4	1.25	5.6

Comparing with the experimental values given in the question, there is very good agreement for the elements with partially filled 4f subshells (i.e. Nd^{3+} and Gd^{3+}), but poor agreement for the other elements (which are all in the iron transition group).

8.12 If we assume that the iron group transition elements all have $L = 0$, then the values of the effective Bohr magneton, p, are as follows:

	S	L	J	g	p
Cr^{3+}	$\frac{3}{2}$	0	$\frac{3}{2}$	2	3.9
Cr^{2+}	2	0	2	2	4.9
Fe^{2+}	2	0	2	2	4.9
Ni^{2+}	1	0	1	2	2.8

These values are in much better agreement with the actual values of p than those calculated in Question 8.11, suggesting that the assumption that $L = 0$ in these metals is correct.

8.13 The solution is similar to Example 8.4. The magnetization is due purely to the Ni^{2+} ions. From the solution to Question 8.4(a) we know that the electron spins are aligned such that there are five with spins parallel to one another and three antiparallel, i.e.

↑↑↑↑↑↓↓↓

Therefore the net magnetic dipole moment per Ni^{2+} ion is 2 μ_B. The magnetization is therefore

$$M = \frac{2\mu_B N_A}{\text{molar volume}} = \frac{(2)(9.27\times10^{-24}\ \text{J T}^{-1})(6.02\times10^{23}\ \text{mol}^{-1})}{4.30\times10^{-5}\ \text{m}^{-3}\ \text{mol}^{-1}}$$

$$= 2.60\times10^{5}\ \text{J T}^{-1}$$

8.14 The magnetic dipole moment of an Fe^{2+} ion is $4\mu_B$ (see Example 8.4), and the volume of 1 mole of iron, V_m, is

$$V_m = \frac{\text{molar mass}}{\text{density}} = \frac{0.056\ \text{kg mol}^{-1}}{7.87\times10^{3}\ \text{kg m}^{-3}} = 7.12\times10^{-6}\ \text{m}^3\ \text{mol}^{-1}$$

So the predicted magnetization is given by

$$M = \frac{\text{dipole moment of 1 mole}}{\text{molar volume}} = \frac{4\mu_B N_A}{V_m}$$

$$= \frac{(4)(9.27\times10^{-24}\ \text{J T}^{-1})(6.02\times10^{23}\ \text{mol}^{-1})}{7.12\times10^{-6}\ \text{m}^3\ \text{mol}^{-1}}$$

$$= 3.14\times10^{6}\ \text{J T}^{-1}\ \text{m}^{-3}$$

8.15 Following Example 8.5, the volume of 1 mole of cobalt, V_m, is

$$V_m = \frac{\text{molar mass}}{\text{density}} = \frac{0.059\ \text{kg mol}^{-1}}{8.9\times10^{3}\ \text{kg m}^{-3}} = 6.63\times10^{-6}\ \text{m}^3\ \text{mol}^{-1}$$

The volume per ion is

$$V_i = \frac{V_m}{N_A} = \frac{6.63\times10^{-6}\ \text{m}^3\ \text{mol}^{-1}}{6.02\times10^{23}\ \text{mol}^{-1}} = 1.10\times10^{-29}\ \text{m}^3$$

So the dipole moment of a single cobalt ion is

$$\mu = M.V_i = (14.0\times10^{5}\ \text{J T}^{-1}\ \text{m}^{-3})(1.10\times10^{-29}\ \text{m}^3) = 1.54\times10^{-23}\ \text{J T}^{-1}$$

or

$$n_{eff} = \frac{\mu}{\mu_B} = \frac{1.54\times10^{-23}\ \text{J T}^{-1}}{9.27\times10^{-24}\ \text{J T}^{-1}} = 1.67$$

8.16 Assuming that the interaction energy ΔE is approximately equal to the mean thermal energy at the Curie temperature, then

$$\Delta E = k_B \theta_C$$

For iron, $\theta_C = 1043$ K, so $\Delta E = 1.44\times10^{-20}$ J = 0.090 eV.
For nickel, $\theta_C = 627$ K, so $\Delta E = 0.054$ eV.
For cobalt, $\theta_C = 1388$ K, so $\Delta E = 0.120$ eV.

8.17 B and B_0 have units of T and μ_0 has units of $T^2\ m^3\ J^{-1}$ (see Section 8.2), so dimensionally we have

$$\frac{[B][B_0]}{[\mu_0]} = \frac{[T]^2}{[T^2\ J^{-1}\ m^3]} = \frac{[J]}{[m^3]} = \frac{[\text{energy}]}{[\text{volume}]}$$

8.18 In Fig. 8.24(a),

$$\left(\frac{BB_0}{\mu_0}\right)_{max} = \frac{\mu_0 M_R B_C}{\mu_0} = M_R B_C$$

In Fig. 8.24(b), (BB_0/μ_0) is maximum at $B_0 = B_C/2$ and $B = \mu_0 M_R/2$, so

$$\left(\frac{BB_0}{\mu_0}\right)_{max} = \frac{M_R B_C}{4}$$

Using these as upper and lower limits, respectively, then we obtain the following values for the energy product:

	B_C (T)	M_R ($J\,T^{-1}\,m^{-3}$)	$(BB_0/\mu_0)_{max}$ lower ($kJ\,m^{-3}$)	$(BB_0/\mu_0)_{max}$ upper ($kJ\,m^{-3}$)
Alnico V	0.07	1.03×10^6	18	72
Co_5Sm	1.0	6.35×10^5	159	634
$Nd_2Fe_{14}B$	1.2	9.52×10^5	286	1143

Comparing with the values in Table 8.4 we can see that all of the actual values of $(BB_0/\mu_0)_{max}$ lie (just) within these limits.

Chapter 9

9.1 First determine the critical magnetic field at 4.2 K. Using eqn (9.1)

$$B_T = B_C\left[1-\left(\frac{T}{T_c}\right)^2\right] = 5.29\times10^{-2}\,T$$

The magnetic field due to the current in the wire is $B = \mu_0 I/2\pi r$.
Maximum current is when the field at the surface equals B_T, i.e. $I_{max} = 2\pi r B_T/\mu_0$ = 527.6 A.
The maximum current density is therefore $J_{max} = I_{max}/\pi r^2 = 4.20\times10^7\ A\ m^{-2}$.

9.2 First use eqn (9.1) and the data at 4.2 K to determine B_c:

$$B_C = B_T\left[1-\left(\frac{T}{T_c}\right)^2\right]^{-1} = 31.4\ T$$

Rearranging eqn (9.1) again, the temperature T corresponding to $B_T = 10$ T is

$$T = T_c\left(1 - \frac{B_T}{B_c}\right)^{1/2} = 16.5 \text{ K}$$

9.3 From eqn (9.2) we have

$$I(t) = I_0 e^{-t/\tau}$$

where $I(t)/I_0 > 0.99$ and $t = 2.5$ years $= 7.88 \times 10^7$ s. So

$$\tau > \frac{t}{\ln\left(\frac{1}{0.99}\right)} = 7.84 \times 10^9 \text{ s}$$

Since σ is proportional to τ, the ratio of conductivities of a superconductor to a normal metal is

$$\frac{\sigma_s}{\sigma_n} = \frac{\tau_s}{\tau_n} > \frac{7.84 \times 10^9 \text{ s}}{10^{-14}} = 8 \times 10^{23}$$

9.4 At the temperature at which the disc becomes superconducting, the magnetic field is expelled and the magnet levitates above the sample.
This cannot be explained using classical electromagnetism because there is no change in the external magnetic field, and therefore no induced current.

9.5 The position of equilibrium is when the gravitational potential energy equals the magnetic potential energy, i.e.

$$mgx = \frac{\mu_0 m_m^2}{32\pi x^3}$$

where x is the height of the magnet. Therefore

$$x = \left(\frac{\mu_0 m_m^2}{32\pi mg}\right)^{1/4} = \left(\frac{(1.26 \times 10^{-6} \text{ H m}^{-1})(0.30 \text{ J T}^{-1})}{(32)(3.14)(0.01 \text{ kg})(9.81 \text{ m s}^{-2})}\right)^{1/4} = 1.04 \times 10^{-2} \text{ m}$$

In the mixed state, not all of the magnetic field is expelled; therefore the magnetic potential energy is smaller and the height of the magnet is reduced.

9.6 Type II superconductor characteristics are shown below:

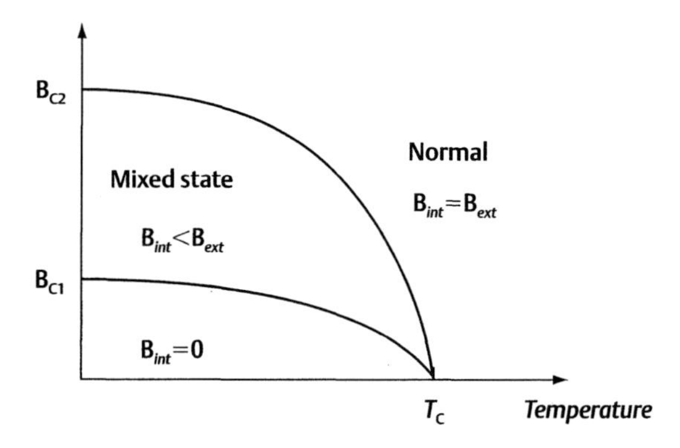

$$B_{c1} = B_{c1}(0)\left[1 - \left(\frac{T}{T_c}\right)^2\right]$$

$$B_{c2} = B_{c2}(0)\left[1 - \left(\frac{T}{T_c}\right)^2\right]$$

where $B_{c1}(0)$ and $B_{c2}(0)$ are the critical fields at $T = 0$ K.

9.7 Using the solution for Question 9.1, assume the critical current density J_c is given by

$$J_c = \frac{I_{max}}{\pi r^2} = \frac{2B_T}{\mu_0 r} = \frac{(2)(21\ \text{T})}{(1.26\times 10^{-6}\ \text{H m}^{-1})(0.5\times 10^{-3}\ \text{m})} = 6.67\times 10^{10}\ \text{A m}^{-2}$$

The actual critical current density is much lower than this calculated value because in a type II superconductor the resistivity is also non-zero if the current causes the filaments of normal material to move (see Section 9.5).

9.8 Using subscripts 1 and 2 to denote the two isotopes, we can write down two simultaneous equations

$$T_{c1} = AM_1^{-\alpha}$$

$$T_{c2} = AM_2^{-\alpha}$$

Eliminating A we obtain

$$\frac{T_{c1}}{T_{c2}} = \left(\frac{M_1}{M_2}\right)^{-\alpha}$$

Therefore

$$\alpha = \frac{\log(T_{c1}/T_{c2})}{\log(M_2/M_1)} = \frac{\log\left(\frac{4.185\text{K}}{4.146\text{K}}\right)}{\log\left(\frac{2034\,\text{u}}{1995\,\text{u}}\right)} = 0.484$$

The equation of motion of a simple harmonic oscillator is

$$M\frac{d^2x}{dt^2} + kx = 0$$

where k is a constant. The equation has solutions of the form

$$x = Ae^{i\omega t}$$

Substituting into the above equation we obtain

$$\omega^2 = \frac{k}{M}$$

Therefore, the experimental data for mercury suggests T_c is proportional to $M^{-0.484}$ and the frequency of vibration for a harmonic oscillator is proportional to $M^{-0.5}$, so this supports the arguement that T_c is related to the vibrations of the ions.

9.9 The supercurrent, I_s, can be expressed using the same formula as a normal current, i.e. from Section 4.2

$$I_s = n_s Aev_s$$

where n_s is the number density of Cooper pairs, v_s is their velocity and A is the cross-sectional area of the wire,

$$A = \pi r^2 = (3.14)(3 \times 10^{-3}\ \text{m})^2 = 2.83 \times 10^{-5}\ \text{m}^2$$

Therefore

$$v_s = \frac{I_s}{n_s Ae} = \frac{1000\ A}{(5.0 \times 10^{27}\ \text{m}^{-3})(2.83 \times 10^{-5}\ \text{m}^2)(1.60 \times 10^{-19}\ C)} = 0.442\ \text{m s}^{-1}$$

The velocity of the individual electrons is given by eqn (4.20). If E_F = 9.37 eV = 1.50×10^{-18} J, then

$$v = \left(\frac{2E_F}{m}\right)^{1/2} = \left(\frac{(2)(1.50 \times 10^{-18}\ \text{J})}{9.11 \times 10^{-31}\ \text{kg}}\right)^{1/2} = 1.81 \times 10^6\ \text{m s}^{-1}$$

The velocity of the Cooper pairs is very small in comparison because each pair is made up of two electrons travelling at almost equal speeds in opposite directions; therefore the net velocity is close to zero.

9.10 In a superconductor the photons destroy some of the Cooper pairs by exciting the paired electrons into two single-electron states. If all of the Cooper pairs are destroyed then the resistivity becomes non-zero.
In a semiconductor the photons excite electrons from the valence band into the conduction band and therefore reduce the resistivity.

9.11 The number of Cooper pairs against temperature is plotted in the following diagram:

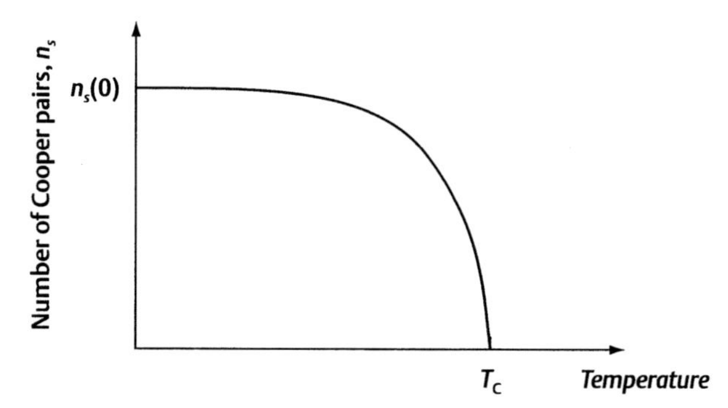

When $n_s(T) = n_s(0)/2$,

$$\frac{n_s(T)}{n_s(0)} = \frac{1}{2} = 1 - \left(\frac{T}{T_c}\right)^4$$

Therefore

$$T = \left(\frac{1}{2}\right)^{1/4} T_c = 0.841\ T_c$$

9.12 The volume of a sphere with diameter equal to the coherence length, ξ, is

$$V = \frac{4\pi}{3}\left(\frac{\xi}{2}\right)^3 = \frac{(4)(3.14)(100\times10^{-9}\ \text{m})^3}{3} = 4.19\times10^{-21}\ \text{m}^3$$

The number of Cooper pairs in this volume is

$$n_s V = (1\times10^{28}\ \text{m}^{-3})(4.19\times10^{-21}\ \text{m}^3) \approx 4\times10^7$$

9.13 Using eqn (9.3) the energy gap is

$$E_g = 3.53 k_B T_c = (3.53)(1.38\times10^{-23}\ \text{J K}^{-1})(90\ \text{K}) = 4.38\times10^{-21}\ \text{J} = 0.0274\ \text{eV}$$

From eqn (9.4) the voltage is

$$V_c = \frac{E_g}{2e} = \frac{4.38\times10^{-21}\ \text{J}}{(2)(1.60\times10^{-19}\ \text{C})} = 0.0137\ \text{V}$$

9.14 Using eqn (9.7) the upper critical field is given by

$$B_{c2} = \sqrt{2}\frac{\lambda}{\xi}B_c = \frac{(1.414)(700\times10^{-9}\ \text{m})(0.1\ \text{T})}{1\times10^{-9}\ \text{m}} = 99.0\ \text{T}$$

9.15 The maximum current in the wire is

$$I_{max} = J_c \pi r^2 = (1\times10^{10}\ \text{A m}^{-2})(3.14)(2.5\times10^{-3}\ \text{m})^2 = 1.96\times10^5\ \text{A}$$

giving a maximum field at the centre of the solenoid of

$$B_{max} = \mu_0 n I_{max} = (1.26\times10^{-6}\ \text{H m}^{-1})(100)(1.96\times10^5\ \text{A}) = 247\ \text{T}$$

9.16 Since the change in the supercurrent is directly proportional to the change in magnetic flux, an error of 1% in the current corresponds to a sensitivity to changes in flux of $\Delta\phi = 0.01\phi_0$, where ϕ_0 is a flux quantum ($= 2.07\times10^{-15}$ T m). The minimum detectable change in the field is therefore

$$\Delta B = \frac{\Delta\phi}{A} = \frac{(0.01)(2.07\times10^{-15}\ \text{T m})}{(3.14)(5.0\times10^{-3}\ \text{m})^2} = 2.64\times10^{-13}\ \text{T}$$

Chapter 10

10.1 Using eqn (10.3),

$$\varepsilon = \frac{C}{C_0} = \frac{0.195\ \mu\text{F}}{0.085\ \mu\text{F}} = 2.29$$

10.2 From Table 10.1, the static dielectric constant of LiCl is 11.95, and the external electric field is

$$\mathcal{E}=\frac{12\text{ V}}{10^{-4}\text{ m}}=1.20\times 10^{5}\text{ V m}^{-1}$$

Substituting in eqn (10.12) gives

$$\mathcal{E}_{loc}=\frac{1}{3}(\varepsilon+2)\mathcal{E}=\frac{1}{3}(11.95+2)(1.20\times 10^{5}\text{ V m}^{-1})=5.58\times 10^{5}\text{ V m}^{-1}$$

10.3 From eqn (10.8), the dipole moment of the atom is

$$p=\alpha\mathcal{E}=(4.3\times 10^{-41}\text{ F m}^{2})(5.0\times 10^{4}\text{ V m}^{-1})=2.15\times 10^{-36}\text{ C m}$$

For a neon atom, the atomic number is 10; therefore the charge q is $(10.0)(1.60\times 10^{-19}\text{ C})=1.60\times 10^{-18}\text{ C}$.
Using eqn (10.6) the displacement of the centroids of positive and negative charge is therefore

$$\delta=\frac{p}{q}=\frac{2.15\times 10^{-36}\text{ C m}}{1.60\times 10^{-18}\text{ C}}=1.34\times 10^{-18}\text{ m}$$

10.4 (a) If the electron transfers completely from the hydrogen to the chlorine, then we have a dipole with charges of $\pm e$ separated by a distance $d=0.128$ nm. Therefore the dipole moment is

$$p=e.d=(1.60\times 10^{-19}\text{ C})(0.128\times 10^{-9}\text{ m})=2.05\times 10^{-29}\text{ C m}$$

(b) If the amount of charge actually transferred is q, then $p=q.d$, i.e.

$$q=\frac{p}{d}=\frac{3.3\times 10^{-30}\text{ C m}}{0.128\times 10^{-9}\text{ m}}=2.58\times 10^{-20}\text{ C}=0.16\ e$$

10.5 (a) Since there are two different types of ions involved we will use eqn (10.14). The number of ions of each type per unit volume is given by

$$N=\frac{\text{Avogadro's number}}{\text{molar volume}}=\frac{6.02\times 10^{23}\text{ mol}^{-1}}{3.71\times 10^{-5}\text{ m}^{3}\text{ mol}^{-1}}=1.62\times 10^{28}\text{ m}^{-3}$$

If we write N_α for $\Sigma_j N_j\alpha_j$, then

$$N\alpha=\sum_j N_j\alpha_j=N(\alpha_{K^+}+\alpha_{Cl^-})$$
$$=(1.62\times 10^{28}\text{ m}^{-3})[(1.48\times 10^{-40}\text{ F m}^{2})+(3.29\times 10^{-40}\text{ F m}^{2})]$$
$$=7.73\times 10^{-12}\text{ F m}^{-1}$$

Rearranging eqn (10.14) we can write

$$\varepsilon=\frac{2N\alpha-3\varepsilon_0}{3\varepsilon_0-N\alpha}=\frac{2(7.73\times 10^{-12}\text{ F m}^{-1})+3(8.85\times 10^{-12}\text{ F m}^{-1})}{3(8.85\times 10^{-12}\text{ F m}^{-1})-(7.73\times 10^{-12}\text{ F m}^{-1})}$$
$$=2.23$$

(For comparison, the measured value is 2.22.)
(b) From eqns (10.4) and (10.5) the total polarization is

$$P = \varepsilon_0(\varepsilon - 1)\mathcal{E}$$
$$= (8.85\times 10^{-12}\ \mathrm{F\ m^{-1}})(4.84 - 1.0)(1.0\times 10^{5}\ \mathrm{V\ m^{-1}})$$
$$= 3.40\times 10^{-6}\ \mathrm{C\ m^{-2}}$$

Also, from eqn (10.15) we can write the total polarization as

$$P = P_i + P_e$$

Assuming that the local field is given by eqn (10.12), i.e. $\mathcal{E}_{loc} = \frac{1}{3}(\varepsilon_s + 2)\mathcal{E}$, and using eqn (10.10) we get

$$P_e = N\alpha\mathcal{E}_{loc}$$
$$= (1.62\times 10^{28}\ \mathrm{m^{-3}})[(1.48\times 10^{-40}\ \mathrm{F\ m^{2}}) + (3.29\times 10^{-40}\ \mathrm{F\ m^{2}})]$$
$$\times\frac{1}{3}(4.84 + 2)(1.0\times 10^{5}\ \mathrm{V\ m^{-1}})$$
$$= 1.76\times 10^{-6}\ \mathrm{C\ m^{-2}}$$

Therefore

$$P_i = P - P_e = 1.64\times 10^{-6}\ \mathrm{C\ m^{-2}}$$

From eqn (10.16) $P_i = NeA$. Therefore the amplitude of the ionic displacement is

$$A = \frac{P_i}{Ne} = \frac{1.64\times 10^{-6}\ \mathrm{C\ m^{-2}}}{(1.62\times 10^{28}\ \mathrm{m^{-3}})(1.60\times 10^{-19}\ \mathrm{C})} = 0.633\times 10^{-16}\ \mathrm{m}$$

10.6 Substituting $x = Ae^{i\omega t}$ into the equation for a forced undamped harmonic oscillator, i.e.

$$\frac{d^2x}{dt^2} + \omega_0^2 x = \frac{e\mathcal{E}}{m}e^{i\omega t}$$

gives

$$-\omega^2 Ae^{i\omega t} + \omega_0^2 Ae^{i\omega t} = \frac{e\mathcal{E}}{m}e^{i\omega t}$$

Rearranging, the amplitude of the vibration is

$$A = \frac{e\mathcal{E}}{m}\frac{1}{\omega_0^2 - \omega^2}$$

So when $\omega = \omega_0$, the amplitude of the vibration becomes infinite.

10.7 (a) Using eqn (10.25)

$$\varepsilon' = \frac{Ne^2}{\varepsilon_0 m} \frac{\omega_0^2 - \omega^2}{(\omega_0^2 - \omega^2)^2 + \gamma^2 \omega^2} + \varepsilon_{opt}$$

The static case is obtained by putting $\omega = 0$, i.e.

$$\varepsilon_s = \frac{Ne^2}{\varepsilon_0 m} \frac{\omega_0^2}{(\omega_0^2)^2} + \varepsilon_{opt}$$

Therefore

$$\varepsilon_s - \varepsilon_{opt} = \frac{Ne^2}{\varepsilon_0 m \omega_0^2}$$

(b) The number of ions per unit volume in sodium chloride is

$$N = \frac{\text{density}}{\text{average mass of an ion}} = \frac{2200 \text{ kg m}^{-3}}{(29.2)(1.66 \times 10^{-27} \text{ kg})} = 4.54 \times 10^{28} \text{ m}^{-3}$$

Substituting the values in the above formula gives

$$\varepsilon_s - \varepsilon_{opt} = \frac{(4.54 \times 10^{28} \text{ m}^{-3})(1.60 \times 10^{-19} \text{ C})^2}{(8.85 \times 10^{-12} \text{ F m}^{-1})(29.2)(1.66 \times 10^{-27} \text{ kg})(3.0 \times 10^{13} \text{ rad s}^{-1})^2}$$
$$= 3.01$$

(For comparison, using the data in Table 10.1 gives $\varepsilon_s - \varepsilon_{opt} = 3.56$.)

10.8 From eqn (10.28)

$$\varepsilon' = \varepsilon_{opt} + \frac{\varepsilon_s - \varepsilon_{opt}}{1 + \omega^2 \tau^2}$$

When $\omega = 0$,

$$\varepsilon' = \varepsilon_{opt} + \frac{\varepsilon_s - \varepsilon_{opt}}{1 + 0} = \varepsilon_s$$

When $\omega = 1/\tau$,

$$\varepsilon' = \varepsilon_{opt} + \frac{\varepsilon_s - \varepsilon_{opt}}{1 + 1} = \frac{\varepsilon_s + \varepsilon_{opt}}{2}$$

When $\omega >> 1/\tau$, then $(\omega\tau)^2$ is large and so the term

$$\frac{\varepsilon_s - \varepsilon_{opt}}{1 + \omega^2 \tau^2}$$

becomes very small. Therefore

$$\varepsilon' \to \varepsilon_{opt}$$

(b) From eqn (10.29), the imaginary part of the dielectric function is

$$\varepsilon'' = \frac{(\varepsilon_s - \varepsilon_{opt})\omega\tau}{1 + \omega^2 \tau^2}$$

Differentiating this expression gives

$$\frac{d\varepsilon''}{d\omega}=\frac{(\varepsilon_s-\varepsilon_{opt})\tau(1+\omega^2\tau^2)-2\omega\tau^2(\varepsilon_s-\varepsilon_{opt})\omega\tau}{(1+\omega^2\tau^2)^2}$$

The maximum power is dissipated when $d\varepsilon''/d\omega = 0$, i.e.

$$(\varepsilon_s-\varepsilon_{opt})\tau(1+\omega^2\tau^2)-2\omega\tau^2(\varepsilon_s-\varepsilon_{opt})\omega\tau=0$$
$$\Rightarrow \omega^2\tau^2=1$$
$$\Rightarrow \omega=\frac{1}{\tau}$$

10.9 Rearranging eqn (5.21), the largest band gap that can absorb visible light is

$$E_g=\frac{hc}{\lambda}=\frac{(6.63\times10^{-34}\text{ J s})(3.00\times10^{8}\text{ m s}^{-1})}{380\times10^{-9}\text{m}}=5.23\times10^{-19}\text{ J}$$
$$=3.27\text{ eV}$$

Therefore pure materials with a band gap larger than 3.27 eV should be transparent to visible light.

10.10 The acceptor state is at an energy $E_a = 0.36$ eV above the valence band edge, and for reasons stated in the question we assume that the Fermi energy is at an energy $E_a/2 = 0.18$ eV above the valence band edge.
Following a similar line of reasoning to that used in Example 5.6, the proportion of boron impurities that are ionized, N_{a1}, is equal to the probability of an electron occupying the acceptor state, which in turn is given by the Fermi–Dirac distribution, so

$$N_{a1}=f(E_a)=[e^{(E_a-E_F)/k_BT}+1]^{-1}$$

where $E_a - E_F = (0.36 - 0.18)$ eV $= 2.88\times10^{-20}$ J. Therefore

$$N_{a1}=\left[\exp\left(\frac{2.88\times10^{-20}\text{ J}}{(1.38\times10^{-23}\text{ J K}^{-1})(300\text{ K})}\right)+1\right]^{-1}=9.51\times10^{-4}$$

10.11 For silicon we have

$$n=(2\times10^{26}\text{ m}^{-3})\exp\left(-\frac{(0.88)(1.6\times10^{-19}\text{ J})}{(1.38\times10^{-23}\text{ J K}^{-1})(1200\text{ K})}\right)\approx4\times10^{22}\text{ m}^{-3}$$

and for diamond

$$n=(2\times10^{26}\text{ m}^{-3})\exp\left(-\frac{(5.3)(1.6\times10^{-19}\text{ J})}{(1.38\times10^{-23}\text{ J K}^{-1})(1200\text{ K})}\right)\approx1\times10^{4}\text{ m}^{-3}$$

A silicon device could not operate at this temperature because the concentration of thermal carriers is of the same magnitude as the concentration of carriers normally introduced by doping. Consequently, the contact potential at a p–n junction would be small (of the order of 0.1 eV), which is comparable with the thermal energy kT.

10.12 Since the polarization is the dipole moment per unit volume, the dipole moment of 1 mole, p_m, is

$$p_m = P_S V_m = (0.26\ \text{C m}^{-2})(3.8\times 10^{-5}\ \text{m}^3) = 9.88\times 10^{-6}\ \text{C m}$$

Since 1 mole of $BaTiO_3$ consists of N_A ions of Ba^{2+}, N_A ions of Ti^{4+} and $3N_A$ ions of O^{2-} (see Example 10.3), the total charge associated with 1 mole is $\pm 6N_A e$. Therefore the dipole moment of 1 mole is

$$p_m = 6 N_A e \delta$$

Rearranging, we have

$$\delta = \frac{p_m}{6 N_A e} = \frac{9.88\times 10^{-6}\ \text{C m}}{(6.0)(6.02\times 10^{23}\ \text{mol}^{-1})(1.60\times 10^{-19}\ \text{C})} = 1.71\times 10^{-11}\ \text{m}$$

10.13 For an insulator of thickness d, the electric field is

$$\mathcal{E} = \frac{240\ \text{V}}{d}$$

The value of d for which this is equal to the breakdown field of $5\times 10^7\ \text{V m}^{-1}$ is

$$d = \frac{240\ \text{V}}{5\times 10^7\ \text{V m}^{-1}} = 4.8\times 10^{-6}\ \text{m}$$

The standard thickness of insulation is typically about 5×10^{-4} m, i.e. a factor of 100 times thicker.

Chapter 11

11.1 The calculation is similar to that in Example 11.1. Using eqns (7.2) and (7.15) the force constant γ is given by

$$\gamma \approx m\left(\frac{k_B\theta_D}{\hbar}\right)^2 = (37.0)(1.66\times 10^{-27}\ \text{kg})\left(\frac{(1.38\times 10^{-23}\ \text{J K}^{-1})(230\ \text{K})}{1.05\times 10^{-34}\ \text{J s}}\right)^2$$

$$= 56.1\ \text{N m}^{-1}$$

and from eqn (7.4) the amplitude of vibration at the melting point is

$$x_{melt} = \left(\frac{2k_B T_{melt}}{\gamma}\right)^{1/2} = \left(\frac{(2)(1.38\times 10^{-23}\ \text{J K}^{-1})(1063\ \text{K})}{56.1\ \text{N m}^{-1}}\right)^{1/2} = 2.29\times 10^{-11}\ \text{m}$$

As a percentage of the interatomic spacing this corresponds to

$$\frac{x_{melt}}{a_0}\times 100\ \% = \frac{2.29\times 10^{-11}\ \text{m}}{3.14\times 10^{-10}\ \text{m}}\times 100\ \% = 7.3\%$$

11.2 From eqn (7.4), the amplitude of the thermal vibrations at temperature T is

$$x_{max} = \left(\frac{2k_B T}{\gamma}\right)^{1/2}$$

where γ is the force constant. Rearranging, the melting point, T_{melt} is given by

$$T_{melt} = \frac{\gamma x_{melt}^2}{2k_B}$$

where x_{melt} is the amplitude of the atomic vibrations at the melting point. If the crystal melts when the amplitude of vibration is 7% of the equilibrium atomic separation, then

$$x_{melt} = 0.07\ a_0 = (0.07)(1.54\times10^{-10}\ \text{m}) = 1.08\times10^{-11}\ \text{m}$$

From Example 11.1, the force constant is given by

$$\gamma \approx m\left(\frac{k_B\Theta_D}{\hbar}\right)^2 = (12.0)(1.66\times10^{-27}\ \text{kg})\left(\frac{(1.38\times10^{-23}\ \text{J K}^{-1})(1850\ \text{K})}{1.05\times10^{-34}\ \text{J s}}\right)^2$$

$$= 1178\ \text{N m}^{-1}$$

Therefore

$$T_{melt} = \frac{\gamma x_{melt}^2}{2\ k_B} = \frac{(1178\ \text{N m}^{-1})(1.08\times10^{-11}\ \text{m})^2}{(2)(1.38\times10^{-23}\ \text{J K}^{-1})} = 4978\ \text{K}$$

(In fact, diamond does not melt—it transforms to graphite at a temperature of about 3400 K.)

11.3 In the liquid, the atoms occupy 64% of the total volume of the crystal, i.e. $0.64\times10^{-6}\ \text{m}^3$.
In the crystal, the atoms occupy 74% of the volume; therefore the total volume of the crystal is

$$\frac{0.64\times10^{-6}\ \text{m}^3}{0.74} = 8.65\times10^{-7}\ \text{m}^3$$

11.4 The volume of the liquid is greater than that of the crystal by a factor of

$$\frac{0.74}{0.64} = 1.16$$

so the density decreases by a factor of 1/1.16 = 0.862.
If the density of the crystal is denoted by ρ_c and the density of the liquid by ρ_l, then $\rho_l = 0.862\ \rho_c$, and the fractional change in density is

$$\frac{\rho_c - \rho_l}{\rho_c} = \frac{\rho_c - 0.862\rho_c}{\rho_c} = \frac{1-0.862}{1} = 0.138$$

Therefore the percentage change in density is $0.138 \times 100\% = 13.8\%$

11.5 If every photon is absorbed, the electrical energy produced per second by the panel is

$$P = (\text{photon flux}) \times (\text{area of panel}) \times (\text{energy per photon})$$

Since the band gap energy is 1.0 eV = 1.6×10^{-19} J,

$$P = (2 \times 10^{21}\ \text{m}^{-2}\ \text{s}^{-1})(10.0\ \text{m}^2)(1.6 \times 10^{-19}\ \text{J}) = 3200\ \text{W}$$

Since the panel is only 25% efficient, only 25% of the total energy of the photons is actually converted into electrical energy. Therefore the actual power produced is

$$P \times \frac{25\ \%}{100\ \%} = 800\ \text{W}$$

Chapter 12

12.1 The length of the polymer is equal to 20 000 C—C bonds, i.e.

$$(20\,000)(0.154 \times 10^{-9}\ \text{m}) = 3.08 \times 10^{-6}\ \text{m}$$

One CH_2 monomer contains one carbon atom and two hydrogen atoms, so the mass of one CH_2 monomer is 14 u. The mass of the polymer is therefore

$$(20\,000)(14\ \text{u}) = 280\,000\ \text{u} = 4.65 \times 10^{-22}\ \text{kg}$$

To estimate the mass of the soot particle, we will assume that the volume of a carbon atom, V_C, is that of a sphere with a radius equal to half the C—C bond length, i.e.

$$V_C = \frac{4}{3}\pi r^3 = \frac{4}{3}(3.14)(7.7 \times 10^{-11}\ \text{m})^3 = 1.91 \times 10^{-30}\ \text{m}^3$$

Similarly, the volume of the soot particle, V_s, is equal to that of a sphere of diameter 2 μm. Therefore

$$V_S = \frac{4}{3}\pi r^3 = \frac{4}{3}(3.14)(1.0 \times 10^{-6}\ \text{m})^3 = 4.19 \times 10^{-18}\ \text{m}^3$$

The number of carbon atoms in the soot particle, N, is therefore approximately equal to

$$N = \frac{V_S}{V_C} = \frac{4.19 \times 10^{-18}\ \text{m}^3}{1.91 \times 10^{-30}\ \text{m}^3} = 2.19 \times 10^{12}$$

(This is an overestimate of N because the atomic spheres do not fill all of the space—see Chapter 2—but it is good enough for our purposes.)
The mass of the soot particle is therefore approximately

$$N.12\ \mathrm{u} = (2.19\times10^{12})(12)(1.66\times10^{-27}\ \mathrm{kg}) = 4.4\times10^{-14}\ \mathrm{kg}$$

So although a long polymer may be comparable in size (at least in one dimension) with a small macroscopic object such as a soot particle, it is typically about 10^8 times smaller in mass.

12.2 If the rubber strip doubles in length, then the extension Δl is equal to the original length, l, i.e. the strain is

$$\epsilon = \frac{\Delta l}{l} = 1.0$$

The stress is

$$\sigma = \frac{F}{A} = \frac{3\ \mathrm{N}}{0.1\times10^{-4}\ \mathrm{m}^2} = 3.0\times10^5\ \mathrm{N\ m}^{-2}$$

From eqn (3.3), Young's modulus is

$$Y = \frac{\sigma}{\epsilon} = \frac{3.0\times10^5\ \mathrm{N\ m}^{-2}}{1.0} = 3.0\times10^5\ \mathrm{N\ m}^{-2}$$

The same stress applied to an aluminium strip produces a strain

$$\epsilon = \frac{\sigma}{Y} = \frac{3.0\times10^5\ \mathrm{N\ m}^{-2}}{7.0\times10^{10}\ \mathrm{N\ m}^{-2}} = 4.3\times10^{-6}$$

12.3 In the diagram below the angle θ is equal to 109.5°/2, so the distance d is

$$d = a\sin\left(\frac{109.5^\circ}{2}\right) = 0.816a$$

For a chain containing N polymers the total length of the chain is therefore $0.816aN$.

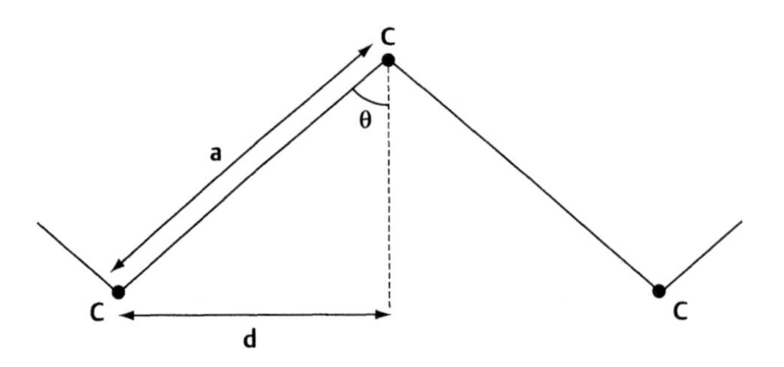

12.4 The solution is similar to Example 12.1 but with $N = 250$.
From eqn (12.2) the natural (unstressed) length of the polymer is

$$d_{unstressed} \approx a\sqrt{N} = a\sqrt{250} = 15.8\ a$$

The maximum elastic length in the stressed case is

$$d_{max} \approx 0.816\, aN = 204\, a$$

The strain is therefore

$$\epsilon = \frac{(204\, a - 15.8\, a)}{15.8\, a} \times 100\,\% = 1191$$

12.5 20° C corresponds to 293 K, so materials with T_g of less than 293 K (i.e. polyethylene, rubber and polypropylene) are in the glassy state, whereas the others are in the rubbery state.

12.6 The total length of the polymer is

$$(25\,000)(0.154 \times 10^{-9}\ \text{m}) = 3.85 \times 10^{-6}\ \text{m}$$

so the number of times the chain must be folded to fit into a single lamella is

$$\frac{3.85 \times 10^{-6}\ \text{m}}{10 \times 10^{-9}\ \text{m}} = 385$$

12.7 If the mass of the cable is m, then the maximum stress in the cable is

$$\sigma = \frac{mg}{A}$$

where A is the cross-sectional area of the cable and g is the acceleration due to gravity. If the maximum length of cable that can be supported under its own weight is L_{max}, then the corresponding mass of the cable is $m_{max} = \rho A L_{max}$ where ρ is the density of the material. Consequently, the breaking stress, σ_B, is given by

$$\sigma_B = \frac{m_{max}\, g}{A} = \frac{\rho\, A L_{max}\, g}{A} = \rho\, L_{max}\, g$$

Therefore, the maximum length is given by

$$L_{max} = \frac{\sigma_B}{\rho\, g}$$

Using the data from Table 12.2, for nylon

$$L_{max} = \frac{1000 \times 10^6\ \text{N m}^{-2}}{(1.15 \times 10^3\ \text{kg m}^{-3})(9.81\ \text{m s}^{-2})} = 8.86 \times 10^4\ \text{m}$$

and for steel

$$L_{max} = \frac{2000 \times 10^6\ \text{N m}^{-2}}{(7.8 \times 10^3\ \text{kg m}^{-3})(9.81\ \text{m s}^{-2})} = 2.61 \times 10^4\ \text{m}$$

12.8 If we denote the mass of the car by m, then the stress in the cable is $\sigma = mg/A$ where A is the cross-sectional area of the cable and g is the acceleration due to gravity. The minimum cross-sectional area A_{min} corresponds to the case where the stress is just equal to the breaking stress, σ_B, i.e. $\sigma_B = mg/A_{min}$. Therefore, using the data in Table 12.2,

$$A_{min} = \frac{mg}{\sigma_B} = \frac{(800\ \text{kg})(9.81\ \text{m s}^{-2})}{3000 \times 10^6\ \text{N m}^{-2}} = 2.62 \times 10^{-6}\ \text{m}$$

Note that this corresponds to a cable with a diameter of less than 2 mm!

Illustration acknowledgements

Fig. 2.2 P. Lacombe, Metallurgical Laboratory, University of Paris-Sud. *from* André Guinier, *The Structure of Matter* (London: Edward Arnold, 1984), p. 126.

Fig. 2.18(b) A. Ourmazd, J. C. Bean, *Physical Review Letters* **55**, 766, (1985). Copyright 1985 by the American Physical Society.

Fig. 2.21 Reprinted with permission from Fig. 3b, Wegscheider et al, "Novel relaxation process in strained Si/Ge superlattices grown on Ge (001)", *Applied Physics Letters*, **57**(15), 1990, pp. 1496–1498. Copyright 1990, American Institute of Physics.

Fig. 2.22 Copyright 2000, Colin Cuthbert Photography.

Fig. 3.16 Fig. 6, Ch V Kopetskii, A. I. Pashkovskii, *Physica status solidi* (a) **21**(741), 1974.

Fig. 8.16 R. W. DeBlois, General Electric Research and Development Center (from F. Keffer, 'The magnetic properties of materials', *Scientific American* Sept 1967, p. 231).

Fig. 8.17 Curtis R. Rodd and Andrew S. Holik, General Electric Company (from J. J. Becker, 'Permanent magnets', *Scientific American* Dec 1970, p. 93).

Fig. 9.5 U.S. Department of Energy/Science Source/Photo Researchers, Inc.

Fig. 9.10 Dr U. Essmann, Fig. 14.5, H. M. Rosenberg *The Solid State*, 3rd edition (Oxford: Oxford University Press, 1988).

Fig. 10.18 Fig. 13.11, H. M. Rosenberg *The Solid State*, 3rd edition (Oxford: Oxford University Press, 1988).

Index

D

E

F

G

H

I

J

K

L

M

T

U

V